MECHANISMS OF ENVIRONMENTAL STRESS RESISTANCE IN PLANTS

MECHANISMS OF ENVIRONMENTAL STRESS RESISTANCE IN PLANTS

Edited by

Amarjit S. Basra and Ranjit K. Basra

Department of Botany
Punjab Agricultural University
Ludhiana, India

harwood academic publishers
Australia • Canada • China • France • Germany • India • Japan • Luxembourg
Malaysia • The Netherlands • Russia • Singapore • Switzerland • Thailand
United Kingdom

Published in The Netherlands by Harwood Academic Publishers.

Amsteldijk 166
1st Floor
1079 LH Amsterdam
The Netherlands

British Library Cataloguing in Publication Data

Mechanisms of environmental stress resistance in plants
1. Plants, Effect of stress on 2. Plants – Hardiness 3. Plant genetic engineering
I. Basra, Amarjit S. II. Basra, Ranjit K.
581:5'222

ISBN 905702036X

CONTENTS

To Sukhmani and Nishchayjit,
Our Joy and Inspiration!

PREFACE

Plant growth and yield are limited in many regions of the world by a wide variety of abiotic and biotic stress factors. Consequently, there is an appreciable deficit between realized and potential crop yields. In the face of mounting population pressure, these limitations will be of increasing severity for global agriculture and forestry in the decades ahead.

In the past, the primary approach for alleviation of environmental stress has relied mainly on modification of the environment through the use of fertilizers, soil amelioration, pesticides and irrigation. For some areas, the widespread use of agrochemicals has resulted in fragile ecosystems, scoring low on sustainability, product quality and environmental protection. Salinization of land and water is an increasing problem, particularly in arid and semi-arid regions where irrigation is a contributory factor. Such an approach is now widely seen as being unsustainable and certainly not a formula to follow indefinitely. A major goal therefore in current plant science is to adapt the plant to the environment, for which it is imperative to understand the mechanisms of plant resistance to environmental stresses and to utilize this knowledge for genetic manipulation of crop and native plants for achieving improved stress resistance. This is one of the most exciting and fast moving areas of plant stress, encompassing diverse component disciplines from molecular and cell biology, through plant physiology and ecology, to the domains of agronomy, plant breeding and biotechnology.

The aim of this book is to bring together in one place an integrated account of mechanisms of resistance to a full range of abiotic and biotic stresses, frequently encountered by plants. The book is divided into two parts. The first and major part of the book consists of chapters focusing on the mechanisms of resistance to abiotic stress factors, including those of water, salt, temperature, solar radiation, nutrient deficiency, metal toxicity and physical forces such as wind. The second part of the book deals with biotic stress and provides a mechanistic account of virus, fungal and insect resistance in plants. A new synthesis is achieved in considering the mechanisms of resistance at various levels of organization: from the level of individual cells and tissues, through whole plants to communities.

The chapters have been written by an international group of experts in their respective fields of interest, who between them have a wealth of research and teaching experience. The coverage is comprehensive and up-to-date. Many outstanding problems are critically analyzed and important questions identified for future research.

Complementing its predecessor volume *Stress-induced Gene Expression in Plants* (Basra, 1994), the present book fills a void in the literature providing a comprehensive view reflecting our present knowledge of overall plant resistance to environmental stress factors based on multi-faceted, multilevel and multidisciplinary approaches. The book is wide ranging in scope and should prove extremely useful to advanced students, teachers and researchers in botany, plant physiology, biochemistry, genetics, molecular biology, plant breeding, biotechnology, agronomy, soil science, horticulture, forestry and environmental sciences.

There remains an enormous number of challenges and our hope is that this book will stimulate new areas of research in this fascinating field. Recent advances in molecular biology and biotechnology are making exploration and manipulation of these mechanisms more feasible and are likely, eventually, to make possible large gains in plant productivity, sustainable use of resources and environmental conservation. Progress in the identification and cloning of stress-resistance genes is gaining momentum. In the coming years, biotechnology can be expected to considerably strengthen the breeding programs, so that the impact of environmental stresses on plants will be reduced.

Finally, we must express our gratitude to all the authors who contributed chapters to this volume, and to all concerned at Harwood Academic Publishers, for without their efforts and enthusiasm, it would not have been possible. Personally, the eternal wisdom of Guru Granth Sahib has been a perennial source of inspiration and peace!

REFERENCE

Basra A.S. (Ed.) (1994) *Stress-Induced Gene Expression in Plants*. Chur, Switzerland: Harwood Academic Publishers.

LIST OF CONTRIBUTORS

E.N. Ashworth
Department of Horticulture
Purdue University
West Lafayette, Indiana 47907
USA

A.S. Basra
Department of Botany
Punjab Agricultural University
Ludhiana–141004
India

R.K. Basra
Department of Botany
Punjab Agricultural University
Ludhiana–141004
India

J. Brown
The University of Arizona
Environmental Research Laboratory
2601 E. Airport Drive
Tucson, AZ 85706
USA

T. Bryngelsson
Department of Plant Breeding Research
The Swedish University of Agricultural Sciences
S–268 31 Savlöv
Sweden

D.B. Collinge
Section for Plant Pathology
Department of Plant Biology
The Royal Agricultural and Veterinary University
Thorvaldsensvej 40
1871 Fredericksberg C Copenhagen
Denmark

P.A. Fay
Division of Biology
Ackert Hall
Kansas State University
Manhattan, KS 66506–4901
USA

R.S.S. Fraser
Society for General Microbiology
Harlborough House
Basingstake Road, Spencers Wood
Reading, RG 7 14E
UK

D.F. Gaff
Department of Ecology and Evolutionary Biology
Monash University, Clayton 3168,
Australia

E.P. Glenn
The University of Arizona
Environmental Research Laboratory
2601 E. Airport Drive
Tucson, AZ 85706,
USA

P.L. Gregersen
Section for Plant Pathology
Department of Plant Biology
The Royal Agricultural and Veterinary University
Thorvaldsensvej 40
1871 Fredericksberg C Copenhagen
Denmark

B. Huang
Department of Horticulture,
Forestry and Recreation Resources
2021 Throckmorton Plant Science Center
Kansas State University
Manhattan, KS 66506–5506
USA

R.R. Johnson
Department of Biology
Colby College
Waterville, ME 04901
USA

R.S. Jones
Mountain State Environmental Services
120, Cloverdale Heights
Charles Town, WV 25414
USA

M.J. Khan
Department of Soil Science
NWFP Agricultural University
Peshawar
Pakistan

A.K. Knapp
Division of Biology
Ackert Hall
Kansas State University
Manhattan, KS 66506–4901
USA

J.R. Mahan
USDA–ARS
Cropping Systems Research Laboratory
Rt. 3, Box 215
Lubbock, Texas 79401
USA

B.L. McMichael
USDA–ARS
Cropping Systems Research Laboratory
Rt. 3, Box 215
Lubbock, Texas 79401
USA

T.M. Murphy
Department of Plant Biology
University of California
Davis, CA 95616
USA

J.N. Pearson
Department of Plant Science
Waite Agricultural Research Institute
The University of Adelaide
Glen Osmond, SA 5064
Australia

Z. Rengel
Soil Science and Plant Nutrition
School of Agriculture
The University of Western Australia
Nedlands, WA 6009
Australia

Z. Ristic
Department of Biology
University of South Dakota
Vermillion, South Dakota 57069
USA

V. Smedegaard-Peterson
Section for Plant Pathology
Department of Plant Biology
The Royal Agricultural and Veterinary University
Thorvaldsensvej 40
1871 Fredericksberg C Copenhagen
Denmark

H. Thomas
Institute of Grassland and Enviornmental Research
Plas Gogerddan, Aberystwyth
Dyfed SY23 3EB, Wales
UK

H. Thordal–Christensen
Section for Plant Pathology
Department of Plant Biology
The Royal Agricultural & Veterinary University
Thorvaldsensvej 40
1871 Fredericksberg C Copenhagen
Denmark

D.F. Wanjura
USDA–ARS
Cropping Systems Research Laboratory
Rt. 3, Box 215
Lubbock, Texas 79401
USA

J.M. Wilson
'Gwyndaf'
Penmon, Beaumaris
Gwynedd, LL58 8SN
USA

1. DROUGHT RESISTANCE IN PLANTS

HENRY THOMAS

Institute of Grassland and Environmental Research, Plas Gogerddan, Aberystwyth, Dyfed SY23 3EB, Wales, UK.

INTRODUCTION

Plants have been coping with water stress for some 400 million years, ever since they first left the seas and colonized dry land. When drought occurs, higher plants have always been obliged to endure it or to adjust their life cycle to avoid it: unlike animals, they cannot move to a better place. Thus a major driving force behind the evolution of land plants has been their need to search for water, to absorb it, to transport it and to retain it. Even so, drought is still the major constraint to crop production (e.g. Boyer, 1982; McWilliam, 1986).

The mechanisms underlying drought resistance, having evolved over such a long period of time, are numerous and still imperfectly understood. The aim of this chapter is to describe the current state of our knowledge, and how it might be put to practical use.

THE DROUGHT ENVIRONMENT

The Components of Drought

To the meteorologist, drought is defined as a period (15d, say) without appreciable rain. In agriculture, a drought is a dry spell that results in a loss of yield below that expected under optimal water supply. To the plant physiologist, drought is more than a lack of rainfall: it is a concurrence of at least seven environmental stresses. These are:

(1) Low soil moisture availability, limiting the supply of water to the roots.
(2) High evaporative load, due to low humidity, high temperature, high insolation and strong winds. The potential loss of water from the leaves exceeds that which can be taken up by the roots, even in well-watered soil. Monteith (1986) has calculated that in equatorial regions, 30–50% of the loss in yield during drought is due to low humidity.
(3) High temperature, causing high respiration and damage to metabolic processes and cell structure (Chapter 7, this volume).
(4) High solar irradiance, leading to photo-inhibition, photo-oxidation, and eventually death of leaves.

(5) Soil hardness increases as soil dries; this can adversely affect root growth and lead to reduced leaf growth and photosynthesis, especially in seedlings (Masle and Passioura, 1987).
(6) Unavailability of nutrients, in particular in the upper soil horizons which dry most rapidly but are most mineral rich (Garwood and Williams, 1967).
(7) Accumulation of salts in the topsoil and around the roots, leading to osmotic and toxic stress (Chapter, 4 this volume).

In practice, the researcher faces the major problem of disentangling the relative contributions of these phenomena to plant growth and adaptation.

Geographical Variation in Drought Patterns

Drought is caused by an imbalance between evaporation, including transpiration, and rainfall. For the purposes of this discussion it is convenient to recognize six climatic regions on the basis of seasonal rainfall (R) and evaporation (E) patterns (Kendrew, 1961; Edmeades *et al.*, 1989). These descriptions are based on monthly means, and give little idea of the reliability of rainfall, or the probabilities of serious drought.

(1) Equatorial climate, having fairly stable E, and two rainy seasons at either side of the solstice (when $R > E$), and two dry seasons (when R is low but not zero).
(2) Tropical climate, with rain in the hottest, sunniest months and a long, dry winter; intermediate between (1) and (3).
(3) Monsoonal climate, occurring in the tropics and to the east of large landmasses, with a very dry winter and spring, highest E in spring, and $R>E$ in summer.
(4) Continental interior climate, a less extreme version of climate (3). R and E tend to run in phase, being at their maxima in summer and minima in winter. In summer, at higher latitudes or altitudes R and E may be similar, but in more extreme climates E will exceed R.
(5) Mediterranean climate, with R concentrated in the winter half of the year, possibly with maxima in autumn and spring, and a reliably dry summer; E low in winter, high in summer.
(6) Maritime temperate climate, with R evenly distributed or rather higher in autumn and winter, and E ranging from zero in winter to moderate in summer. Droughts occur only in some years, but may be serious since crops are less well adapted.

PRINCIPLES OF CROP WATER RELATIONS

The Role of Water in the Plant

Water is essential for life on earth. In plants it performs a number of functions, which show different sensitivities to water stress. Briefly, these roles are:

(1) Acting as a "skeletal" material which has very little metabolic cost. Plants have large vacuoles, consisting mainly of water, which permit a small amount of dry matter to be spread over a large area or volume. This increases the effectiveness with which roots explore the soil and with which foliage exploits the atmosphere and solar energy.

(2) As a medium for metabolism, since many gases, salts and organic compounds are soluble in water. Minerals are taken up in solution from the soil and transported to the leaves in the xylem flow; gas exchange occurs through the film of water in and on mesophyll cells; and metabolites are moved through the plant in solution in the phloem. The tensile strength of water allows it to be drawn, without breaking in the xylem vessels, to the leaves of the tallest trees.

(3) To reduce fluctuations in temperature. Only about 1% of the solar radiation reaching a plant canopy is used in photosynthesis, and a large proportion (about half) of the remaining radiative load is consumed in evaporating water from the leaves (transpiration), thus cooling them and preventing overheating.

Further details of the physical and chemical properties of water in plants are to be found in Milburn (1979).

Water Relations and Growth of Unstressed Plants

Plants have an enormous demand for water. Consider a closely mown grass sward (the meteorologists' standard crop) which accumulates dry matter at the rate of 0.1 t ha^{-1} d^{-1}, under a moderate evaporation load of 3 mm d^{-1}. It will absorb c. 0.5 t ha^{-1} d^{-1} of water in its newly produced tissues but, in doing so, it will transpire 30 t ha^{-1} d^{-1} of water through its leaves into the atmosphere. The water moves from the soil, into the roots and up to the leaves in response to gradients in water potential (ψ), which is a measure of free energy per unit volume of water (Slatyer, 1967). When most of the water passing through the plant is used in transpiration (particularly in sunny weather), the flow rate is linearly proportional to the gradient in ψ between soil and leaf, $\Delta\psi$ (Boyer, 1985).

The water in a plant occupies two distinct compartments: (a) the apoplast, which includes the cell walls, the intercellular spaces and the xylem lumen, all outside the plasmalemmas; and (b) the symplast, lying inside the plasmalemma. The symplast in turn includes two sub-compartments, namely (a) the vacuole, separated by the tonoplast from (b) the cytoplasm and organelles. The symplasts must draw water from the apoplastic transport stream in the xylem in order to keep mature cells hydrated and to expand young growing cells. Potential gradients across the semipermeable plasmalemma are generated by solutes in the cytoplasm and vacuole, which lower the osmotic (solute) potential ψ_s and suck water in from the apoplast. The typical value of ψ_s in mesophytic plants is around –1 MPa. Continued absorption of this water would build up pressure in the protoplast and burst it were it not for the cell walls, which resist this force and push back. The pressure on the walls, which is positive in sign, is known as turgor potential or pressure potential (ψ_P). The relationship between these potentials is defined by the equation $\psi = \psi_s + \psi_P$, which predicts that at night, for example, when bulk ψ is near zero and ψ_s is –1MPa, turgor will be 1MPa.

During the day, when transpiration exceeds water uptake, the water content of the tissues declines, as do the three potentials; the general form of the relationship for a mesophytic leaf is shown in Figure 1.1.

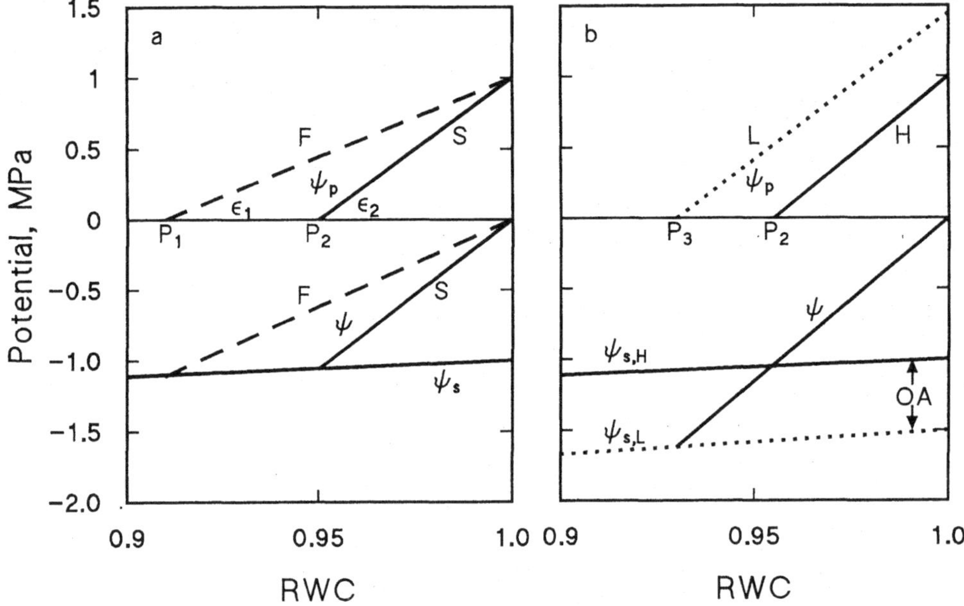

Figure 1.1. Relationships between relative water content (RWC), turgor (ψ_P), water potential (ψ), osmotic potential (ψ_s) and bulk modulus of elasticity ($\in$). A fully hydrated plant (e.g. well watered at dawn) has RWC = 1, ψ = 0, and $\psi_P = -\psi_s$. *(a)* Relationship for a leaf of a well-watered mesophyte, having ψ_s = 1 MPa (=10 bars = 14.7 psi $\simeq$ 10 atmospheres $\simeq$ 400 mOsmoles kg^{-1}). As stress increases, RWC decreases and (if there is no osmotic adjustment) ψ_s decreases approximately as ψ_s/RWC. Two values of $\in$ are shown, for a flexible leaf (F) with $\in_1$ = 10 MPa, and for a stiff leaf (S) with $\in_2$ = 20 MPa. The flexible leaf has a weaker attraction.for water (i. e. develops a weaker gradient in y between leaf and environment) at a given RWC, but loses turgor more slowly and reaches incipient plasmolysis (P1) at a lower RWC. As plants acclimate to drought, tends to increase and ys to decrease. *(b)* Relationship for a stiff-leaved plant exhibiting osmotic adjustment (OA) of 0.5 MPa. The disadvantage of the rapid loss of turgor with RWC is partly counteracted by OA from a high (H) to a low (L) ys, which delays incipient plasmolysis from P2 to P3. Also, it allows a higher turgor to be achieved should the leaf rehydrate overnight or after rain.

During the day, when transpiration exceeds water uptake, the water content of the tissues declines, as do the three potentials; the general form of the relationship for a mesophytic leaf is shown in Figure 1.1.

While the above analysis serves as a first approximation for leaves or shoots consisting mainly of mature tissue, the responses of growing cells are complicated by their demand for water to expand their cells. The biophysical characters contributing to this process can be summarized in two equations first postulated by Lockhart (1965), and developed further by Boyer (1985) and Cosgrove (1986), which explain the effects of potential gradients, wall characters and water supply on the relative volume increase (RVI) of a growing cell:

$$RVI = C\,(\sigma\Delta\psi_s - \psi_p) = \phi\,(\psi_p - Y) \text{..........(Equations 1, 2)}$$

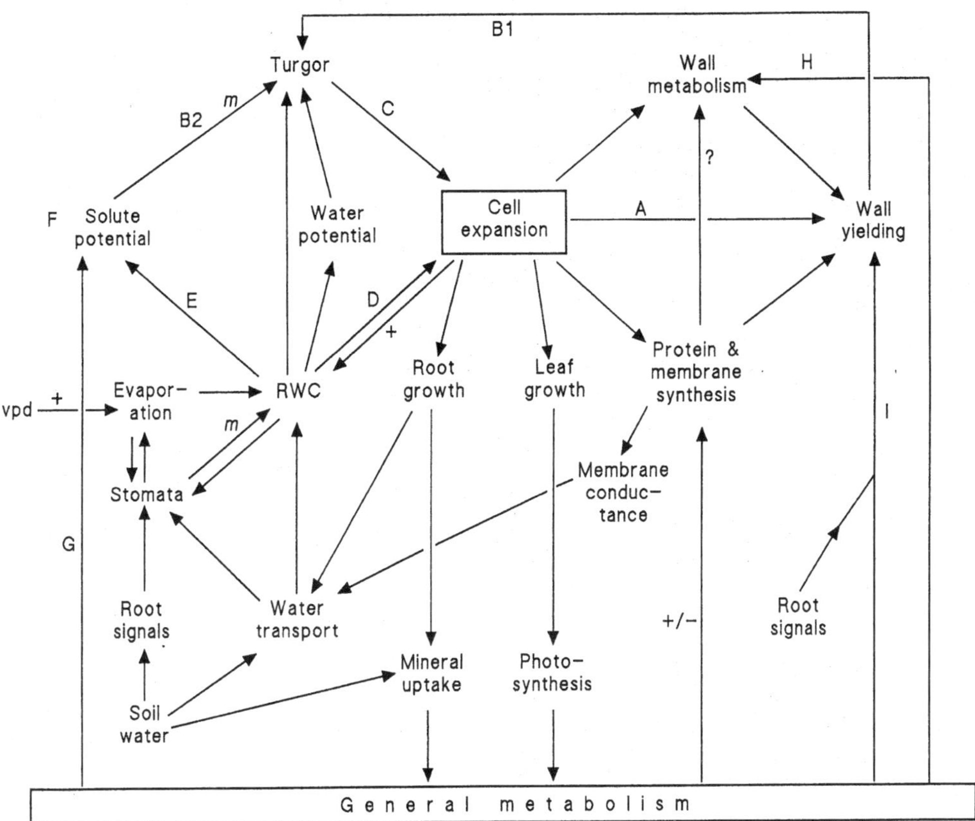

Figure1.2. Outline of inter-relationships between water relations characters that control the expansion of cells, modified after Hsiao, Silk and Jing (1985). For explanation of processes *A–I*, see text. Drought reduces (or makes more negative) most processes or states; exceptions are indicated as follows: + increased, *m* maintained.

The first expression shows that the driving force for the cell to expand in volume by drawing in water is the difference between pressure potential and the symplast-environment osmotic gradient. It is constrained, however, by the volumetric conductance of the cell membrane, C, a measure of how easily water can pass through it (units s^{-1} MPa^{-1}). The variable σ is the reflection coefficient, which quantifies the effectiveness of the plasmalemma at filtering out solute particles, and is set to unity for a perfect semipermeable membrane. This is an oversimplification because membrane permeability is controlled by aquaporins, special channels which facilitate water transport while excluding ions and metabolites (Chrispeels and Maurel, 1994).

The second expression shows that above a certain yield threshold (Y, MPa) in pressure potential, increase in volume is driven by turgor, but restrained by the plastic extensibility (irreversible stretchiness) of the cell wall (ϕ, s^{-1} MPa^{-1}).

All these processes respond to each other (Hejnowicz and Sievers, 1995), to the environment and to general plant metabolism in a dynamic way (Figure 1.2). There is still much uncertainty over quantifying some of the biophysical components and their effects on growth (Hsiao, *et al.*, 1985; Pritchard and Tomos, 1993) because it is difficult to carry out univariate experiments and to measure phenomena on such a small scale. It seems likely that as the cell wall irreversibly yields and stretches under pressure (Figure 1.2, process [A]), the cell expands and therefore loses ψ_P [B1], the driving force for growth, which slows [C]. The cell also draws in water [D], ultimately from the xylem stream, so that the cell sap becomes diluted [E] and the osmotic gradient ($\Delta\psi_s$) is reduced [F]. Again, turgor and growth are reduced [B2, C]. This can be counteracted by accumulation of solutes [G] (such as potassium, amino acids and sugars) (a) because growth is reduced more than imports, (b) because the degree of polymerization of solutes is reduced, or (c) because imports are increased. The extensibility of the cell walls may depend on the synthesis of cell wall material [H], on the "loosening" of the fibres by hormone-stimulated proton extrusion [I], and on the activity of enzymes such as xyloglucan endotransglycosylase (XET) and peroxidases (Fry, 1995).

RESPONSES TO DROUGHT

Adaptation and Acclimation

Adaptation in a species, population or plant can be regarded as modification of structure or metabolism in response to change in the environment that improves prospects of survival either of the individual or of its progeny. Adaptation may occur in two ways. Firstly, plants may modify their phenology, structure or metabolism to alleviate the effects of stress; the changes may take place over the space of a few seconds, or gradually during an entire growing season. This process is usually referred to as acclimation. Secondly, in populations, environmental changes may cause some genotypes to be favored over others, leading eventually to evolution and speciation (Hoffman and Parsons, 1991). Breeding for stress resistance is merely an accelerated version of this process, but is directed more towards human ends (i.e. maintenance of yield).

Not all changes in response to drought are acclimatory — they may be the result of disruption or imbalance in metabolic processes — and their characterization and significance have been the subject of much discussion (Hanson and Hitz, 1982; Hanson, *et al.*, 1986).

Those characteristics that contribute to stress resistance (e. g. hairy or glaucous leaves) and which are expressed whatever the environment, are termed "constitutive".

Life Cycle or Phenology

In climates with pronounced wet and dry seasons, most annuals have evolved to complete their life cycles before drought becomes too severe; this method of drought resistance was termed "drought escape" by May and Milthorpe (1962). A short life cycle and rapid maturity have been identified as valuable traits for drought resistance of annuals in such environments. In Israel, for example, late maturing sorghum varieties tend to be sown in wetter areas, and early varieties in drier areas (Blum, 1979). In a varietal comparison of wheat grown in Mexico, Fischer and Maurer (1978) quantified the contribution of drought escape to yield at 3–8 g m^{-2} d^{-1}. Similar advantages of early over late maturing crops have been described in pearl millet (*Pennisetum americanum* [L.] Leeke) by Bidinger *et al.*, (1987), and in soybean (*Glycine max* L.) by Muchow and Sinclair (1986).

The severity and timing of drought may vary considerably from year to year, and one way to accommodate this might be to select for developmental plasticity, so that in wetter years crops might grow for longer and accumulate more harvestable crop. Such advantages have been shown in cowpea (*Vigna unguiculata*; Lawn, 1986), and in pearl millet subjected to mid-season drought (Bidinger *et al.* 1987). On the other hand, Bremner and Preston (1990) found that the strong determinate habit of sunflower (*Helianthus annuus* L.) gave it a yield advantage over sorghum (*Sorghum bicolor* L.) during terminal drought. Bidinger *et al.* (1987) concluded that "if probabilities of the occurrence of stress at different times in the crop cycle are not known, it is not possible to use phenology to capitalize on drought escape potential". If plasticity is not desirable, matching phenology to drier seasons will give lower but more stable yields, while matching to wetter years would increase risks but give higher potential yields (Baker, 1989).

When sporadic rainfall occurs during mainly dry seasons, plant responses are more complicated because different developmental stages vary in their sensitivity to drought and in their contribution to yield. This is most easily determined by imposing experimental irrigation or rainfall exclusion at different stages of the life cycle (e.g. Harris and Williams, 1981). For example, irrigation of crops such as peas (*Pisum sativum* L.) or strawberries (*Fragaria x ananassa*) during the vegetative stage may induce excessive vegetative growth at the expense of flower and fruit production.

Drought during flowering may be particularly damaging, since it reduces pollen viability and germination (Ekanayanke *et al.*, 1990), disrupt synchrony of ripening in gynoecium and androecium in maize (*Zea mays* L.; Struik *et al.*, 1986; Edmeades *et al.*, 1989), and delays ontogeny and therefore the onset of flowering (Clarkson and Russell, 1976; Ludlow and Ng, 1977; Manjarrez-Sandival *et al.*, 1989; Bremner and Preston, 1990).

The early stages of development after fertilization appear to be the most sensitive in some crops, such as soybean (Muchow and Sinclair, 1986) and cotton (*Gossypium hirsutum*) (Kock *et al.*, 1990), although Husain *et al.*, (1990) concluded that *Vicia faba* was equally sensitive to drought at all developmental phases. Once flowering has begun, ontogeny may be accelerated by stress (Clarkson and Russell, 1976; Manjarrez-Sandoval *et al.*, 1989).

Drought may not always decrease yields, however: mild stress during flowering improved flower production in *V. faba* (Grashoff, 1990) and lint yields in cotton (Kock *et al.*, 1990).

Perennial forages are more sensitive to drought when flowering than when vegetative (Thomas and Evans, 1990). Those from dry climates may complete flowering and set seed before the onset of severe drought, and then survive in a quiescent state until the return of the autumn rains, when their ability to recover rapidly is an important agronomic trait. Such plants must be able to accumulate reserves of carbohydrate (Volaire, 1995) and nitrogen such as vegetative storage proteins (Bray, 1994) during early drought to be able to survive the summer.

We now turn to the beginning of the life cycle, the deposition of seed and subsequent germination and establishment. In some species (groundnut [*Arachis hypogaea*], subterranean clover [*Trifolium subterraneum*] , *Cymbalaria*), the fruits are forced into the ground or into cracks between rocks; the awns of *Avena fatua* screw the caryopsis into the ground; many legumes such as clovers (*Trifolium*) and medics (*Medicago*) produce both soft (rapidly germinating) and hard (slow germinating) seed, which improves the chances of establishment in an unpredictable environment.

The young seedling is very sensitive to drought, with legumes more sensitive than grasses, possibly because their roots are less able to penetrate soil cracks (Humphreys, 1981) or to produce adventitious roots if the seminal root dies (Bassiri *et al.*, 1988). Genetic differences in drought resistance of mature plants may be manifested in seedlings, and this may provide useful opportunities for screening and selection (Sullivan and Ross, 1979; Morgan, 1988). Priming seeds, by rehydrating them in osmotica that prevent germination before sowing, may improve establishment under water stress (Bradford, 1986).

Control of Water Uptake by Roots

Plants of many species respond to drought by increasing the proportion of assimilates diverted into root growth, and thus increasing the root/shoot ratio and the volume of soil water available to the plant (O'Toole and Bland, 1987). This could be due to differential sensitivity of roots and shoots to endogenous ABA, and to greater osmotic adjustment in roots (Sharp and Davies, 1989). Under some circumstances (Sharp and Davies, 1979), total root mass may be increased under drought.

As the soil dries there is a slowing of water movement into the root (Klepper and Rickman, 1990). This reduced flow implies an increase in resistance, the nature of which has been discussed by Boyer (1985) and Passioura (1988a); the consensus is that the resistance lies not in the bulk soil but in the root itself (perhaps due to embolism in the xylem: Milburn, 1973) or in the root/soil interface (perhaps due to root shrinkage away from the soil: Huck *et al.*, 1979).

The optimum pattern of growth of the root system depends on a combination of rainfall distribution through the year, soil depth, soil physical characters (texture, water holding capacity), shoot water demand in the developing crop, and plant-to-plant competition. Many root characters affect water uptake, and include total mass, vertical distribution, maximum depth explored, branching, diameter, root hairs, and possibly benign infection by symbiotic mycorrhizae. The genetic variation in these characters and in their phenotypic plasticity has been reviewed by O'Toole and Bland (1987).

In ecosystems, the various species may occupy different environmental niches partly by rooting most densely in particular soil strata (e.g. Sydes and Grime, 1984). Such compatibility can be exploited in arable agricultural systems, where root exploration of different soil horizons by companion species may lead to greater water use efficiency than shown by either species in monoculture (Ofori and Stern, 1987).

Where soil water deficits are never a problem, a large root system in a crop is a waste of resources that would be better diverted into producing more economic product (O'Toole and Bland, 1987), although a dense surface root mass may prevent invasion by weeds. In drier climates where soils are deep, have a large soil moisture holding capacity and are fully recharged by rain every year, then a deep root system could be an effective use of resources. In perennial plants such as grasses (Troughton, 1957) and trees (Pereira and Pallardy, 1989), the root system may be laid down outside the main shoot growing season, and is already in place should drought occur. Among forage crops, a deep root system is exhibited by tall fescue (*Festuca arundinacea* [L.] Schreb: Garwood and Sinclair, 1979), lucerne (alfafa, *Medicago sativa* L.) and some white clover (*Trifolium repens* L.) populations (Woodfield and Caradus, 1987). In general, it seems that only a few deep roots may be needed to supply water to a droughted plant (McWilliam and Kramer, 1968; Sharp and Davies, 1985; Passioura, 1988a). The effectiveness of roots in taking up water seems to decline with distance from the tip (Hansen, 1974) and it may be that newly produced roots will be more effective than ones formed earlier.

A more efficient use of resources, particularly by annual crops, is for the rooting system to be phenotypically plastic, and explore deep in the soil only when stimulated by drought. For example, Hurd (1974) found the most effective wheat variety for semi-arid conditions was one which had both a high root mass in soil horizons above 30cm, and the ability to explore more deeply if necessary. Deep rooting has contributed to drought resistance in rice (*Oryza sativa* L.) (Mambani and Lal, 1983), sorghum (Boyer *et al.*, 1980), and maize (*Zea mays* L.) (Lorens *et al.*, 1987), but not in potato (*Solanum tuberosum* L.) (Steckel and Gray, 1979).

Two disadvantages of deep roots are (a) that they supply mainly water but few nutrients, since most available minerals are concentrated in the topsoil (Garwood and Williams 1967); and (b) that in acid soil they may be poisoned by an excess of aluminium ions, a phenomenon that may be alleviated by breeding (Goldman *et al.*, 1989)

In shallow soils or under terminal droughts where the only water available is that stored in the soil following the last rains, a smaller, slower growing root system with narrow xylem vessels (Richards and Passioura, 1989) may regulate the water supply early in the season so that some water remains for the crop to complete its life cycle.

Control of Water Loss Through Leaves

Stomatal conductance appears to be controlled by three environmental response mechanisms (Schulze, 1986). The first involves evaporation from the epidermis, particularly in the region of the stomata where the cuticle is thinner and presumably more permeable (Maier-Maercher, 1983). There is some dispute as to whether stomatal changes are best explained by variation in relative humidity (Schulze, 1986) or vapor pressure deficit (Aphalo and Jarvis, 1991).

The second mechanism is in response to bulk leaf water status, the relationship between conductance and ψ depending on the nature of the drought stress (Schulze and Kuppers, 1979; Gollan *et al.*, 1985; Turner *et al.*, 1985). Stomata do close, however, when bulk turgor is lost; this appears to be due to reallocation of preformed abscisic acid (ABA) from chloroplasts to the guard cell apoplast where it modifies ion metabolism of the plasmalemma and stimulates closure (Hartung *et al.*, 1990).

The third mechanism for stomatal closure is a response to chemical signals (probably ABA), which are generated by roots experiencing dry soil and transmitted through the xylem to the guard cell apoplast. This can result in partial stomatal closure even before there is any drought-induced change in bulk ψ (Zhang and Davies, 1989).

The relative importance of these mechanisms in different species is discussed by Tardieu (1993).

During acclimation, stomata may become less sensitive to stress, and transpiration is therefore maintained (McCree, 1974; Brown *et al.*, 1976; Henson *et al.*, 1983).

The result of stomatal closure is, of course, to reduce water vapor diffusion out of the leaf, but the reduction is often not so great as one might at first expect. Jarvis and McNaughton (1986) have shown that for forage and arable crops in the field, evaporation depends mainly on characteristics of the bulk atmosphere rather than on stomatal aperture. Stomatal closure will, however, help conservation of water within the leaf tissues, and hence delay desiccation (Blum *et al.*, 1981).

Plants exhibit a number of other mechanisms that help reduce water loss. Epicuticular wax both decreases water loss through the cuticle (Chatterton *et al.*, 1975) and may also reflect some insolation which reduces leaf temperature and therefore vapor pressure deficit (vpd) gradients and transpiration load (Johnson *et al.*, 1983). The reflectivity of the bloom or glaucousness depends both on the quantity of the wax and on its configuration (smooth or fluffy). Genetic variation in waxiness has been associated with drought resistance in wheat (Johnson *et al.*, 1983), barley (Baenzinger, 1983), *Eragrostis* (Tischler and Voigt, 1990) and tomato (*Lycopersicon esculentum.*) (Jain *et al.*, 1989) but not in maize (Peters, 1987). In many of these species, there was phenotypic plasticity shown by an increase in glaucousness during drought.

Leaves may also reduce transpiration rate and insolation load by rolling, which reduces exposed leaf area and isolates abaxial stomata from the environment. Other morphological and leaf-movement mechanisms of alleviating evaporative stress have been described by Turner (1979) and Ludlow and Muchow (1990).

The rate of water loss from excised leaves, which is a measure of stomatal conductance, has shown promise as a selection criterion for drought resistance in wheat (Clarke and Townley-Smith, 1986).

Changes in Water Relations and Growth of Cells and Organs

Rate of imposition of stress

The growth of organs depends on the rate at which new cells are produced, and the rate at which these cells can expand. Both processes are sensitive to turgor, but in proportions that probably depend on tissue, species and stress (Kirkham *et al.*, 1971; Meyer and Boyer, 1972; Clough and Milthorpe, 1975). The pattern of cell division and extension in leaves is very different in dicotyledons and monocotyledons (Aspinall, 1986; Dale, 1988). In broad leaves, division occurs all over the young leaf, particularly near the base, and as the leaf ages, the number of dividing cells declines (Clough and Milthorpe, 1975) while expansion continues. In the Gramineae, nearly all division and extension occurs at the base of the young leaf, which is enclosed in the older leaf sheaths and therefore insulated from the external environment. One of the first symptoms of drought is the reduction of cell expansion and therefore leaf growth; in grasses, as drought becomes more severe, the expansion zone becomes increasingly shorter from the distal end (Durand *et al.*, 1995).

There has been much conflicting evidence in the literature on the significance of many of the changes observed in biophysical and metabolic acclimation to drought. In many cases, this may be because of the time-scale of the experiments. For example, work in which soil moisture potential is changed rapidly (a few seconds for osmotic shock to a few days drying in a pot-grown plant) may add more confusion than enlightenment, since it shows you what a plant *can* do, not what it *does* do. Comparisons of physiological responses in controlled environments and the field, or of rates of droughting are given by Watts (1974), Jones *et al.*, (1980) and Thomas (1986).

Osmotic adjustment

An acclimatory character that has received much attention over the last 15 years (in part because it is easily measured on bulk tissue) is osmotic adjustment —the lowering of the solute potential in response to water stress. This may occur in two ways: passive adjustment as a result of a lowering of the tissue relative water content, or active adjustment by accumulation of salts and metabolites or by developmental reduction in cell volume. In general, under water stress, photosynthesis is affected less than the consumption of assimilates during growth (see below), and so sugars (but often not starch) and other metabolic products accumulate, although their relative contribution to osmotic adjustment depends on species, tissue and rate of droughting (Wardlaw, 1969; Munns and Weir, 1981; McCree *et al.*, 1984; Van Volkenberg and Boyer, 1985). Osmotic adjustment can be controlled by interconversion between polysaccharides (starch, fructans) and oligosaccharides (sucrose, glucose), since osmotic potential depends on the number of molecules. Hendry (1993) has hypothesized that many temperate species such as grasses store fructans rather than starch because, during their evolutionary history, they were forced to adapt to drought and other stresses.

The sugars, and probably mineral ions too, will tend to accumulate in the vacuole, so in order to prevent osmotic gradients across the tonoplast concentrations of organic molecules in the cytoplasm are increased. The most abundant of these "compatible solutes" are proline and the betaines; their advantage is that they cause relatively little disruption to metabolic processes and may also protect protein configurations during dehydration (Storey and Wyn Jones, 1978; Wyn Jones and Pritchard, 1989). Proline is very labile, and accumulates in growing cells after only a few weeks of drought, while in mature tissues it appears to be a symptom of impending death (Turner, 1990; Thomas, 1991).

The obvious advantage of osmotic adjustment to the plant is that $\Delta\psi_s$ across cells and tissues is maintained against a fall in ψ during drought, and hence any drop in turgor is prevented or reduced (Figure 1.3A). Although turgor maintenance will not help maintain cell growth if there are other constraints (see below), it *will* prevent damage due to plasmolysis or loss of cell volume.

Osmotic adjustment has been associated with drought resistance with some success (e.g. Morgan, 1984), although this approach has been criticized by Munns (1988). Osmotic adjustment has been positively associated with maintenance of grain yield in many species: barley (Singh *et al.*, 1972; Blum, 1989); wheat (Morgan, 1977, 1984; Morgan and Condon, 1986); sorghum and millet (Blum and Sullivan, 1986; Wright *et al.*, 1983); *Brassica* spp (Kumar *et al.*, 1984); and cotton (Timpa *et al.*, 1986). Further examples can be found in the reviews

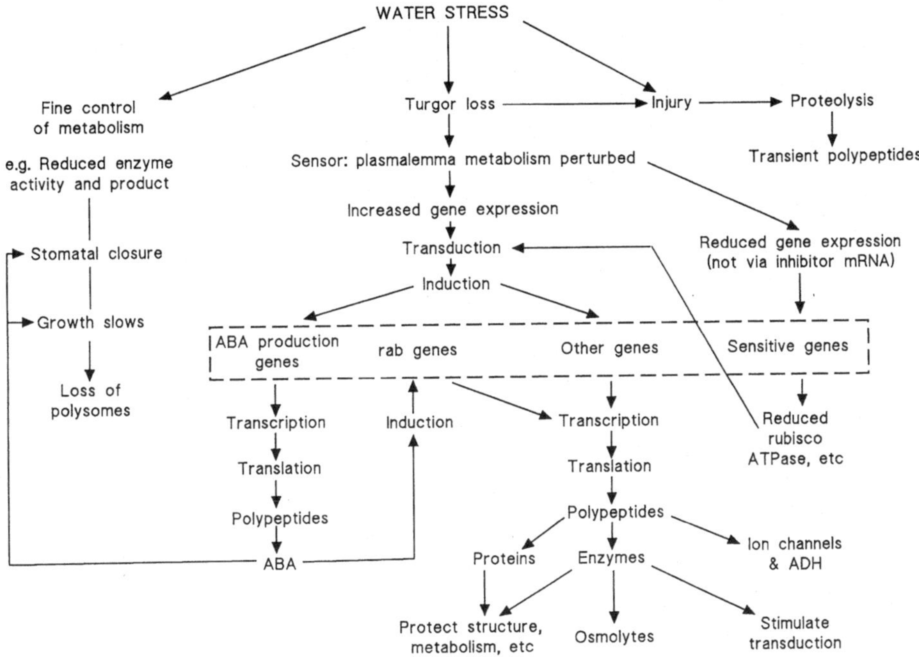

Figure.1.3. Summary of changes at the molecular and biochemical level following exposure to water stress. Skriver and Mundy (1990) and Bray (1994), and from sources quoted in the text. Genes responsive to abscisic acid are shown as rab.

by Morgan (1984), Singh (1989), and Ludlow and Muchow (1990). To set against these relationships are negative correlations between dry matter production and osmotic adjustment in cotton (Quisenberry *et al.*, 1984); barley (Grumet, Albrechtsen and Hanson, 1987); wheat (Johnson *et al.*, 1984); and ryegrass and fescue (Thomas and Evans, 1989; Amin, 1990).

These conflicting results may be a consequence of the correlation of osmotic adjustment with a range of growth and metabolic processes (summarized by Singh, 1989; Ludlow and Muchow, 1990), from whom the following list is largely taken.

Positive correlation with osmotic adjustment: root exploration for water at depth, leaf water conductance (stomata stay open), leaf area duration (less senescence), harvest index (due to reduced abortion and improved assimilate supply), survival of severe stress (through maintained turgor), transpirational cooling, possible retention of assimilates in assimilating leaves (and hence reduced allocation for tillering, Thomas and Evans, 1990).

Negative correlation with osmotic adjustment: water loss from excised tissue, leaf rolling, ABA production, electrolyte leakage from membranes.

Thus, Santamaria *et al.*, (1990) found that sorghum lines differing in osmotic adjustment had very similar total above-ground yield under drought, but that the better adjusters had higher harvest indices, and hence greater yields.

Constraints to cell expansion

When plants are subjected to drought, even before there is any loss in turgor, the walls of growing cells in leaves and stems of plants generally become less plastic and expansion declines, with a resulting reduction in organ growth. This may be due to reduction in ϕ or increase in Y (Equation 2) (Davies and Van Volkenberg, 1983; Van Volkenburgh and Boyer, 1985; Matthews *et al.*, 1987). Cell wall metabolism, like stomatal opening, is moderated by growth-retarding substances (phytohormones) such as ABA originating in the root system and transported to the meristems (Passioura, 1988b; Zhang and Davies, 1989). Root cell walls seem to be less sensitive to water stress, and hence root growth may be maintained even when shoot growth has stopped (Sharp and Davies, 1989; Pritchard *et al.*, 1990).

Changes in membrane and tissue conductances also occur during stress; in young stems of soybean, for example, water stress decreased conductance of cortical cells (Nonami and Boyer, 1990). As drought progresses, the conductance of the whole plant decreases, perhaps due to cavitation or embolisms in the xylem vessels (Grace, 1993).

Mature cells are not plastic, but are elastic. In most plants, as they adapt to drought, the bulk modulus of elasticity ($\in$) of the tissue increases; i.e. the tissue becomes stiffer. In theory, a flexible leaf (low $\in$) would delay the loss of turgor as RWC declines; thus, Cutler *et al.*, (1977) found that in rice the production of small cells during drought increased elasticity and obviated the “need” for osmotic adjustment. Alternatively, a stiff leaf (high $\in$) would enable a low ψ to develop for only a small loss of relative water content and cell volume (Figure 1.2B) and hence help maintain water potential gradients and water uptake; it would be able to maintain turgor if osmotic adjustment occurred. Although during drought cells may become smaller (Jones *et al.*, 1980) they do not necessarily make the tissue more flexible, since the walls may also become thicker.

Carbon Assimilation and Metabolism

Stomatal and non-stomatal limitations

As drought develops photosynthesis declines, although not so rapidly as the decline in leaf growth (Jones *et al.*, 1980; McCree, 1986). Respiration measured on a single leaf basis tends to decline more slowly than photosynthesis, and in the earlier stages of drought may actually increase (Levitt, 1980). Measured on a canopy basis, however, respiration as a proportion of photosynthesis initially declines in parallel with leaf growth but then increases as senescent leaf mass accumulates (McCree, 1986), presumably reflecting the metabolic demands of disassembling and exporting protoplast degradation products.

The photosynthetic mechanism in the chloroplasts is particularly robust, and during the early stages of drought the main limitation to photosynthesis is due to stomatal closure (Schulze, 1986; Chaves, 1991; Davies and Pereira, 1992). Eventually, photosynthesis declines even when measured under very high ambient CO_2 concentrations, implying that the photosynthetic apparatus itself is being affected (Schulze, 1986). Kaiser (1987) tabulated the probable effects of increasing dehydration as follows:

RWC 100–70%: Reduction in photosynthesis due to stomatal closure, which is rapidly reversible.

RWC 70–35%: In bright light, photosynthetic capacity is reduced, and recovers only slowly on rehydration. The main cause could be photoinhibition, since carboxylation, the Calvin cycle and photorespiration are all reduced; electron transport appears to be rather more robust.

RWC < 30%: Irreversible decrease in photosynthetic capacity, leading to death, due to membrane damage in the chloroplast.

Chaves (1991) has drawn attention to methodological problems in evaluating the relative importance of non-stomatal inhibition. An early adaptive response to water stress, such as occurs in osmotically shocked leaf discs of spinach, is the increased partitioning of assimilates into sucrose rather than into starch, even before there is an effect on photosynthetic rate (Quick *et al.*, 1989); this could lead to osmotic adjustment in the mesophyll cells. One cause of reduction in photosynthetic potential seems to be loss of stromal volume, which under moderate stress may be maintained by osmotic adjustment in the chloroplasts (Gupta and Berkowitz, 1988).

In the field, water stress usually coincides with high light and temperature, and the combination of these factors can lead to photoinhibition (Björkman and Powles, 1984) although this does not appear to be of agronomic significance in sorghum (Ludlow and Powles, 1988).

Water use efficiency

The ratio between production and water use is conventionally referred to as water use efficiency (WUE, or W), although the term has been criticized by Monteith (1986) and Stanhill (1986). Its precise definition varies according to the system under investigation. For crop studies in the field

$$W = D/E \qquad \text{Equation 3}$$

where D = dry weight gain of the entire plant, the above-ground parts, or the harvestable yield, and E = total evaporation from the crop and soil, or merely transpiration through the crop. At a more fundamental level, W can be regarded as the ratio of carbon fixed to water transpired, expressed as weight or mol. An ingenious method of retrospectively deriving the WUE of C_3 crops is to measure the change in discrimination of photosynthesis against naturally-occurring $^{13}CO_2$ during drought, by measuring the ratio ^{13}C to ^{12}C on dried tissue samples (Farquhar and Richards, 1984). This has been used to sample for high WUE, though not necessarily drought resistance, in cereals (Condon *et al.*, 1990) and grasses (Johnson, 1994). Interpretation can be difficult in crops in which only part of the plant is harvested, especially when carbon reserves formed before drought are subsequently translocated to the harvested parts.

There are considerable differences in W over the world: Stanhill (1986) calculated the range of potential W (photosynthesis / transpiration, g kg^{-1}) for typical locations as follows:

Temperate	30 (winter) to 12 (summer)
Desert and semi-arid	13 (winter) to 5 (summer)
Tropical and equatorial	8

The low values in summer and in the warmer climates are due to high values of vpd (vapor pressure deficit) and hence greater evaporative load. The importance of vpd was demonstrated by Turner *et al.*, (1984) who showed that, for a range of species, increasing the vpd gradient by a factor of 2.5 increased transpiration by 115–196% but decreased photosynthesis by 9–42%. Another example is the WUE of tomatoes growing in Peru, which was three times greater in the humid winter than in the low humidity of summer (Trebejo and Midmore, 1990).

Management may help improve WUE by reducing water wasted by evaporation from the soil and therefore increasing the "useful" water available for transpiration through the crop. The aim is to increase the ground cover by leaves as soon as possible in the season: early growth at low spring temperatures, fertilizer regime and weed control may all make an important contribution (French and Schultz, 1984 a,b).

At the end of a drought when the soil surface is dry, all evaporation will be transpiration, and there is the possibility of reducing it by reducing leaf area. This will only conserve water when the LAI is below 3 (Ritchie and Burnett, 1971; Johns and Lazenby, 1973), and is appropriate only to grain crops which can draw on assimilates translocated from dying leaves and stems, or to perennial crops entering a period of dormancy, quiescence or suspended growth. In forage crops, loss of photosynthetic leaf area (along with stomatal closure) is the major cause of the decline in canopy assimilation rate during drought (Jones *et al.*, 1980). Drought may even suspend the ageing process, so that on re-watering photosynthesis exceeds that in unstressed leaves of the same chronological age (Ludlow, 1975).

The largest and most consistent metabolic process contributing to differences in WUE is the photosynthetic pathway. C_4 plants, because of their efficient CO_2 "scavenging" mechanism and ability to utilize very high irradiance, can maintain photosynthesis at lower stomatal conductances than C_3 plants; the net result is that they use water about twice as efficiently (Stanhill, 1986). The ways in which stomata may respond to different environments in order to maximize WUE have been described and modelled by Cowan (1977) and Jones (1980). Cowan's model predicts that under well-watered conditions, plants can afford to be profligate in their water use, and maintain high stomatal conductances all day. Under dry conditions it is best for the stomata to close around the middle of the day when the evaporative load is greatest, and to open for assimilation (and transpiration) in the morning and evening, when light is adequate but stress is low. Such a response has often been observed in the field in dry environments (Slatyer, 1967).

Jones' model responses apply to the whole life cycle of the plant in relation to water supply through the season. The two extreme patterns of water use are termed (a) optimistic (or water spenders, Levitt, 1980), which operate little control over water loss and rapidly exhaust soil water reserves, and (b) pessimistic (or water savers), which have constitutively low rates of water uptake (see discussion of root growth above) and transpiration. Water spenders will tend to be most effective in conditions where stress is unlikely, and will be more aggressive under competitive situations. Water savers may only be at an advantage during terminal drought, when it is necessary to meter out water through the season (Richards and Passioura, 1989); in mixed communities they could be at a competitive disadvantage if they leave water in the soil which is taken up by more profligate plants. In nature, stomatal responses to stress mean that plants lie between these extremes.

Assimilate distribution

Sugars manufactured in the leaves may stay there or be transported (as sucrose) to other parts of the plant via the phloem. They may be consumed in respiration, be used in manufacturing more plant tissue, contribute to osmotic adjustment, or be diverted into storage. The ways in which plants may change their priorities in response to drought have been reviewed by Daie (1988), and a more general review of partitioning is given by Wardlaw (1990).

During the early stages of drought, as we have seen above, growth may be more reduced than photosynthesis, and assimilates will accumulate; later, if assimilation is further reduced, there may be conflicting demands for assimilate, which may be met by mobilizing reserves (Huber *et al.*, 1984). Transport in the phloem appears to be relatively unaffected during drought (Wardlaw, 1969), and so distribution will depend on loading at the source and demand at the sink.

If there are gradients of apoplastic water potential in the plant, then assimilates may be diverted to the parts with the lowest potential (Lang and Thorpe, 1986). An example of the nature of changes in partitioning during drought, and also the influence of genetic diversity, is provided by Nicolas *et al.* (1985). They imposed a drought stress just after anthesis on two wheat varieties, one resistant and the other susceptible to drought. Carbon partitioning responded to stress more quickly than did photosynthesis. Under moderate drought the roots were strong competitors for available assimilates, especially in the resistant variety, and adjusted osmotically mainly due to increased potassium. Although root respiration declined it became more efficient, as less took place through the alternative (cyanide-resistant) pathway, a sign of short-term adaptation. Later, the grains became stronger competitors than roots for carbon, but imported less in droughted than in watered plants; as a consequence, reserves accumulated in the stems.

Harvest index (HI), is the proportion of total plant weight or above-ground parts diverted into the economic part, usually grain. Passioura (1977) suggested carrying out direct selection for HI under drought, but there are few records of success (Morgan, 1989). Furthermore, there is obviously little scope for modifying HI in crops where the economic part is the foliage (e.g. forage, cabbage).

Assimilates produced before anthesis and stored in the stem can be used in grain filling, and there is genetic variation for this in wheat and barley (Austin *et al.*, 1977, 1980). The ability of plants growing under terminal drought to use stored assimilates to maintain grain growth has been assessed even in wet soil by killing the leaves with a chemical desiccant about 2 weeks after anthesis (Blum *et al.*, 1983).

Drought and Mineral Nutrition

Drought and mineral nutrition may interact in a number of ways, as a consequence of (1) reduced transport of ions through the soil to the roots; (2) modified uptake of ions by roots; (3) changing demand of shoots and roots for minerals; (4) reduced transport through the plant; (5) deficiencies or accumulation of ions which may disrupt metabolism or induce acclimatory responses.

Availability, absorption and mineral content

The flux of ions through the soil to the root surface is the sum of two processes: diffusion and mass flow, of which the former appears to be the most important (Nye and Tinker, 1977). During drought the ion most likely to be unavailable to plants is phosphate, because (1) it is tightly adsorbed onto clay particles, so that only a small proportion of total soil phosphate is in solution; and (2) its diffusion coefficient is 10^4 to 10^7 times lower than that of potassium, nitrate and ammonium.

The effect of soil water stress on mineral uptake has been reviewed by Pitman (1981). The influence of transpiration on mineral uptake appears to be important mainly under conditions of high nutrient status when root uptake is non-limiting (Lazaroff and Pitman, 1966). There is also a basal rate of uptake that is flow-independent (Baker and Weatherley, 1969). Tanner and Beevers (1990) found that varying the transpiration rate of maize by a factor of 2–3 had little effect on shoot mineral content, and suggested that minerals are transported by a non-transpirational xylem flow. During drought, growth rates and mineral uptake may be reduced to a similar extent, so that plant mineral content may be relatively stable (Pitman, 1981).

Nitrogen and drought

During drought in the field, nitrogen content and uptake may be increased, reduced or unaffected (Clark, 1981). Even when plants are grown in nutrient solution, a much simpler experimental system, osmotic stress may decrease nitrate uptake (Meyer and Gingrich, 1966) or increase it (Talouizite and Champigny, 1988). High applications of nitrogen fertilizer have lead to greater vigor and deeper water extraction in droughted wheat (Barraclough, 1989), and to greater water use efficiency in forage grasses (Lemaire and Denoix, 1987).

Once inside the plant, nitrate is converted to nitrite (catalyzed by nitrate reductase, NR), which is converted to ammonia (enzyme: nitrite reductase, NiR), and thence via glutamine (enzyme: glutamine synthetase) and glutamate (enzyme: glutamate synthase) to amino acids and proteins. Conversion to glutamine may occur in roots or leaves, depending on species. Hanson and Hitz (1982) proposed two design principles to explain changes in N metabolism during drought:

(1) to achieve a balance between nitrogen and carbon metabolism, by decreasing enzyme synthesis;
(2) to maintain potentially toxic nitrite and ammonia at low concentrations.

Thus, the activity of NR, the first enzyme in the chain, is more sensitive to water stress than NiR (Huffaker *et al.*, 1970), reducing formation and accumulation of nitrite. The next enzymes, glutamine synthetase and glutamate synthase, are also robust, so that accumulation of ammonia is prevented (Hanson and Tully, 1979). Photorespiration is less sensitive to drought than many other metabolic processes, and will therefore continue to consume glutamine (Lawlor and Fock, 1977).

In legumes, nitrogen fixation is sensitive to even a small drop in ψ (Sprent, 1973; Engin and Sprent, 1973; Guerin *et al.*, 1990; Paraja-singham and Knievel, 1990), and more affected by drought than photosynthesis, N uptake or nodule respiration (Sinclair *et al.*, 1987; Kirda *et al.*, 1989; Davey and Simpson, 1990).

Low nitrogen supply may affect biophysical processes, reducing hydraulic conductance in sunflower (Radin and Boyer, 1982) and cell wall extensibility in tomato (Radin, 1983).

Nitrate may also contribute to osmotic adjustment (Talouizite and Champigny, 1988), particularly under low light when carbohydrates and organic acids are in short supply (Blom-Zandstra and Lampe, 1985).

Phosphorus and drought

Phosphorus uptake is limited by drought not only by restricted availability in the soil, but also because the absorptive power of the roots declines with drying (Dunham and Nye, 1976). In barley, Day *et al.* (1978) found that P uptake was closely negatively correlated with the number of days the top-soil was dry, and it was likely that P deficiency limited yield. Once in the root, P is rapidly converted to nucleotides and other organic forms (Milthorpe and Moorby, 1979).

Mycorrhizae, commensual fungi whose hyphae both explore the soil and colonize roots, can have a profound effect on P uptake, especially in P-deficient soils. During drought, vesicular-arbuscular mycorrhizal (VAM) infection in crops may contribute to maintenance of growth, water use, stomatal conductance and ψ (Fitter, 1985, 1988).

Drought-induced phosphorus deficiency appears to impinge most seriously on metabolism of the adenylated nucleotides, ADP and ATP (Stonov and Petinov, 1980). Turner and Wellburn (1985) suggested that this could be a cause of the early reduction in growth that occurs before loss of turgor. On relief of water stress, levels of P_i in leaves of *Capsicum* and *Lolium* do not recover as rapidly as does growth (Turner, 1985; Thomas, 1991).

Phosphorus deficiency also affects the water relations of plants. Stomatal conductance can be greatly reduced, for example in *Festuca* (Pitcairn and Grace, 1982), *Trifolium* (Fitter, 1988), and *Coix* and *Chloris* (Saneoka, 1990). This may be because P deficiency increases sensitivity of stomata to ABA (Radin, 1984). P deficiency rapidly reduces root hydraulic conductance in cotton, leading to low ψ, reduced leaf expansion and eventually lower yield (Radin and Eidenbock, 1984).

Potassium and drought

Potassium is particularly important in plants as an osmoticum: it may make a 30–50% contribution to ψ_s, particularly in older leaf tissue (e.g. Munns *et al.*, 1979; Thomas, 1991). After prolonged drought in the field, potassium (and often sodium and calcium) accumulate in leaves of ryegrass (Nielsen, results presented by Pitman, 1981) and barley (Leigh and Johnson, 1983), and could have a role in osmotic adjustment. If potassium is deficient, its role as an osmoticum in leaves may be partly taken over, for example, by sugars and citrate in sunflower (Lindhauer, 1987), or by calcium and sodium in barley (Leigh and Johnson, 1983). In roots of maize subjected to osmotic shock, Sharp *et al.*, (1990) found that hexoses rather than potassium contributed to osmotic adjustment as cell volume decreased. Exposure of wheat roots to high levels of potassium can decrease cell wall extensibility (Pritchard *et al.*, 1990).

Potassium plays a key role in stomatal opening. A reduced supply of potassium therefore reduces stomatal conductance to CO_2 much more than it reduces internal conductances (Terry and Ulrich, 1973), because potassium is lost from the guard cells (Ehret and Boyer, 1979). Non-stomatal inhibition of photosynthesis during water stress may also be sensitive to potassium (Pier and Berkowitz, 1987).

Cell Metabolism

Recent research into the effect of dehydration at the cellular level has been drawn together in a recent volume (Close and Bray, 1993). The following account concentrates on events most likely to contribute to drought resistance in whole plants.

Proteins and enzymes

The sensitivity of protein synthesis to stress has been reviewed by Bewley (1981) and by Rhodes (1987). Rapid drought reduces the numbers of polysomes in cells, and hence protein assembly in maize (Hsiao, 1970) and in a number of other crop species (Aspinall, 1986). In wheat, however, water stress down to – 3MPa had no effect on polysome populations of mature leaves, but reduced those in growing tissues, indicating that polysome loss is a consequence of reduced growth (Scott *et al.*, 1979). Poikilohydric plants such as mosses also lose polysomes, but they conserve ribosomes, mRNA, and the protein synthesis complex in an inactive state, which is stable and responds rapidly to rehydration (Bewley, 1981).

Once formed, many protein molecules have only a short life because their turnover is so rapid. Protein degradation products, such as amino acids, will therefore accumulate during drought and contribute to osmotic adjustment (e.g. Munns *et al.*, 1979), or be stored and used as substrates for subsequent recovery (Blum and Ebercon, 1976).

These net changes mask the different responses of enzymes and other protein to desiccation *in vitro*, (Stewart, 1989) and *in vivo* (Dhindsa and Cleland, 1975; Bewley *et al.*, 1983). The enzymes of carbon and mineral metabolism in well-watered plants are inhibited to various extents as drought takes hold. Others, however, are more stable, or may even increase in activity, and it is likely that they are responsible for scavenging potentially damaging free radicals that would otherwise accumulate when plants are stressed (Luna *et al.*, 1985; Smirnoff, 1993).

Other stimulated enzymes may be responsible for the synthesis and transport of compatible solutes and osmoprotectants, and new proteins themselves may help stabilize existing cellular proteins (Skriver and Mundy, 1990).

Membrane stability

Much of the living cell is composed of membranes, which are not inert but are dynamic structures having a turnover time of only a few hours in an expanding cell (Steer, 1988). The ability of membranes to tolerate dehydration has been the subject of research for over 100 years (Levitt, 1980). When cells are severely dehydrated, turgor is lost, the protoplast shrinks, and the cell wall, if it is thin, may stay attached to the protoplast and collapse with it (cytorrhisis). If the wall is thick the protoplast may become completely detached as it shrinks (plasmolysis).

If the plasmalemma is damaged the cell contents may leak out and the cell will die. A number of models have been proposed to explain the nature of the damage to the plasmalemma: Levitt (1981) describes Iljin's proposal that the damage is mechanical, and his own that sulphydryl bonds are oxidized into S–S bonds. More recent theories consider that the phospholipid bilayer is disrupted to become either a leaky collection of hexagonal platelets or has undergone phase separation from a liquid-crystalline phase to a gel phase (Stewart, 1989).

Plants exhibit a number of mechanisms to reduce the effects of drought on membranes. Osmotic adjustment, which helps reduce loss of turgor and cell volume, has been discussed above. Changes in the physical and chemical constitution of membranes, such as increase in sterol/phospholipid ratio, or in levels of lipids and phospholipids, contribute to their stability and hence to desiccation resistance (Douglas and Paleg, 1981; Svenningsson and Liljenberg, 1986; Gantet *et al.*, 1990; Premachandra *et al.*, 1991).

Membranes, like proteins, are susceptible to damage by free radicals. Mechanisms that protect membranes during drought include (1) enzymic degradation of superoxide by superoxide dimutase, catalase and peroxidase; (2) inhibition of lipid peroxidation by alpha-tocopherol and other anti-oxidants; and (3) scavenging of free radicals by anions, sugars, amino acids and proline (Stewart, 1989).

Gene Expression

Although plants respond to different environmental signals in many different ways, there is a common sequence of processes (shown in Fig.1.3) which occur in three stages (Smith, 1990), namely:

(1) Perception of environmental signals (biophysical or ionic) by specific receptor or sensor systems. Sensors of drought include compression or stretching of the plasmalemma in response to changes in turgor or protoplast volume, and changes in mineral fluxes.
(2) Transduction of the signals by a chain of processes that operate either directly on metabolism or by modifying gene expression. Metabolic changes can be very rapid and can be regarded as "fine control", whilst changes (enhancement or suppression) in gene expression are slower and more energy-demanding ("coarse control"). Since acclimation to drought has many similarities to embryo development and dehydration in the maturing seed, a number of genes occurring in vegetative tissues are homologous with "late embryogenesis abundant" (*lea*) genes (Dure and Verma, 1993), especially in resurrection plants (Alamillo *et al.*, 1994). The subject of alteration in gene expression in response to water deficit has been reviewed by Bray (1994), and in response to ABA by Chandler and Robertson (1994).
(3) Evocation of a physiological response, making the phenotype more fit for the current environment. Our knowledge of acclimatory processes at the molecular level is still rudimentary and not very helpful, most work having been done on rapidly stressed (shocked) tissues rather than on plants that have gradually acclimated to drought. If genes contributing to particular acclimatory processes can be identified, however, the contribution to breeding and

physiology could be enormous. Such studies are further advanced in the fields of salt and heat tolerance than in drought resistance.

GENETIC DIVERSITY AND VARIETY IMPROVEMENT

Operational Classification of Drought-response Strategies

All of the acclimatory processes described above contribute to three main strategies of coping with drought: drought escape, dehydration avoidance and dehydration tolerance. The classification presented below is based on May and Milthorpe (1962), Levitt (1980), McWilliam (1989) and Ludlow and Muchow (1990).

(I) DROUGHT ESCAPE OR AVOIDANCE: the ability (constitutive or facultative) to complete growth before drought is severe, and to survive the dry season either as seeds (cereals, annual forages) or in a dormant, quasi-dormant or quiescent state (trees, shrubs, perennial forage).

(II) DROUGHT RESISTANCE: the ability to grow and yield during dry conditions.

(1) *Avoidance or delay of dehydration*: the maintenance of relative water content (RWC), protoplast volume or turgor during drought; growth may or may not be reduced, but most metabolic processes continue.

(a) Control of water and mineral uptake from soil, root growth, distribution and conductance matched to probability of drought and phenology of crop.
(b) Control of water loss through leaves. The balancing of water conservation against reduced evaporative cooling and C-fixation may be critical. Mechanisms include: stomatal closure in response to vapor pressure deficit (vpd), bulk ψ or root signalling; high cuticular resistance; high reflectance of insolation from wax or indumentum; leaf movements; leaf shedding; change from C_3 to CAM metabolism.
(c) Osmotic adjustment to maintain turgor, if not leaf growth.
(d) Changes in cell-wall characters, cell size.

(2) *Tolerance of dehydration*: the ability to endure periods of low RWC, volume or turgor, and to recover afterwards.

(a) Osmotic adjustment, changes in cell size and wall rheology to prevent plasmolysis and cytorrhisis.

(b) Membrane stability: changes in phospholipid density and permeability, osmoprotectants, superoxide scavengers.
(c) Robustness of photosynthetic apparatus
(d) Storage of fixed carbon, nitrogen etc. for rapid recovery.
(e) Ability to transport reserves to economic component under severe stress.
(f) Maintenance of functioning leaf tissue at low RWC and high temperature for rapid recovery.

Diversity in Response

Most plants possess a number of mechanisms, and species of similar drought resistance may achieve it in different ways (Table 1). For example, sorghum has a determinate growth habit (which limits its ability to respond to sporadic rainfall) and is a dehydration avoider and not able to tolerate low RWC (Ludlow and Muchow, 1990) whereas millet is more plastic in phenology and can tolerate lower RWC (Bidinger *et al.,* 1987). In Australian grasslands, siratro maintains RWC by early stomatal closure while companion grass species lose RWC but are able to maintain turgor by osmotic adjustment (Ludlow, 1989). In temperate grassland, tall fescue delays dehydration because it has deep roots and can roll its leaves, while cocksfoot is more dehydration tolerant and has fine surface roots that can take advantage of intermittent rain (Garwood *et al.*, 1979; Lemaire and Denoix, 1987; Thomas, unpublished). Trees such as olive, carob and laurel have leaves which are similar (hard, shiny and evergreen), but each species employs a quite different strategy in response to drought (Lo Gullo and Salleo, 1988).

The following rankings of cereal crops, in order of decreasing drought resistance, is given by Fischer (1989). Cool season: barley > triticale (complete) > durum wheat > bread wheat > triticale (substituted) > oats. Warm season: millet > sorghum > maize > rice. Recommendations for Haryana State in India (quoted by Singh, 1989) are *Eruca sativa* (requires only 150 mm rain) > mustard > gram > lentil > barley > wheat (350 mm rain).

In temperate regions, vegetable crops with a low requirement for water include carrots, parsnips, onions, radish, swede and turnip, while more water is required for leafy crops such as celery, lettuce, spinach and cabbage.

Among forage crops, which are less specialized than vegetables and cereals, a much greater range of unimproved species and bred varieties is available (Duke, 1981; Skerman and Riveros, 1990). Humphreys (1981) has rated the following tropical forage species as particularly drought resistant. Grasses: *Cenchrus ciliaris, Panicum antidotale, P. coloratum makarikariense, Sorghum almum, Urochloa mosambicensis.* Legumes: *Stylosanthes guianensis intermedia, S. hamata, S. humilis, S. scabra.* Of the temperate forage grasses that have been investigated (Hubbard, 1968; Garwood *et al.*, 1979; Norris and Thomas, 1982) the ranking, in decreasing order of drought resistance is *Bromus madritensis* >

Table 1.1 Summary of contributions of various physiological traits to drought resistance, particularly in the context of crop improvement. Based on Ludlow and Muchow (1990), and including information from other published reports quoted in this section. Symbols: + advantageous, o neutral, – deleterious, ~ depends on circumstances; normal type, conjectural; **bold**, reasonably certain. Recommended traits: number of authors (Blum 1988, and in Baker 1989) recommending trait [+] or not [–]; * indicates trait particularly recommended.

Crop:	*Grain*			*Forage*				*Recommended trait*
Climate: temperate (T), dry (D):	*T*	*D*	*D*	*T*	*D*	*D*	*D*	
Sporadic rain (S), terminal drought (T):	*S*	*S*	*T*	*S*	*S*	*T*	*T*	
Annual (A), perennial (P):	*A*	*A*	*A*	*AP*	*AP*	*A*	*A*	
Rapid ground cover	+	+	+	+	+	+	+	+++
Early flowering	–	+	+	+	+	?	+	++++*
Summer dormancy	**o**	**o**	**o**	–	~	*o*	+	*0*
Phenotypic plasticity	+	+	–	+	+	+	+	+++–*
Roots: deep, rapid growth	+	+	+	+	+	+	+	+++++[f*]
dense surface[a]	+	~	~	+	+	~	+	*0*
low conductance	–	–	+	–	–	?	+	+(+)
Leaf rolling at high RWC	–	–	–	–	+	?	+	+[e]
at lowRWC	?	?	?	+	+	?	+	+[2]
Leaf movements, helionasty	+	+	–	?	+	+	+	*0*
Leaf shedding	–	–	+	–	–	?	–	–[f]
Leaf reflectance	+	+	+	+	+	+	+	++++
Stomata close at high RWC	–	–	+	–	–	+[b]	+[b]	–+[f]
at low RWC	+	+	+	+	+	+	+	–+[d]
Sunken stomata	?	?	?	+	?	?	?	*0*
C_3 metabolism	+	~	~	+	~	~	~	*0*
C_4 metabolism	–	~	~	–	~	~	~	*0*
Osmotic adjustment	+	+		?	+	+	+	+++[f] +*
Osmoprotectants	+	+	*o*	+	+	*o*	+	*0*
Membrane stability	+	+	+	+	+	+	+	(+)+[f]
Robustness of photosynthetic mechanism	+	+	*o*	+	+	*o*	*o*	++*o*
Storage of reserves for recovery	+	+	*o*	+	+	*o*	+	*0*
Translocation of harvestable parts	+	+	+	**o**	*o*	+	+	++++
Survival of green leaf function	+	+	~	+	+	~	+	+++*
Survival of axillary buds	*o*	+	**o**	+	+	**o**	+	+[e]
Recovery when rains return	+	+	**o**	+	+	+	+[e]	+[e]

[a] Improves water and nutrient uptake, keeps out weeds. [b]– for dehydration avoiders. [c] Only if start of wet season. [d] Noon closure may improve WUE. [e] Vegetative growth of forages. [f] Fast-growing trees (Pereira and Pallardy, 1989).

Dactylis glomerata > *Festuca arundinacea* > *Lolium perenne* > *Phleum pratense* > *L. multiflorum* × *perenne L. multiflorum* > *Poa trivialis*. The physiology and ecology of rangeland plants, including drought resistance, is covered in the volumes edited by Jones and Lazenby (1988) and Tueller (1988).

There is also considerable diversity within species, as the result of both natural adaptation and breeding, and as we have seen above much has been learnt about the physiological basis of drought resistance from comparisons of different populations or ecotypes. Blum (1988) has described crop evolution as being "nearly synonymous with adaptation to various stress environments".

The genetic diversity within landraces of cereals, as opposed to the near homozygous state of advanced cultivars, has been found to contribute to maintenance of yield ("populational buffering", Allard and Bradshaw, 1964) in unpredictably dry environments (e.g. Sojka, 1985; Acevedo and Cecarelli, 1989).

Selection and Breeding for Yield

The whole subject of breeding for stress environments has been comprehensively covered by Blum (1988), and the opinions of a consortium of breeders and physiologists on breeding for drought resistance in cereals are given in the volume edited by Baker (1989). In addition, many traits and strategies appropriate to improving a range of crop species have been critically evaluated by Turner (1986), Ludlow (1989) and Ludlow and Muchow (1990).

The most extensively used empirical measure of drought resistance is through evaluating the "genotype by environment interaction", or G×E (Finlay and Wilkinson, 1963). Briefly, in a variety trial the "environment" is equated to the mean of all varieties under the conditions of a particular site, management procedure or year. Some control over the nature of the "environments" can be imposed by using a set of reference varieties and by ensuring that the actual environments vary in only one respect, in this case, water supply. The yield of individual genotypes or populations is then plotted or regressed against the "environment" (Figure 1.4). The yield potential of a variety is its yield in the best "environment". The yield stability is a measure of the maintenance of yield of a particular variety as the quality of the "environment" declines; thus, when G is regressed on E, an average variety has a regression coefficient (b) of 1, a stable variety has $b < 1$ and an unstable variety $b > 1$ (Figure 1.4). Modified versions of this technique have been developed by Fischer and Maurer (1978) and by Bidinger *et al.*, (1987b). In forages, the situation is complicated by necessary management activities, which introduce complex interactions between timing of drought, heading, and defoliation height and frequency.

Mass selection for improved yield under drought may be conducted in three ways. Firstly, selection for yield potential in near optimal environments may be employed because it gives the plants a head start as drought approaches (Figure 1.4, line A). A recent example of this "residual effect of yield potential" is the successful use in dry Australian environments of CIMMYT wheat germplasm selected under near-optimal conditions (Edmeades *et al.*, 1989). Another advantage of selecting for yield potential is that it is easier to

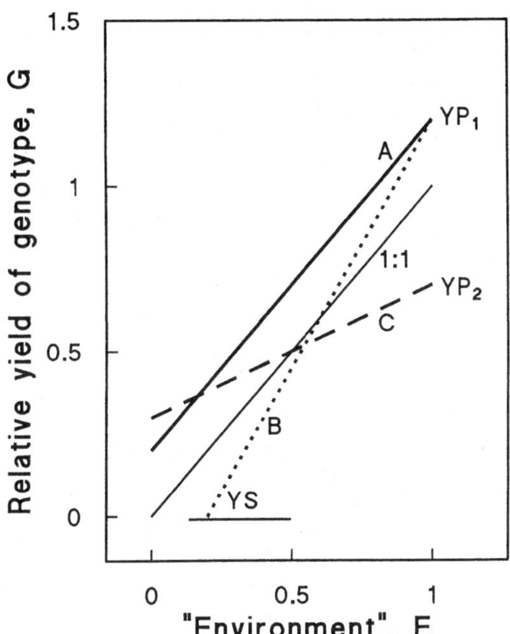

Figure 1.4. Genotype by environment interactions (GxE) showing yield potential (YP) and yield stability (YS). E is the mean over all varieties for each environment in a trial. Genotype A has high YP and average YS (parallel to the 1:1 line), B has high YP and low YS (yield declines rapidly as the environment worsens), while C has low YP but is very stable and yields well in poor environments.

detect expression of genetically-controlled characters under good growing conditions (Mederski and Jeffers, 1973). This method is likely to be most appropriate where management quality is high and stresses are rare or moderate. In perennial crops, the ability to regrow when it rains again is important (Kemp and Culvenor, 1994); in temperate forages, it is positively correlated with potential yield (Thomas and Norris, 1981), whereas in Mediterranean environments it is associated with low potential summer growth (Silsbury, 1961; Volaire, 1995).

Secondly, selection for yield stability can be conducted in a number of sites and managements in the hope that some kind of general stress resistance emerges. Rumbaugh *et al.* (1984) concluded that the best compromise was to select under moderate, controlled stress conditions.

Thirdly, selection may be conducted in environments where drought is the main stress factor, using sites having different rainfall totals and distribution. Further experimental treatments such as irrigation and rain-out shelters can be imposed, but irrigation does not fully simulate a wetter climate, nor shelters a drier (Monteith, 1986).

These categories are not mutually exclusive; Byrne *et al.* (1995) have concluded that, in their maize breeding programme for drought-prone tropical areas, the best combination would be to combine imposed drought at one site with multi-location testing.

The imposition of drought regimes must obviously be related to the range of environments that the crop is to experience. A broad outline of these environments was given at the beginning of this chapter, and an example of more precise climatic description (for Sub-Saharan Africa) is presented by Peacock and Sivakumar (1987).

In crop improvement, the choice of germplasm may be very wide, and ranges from existing elite varieties through landraces to wild relatives and (if genetic manipulation ever becomes practical) unrelated species. The consensus of opinion (Blum, 1988; Baker, 1989) is that, despite the risk of continually narrowing the genetic base, for grain crops it is generally most effective to work with varieties or landraces that are already high-yielding. There are three reasons for this:

(1) High yield has been achieved over centuries or millennia, and it would be foolish to dilute this very complex attribute by crossing with generally inferior germplasm which might include many undesirable traits. Some success has, however, been achieved in transferring salt tolerance from exotic species into crop plants (Chapter 4, this volume).
(2) Such crossing and introgression can be technically difficult, although trans- genic techniques are proving increasingly useful.
(3) Elite varieties or landraces may already contain an as yet unrealized pool of drought resistance.

Forage varieties are much less specialized than the cereals and are generally merely improved ecotypes; consequently item (1) above is not really a limitation to breeding. Genotypes from Mediterranean-type environments do, however, practice drought avoidance (see above), which makes them agronomically undesirable for more temperate climates.

Ideotype Breeding and Identification of Traits

Donald (1968) suggested breeding should be directed towards the concept of a model plant or ideotype, rather than towards high yield alone, but selection for physiological or biochemical attributes has so far been largely unsuccessful (Gifford *et al.*, 1984). Notable exceptions include the work of Richards and Passioura (1989) on water-saving root systems, and that of Morgan (1984) and Morgan and Condon (1986) on osmotic adjustment in wheat. Ramusson (1987) reviewed the roles of traditional (yield) and ideotype breeding in barley, and concluded that the relative research efforts should be in a 3:1 ratio. Blum (1988) has pointed out that an incomplete understanding of physiological criteria could lead the breeder in search of a "mirage", but with "empirical and judicious testing" could prove commercially effective. Breeding is, of course, a numbers game, and the breeder does not have to identify promising germplasm (or eliminate unpromising) with absolute certainty—he just has to win on averages.

Appropriate traits can be identified with varying levels of certainty in a number of ways. Firstly, species, populations, varieties or genotypes with "known" levels of drought resistance in field trials can be compared. Secondly, experimental populations can be produced by divergent selection using the following sequence: (a) assess genotypic variation for the target character in a population; (b) identify plants with extreme expression of the character; (c) produce seed of the high lines and low lines; (d) compare the drought resistance of the contrasting lines. Ideally, the contrasting lines should have an identical but heterogeneous genetic background, and differ only in the target character (Richards and Passioura, 1989). Alternatively, an isogenic background may be used (Grumet *et al.*, 1987; Innes *et al.*, 1984). In outbreeding crops, which suffer considerably from inbreeding depression, it may be possible to compare only divergent breeding populations (Thomas and Evans, 1989; Gay, 1994).

Use of the Newer Technologies

One of the biggest problems in improving drought resistance is that it is a multi-character phenomenon. Also, it is likely that many component traits are themselves of a quantitative nature and controlled by a number of genes: in sorghum, for example, osmotic adjustment appears to be under the control of two major genes and several minor ones (Basnayake *et al.*, 1995), whereas in wheat only a single gene locus seems to be involved (Morgan, 1991). However, as we move from a higher to a lower level of complexity, such as from development through structure and physiology to biochemistry, the number of component processes becomes smaller, and there is more likelihood that a trait at that level is controlled by a single gene.

The relatively new techniques collectively known as biotechnology provide tool-kits that will play an important role in a number of crop improvement activities (Rains, 1989; Hughes *et al.*, 1989; Lal and Lal, 1990; Nguyen 1993; Bray, 1994).

Cell and tissue culture

By using tissue, cell or protoplast culture, it is possible to screen many accessions of germplasm for expression of genes without the confounding effects of cell age, cell type, and differentiation. A major defect in the present context is that it is impossible to impose "real" water stress on such cultures, and so osmotic stress must be used instead; the two stresses may have little in common (Hurkman and Tanaka, 1988). For example, Bhaskaran *et al.*, (1985) and Newton *et al.*, (1986) found no correlation between callus proline production under osmotic stress and resistance to drought.

Another application of culture techniques is inducing additional genetic diversity through somaclonal variation, the genome being unstable under these conditions.

Marker-assisted selection, gene maps and genetic dissection of traits

The application of molecular biology has so far proven most effective in two areas. Firstly, where single metabolic events and therefore single gene loci are involved it may be possible to modify the genotype (Hanson *et al.*, 1986). For example, the rapid development of transformation systems (Bettany, 1995) then enables single genes presumed to code for a desirable trait to be moved to the target plant from unrelated species, even animals or bacteria.

The second area is in the use of molecular markers to follow the movement of traits in a breeding programme (Paterson *et al.*, 1991). This enables us to overcome a major constraint in crop improvement: the measurement of traits on large numbers of individual plants or populations, which is made difficult because of spatial variability and the time labour and land area required. In the past, it was occasionally possible to use visual markers (easily-detected characters genetically linked with desirable traits) to identify promising plants. Such markers are rare, but the development of restriction length fragment polymorphism (RFLP) analysis has provided numerous DNA markers (quantitative trait loci, or QTLs). These can be associated with particular traits and gene loci, and used to map the positions of genes on chromosomes accurate to about 0.01 of a chromosome (Paterson *et al.*, 1991). For example, a gene regulating drought-induced ABA production has been located to a region on chromosome 5A of wheat (Quarrie *et al.*, 1994). QTL analysis has greatly accelerated traditional breeding programmes; in rice, for example, more linkages were quantified in 2 years than conventional crossing achieved in 60 years (O'Toole, 1989). A further advantage of this technique is that findings on one species can be extrapolated to others because of synteny-the strong similarity of homologous chromosomes in species derived from a common ancestor.

One uncertainty in improving drought resistance is that it is not always certain which traits are relevant. QTL analysis may then be used in reverse by detecting markers occurring in plants or populations whose drought resistance or susceptibility has been determined experimentally, and discovering which traits have previously been mapped to these markers.

Physical mapping of trait

QTL analysis, like traditional genetic analysis, works on probabilities and requires powerful statistical analysis-of-variance algorithms to determine the most likely, not the actual, location of a trait in a genome. A more direct approach is to prepare a physical map of drought resistance traits by "chromosome painting", using genomic *in situ* hybridization (GISH) and similar techniques (Humphreys and Ghesquière, 1994). Hybrids are made between related species (species ***A***, say, being the most agronomically desirable, and species ***B*** containing useful drought resistance traits). The hybrids are then repeatedly backcrossed to ***A*** to produce introgressed lines whose genome is mostly that of ***A***, but contains one or more small recombinant segments of genome ***B***. Elite plants, those having drought resistance or presumed constituent traits, can be identified experimentally. Squash preparations are made of mitotic cells of these élite plants and the recombinant segments stained or "painted" using a fluorescent probe prepared from the DNA of parent *B*. The segments can then be detected under the microscope by their characteristic fluoresecnce. It is likely that the transferred traits are somewhere in an introgressed segment, and future analysis is carried out much as for QTL studies. In future it should be possible to physically remove introgressed segments by microdissection and, if they are small enough, determine their DNA sequences.

CONCLUSION

The most effective way forward in understanding the mechanisms of drought resistance, and applying our knowledge to crop improvement is through collaboration between disciplines, such as cytology, molecular biology, physiology and breeding. It is reassuring and encouraging to see the increasing number of such multidisciplinary projects and teams that have been established over the last few years.

REFERENCES

Acevedo E. and Ceccarelli S. (1989) Role of the physiologist-breeder in a breeding programme for drought-resistance conditions. In *Drought Resistance in Cereals*, edited by F.W.G. Baker, pp. 117–140. Wallingford, Oxon: CAB International.

Alamillo J.M., Roncarti R., Heino P., Velasco R., Nelson D., Elster R. *et al.* (1994) Molecular analysis of desiccation tolerance in barley embryos and in the desiccation plant *Craterostigma plantagineum. Agronomie,* **2**, 161–167.

Allard R.W. and Bradshaw A.D. (1964) Implications of genotype-environmental interactions in plant breeding. *Crop Science*, **4**, 503–507.

Amin R.Md. (1990) Water relations of diverse *Lolium perenne* populations. M.Sc. Thesis, University College of Wales, Aberystwyth.

Aphalo P.J. and Jarvis P.G. (1991) Do stomata respond to relative humidity? *Plant, Cell and Environment,* 127–132.

Aspinall D. (1986) Metabolic effects of water and salinity stress in relation to expansion of the leaf surface. *Australian Journal of Plant Physiology*, **13**, 59–74.

Austin R.B., Edrich J.A., Ford M.A. and Blackwell R.D. (1977) The fate of dry matter, carbohydrates and ^{14}C lost from the leaves and stems of wheat during grain filling. *Annals of Botany,* **41**, 1309–1321.

Austin R.B., Morgan C.L., Ford M.A. and Blackwell R.D. (1980) Contributions to grain yield from pre-anthesis assimilation in tall and dwarf barley phenotypes in two contrasting seasons. *Annals of Botany*, **45**, 309–319.

Baenzinger P.S., Wesenberg D.M. and Sicher R.C. (1983) The effects of genes controlling barley leaf and sheath waxes on agronomic performance in irrigated and dryland environments. *Crop Science,* **23**, 116–120.

Baker D.A. and Weatherley P.E. (1969) Water and solute transport by exuding root systems of *Ricinus communis*. *Journal of Experimental Botany*, **20**, 485–496.

Baker F.W.G. (1989) Conclusions. In *Drought Resistance in Cereals*, edited by. F.W.G. Baker, pp. 213–220. Wallingford, Oxon: CAB International.

Barraclough P.B., Kuhlmann H. and Weir. A.H. (1989) The effects of prolonged drought and nitrogen fertilizer on root and shoot growth and water uptake by winter wheat. *Journal of Agronomy and Crop Science*, **163**, 352–360.

Basnayake J., Cooper M, Ludlow M.M., Henzell R.G. and Snell P.J. (1995) Inheritance of osmotic adjustment to water stress in three grain sorghum crosses. *Theoretical and Applied Genetics*, **90**, 675–682.

Bassiri M., Wilson A.M. and Grami B. (1988) Dehydration effects on seedling development of four range species. *Journal of Range Management*, **41**, 383–386.

Bewley J.D. (1981) Protein synthesis. In *The Physiology and Biochemistry of Drought Resistance in Plants*, edited by L.G. Paleg and D. Aspinall, pp.261–282. Sydney: Academic Press.

Bewley J.D., Larsen K.M. and Papp J.E.T. (1983) Water-stress-induced changes in the pattern of protein synthesis in maize seedling hypocotyls: a comparison with the effects of heat shock. *Journal of Experimental Botany,* **34**, 1126–1133.

Bhaskaran S., Smith R.H. and Newton R.J. (1986) Physiological changes in cultured sorghum cells in response to induced water stress. I. Free proline. *Plant Physiology*, **79**, 266–269.

Bidinger F.R., Mahalakshmi V. and Rao G.D.P. (1987a,b) Assessment of drought resistance in pearl millet *Pennisetum americanum* (L.) Leeke]. I. Factors affecting yields under stress. II. Estimation of genotype response to stress. *Australian Journal of Agricultural Research,* **38**, 37–48,49–59.

Björkman O. and Powles S.B. (1984) Inhibition of photosynthetic reactions under water stress: interaction with light level. *Planta*, **161**, 490–504.

Blom-Zandstra M. and Lampe J.E.M. (1985) The role of nitrate in the osmoregulation of lettuce (*Lactuca sativa* L.) grown at different light intensities. *Journal of Experimental Botany,* **36**, 1043–1052.

Blum A. (1979) Genetic improvement of drought resistance in crop plants: a case for sorghum. In *Stress Physiology in Crop Plants*, edited by. H. Mussell and R.C. Staples, pp.430–445. New York: John Wiley.

Blum A. (1988) *Plant Breeding for Stress Environments.* Boca Raton: CRC Press.

Blum A. (1989) Osmotic adjustment and growth of barley genotypes under drought stress. *Crop Science*, **29**, 230–233.

Blum A. and Ebercon A. (1976) Genotypic responses in sorghum to drought stress. III. Free proline accumulation and drought resistance. *Crop Science*, **16**, 428–431.

Blum A., Gozlan G and Mayer J. (1981) The manifestation of dehydration avoidance in wheat breeding germplasm. *Crop Science,* **21**, 495–499.

Blum A., Gozlan G and Mayer J. (1983) Chemical desiccation of wheat plants as a simulator of post-anthesis stress. II. Relations to drought stress. *Field Crops Research,* **6**, 149–155.

Blum A. and Sullivan C.Y. (1986) The comparative drought resistance of landraces of sorghum and millet from dry and humid regions. *Annals of Botany,* **57**, 835–846.

Boyer J.S. (1982) Plant productivity and environment. *Science,* **218**, 443–448.

Boyer J.S. (1985) Water transport. *Annual Review of Plant Physiology,* **36,** 473–516.

Boyer J.S., Johnson R.R. and Saupe S.G. (1980) Afternoon water deficits and grain yields in old and new soybean cultivars. *Agronomy Journal,* **72**, 981–786.

Boyer J.S. and Nonami H. (1990) Xylem hydraulics, turgor and wall properties during growth. In *Importance of Root to Shoot Communication in the Responses to Environmental Stress*, edited by. W.J. Davies and B. Jeffcoat, pp. 45–52 Bristol: British Society for Plant Growth Regulation.

Bradford K.J. (1986) Manipulation of seed water relations via osmotic priming to improve germination under stress conditions. *Hortscience,* **21**, 1105–1112.

Bray E.A. (1994) Alterations in gene expression in response to water deficit. In *Stress-Induced Gene Expression in Plants*, edited by. A.S. Basra, pp 1–23. Chur: Harwood Academic Publishers.

Bremner P.M. and Preston G.K. (1990) A field comparison of sunflower (*Helianthus annuus*) and sorghum (*Sorghum bicolor*) in a long drying cycle. II. Plant water relations, growth and yield. *Australian Journal of Agricultural Research*, **41**, 463–478.

Brown K.W., Jordan W.R. and Thomas J.C. (1976) Water stress induced alterations of the stomatal response to decreases in leaf water potential. *Physiologia Plantarum*, **37**, 1–5.

Byrne P.F., Bolanos J., Edmeades G.O. and Eaton D. (1995) Gains from selection under drought versus mulit-location testing in related tropical maize populations. *Crop Science*, **35**, 63–69.

Chandler P.M. and Robertson M. (1994) Gene expression regulated by abscisic acid and its relation to stress tolerance. *Annual Review of Plant Physiology and Plant Molecular Biology*, **45**, 113–141.

Chatterton N.J., Hanna W.W., Powell J.B. and Lee D.R. (1975) Photosynthesis and transpiration of bloom and bloomless sorghum. *Canadian Journal of Plant Science*, **55**, 641–643.

Chaves M.M. (1991) Effects of water deficits on carbon assimilation. *Journal of Experimental Botany*, **42**, 1–16.

Chrispeels M.J. and Maurel C. (1994) Aquaporins: the molecular basis of facilitated water movement through living plant cells. *Plant Physiology*, **105**, 9–13.

Clark R.B. (1981) Effects of light and water stress on mineral element composition of plants. *Journal of Plant Nutrition*, **3**, 835–885.

Clarke J.M. and Townley-Smith T.F. (1986) Heritability and relationship to yield of excised-leaf water retention in durum wheat. *Crop Science*, **26**, 289–292.

Clarkson N.M. and Russell J.S. (1976) Effect of water stres on the phasic development of annual *Medicago* species. *Australian Journal of Agricultural Research*, **27**, 227–234.

Clough B.F. and Milthorpe F.L. (1975) Effects of water deficit on leaf development in tobacco. *Australian Journal of Plant Physiology*, **2**, 291–300.

Condon A.G., Farquhar G.D. and Richards R.A. (1990) Genotype variation in carbon isotope variation and transpiration efficiency in wheat. *Australian Journal of Plant Physiology*, **17**, 9–22.

Cosgrove D. (1993) Wall extensibility; its nature, measurement and relationship to plant cell growth. *New Phytologist*, **124**, 1–23.

Cowan I.R. (1977) Stomatal behaviour and the environment. *Advances in Botanical Research*, **4**, 117–228.

Cutler J.M., Rains D.W. and Loomis R.S. (1977) The importance of cell size in the water relations of plants. *Physiologia Plantarum*, **40**, 255–260.

Daie J. (1988) Mechanism of drought-induced alterations in assimilate partitioning and transport in crops. *CRC Critical Reviews in Plant Sciences*, **7**, 117–137.

Dale J.E. (1988) The control of leaf expansion. *Annual Review of Plant Physiology and Plant Molecular Biology*, **39**, 267–295.

Davey A.G. and Simpson R.J. (1990) Nitrogen fixation by subterranean clover at varying stages of nodule dehydration. I. Carbohydrate status and short-term recovery of nodulated root respiration. *Journal of Experimental Botany*, **41**, 1175–1187.

Davies W.J. and Pereira J.S. (1992) Plant growth and water use efficiency. In *Crop Photosynthesis: Spatial and Temporal Determinants*, edited by. N.R. Baker and H.Thomas. pp. 213–234. Amsterdam: Elsevier.

Davies W.J. and Van Volkenburgh E. (1983) The influence of water deficit on the factors controlling the daily pattern of growth of *Phaseolus* trifoliates. *Journal of Experimental Botany*, **34**, 987–999.

Day W., Legg B.J., French B.K., Johnson A.E., Lawlor D.W. and Jeffers. W.De C. (1978) A drought experiment using mobile rain shelters: the effect of drought on barley yield, water use and nutrient uptake. *Journal of Agricultural Science*, **91**, 599–623.

Dhindsa R.S. and Cleland R.E. (1975) Water stress and protein synthesis. I.Differential inhibition of protein synthesis. *Plant Physiology*, **55**, 778–781.

Donald C.H. (1968) The breeding of crop ideotypes. *Euphytica*, **17**, 385–403.

Douglas T.J. and Paleg L.G. (1981) Lipid composition of *Zea mays* seedlings and water stress-induced changes. *Journal of Experimental Botany*, **32**, 499–508.

Duke J.A. (1981) *Handbook of Legumes of World Economic Importance*. New York: Plenum Press.

Dunham R.J. and Nye P.H. (1976) The influence of soil water content on the uptake of ions by roots. III. Phosphate, potassium, calcium and magnesium uptake and concentration gradients. *Journal of Applied Ecology*, **13**, 967–984.

Durand J.L., Onillon B., Schnyder H. and Rademacher I. (1995). Drought effects on cellular and spatial parameters on leaf growth in tall fescue. *Journal of Experimental Botany*, **46**, 1147–1155.

Dure L. and Verma D.P.S. (1993) The *lea* proteins of higher plants. In *Control of Plant Gene Expression*, edited by D.P.S. Verma, pp. 325–335. Boca Raton: CRC Press.

Edmeades G. O., Bolanos J., Lafitte H. R., Rajaram S., Pfeiffer W. and Fischer R.A. (1989) Traditional approaches for breeding for drought resistance in cereals. In *Drought Resistance in Cereals*, edited by. F.W.G. Baker, pp.27–52. Wallingford, Oxon: CAB International.

Ehret D.L. and Boyer J.S. (1979) Potassium loss from stomatal guard cells at low water potentials. *Journal of Experimental Botany*, **30**, 225–234.

Ekanayake I.J., Steponkus P.L. and de Datta S.K. (1990) Sensitivity of pollination to water deficits in upland rice. *Crop Science*, **30**, 310–315.

Engin M. and Sprent J.I. (1973) Effects of water stress on growth and nitrogen-fixing activity of *Trifolium repens*. *New Phytologist*, **72**, 117–126.

Farquhar G.D. and Richards R.A. (1984) Isotopic composition of plant carbon correlates with water-use efficiency of wheat genotypes. *Australian Journal of Plant Physiology*, **11**, 539–552.

Finlay K.W. and Wilkinson G.N. (1963). The analysis of adaptation in a plant breeding programme. *Australian Journal of Agricultural Research*, **14**, 742–754.

Fischer R.A. (1989) Cropping systems for greater drought resistance. In *Drought Resistance in Cereals*, edited by F.W.G. Baker, pp. 201–212. Wallingford, Oxon.:CAB International.

Fischer R.A. and Maurer R. (1978) Drought responses in spring wheat cultivars. I. Grain yield responses. *Australian Journal of Agricultural Research*, **29**, 897–912.

Fitter A.H. (1985) Functioning of vesicular-arbuscular mycorrhizas under field conditions. *New Phytologist*, **99**, 257–265.

Fitter A.H. (1988) Water relations of red clover *Trifolium pratense* L. as affected by VA mycorrhizal infection and phosphorus supply before and during drought. *Journal of Experimental Botany*, **39**, 595–603.

French R.J. and Schulz J.E. (1984a,b) Water use efficiency of wheat in a Mediterranean-type environment. I. The relation between yield, water use and climate. II. Some limitations to efficiency. *Australian Journal of Agricultural Research*, **35**, 743–764, 765–775.

Fry S.C. (1995) Polysaccharide-modifying enzymes in the plant cell wall. *Annual Review of Plant Physiology and Plant Molecular Biology*, **46**, 497–520.

Gantet P., Hubac C. and Brown S.C. (1990) Flow cytometric fluorescence anisotropy of liphophilic probes in epidermal and mesophyll protoplasts from water-stressed *Lupinus albus* L. *Plant Physiology*, **94**, 729–737.

Garwood E.A. and Sinclair J. (1979) Water use by six grass species. 2. Root distribution and use of soil water. *Journal of Agricultural Science*, **93**, 25–35.

Garwood E.A., Tyson K.C. and Sinclair J. (1979) Water use by six grass species. 1. Dry matter yields and response to nitrogen. *Journal of Agricultural Science*, **93**, 13–24.

Garwood E.A. and Williams T.E. (1967) Growth, water use and nutrient uptake from the subsoil by grass swards. *Journal of Agricultural Science*, **69**, 125–130.

Gay A.P. (1994) Breeding for leaf water conductance, its heritability and its effect on water use in *Lolium perenne*. *Aspects of Applied Biology*, **38**, 41–46.

Gifford R.M., Thorne J.H., Hitz W.D. and Giaquinta R.T. (1984) Crop productivity and assimilate partitioning. *Science*, **225**, 801–807.

Goldman I.L., Carter T.E. and Patterson R.P. (1989) Differential genotypic responses to drought stress and subsoil aluminum in soybean. *Crop Science*, **29**, 330–334.

Gollan T., Turner N.C., and Schulze E.D. (1985) The responses of stomata and leaf gas exchange to vapor pressure deficits and soil water content. II. In the sclerophyllous woody species *Nereum oleander*. *Oecologia*, **65**, 356–362.

Gonzalez E.M., Gordon A.J., James C.L. and Arrese-Igor C. (1995) The role of sucrose synthase in the response of soy-bean nodules to drought. *Journal of Experimental Botany,* **46**, 1515–1523.

Grace J. (1993) Consequences of xylem cavitation for plant water deficits. In *Water Deficits: Plant Responses from Cell to Community*, edited by J.A.C. Smith, and H. Griffiths, pp.109–128. Cambridge: Bios.

Grashoff C. (1990) Effect of pattern of water supply on *Vicia faba* L. *Netherlands Journal of Agricultural Science*, **38**, 131–143.

Grumet R., Alberchtsen R.S. and Hanson A.D. (1987) Growth and yield of barley isopopulations differing in solute potential. *Crop Science*, **28**, 991–995.

Guerin V., Trinchant J.C. and Rigaud J. (1990) Nitrogen fixation (C_2H_2 reduction) by broad bean (*Vicia faba* L.) nodules and bacteroids under restricted conditions. *Plant Physiology,* **92**, 595–601.

Gupta A.S. and Berkowitz G.A. (1988) Chloroplast osmotic adjustment and water stress effects on photosynthesis. *Plant Physiology,* **88**, 200–206.

Hansen G.K. (1974) Resistance to water transport in young wheat plants. *Acta Agriculturae Scandinavica,* **24**, 37–47.

Hanson A.D. and Hitz W.D. (1982) Metabolic responses of mesophytes to plant water deficits. *Annual Review of Plant Physiology,* **33**, 163–203.

Hanson A.D., Hoffman N.E. and Samper C. (1986) Identifying and manipulating metabolic stress-resistance traits. *Hortscience,* **21**, 1313–1317.

Hanson A.D. and Tully R.E. (1979) Amino acids translocated from turgid and water-stressed barley leaves. II. Studies with ^{13}N and ^{14}C. *Plant Physiology,* **64**, 467–471.

Harris E. and Williams J.B. (Eds) (1981) *Irrigation.* London: Her Majesty's Stationery Office.

Hartung W., Slovik S. and Baier M. (1990) pH changes and redistribution of abscisic acid within the leaf under stress. In *Importance of Root to Shoot Communication in the Responses to Environmental Stress,* edited by W. J. Davies and B. Jeffcoat, pp.215–236. Bristol: British Society for Plant Growth Regulation.

Hejnowicz Z. and Sievers A. (1995) Tissue stress in organs of herbaceous plants. *Journal of Experimental Botany,* **46**, 1035–1043.

Hendry G. (1993) Evolutionary origins and natural functions of fructans. *New Phytologist,* **123**, 3–14.

Henson I.E., Alagarswarmy G., Mahalakshmi V. and Bidinger F.R. (1983) Stomatal response to water stress and its relationship to bulk leaf water status and osmotic adjustment in pearl millet (*Pennisetum americanum* [L.]Leeke). *Journal of Experimental Botany,* **34**, 442–450.

Hoffman A.A. and Parsons P.A. (1991) *Evolutionary Genetics and Environmental Stress.* Oxford: Oxford University Press.

Hsiao T.C. (1970) Rapid responses in levels of polysomes in *Zea mays* in response to water stress. *Plant Physiology,* **46**, 281–285.

Hsiao T.C., Silk W.K., and Jing J. (1985) Leaf growth and water deficits: biophysical effects. In *Control of Leaf Growth,* edited by. N.R. Baker, W.J. Davis and C.K. Ong, pp-239–266. Cambridge: Cambridge University Press.

Hubbard C.E. (1968) *Grasses.* London: Penguin.

Huber S.C., Rogers H. and Mowry F.L. (1984) Effects of water stress on photosynthesis and carbon partitioning in soybean (*Glycine max* [L.] Merr.) plants grown in the field at different CO_2 levels. *Plant Physiology,* **76**, 244–249.

Huck M.G., Klepper B. and Taylor H.M. (1979) Diurnal variations in root diameter. *Plant Physiology,* 529–530.

Huffaker R.C., Radin T., Kleinkopf G.E. and Cox E.L. (1970) Effects of mild water stress on enzymes of nitrate assimilation and of the carboxylative phase of photosynthesis in barley. *Crop Science,* **10**, 471–474.

Hughes S.G., Bryant J.A. and Smirnoff N. (1989) Molecular biology: application to studies of stress tolerance. In *Plants Under Stress.* edited by H.G. Jones, T.J. Flowers and M.B. Jones., pp.131–156. Cambridge: Cambridge University Press.

Humphreys L.R. (1981) *Environmental Adaptation of Tropical Pasture Plants.* London: Macmillan.

Humphreys M.W. (1989) The controlled introgression of *Festuca arundinacea* genes into *Lolium multiflorum. Euphytica*, **42**, 105–116.

Humphreys, M.W. and Ghesquière M. (1994) Assessing success in gene transfer between *Lolium multiflorum and Festuca arundinacea. Euphytica* **77,** 283–289.

Hurd E.A. (1974) Phenotype and drought tolerance in wheat. *Agricultural Meteorology,* **14**, 39–55.

Hurkman W.J. and Tanaka C.K. (1988) Polypeptide changes induced by salt stress, water deficit and osmotic stress in barley roots: A comparison using two-dimensional gel electrophoresis. *Electrophoresis,* **9**, 781–787.

Husain M.M., Reid J.B., Othman H. and Gallagher J.N. (1990) Growth and water use of faba beans (*Vicia faba*) in a sub-humid climate. I. Root and shoot adaptations to drought stress. *Field Crops Research,* **23**, 1–17.

Innes P. Blackwell R.D. and Quarrie S.A. (1984) Some effects of genetic variation in drought-induced abscisic acid accumulaton on the yield and water use of spring wheat. *Journal of Agricultural Science,* **102**, 341–351.

Jain V., Madan S. and Gupta S.K. (1989) Changes in the wax content of leaves of rainfed and Mexican wheat cultivars. *Acta Botanica Indica,* **17**, 59–61.

Jarvis P.G. and McNaughton K.G. (1986) Stomatal control of transpiration. *Advances in Ecological Research,* Volume **15**, pp.1–49. London: Academic Press.

Johns G.G. and Lazenby A. (1973) Effect of irrigation and defoliation on the herbage production and water use efficiency of four temperate pasture species. *Australian Journal of Agricultural Research,* **24**, 797–808.

Johnson D.A., Richards R.A and Turner N.C. (1983) Yield, water relations, gas exchange and surface reflectances of near-isogenic wheat lines differing in glaucousness. *Crop Science,* **23**, 318–325.

Johnson R.C. (1994) Genetic variation and physiological mechanisms for water-use efficiency in temperate grasses and legume germplasm. *Aspects of Applied Biology,* **38**, 71–78.

Johnson R.C., Nguyen H.T. and Croy L.I. (1984) Osmotic adjustment and solute accumulation in two wheat genotypes differing in drought resistance. *Crop Science,* **24**, 957–962.

Jones H.G. (1980) Interaction and integration of adaptive responses to water stress: the implications of an unpredictable environment. In *Adaptation of Plants to Water and High Temperature Stress*, edited by N.C. Turner and P.J. Kramer, pp.353–365. New York: John Wiley.

Jones M.B. and Lazenby A. (1988) *The Grass Crop*. London: Chapman and Hall.

Jones M.B., Leafe E.L. and Stiles W. (1980) Water stress in field-grown perennial ryegrass. I. Its effect on growth, canopy photosynthesis and transpiration. *Annals of Applied Biology,* **96**, 87–101.

Kaiser W.M. (1987) Effects of water deficit on photosynthetic capacity. *Physiologia Plantarum,* **71**, 142–149.

Kemp D.R. and Culvenor R.A. (1994) Improvong thr grazing and drought tolerance of temperate perennial grass. *New Zealand Journal of Agricultural Research,* **37**, 365–378.

Kendrew W.G. (1961) *The Climates of the Continents*. Oxford: Oxford University Press.

Kirda C., Danso S.K.A. and Zapata, L. (1989) Temporal water stress effects on nodulation, nitrogen accumulation and growth of soybean. *Plant and Soil,* **120**, 49–55.

Kirkham M.B., Gardner W.R. and Gerloff G.C. (1972) Regulation of cell division and cell enlargement by turgor pressure. *Plant Physiology,* **49**, 961–962.

Klepper B and Rickman R.W. (1990) Modeling crop root growth and function. *Advances in Agronomy,* **44**, 113–132.

Kock J.de, Bruyn L.P. de and Human J.J. (1990) The relative sensitivity to plant water stress during the reproductive phase of upland cotton (*Gossypium hirsutum* L.). *Irrigation Science,* **11**, 239–244.

Kumar A., Singh P., Singh D.P., Singh H. and Sharma H.C. (1984) Differences in osmoregulation in *Brassica* species. *Annals of Botany,* **54**, 537–541.

Lal R. and Lal S. (1990) *Crop Improvement Utilizing Biotechnology*. Boca Raton: CRC Press.

Lang A. and Thorpe M. (1986) Water potential, translocation and assimilate partitioning. *Journal of Experimental Botany,* **37**, 495–503.

Lawlor D.W. and Fock H. (1977) Photosynthetic assimilation of $^{14}CO_2$ by water-stressed sunflower leaves at two O_2 concentrations and the specific activity of products. *Journal of Experimental Botany,* **28**, 320–328.

Lawn R.J. (1982) Response of four grain legumes to water stress in south-eastern Queensland. I. Physiological response mechanisms. *Australian Journal of Agricultural Research,* **33**, 481–496.

Lazaroff N. and Pitman M.G. (1966) Calcium and magnesium uptake by barley seedlings. *Australian Journal of Biological Science,* **19**, 991–1005.

Leigh R.A. and Johnson A.E. (1983) The effects of fertilizers and drought on the concentrations of potassium in the dry matter and tissue water of field-grown spring barley. *Journal of Agricultural Science,* **101**, 741–748.

Lemaire G. and Denoix A. (1987) Croissance estivale en matiere seche de peuplements de fetuque elevee (*Festuca arundinacea* Schreb.) et de dactyle (*Dactylis glomerata* L.) dans l'Ouest de France. II. Interaction entre les niveaux d'alimentation hydrique et de nutrition azotee. *Agronomie,***7**, 381–389.

Levitt J. (1980) *Responses of Plants to Environmental Stresses*, Vol.2. New York: Academic Press.

Lindhauer M.G. (1987) Solute concentrations in well-watered and water-stressed sunflower plants differing in K nutrition. *Journal of Plant Nutrition,* **10**, 1965–1973.

Lockhart J.A. (1965) An analysis of irreversible plant cell elongation. *Journal of Theoretical Biology,* **8**, 264–275.

Lo Gullo M.A. and Salleo S. (1988) Different strategies of drought resistance in three Mediterranean sclerophyllous trees growing in the same environmental conditions. *New Phytologist,***108**, 267–276.

Lorens G.F., Bennett J.M. and Loggale L.B. (1987) Differences in drought resistance between two corn hybrids. I. Water relations and root length density. *Agronomy Journal,* **79**, 802–807.

Ludlow M.M. (1975) Effect of water stress on the decline of leaf net photosynthesis with age. In *Environmental and Biological Control of Photosynthesis,* edited by. R. Marcelle, pp. 123–133. The Hague: Junk.

Ludlow M.M. (1989) Strategies of response to water stress. In *Structural and Functional Responses to Environmental Stresses,*, edited by K.H. Kreeb, H. Richter and T.M. Hinckley, pp. 269–281. The Hague: SPB Academic Publishing.

Ludlow M.M., and Muchow R.C. (1990) A critical evaluation of traits for improving crop yields in water-limited environments. *Advances in Agronomy,* **43**, 107–153.

Ludlow M.M. and Ng T.T. (1977) Leaf elongation rate in *Panicum maximum* var. *trichoglume. Australian Journal of Plant Physiology,* **3**, 401–411.

Ludlow M.M. and Powles S.B. (1988) Effects of photoinhibition induced by water stress on growth and yield of grain sorghum. *Australian Journal of Plant Physiology,* **15**, 179–194.

Luna M., Maurizio M., Felici M, Artemi F. and Sermanni G.G. (1985) Selective enzyme inactivation under water stress in maize (*Zea mays* L.) and wheat (*Triticum aestivum* L.) seedlings. *Experimental and Environmental Botany,* **25**, 153–156.

Maier-Maercker U. (1983) The role of peristomatal transpiration in the mechanism of stomatal movement. *Plant, Cell and Environment,* **6**, 369–380.

Mambani B and Lal R. (1983) Response of upland rice varieties to drought stress. II. Screening rice varieties by means of variable moisture regimes along a toposequence. *Plant and Soil* , **73**, 73–94.

Manjarrez-Sandoval P., Gonzalez-Hernandez V.A., Mendoza-Onofre L.E. and Engelman E.M. (1989) Drought stress effects on the grain yield and panicle development of sorghum. *Canadian Journal of Plant Science,* **69**, 631–641.

Masle J., and Passioura J.B. (1987) Effects of soil strength on the growthy of wheat seedlings. *Australian Journal of Plant Physiology,* **14**, 643–656.

May L.H. and Milthorpe F.L. (1962) Drought resistance of crop plants. *Field Crop Abstracts,* **15**, 171–179.

McCree K.J. (1974) Changes in the stomatal response characteristics of grain sorghum produced by water stress during growth. *Crop Science,* **14**, 273–278.

McCree K.J. (1986) Whole plant carbon balance during osmotic adjustment to drought and salinity stress. *Australian Journal of Plant Physiology,* **13**, 33–44.

McCree K.J., Kallsen C.E. and Richardson S.G. (1974) Carbon balance of sorghum plants during osmotic adjustment to water stress. *Plant Physiology,* **76**, 898–902.

McWilliam J.R. (1986) The national and international importance of drought and salinity effects on agricultural production. *Australian Journal of Plant Physiology*, **13**, 1–14.

McWilliam J.R. (1989) The dimensions of drought. In *Drought Resistance in Cereals,* edited by. F.W.G. Baker, pp.1–12. Wallingford, Oxon.: CAB International.

McWilliam J.R. and Kramer P.J. (1968) The nature of the perennial response in Mediterranean grasses. I. Water relations and summer survival in *Phalaris. Australian Journal of Agricultural Research,* **19**, 381–385.

Mederski H.J. and Jeffers D.J. (1973) Yield response of soybean varieties grown at two soil moisture stress levels. *Agronomy Journal,* **65**, 410–412.

Meyer R.F. and Boyer J.S. (1972) Sensitivity of cell division and cell elongation to low water potentials in soybean hypocotyls. *Planta,* **108**, 77–87.

Meyer R.E. and Gingrich J.R. (1966) Osmotic effects on wheat using a split root solution culture system. *Agronomy Journal,* **58**, 377–381.

Milburn J.A. (1973) Cavitation studies on whole *Ricinus* plants by acoustic detection. *Planta,* **112**, 333–342.

Milburn J.A. (1979) *Water Flow in Plants*. London: Longman.

Milthorpe F.L. and Moorby. J. (1979) *An Introduction to Crop Physiology*. Cambridge: Cambridge University Press.

Monteith J.L. (1986) Significance of the coupling between saturation vapour pressure deficit and rainfall in monsoon climates. *Experimental Agriculture*, **22**, 329–338.

Morgan J.M. (1977) Differences in osmoregulation between wheat genotypes. *Nature,* **270**, 234–235.

Morgan J.M. (1984) Osmoregulation and water stress in higher plants. *Annual Review of Plant Physiology*, **35**, 299–319.

Morgan J.M. (1988) The use of coleoptile responses to water stress to differentiate wheat genotypes for osmoregulation, growth and yield. *Annals of Botany*, **62**, 193–198.

Morgan J.M. (1989) Physiological traits for drought resistance. In *Drought Resistance in Cereals*, edited by F.W.G. Baker, pp. 53–64. Wallingford, Oxon: CAB International.

Morgan J.M. (1991) A gene controlling differences in osmoregulation in wheat. *Australian Journal of Plant Physiology*, **18**, 249–257.

Morgan J.M. and Condon, A.G. (1986) Water use, grain yield and osmoregulation. *Australian Journal of Plant Physiology*, **13**, 523–532.

Muchow R.C. and Sinclair T.R. (1986) Water and nitrogen limitations in soybean grain production. II Field and model analyses. *Field Crops Research*, **15**, 143–156.

Munns R. (1988) Why measure osmotic adjustment? *Australian Journal of Plant Physiology*, **15**, 717–726.

Munns R., Brady C.J. and Barlow E.W.R. (1979) Solute accumulation in the apex and leaves of wheat during water stress. *Australian Journal of Plant Physiology,* **6**, 379–389.

Munns R. and Weir R. (1981) Contribution of sugars to osmotic adjustment in elongating and expanded zones of wheat leaves during mild water defecits at two light levels. *Australian Journal of Plant Physiology* , **8**, 93–105.

Newton R.J., Bhaskaran S. Puryear J.D. and Smith R.H. (1986) Physiological changes in cultured sorghum cells in response to induced water stress. II. Soluble carbohydrates and organic acids. *Plant Physiology*, **81**, 626–629.

Nguyen H.T., Joshi C.F. and Kuo C.G. (1993) Molecular strategies for the genetic dissection of water and high-temperature stress adaptation in cereal crops. In *Adaptation of Food Crops to Temperature and Water Stress,* pp. 3–19. Taipei: AVRDC.

Nicolas M.E., Lambers H., Simpson R.J. and Dalling M.J. (1985) Effect of drought on metabolism and partitioning of carbon in two wheat varieties differing in drought tolerance. *Annals of Botany,* **55**, 727–742.

Nonami H. and Boyer J.S. (1990) Wall extensibility and cell hydraulic conductivity decrease in enlarging stem tissues at low water potentials. *Plant Physiology,* **93**, 1610–1619.

Norris I. B. and Thomas H. (1982) Variation in growth of varieties and ecotypes of *Lolium* , *Dactylis* and *Festuca* subjected to contrasting soil moisture regimes. *Journal of Applied Ecology,* **19**, 881–889.

Nye P.H. and Tinker P.B. (1977) *Solute Movement in the Soil-root System.* Oxford: Blackwell Scientific Publications.

Ofori F. and Stern W.R. (1987) Cereal-legume intercropping systems. *Advances in Agronomy,* **41**,. 41–90.

O'Toole J.C. (1989) Breeding for drought resistance in cereals: emerging new technologies. In *Drought Resistance in Cereals,* edited by F.W.G. Baker, pp. 81–94. Wallingford, Oxon.: CAB International.

O'Toole J.C. and Bland W.L. (1987) Genotypic variation in crop plant root systems. *Advances in Agronomy* , **41**, 91–145.

Pararajasingham S. and Knievel D.P. (1990) Nitrogenase activity of cowpea *(Vigna unguiculata* [L.]Walp.) during and after drought stress. *Canadian Journal of Plant Science,* **70**, 163–171.

Passioura J.B. (1977) Grain yield, harvest index and water use of wheat. *Journal of the Australian Institute of Agricultural Science,* **43**, 117–120.

Passioura J.B. (1988a) Water transport in and to roots. *Annual Review of Plant Physiology and Plant Molecular Biology,* **39**, 245–265

Passioura J.B. (1988b) Root signals control leaf expansion in wheat seedlings growing in drying soil. *Australian Journal of Plant Physiology,* **15**, 687–693.

Paterson A.H., Tanksley S.D. and Sorrells M.E. (1991) DNA markers in plant improvement. *Advances in Agronomy* , **46**, 39–90.

Peacock J.M. and Sivakumar M.V.K. (1987) An environmental physiologists' approach to screening for drought resistance in sorghum with particular reference to sub–Saharan Africa. In *Food Grain Production in Semi-arid Africa,* edited by. J.M.Menyonga, T.Bezuneh and A.Youdeowei, pp. 101–120. Burkina Faso: OAU/STRC-SAFGRAD.

Pereira J.S. and Pallardy S. (1989) Water stress limitations to tree productivity. In *Biomass Production by Fast-growing Trees,* edited by. J.S. Pereira and J.J. Landsberg, pp.37–56. Dordrecht: Kluwer.

Peters L.L. (1987) Epicuticular wax and drought resistance of several maize genotypes. *Dissertation Abstracts International, B* 47, 12,I,4707B.

Pier P.A. and Berkowitz G.A. (1987) Modulation of water stress effects on photosynthesis by altered leaf K^+. *Plant Physiology,* **85**, 655–661.

Pitcairn C.E.R. and Grace J. (1982) The effect of wind and a reduced supply of phosphorus and nitrogen on the growth and water relations of *Festuca arundinacea* Schreb. *Annals of Botany* , **49**, 649–660.

Pitman M.G. (1981) Ion uptake. In *The Physiology and Biochemistry of Drought Resistance in Plants,* edited by L.G. Paleg and D. Aspinall, pp. 71–96. Sydney: Academic Press.

Premachandra S., Saneoka H., Kanaya M. and Ogata S. (1991) Cell membrane stability and leaf surface wax content as affected by increasing water defícts in maize. *Journal of Experimental Botany,* **42**, 167–171.

Pritchard J., Hetherington P.R., Fry S.C. and Tomos A.D. (1993) Xyloglucan endotransglycosylase activity, microfibril orientation and the profiles of cell wall properties along growing regions of maize roots. *Journal of Experimental Botany,* **44**, 1281–1289.

Pritchard J., Wyn Jones R.G. and Tomos A.D. (1990) Measurement of yield threshold and cell wall extensibility of intact wheat roots under different ionic, osmotic and temperature treatments. *Journal of Experimental Botany,* **41**, 669–675.

Quarrie S.S., Gulli M., Calestani C., Steed A. and Marmiroli N. (1994) Location of a gene regulating drought-induced abscisic acid production on the long arm of chromosome 5A of wheat. *Theoretical and Applied Genetics,* **89**, 794–800.

Quick P., Siegl G., Neuhaus E., Feil R. and Stitt M. (1989) Short term water stress leads to stimulation of sucrose synthesis by activating sucrose-phosphate synthase. *Planta,* **177**, 535–546.

Quisenberry J.E., Cartwright G.B. and McMichael B.L. (1984) Genetic relationship between turgor maintenance and growth in cotton germplasm. *Crop Science,* **24**, 470–483.

Radin J.W. (1983) Control of plant growth by nitrogen: differences between cereals and broad-leaf species. *Plant, Cell and Environment,* **6**, 65–78.

Radin J.W. (1984) Stomatal responses to water stress and to abscisic acid in phosphorus-deficient cotton plants. *Plant Physiology,* **76**, 392–394.

Radin J.W. and Boyer J.S. (1982) Control of leaf expansion by nitrogen nutrition in sunflower plants. *Plant Physiology* , **69**, 771–775.

Radin J.W. and Eidenbock M.P. (1984) Hydraulic conductance as a factor limiting leaf expansion of phosphorus-deficient cotton plants. *Plant Physiology*, **75**, 372–377.

Rains D.W. (1989) Plant tissue and protoplast culture: applications to stress physiology and biochemistry. In *Plants Under Stress.* edited by. H.G. Jones, T.J. Flowers and M.B. Jones, pp. 181–196. Cambridge: Cambridge University Press.

Rasmusson D.C. (1987) An evaluation of ideotype breeding. *Crop Science*, **27**, 1140–1146.

Reed R.H. (1984) Use and abuse of osmo-terminology. *Plant, Cell and Environment*, **7**, 165–170.

Rhodes D. (1987) Metabolic responses to stress. In *The Biochemistry of Plants*, Vol. 12, edited by D.D. Davies, pp. 201–242. San Diego: Academic Press.

Richards R.A. and Passioura J.B. (1989) A breeding program to reduce the diameter of the major xylem vessel in the seminal roots of wheat and its effect on grain yield in rain-fed environments. *Australian Journal of Agricultural Research*, **40**, 943–950.

Ritchie J.T. and Burnett E. (1971) Dryland evaporative flux in a dryland climate. II. Plant influences. *Agronomy Journal*, **63**, 56–62.

Rumbaugh M.D., Asay K.H. and Johnson D. A. (1984) Influence of drought stress on genetic variances of alfafa and wheatgrass seedlings. *Crop Science*, **23**, 297–303.

Saneoka H., Fujita K. and Ogata S. (1990) Effect of phosphorus on drought tolerance in *Chloris gayana* Kunth and *Coix lachryma-jobi* L. *Soil Science and Plant Nutrition*, **36**, 267–274.

Santamaria J.M., Ludlow M.M. and Fukai S. (1990) Contribution of osmotic adjustment to grain yield in *Sorghum bicolor* (L.)Moench under water-limited conditions. I. Water stress before anthesis. *Australian Journal of Agricultural Research*, **41**, 51–65.

Schulze E.D. (1986) Carbon dioxide and water vapour exchange in response to drought in the atmosphere and in the soil. *Annual Review of Plant Physiology*, **37**, 247–274.

Schulze E.D. and Kuppers M. (1979) Short-term and long-term effects of plant water deficits on stomatal response to humidity in *Corylus avellana* L. *Planta*, **146**, 319–326

Scott N.S., Munns R. and Barlow E.W.R. (1979) Polyribosome content in young and aged wheat leaves subjected to drought. *Journal of Experimental Botany*, **30**, 905–911.

Sharp R.E. and Davies W.J. (1979) Solute regulation and growth by roots and shoots of water-stressed maize plants. *Planta*, **147**, 43–49.

Sharp R.E. and Davies W.J. (1985) Root growth and water uptake by maize plants in drying soil. *Journal of Experimental Botany*, **36**, 1441–1452.

Sharp R.E. and Davies W.J. (1989) Regulation of growth and development of plants with a restricted supply of water. In *Plants Under Stress*, edited by H.G. Jones, T.J. Flowers and M.B. Jones, pp. 71–93 Cambridge: Cambridge University Press.

Sharp R.E., Hsiao T.C. and Silk W.K. (1990) Growth of the maize primary root at low water potentials. II. The role of growth and deposition of hexose and potassium in osmotic adjustment. *Plant Physiology*, **93**, 1337–1346.

Silsbury J.H. (1961) A study of dormancy, survival and other characters in *Lolium perenne* at Adelaide, SA. *Australian Journal of Agricultural Research*, **12**, 1–9.

Sinclair T.R., Muchow R.C., Bennett J.M. and Hammond L.C. (1987) Relative sensitivity of nitrogen and biomass accumulation to drought in field-grown soybean. *Agronomy Journal*, **79**, 986–991.

Singh D.P. (1989) Evaluation of specific dehydration tolerance traits for improvement of drought resistance. In *Drought Resistance in Cereals*, edited by F.W.G. Baker, pp. 165–175. Wallingford, Oxon: CAB International.

Singh T.N., Aspinall D. and Paleg L.G. (1972) Proline accumulation and varietal adaptability to drought in barley: a potential metabolic measure of drought resistance. *Nature New Biology*, **236**, 188–190.

Skerman P.J. and Riveros, F. (1990) *Tropical Grasses.* Rome: F.A.O.

Skriver K. and Mundy J. (1990) Gene expression in response to abscisic acid and osmotic stress, *The Plant Cell*, **2**, 503–512.

Slatyer R.O. (1967) *Plant-Water Relationships.* London: Academic Press.

Slonov L.Kh. and Petinov N.S. (1980) Nucleotide content and ATPase activity in hemp leaves as a function of water supply. *Soviet Plant Physiology,* **27**, 811–815.

Smirnoff N. (1993) The role of active oxygen in the response of plants to water deficit and desiccation. *New Phytologist,* **125**, 27–58.

Smith H. (1990) Signal perception, differential expression within multigene families and the molecular basis of phenotypic plasticity. *Plant, Cell and Environment,* **13**, 585–594.

Sojka R.E. (1985) Field evaluation of drought resistance in small-grain cereals. In *Progress in Plant Breeding-1.* edited by G.E. Russel. pp. 165–191. London: Butterworths.

Sprent J.I. (1973) The effects of water stress on nitrogen-fixing root nodules. IV. Effects on whole plants of *Vicia faba* and *Glycine max. New Phytologist,* **71**, 603–611.

Stanhill G. (1986) Water use efficiency. *Advances in Agronomy,* **39**: 53–85.

Steckel J.R.A. and Gray D. (1979) Drought tolerance in potatoes, *Journal of Agricultural Science,* **92**, 375–381.

Steer M.W. (1988) Plasma membrane turnover in plant cells. *Journal of Experimental Botany,* **39**, 987–996.

Stewart G.R. (1989) Desiccation injury, anhydrobiosis and survival. In *Plants Under Stress,* edited by H.G. Jones, T.J. Flowers and M.B. Jones, pp. 115–130. Cambridge: Cambridge University Press.

Storey R. and Wyn Jones R.G. (1978) Salt stress and comparative physiology in the gramineae. I. Ion relations of two salt-and water-stress cultuvars, California Mariout and Arimar. *Australian Journal of Plant Physiology,* **5**, 801–816.

Struik P.C., Doorgeest M. and Boonman J.G. (1986) Environmental effects on flowering characteristics and kernel set of maize (*Zea mays* L.). *Netherlands Journal of Agricultural Science,* **34**, 469–484.

Sullivan C.Y. and Ross W.M. (1979) Selecting for drought and heat resistance in grain sorghum. In *Stress Physiology in Crop Plants,* edited by H. Mussell and R.C. Staples. pp. 263–281. New York: John Wiley.

Svenningsson H. and Liljenberg C. (1986) Membrane lipid changes in root cells of rape *(Brassica napus)* as a function of water-deficit stress. *Physiologia Plantarum,* **68**, 53–58.

Sydes C.L. and Grime J.P. (1984) A comparative study of root development using a simulated rock crevice. *Journal of Ecology,* **72**, 937–946.

Talouizite A. and Champigny M.J. (1988) Response of wheat seedlings to short-term drought stress with particular respect to nitrogen nutrition. *Plant, Cell and Environment,* **11**, 149–155.

Tanner W. and Beevers H. (1990) Does transpiration have an essential function in long-distance ion transport in plants? *Plant, Cell and Environment,* **13**, 745–750.

Tardieu F. (1993) Will progress in understanding soil root relations and root signalling substantially alter water-flow models? *Philosophical Transactions of the Royal Society, London,* **341**, 57–76.

Terry N. and Ulrich A. (1973) Effects of potassium deficiency on the photosynthesis and respiration of leaves of sugar beet. *Plant Physiology,* **51**, 783–786.

Thomas H. (1986) Effects of rate of dehydration on leaf water stress and osmotic adjustment in *Dactylis glomerata L., Lolium perenne* L. and *L. multiflorum* Lam. *Annals of Botany,* **57**, 225–235.

Thomas H. (1990) Osmotic adjustment in *Lolium perenne* L., its heritability and the nature of solute accumulation. *Annals of Botany,* **64**, 581–587.

Thomas H. (1991) Accumulation and consumption of solutes in swards of *Lolium perenne* during drought and after rewatering. *New Phytologist,* **118**, 35–48.

Thomas H. and Evans C. (1989) Effects of divergent selection for osmotic adjustment on water relations and growth of plants of *Lolium perenne. Annals of Botany,* **64**, 581–587.

Thomas H and Evans C. (1990) Influence of drought and flowering on growth and water relations of perennial ryegrass populations. *Annals of Applied Biology,* **116**, 371–382.

Thomas H. and Norris I.B. (1981) Evaluation of drought resistance of grass and clover populations grown in containers. In *Plant Physiology and Herbage Production,* edited by C.E. Wright, pp. 217–219. Hurley: British Grassland Society.

Timpa J.D., Burke J.J., Quisenberry J.E. and Wendt C.W. (1986) Effects of water stress on the organic acid and carbohydrate composition of cotton plants. *Plant Physiology,* **82**, 724–728.

Tischler C.R. and Voigt R.W. (1990) Variability in leaf characteristics and water loss in the weeping lovegrass complex. *Crop Science,* **30**, 111–117.

Tomos A.D. (1990) Growth - a role for plant growth regulators? In *Importance of Root to Shoot Communication in the Responses to Environmental Stress*, edited by. W.J. Davies and B. Jeffcoat, pp. 53–70. Bristol: British Society for Plant Growth Regulation.

Trebejo I. and Midmore D.J. (1990) Effect of water stress on potato growth, yield and water use in a hot and a cool temperate climate. *Journal of Agricultural Science*, **114**, 321–334.

Troughton A. (1957) *The Underground Organs of Herbage Grasses* Farnham Royal: Commonwealth Agricultural Bureaux.

Tueller P.T. (Ed.) (1988) *Vegetation Science Applications for Rangeland Analysis and Management* Dordrecht: Kluwer Academic Publishers.

Turner L.B. (1985) Changes in the phosphorus content of *Capsicum annuum* leaves during water stress. *Journal of Plant Physiology*, **121**, 429–439.

Turner L.B. (1990) The extent and pattern of osmotic adjustment in white clover *(Trifolium repens*) L. during the development of water stress. *Annals of Botany*, **66**, 721–727.

Turner L.B. and Wellburn A.R. (1985) Changes in adenylate nucleotide levels in the leaves of *Capsicum annuum* during water stress. *Journal of Plant Physiology*, **120**, 111–122.

Turner N.C. (1979) Drought resistance and adaptation to water deficits in crop plants. In *Stress Physiology in Crop Plants*, edited by. H. Mussell and R.C. Staples, pp. 343–372. New York: John Wiley.

Turner N.C. (1986) Crop water deficits: a decade of progress. *Advances in Agronomy*, **39**, 1–51.

Turner N.C., Schulze E.D. and Gollan T. (1984) The responses of stomata and leaf gas exchange to vapor pressure deficits and soil water content. I. Species comparison at high soil water contents. *Oecologia*, **63**, 338–342.

Turner N.C., Schulze E.D. and Gollan T. (1985) The responses of stomata and leaf gas exchange to vapor pressure deficits and soil water content. II. In the mesophytic herbaceous species *Helianthus annuus*. *Oecologia*, **65**, 348–355.

Van Volkenburgh E. and Boyer, J.S. (1985) Inhibitory effects of water deficit on maize leaf elongation. *Plant Physiology*, **77**, 190–194.

Volaire F. (1995) Growth, carbohydrate reserves and drought survival strategies of contrasting *Dactylis glomerata* populations in a Mediterranean environment. *Journal of Applied Ecology*, **32**, 56–66.

Wardlaw I.F. (1969) The effect of water stress on translocation in relation to photosynthesis and growth. II. Effect during leaf development in *Lolium temulentum* L. *Australian Journal of Biological Sciences*, **22**, 1–16.

Wardlaw I.F. (1990) The control of carbon partitioning in plants. *New Phytologist*, **116**, 341–381.

Watts W.R. (1974) Leaf extension in *Zea mays*. III. Field measurements of leaf extension in response to temperature and leaf water potential. *Journal of Experimental Botany*, **25**, 1085–1096.

Woodfield D.R. and Caradus J.R. (1987) Adaptation of white clover to moisture stress. *Proceedings of the New Zealand Grassland Association*, **48**, 143–149.

Wright G.C., Smith R.G.C. and Motgan J.M. (1983) Differences between two grain sorghum genotypes in adaptation to drought stress. III. Physiological responses. *Australian Journal of Agricultural Research*, **34**, 637–651.

Wyn Jones R.G. and Pritchard J. (1989) Stresses, membranes and cell walls. In *Plants Under Stress*. edited by. H.G. Jones, T.J. Flowers and M.B. Jones, pp.95–114. Cambridge: Cambridge University Press.

Zhang J and Davies W.E. (1989) Sequential response of whole plant water relations to prolonged soil drying and the involvement of xylem sap ABA in the regulation of stomatal behaviour of sunflower plants. *New Phytologist*, **113**, 167–174.

2. MECHANISMS OF DESICCATION TOLERANCE IN RESURRECTION VASCULAR PLANTS

D.F. GAFF

Department of Ecology and Evolutionary Biology, Monash University, Clayton 3168, Australia

INTRODUCTION

The importance of water deficit in plant protoplasm is not confined to the stress imposed on plants by drought: protoplasmic water deficit is a major factor involved in the effects of other environmental stresses acting on plants, freezing and soil salinity. When shoots freeze, most commonly ice crystals nucleate in xylem sap, then continue to grow in the *intercellular* spaces at the expense of water osmosing from cells. Resulting water stress can be considerable, as protoplasts in many species may shrink to less than one third their former volume (Levitt, 1980).

Where soil becomes saline, its water potential (i.e the free energy of water in a system) is lowered by the osmotic potential of the dissolved salt. Consequently the water potentials of tissues throughout plants on the site are also lowered (indicative of water stress) as they adjust thermodynamically to the modified soil-to-air water potential gradient. If vacuolar uptake of salt and/or accumulation of organic solutes occur, the increase in the amount of osmotically active solute in cells will allow cell volumes to recover, but cell water potentials are determined by each cell's position in the soil-plant-air gradient and so they remain low.

Plant water potentials in halophytic communities are about 2.3 MPa lower than in neighbouring non-saline communities, a considerable water stress. If a soil solution dries until salt crystallizes, soil water potential falls to –40 MPa. Normal plants do not survive water potentials below –20 MPa. A limited number of higher plants "resurrection plants" (discussed below) tolerate water potentials well below –40 MPa. They therefore have one attribute which is a necessary component of overall salt resistance. Relative salt resistances of resurrection grasses range from "salt-tolerant" to "salt sensitive" on the Maas and Hoffman (1977) scale (Table 2.1). When ranked on the basis of regrowth ability under salinized conditions, species with high control dry matter increments out-perform more salt tolerant species (Table 2.2).

Clearly, desiccation tolerance must be complemented by other attributes (particularly the ability to exclude salt from root tissues) to achieve high levels of salt resistance.

Table 2.1. Soil salinities giving yield reductions of 50% of control shoot dry matter production of resurrection grasses.

Species	*mM NaCl*
Sporobolus stapfianus	175–215
S. festivus	170
Tripogon jacquemonti	155
S. aff. fimbriatus	150
S. pellucidus	100
Microchloa caffra	90
Eragrostis nindensis	70
S. lampranthus	35

Based on wood and Gaff (1989) and Collins (1989)

Table 2.2. Effect of soil salinity on shoot regrowth ratio, i.e. the ratio of shoot dry weights for shoots after 6 weeks soil shoots salinization relative to shoots removed just before salinization.

Species	*0 NaCl*	*100 mM NaCl*
*Sporobolus pyramidalis**	9.5	6.1
S. aff. fimbriatus	4.1	2.4
S. lampranthus	5.3	2.1
S. festivus	2.7	2.0
S. pellucidus	2.8	1.4
S. stapfianus	1.0	1.0

Based on Wood and Gaff (1989)
* a desiccation-sensitive species.

Although water stress is not directly involved in other environmental stresses as in freezing and salinity, some responses to water stress are shared, e.g. endogenous accumulation of abscisic acid (ABA) induced by stress and improvement of stress resistance by exogenous ABA. The study of desiccation tolerance therefore has importance in the topic of stress tolerance in general.

RESURRECTION PLANTS

Most angiosperm species have reproductive cells whose protoplasm is able to withstand air-dryness. With the notable exception of grasses, pollen grains, survive water potentials below –100 MPa (Visser, 1955). Embryos, again with some exceptions, also survive water potentials below this very low value. The great majority of angiosperms are unable to express this ability for desiccation tolerance in the non-reproductive tissues; few crop plants, even after hardening to environmental stress, can survive –20 MPa (Iljin,1931; Levitt *et al.*, 1960). The resur-

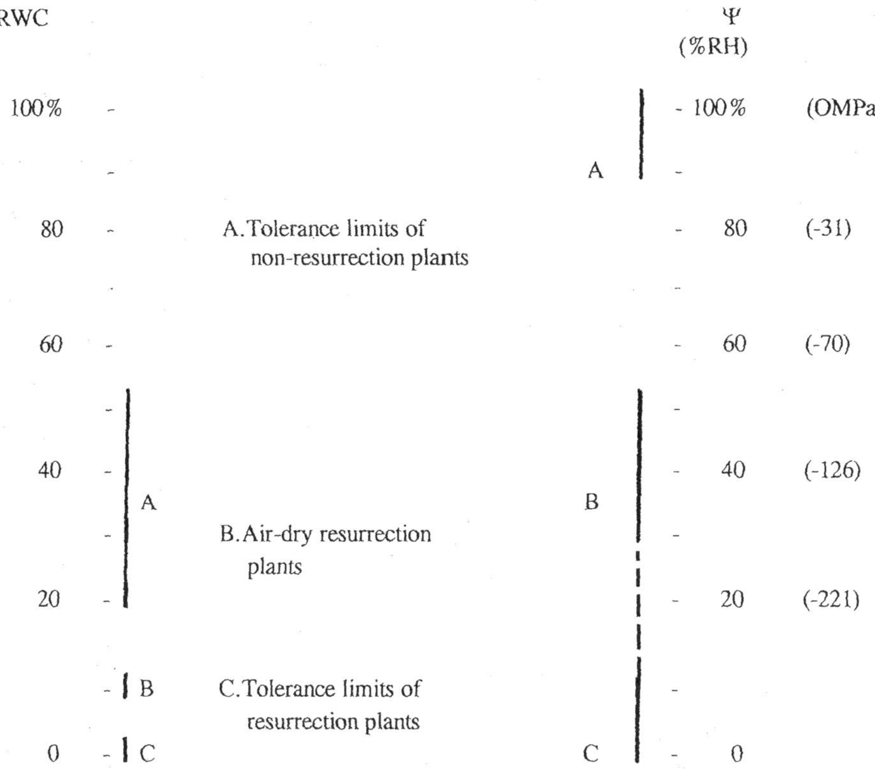

Figure 2.1 A comparison of the usual ranges of protoplasmic drought tolerance limits for (A) non-resurrection plants and (C) resurrection plants, and for water stress levels in live resurrection plants air-dry in the field (B) when expressed on a scale of relative water content (RWC), left, or on a scale of water potential (ψ) given as equivalent relative humidity of air at 28°C (corresponding MPa are in parentheses), right.

rection plants, however, are able to express the desiccation tolerance of the reproductive tissues in their leaves, stems and roots. Of the 160,000 or so species of angiosperms, almost 100 have leaves which survive water potentials below –100 MPa; most of these survive below –500 MPa (Vieweg and Ziegler, 1969; Gaff, 1977,1989; Gaff and Latz, 1978).

The water potential scale refers to the Gibbs free energy associated with water in a particular system. Water stress in plants can also be expressed in terms of relative water content (RWC), i.e. the water content of a tissue as a percent of the water content of the tissue when it is fully turgid. The difference between "normal" plants and "resurrection" plants is not so marked on this scale (Figure 2.l). Normal plants die in the range of 20–50% RWC. In comparison, resurrection plants dry in the field to about 4–13% RWC, without injury. The difference in RWC between the two ranges is small.

The volume/concentration aspects of water stress, are best expressed by the RWC parameter, and may well be more basic to the biochemical responses of plants to water stress (c.f. Flowers and Ludlow, 1986). The water potential however determines the direction of water flow and any equilibrium between plant and air; consequently it is of basic ecological importance. The water potentials of field-dry (living) resurrection plants represent equilibria with air of 30–50% relative humidity; the water potentials at the limits of protoplasmic drought tolerance for "normal" plants represent equilibria with air of 85–98% relative humidity. The small difference in RWC equates to a large difference in water potential, which the plant must cope with to survive air-dryness.

The phenomenon of desiccation tolerance is found in more monocotyledonous species than dicotyledonous species. The fact that the grass family (Poaceae) contains more reported resurrection plants (39 species) than any other angiosperm family holds promise that a fuller understanding of the mechanism of desiccation tolerance may lead eventually to practical application of the phenomenon in cereal crops. The following section examines the processes that may occur in drying resurrection plants in relation to their tolerance of protoplasmic desiccation.

RESPONSES DURING DRYING

Leaves of many resurrection species do not display desiccation tolerance when they are dried detached from the plant, but die at similar water potentials to the death point for normal crop plants. This suggests that they are subject to similar processes of injury to those in non-resurrection plants, and that a protective process occurs in intact resurrection plants. The patterns of injury in relation to water potential, to time of drying and to the time during which leaves are held dry, suggest that injury arises from a number of processes with contrasting characteristics.

RAPID INJURY WITH DECREASING WATER POTENTIAL

When dried too rapidly for any hardening process to occur, leaves of resurrection plants and of non-resurrection plants die once the lethal level of water stress (for that species and tissue) is reached. Injury is rapid compared to the speed of drying, and in most resurrection angiosperms may occur at any water potential below –5 MPa, depending on the species and a plant's prior treatment. The speed of the injury over such a wide range of leaf water potential is suggestive of a rapid physical or physiochemical process triggered at the injurious water potential or cell volume and may involve the following processes of injury.

Mechanical Injury

Iljin (1931) proposed that the protoplasm tore under tensions set up by contraction of the drying protoplast beyond the point of zero turgor and by adhesion of the protoplast to the cell wall. There is evidence that such "negative turgors" do arise (Grieve, 1961), but the extent to which they cause injury is debatable. Electron microscope studies of desiccation-sensitive plants, injured by water stress, reveal degeneration of cell organelles as discrete entities, but no massive tears in the protoplasm (Schnepf, 1961). The plasmalemma of drying resurrection plant cells often appears torn, but this is probably an artefact of differential swelling of dry cell walls and dry protoplasm during fixation.

Degradation of Lipoprotein Membranes

Water stress induces deterioration of the membrane systems in desiccation-sensitive protoplasm: breaks form in membranes bounding plastids, mitochondria and nuclei, as well as in the plasmalemma, and dictyosome membranes (Schnepf, 1961; Nir *et al.*, 1969; Buttrose and Swift, 1975). Leakiness of membranes, at low water potential, to solutes support the view that membranes are degraded during water stress (Simon, 1974). Some leakage of potassium ions occurs in viable leaves of resurrection plants as they rehydrate from air-dryness (Gaff, 1980), but twice as much leakage took place from killed leaves. It seems that partial leakiness of the plasmalemma need not be fatal. Monocotyledonous resurrection plants survive loss of many cristae, loss of most thylakoids, and breaks in outer membranes of mitochondria and chloroplasts (Gaff *et al.*, 1976; Hallam and Gaff, 1978; Rajeswari *et al.*, 1993). The phenolic and hydrolytic contents of vacuoles and lysozymes, theoretically would require that these organelles are kept compartmented from the rest of the cell to avoid disruption of metabolism.Heber and Santarius (1964) proposed that increased concentration of ions in drying cells has a destabilizing effect on the lipoprotein membranes of the cell, once the toxicity thresholds of the ions are exceeded.

Repulsive forces between membranes in close proximity in drying cells have also been suggested to cause tension in the liquid between. Resulting strains in the membranes are postulated to cause physical damage to the membranes (Wolfe and Steponkus 1983). These last two hypotheses of injury during water stress are consistent with the type of membrane damage observed under the electron microscope.

Protein Denaturation

Levitt (1963) proposed the hypothesis that injury under water stress was produced by denaturation of proteins due to conversion of native protein sulphydryl to disulphide; on re-imbibition this leads to unfolding of the polypeptide chain, with a consequent loss of activity of enzymes, or irreversible permeabilization of membranes. Enzyme activity in extracts of dry plants of the resurrection fern, *Selaginella lepidophylla* was markedly less than control extracts in the case of photosynthetic enzymes but was less reduced in the case of respiratory enzymes (respiration appears to be important in recovery on rehydration, but photosynthesis initially is not) (Harten and Eckmeier, 1986). A rise in the content of reactive sulphydryl in structural-protein in cabbage leaves, dried to sublethal water potentials, was consistent with partial denaturation of membrane protein, while at lethal stress, a third of the structural protein sulphydryl was converted to disulphide (Gaff, 1966). Small amounts of sulphydryl were converted to disulphide in the resurrection plant *Xerophyta viscosa*, but not in two other resurrection species or in *Sporobolus pyramidalis*, a desiccation-sensitive grass (Daniel and Gaff, 1980a).

Biochemical processes may contribute to tissue injury, if drying is not particularly rapid. Injury mechanisms that have been proposed are:

Decreased Protein Content

Protein contents usually fall as water stress becomes intense in desiccation-sensitive plants (see Levitt, 1980). Protein synthesis may be sensitive to water stress, e.g. most polysomes revert to monosomes in *Zea mays* in 4h at 0.5 MPa (Hsiao, 1970). Reduced protein synthesis was attributed by Maranville and Paulsen (1972) partly to diminished RNA synthesis and partly to four-fold increases in RNase activity.

Net loss of structural protein was found in leaves of drying resurrection plants, but soluble protein contents were often maintained or, in some cases, increased (Gaff and McGregor, 1979). In *Xerophyta villosa*, levels of soluble and of insoluble protein were similar in live air-dry leaves and in dead air-dry leaves, i.e. in this case, injury was not due to hydrolysis of protein.

Ammonia Toxicity

Savitskaya (1965) considered ammonia accumulation might cause stress injury. Ammonia contents in water stressed barley rose to 10–25 ppm, contents that seem low for injury (c.f. toxicity levels in cucumber is 200 ppm; Matsumoto *et al.*, 1968).

Impaired Respiration

Respiration rates decline at sever water stress (Stuart, 1968), and ATP supply may fall (Zholkevich and Rogacheva, 1967). Trends in ATP content in resurrection plants are highly variable, falling in some species and rising in others. In desiccation-tolerant angiosperms, appreciable ATP was always present in the air-dry leaves (Gaff and Ziegler, 1989); the desiccation tolerant moss *Tortula ruralis*, however, revived from about –40 MPa although ATP was virtually absent. At least in the short term, a high ATP content was evidently not essential for recovery on rehydration.

Photo-oxidation

In *Selaginella lepidophylla*, water stress at high temperatures exacerbates photoinhibition, even though rolling up of the fronds into a tight ball provides protection from intense light at full dehydration (Lebkuecher and Eickmeier, 1991). Protection is extended by zeaxanthin-associated mechanisms (Eickmeier *et al.*, 1993). Levels of H_2O_2 increased during drying and rehydration of detached leaves of the resurrection grass *Sporobolus stapfianus;* activity of enzymes of the glutathione-ascorbate cycle recorded in extracts from dry and rehydrated leaves of this grass and the resurrection dicot *Boea hygroscopica* indicates that some protective antioxidant action may be operating. (Sgherri *et al.*, 1994 a,b).

SLOW INJURY DURING STORAGE AT LOW WATER POTENTIALS

Whereas the above processes may injure either desiccation-sensitive or desiccation-tolerant species, other injury processes affect desiccation-tolerant tissue at stress levels beyond those survived by sensitive species.

In Air of High Relative Humidity

Resurrection plant tissue which survives well below –100 MPa for long periods, will deteriorate and die if kept for several days at only –20 MPa. Injury is slower at more intense water stress: most species die after 46 days storage at 76% relative humidity, but last 79 days at 52% RH (Gaff, 1977). Pollen and seed are subject to this injury also. Slow proliferation of fungi and bacteria may be implicated above 70% RH. Injury, however, persists in pathogen-free wheat caryopses (Harrison and Perry, 1976). Possibly, biochemical processes of injury are operating at low speed, as the general rate of metabolism slows at very low tissue water contents. These may include partial loss of DNA and RNA (Roberts and Osborne, 1973).

In Air of Low Relative Humidity

Slow injury takes place during prolonged storage *after* equilibration of seed, pollen and leaves of resurrection plants to very low water potentials (below −200 MPa) (Pruzsinszky, 1960; Gaff, 1977; Powell and Mathews, 1972). Unlike storage injury, speed of injury increases as stress intensifies (i.e. as water potential decreases to more negative values). Gradual oxidation of lipids may be occurring in the absence of effective respiration (Koostra, 1973).

INDUCTION OF DESICCATION TOLERANCE

The overall drought resistance of most plants can be improved by a period of moderate stress (either water stress or cold stress). In some species, only mechanisms of drought avoidance are improved, in others protoplasmic drought tolerance increases, either alone or in combination with the drought avoidance. Although the increased tolerance may be an appreciable improvement from an agricultural point of view, it is still only small compared with the desiccation tolerance of resurrection plants. Three crop species improved protoplasmic drought tolerance by 3 to 9 MPa during 2 to 3 weeks of low soil water (Levitt *et al.*, 1960). Pea plants were able to improve drought tolerance by 3 MPa as the root medium water potential fell to −5 MPa, but only in a particular stage of development (2nd to 4th week after planting) (Lee-Stadelmann and Stadelmann, 1979). Current evidence indicates that during drying, the potential operation of the injurious processes, mentioned above, is opposed in the resurrection plants by a "hardening" or "acclimation" process which induces full desiccation tolerance before plants reach air-dryness. The evidence for this is:

(a) When leaves are detached from hydrated resurrection plants, in most species they show only the drought tolerance levels of normal crop plants (Gaff, 1980).
(b) With the resurrection plants *Coleochloa setifera*, *Sporobolus stapfianus* and *Xerophyta villosa*, detached leaves survive drying if their source plant was moderately stressed when they were removed (Bartley, 1978; Tymms, 1979; Gaff and Loveys, 1993).
(c) Leaves detached from the resurrection plant, *Borya constricta* survive complete dehydration after detachment from hydrated plants, if they are equilibrated for two days to air of 90 to 98% relative humidity before complete drying (Gaff and Churchill, 1976). The induction of desiccation tolerance in *Borya* required two days on the water potential range −3 to −15

MPaat temperatures between 18 and 35°C. Light was not required. Tolerance-induction was accompanied by loss of chlorophyll and loss of grana, and vesicularization of thylakoids (Gaff *et al.*, 1976; Hetherington *et al.*, 1982). Leaves which did not undergo these changes failed to survive desiccation. The failure of detached leaves of many resurrection plants to survive desiccation raises the possibility that a hormonal link with the rest of the plant is necessary to trigger tolerance induction. Presumably *Borya*, under suitable conditions, can synthesize any such hormone in detached leaves.

Exogenous abscisic acid (ABA) applied to fully hydrated detached leaves induced desiccation tolerance without the two-day period of water stress (Gaff and Loveys, 1984). Moreover, endogenous ABA increased with water stress to several times compared with hydrated controls of *Borya* and of *Myrothamnus flabellifolia*, a resurrection plant in which exogenous ABA also stimulated desiccation tolerance. It seems likely then that ABA is implicated in the normal mechanism for triggering desiccation tolerance in these species. Exogenous ABA also stimulated desiccation tolerance in cultured callus of the resurrection plant *Craterostigma plantagineum* (Bartels *et al.*, 1990).

However, exogenous ABA did not improve protoplasmic drought tolerance in detached leaves of this species dried in the dark (Marr, 1990). Detached hydrated leaves of the resurrection grass *Sporobolus stapfianus* show little or no stimulation of desiccation tolerance when treated with ABA, even though the content of ABA rises markedly with water stress (Gaff and Loveys, 1993). At this stage then it appears that ABA increases with water stress in resurrection plants as in non-resurrection plants (Wright and Hiron, 1969), but that the response in inducing desiccation tolerance in the former group is species-dependent.

THE MECHANISM OF DESICCATION TOLERANCE

Data available to date on resurrection plants suggest that the following factors may be important in the mechanism of desiccation tolerance:

Accumulation of Protective Substances

In proposing that injury in water stressed cells stemmed from disruption of membrane structure by high concentrations of ions, Heber and Santarius (1964) also suggested that sufficient concentrations of solutes such as sugars could protect membranes from these effects. The literature on protection by sugars in dry biological systems has been reviewed by Crowe *et al.* (1992). Hydrolysis of starch accompanied by accumulation of sugars was noted in drying leaves of the resurrection plant *Xerophyta schnitzleinia* (Owoseye and Sanford, 1972) and is one of the earliest effects of water stress on resurrection plants (Gaff *et al.*, 1976; Hallam and Luff, 1980). On air-dryness, most starch is hydrolyzed. Starch hydrolysis is common during water stress in non-resurrection plants as well. Sugar contents are marginally higher in resurrection plants than in other stressed plants, but sugar to anion ratios are markedly higher in the former than the latter (Schwab and Heber, 1984; Schwab and Gaff, 1986).

Sucrose constitutes the bulk of the sugar accumulated during water stress in both resurrection and non-resurrection species (Kaiser *et al.*, 1985; GHasempour-Mahidashti, 1995). Evidence for this hypothesis is strengthened by the observation that sucrose (specifically) partially protected three respiratory enzymes, *in vitro*, from inhibition by high cation concentrations. Moreover, preincubation of the enzymes with sucrose (before high ion concentrations) stimulated activity to above control levels (Schwab and Gaff, 1990). A specific sucrose interaction with the proteins of drying resurrection plants, but not those of desiccation-sensitive tissues, could explain why tissues such as sugar beet are not desiccation-tolerant. It should be noted however, that sugars and amino acids, and even proline, do not accumulate during drying in the desiccation-tolerant moss *Tortula ruralis* (Bewley and Krochko, 1982).

The disaccharide trehalose is particularly effective in stabilizing biological material during freezing and dehydration; partly due to the ability of trehalose solutions to avoid crystallizing as they dehydrate (Colaco *et al.*, 1992). High sugar contents assist the production of a vitreous phase in dehydrating protoplasm and thereby stabilizes desiccated tissue (Williams and Leopold, 1989).

Protein Synthesis

Soluble protein content is often higher in air-dry resurrection plants than in hydrated ones, suggesting that protein synthesis is active during drying (Gaff and McGregor, 1979). Studies of the desiccation-tolerant *Xerophyta villosa* revealed a complex response in the polysome content during drying. As in desiccation-sensitive plants (Hsiao, 1970; Brandle *et al.*, 1977), the polysome content declined during mild stress. However with further drying, polysome content in *X. villosa* increased to a value 50% greater than in the unstressed control plants; also incorporation of labelled amino acids increased at 60% RWC, in the range of water stress where desiccation tolerance is induced (Tymms *et al.*, 1982). Subsequently, polysome content fell to zero as RWC approached 20%. Similar trends in glucose-6-phosphate dehydrogenase activity occur in *Sporobolus stapfianus* leaves when they dry attached to the plant (and are viable) but not when they are drying detached (and die) (Gaff, Wallek and Duivenvorden, unpublished). Presumably, increased protein synthesis contributes to the greater activity of the enzyme in air-dry leaves, when dried attached (70% of control levels), compared to leaves dried detached (26%).

Qualitative changes in proteins occur during drying of resurrection plants and may well play a crucial role in survival of the tissue. Levitt(1962) proposed that protoplasmic hardening to water stress involved, among other factors, synthesis of stress-stable proteins, in which the native structure was stabilized by intramolecular disulphide bonds. Electrophoretic patterns of peroxidase activity in soluble protein show the disappearance of bands of activity and the appearance of new bands during desiccation of *Xerophyta viscosa* (desiccation-tolerant) (Daniel and Gaff, 1980b).

A new band of glucose-6-phosphate dehydrogenase activity, which is stress-stable, arises with air-drying of attached leaves of *Sporobolus stapfianus* (Gaff, Wallek and Duivenvorden, unpublished). Over 20 new proteins were discerned in extracts of dry viable leaves of *Craterostigma plantagineum* compared with extracts from fully hydrated leaves (Bartels *et al.*, 1990). Resurrection plants may remain viable in the air-dry state for periods of two years, or more, depending on the ambient conditions of temperature, insolation and humidity. Although various degrees of degradation of fine structure are apparent in the different species, an essential core of organization must remain as a basis for recovery of normal metabolism in the rehydrated tissue. Messenger RNA is retained in air-dry leaves of *Xerophyta villosa* and *Craterostigma plantagineum* (Tymms, 1979; Bartels *et al.*, 1990). The desiccation-tolerant moss, *Tortula ruralis* even retains polysomes when air-dried rapidly, and rapidly resumes protein synthesis on rehydration, at least in part based on retained polysomes (Bewley, 1972; Oliver and Bewley, 1984; Oliver 1991). In desiccation-tolerant angiosperms, respiration recovers rapidly (0.5 to 1 h); RNA and protein synthesis recover gradually until by about 16 hours rehydration control rates are attained (Gaff and McGregor, 1979). In the early stage of recovery, incorporation of amino acids into protein is higher into low molecular weight proteins than in control plants of *X. villosa* (Tymms, 1979).

GENETICS OF DESICCATION TOLERANCE

The desiccation tolerance of plant embryos and pollen in general imply that the genetic information for tolerance is present in most angiosperm plants, but it is rarely expressed in foliage. Several transcripts, which are related to the induction of desiccation tolerance in drying leaves of *Craterostigma plantagineum* and in ABA-treated callus of that species, have recently been cloned (Piatkowski *et al.*, 1990). Five cDNA clone families were characterized; predicted proteins included ones homologous to dehydrins and LEA proteins, both found in seed, and ELIP in thylakoids (Michel *et al.*, 1994). Genes encoding transketolases were upregulated during rehydration, related to sugar metabolism (Bernacchia *et al.*, 1995). A further clone, derived from barley embryos at the stage when desiccation tolerance is attained, had homology to mammalian genes encoding for aldose reductase involved in the synthesis of sorbitol (Bartels *et al.*, 1991). This clone encoded for an active aldose reductase in plants, expressed during embryo maturation and modulated by ABA and gibberellic acid. This may be involved in synthesis of protective solutes during water stress.

Desiccation tolerance is not regulated by ABA in the resurrection grass *Sporobolus stapfianus*. Protein synthesis persists to extreme water stress. Novel leaf-proteins are detected in 2 phases in drying plants: from 85% to 60% RWC and 60% to below 30% RWC (Kuang *et al.*, 1995). These protein changes reflect qualitative alterations in transcription (Gaff *et al.*, 1993). Differential screening of a cDNA library yielded water families of cDNA clones correlated with induction of desiccation tolerance in drying plants of this grass (Blomstedt *et al.*,1995). Homologies of the clone families with registered database DNA sequences support the hypotheses that the following processes are probably important for desiccation tolerance: (a) protein turnover, (b) antioxidant activity, and (c) prevention of excessive accumulation of ions in the cytosol and nucleoplasm.

CONCLUSION

It is clear that much work remains to be done in this field. On the one hand, basic screening of more desiccation-tolerant grasses for salt tolerance is desirable. On the other, molecular biology promises new insights into the mechanism of desiccation tolerance in plants to complement physiological studies. If the genetic control of desiccation tolerance is not too complicated, it should be possible eventually to genetically engineer crop and pasture plants for this characteristic. While desiccation tolerance is a factor in salt resistance, it cannot be effective alone; a greater understanding of the mechanism of salt exclusion by roots is also needed.

ACKNOWLEDGEMENTS

The author is grateful to the Meat Research Corporation and the Australian Research Commission for funding during the preparation of this Chapter (Grants UMON.004 and A19230441, respectively) and during some of the research cited here.

REFERENCES

Bartels D., Engelhardt K., Roncarati R., Schneider F., Rotter M and Salamini R. (1991) An ABA and GA modulated gene expressed in barley embryo encodes an aldose reductase related protein. *The EMBO Journal*, 10, 1037–1043.

Bartels D., Schneider K., Terstappen G., Piatkowski D. and Salamini F. (1990) Molecular cloning of abscisic acid- modulated genes which are induced during desiccation of the resurrection plant *Craterostigma plantagineum*. *Planta*, **181**, 27–34.

Bartley M. (1978) Some aspects of the ultrastructure and physiology of the southern African sedge *Coleochloa setifera* (Ridley) Gilly under water deficit stress. B.Sc. (Hons) Report, Monash University, Clayton, Australia.

Bernacchia G., Schwall G., Lottspeich F., Salamini F. and Bartels D. (1995) The transketolase gene family of the resurrection plant *Craterostigma plantagineum*: Differential expression during the rehydration phase. *The EMBO Journal*, **614**, 610–618.

Bewley J.D. (1972) The conservation of polysomes in the moss *Tortula ruralis* during total desiccation. *Journal of Experimental Botany*, **23**, 692–698.

Bewley J.D. and Krochko J.E. (1982) Desiccation-tolerance. In Physiological Plant Ecology II; Water Relations and Carbon Assimilation, edited by O.L. Lange, P.S. Nobel, C.B. Osmond and H. Ziegler. Encyclopedia of Plant Physiology, New Series 12B, Heidelberg/New York: Springer-Verlag.

Blomstedt C.K., Gaff D.F., Gianello R.D., Hamill J.D. and Neale A.D. (1995) Characterization of genes involved in desiccation tolerance of resurrection grass *Sporobolus stapfianus*. *Australian Society of Plant Physiologists 35th Annual General Meeting Sept.* 26–29, 1995 Sydney, Abstracts, 138.

Brandle J.R., Hinckley T.M., and Brown G.N. (1977) The effect of dehydration-rehydration cycles on protein synthesis. *Physiologia Plantarum*, **40**, 1–5.

Buttrose M.S. and Swift J.G. (1975) Effects of killing by heat or desiccation on membrane structure in pea roots. *Australian Journal of Plant Physiology*, **2**, 225–233.

Colaço C., Sen S., Thangavelu M., Pinder S. and Roser B. (1992) Extraordinary stability of enzymes dried in trehalose: simplified molecular biology. Bio/Technology, **10**, 1007–1011.

Collins J.P. (1989) Salt tolerance of desiccation tolerant grasses. B.Sc. (Hons) Report, Monash University, Australia.

Crowe J.H., Hoekstra F.A. and Crowe L.M. (1992) Anhydrobiosis. *Annual Review of Physiology*, **54**, 579–599.

Daniel V. and Gaff, D.F. (1980a) Sulphydryl and disulphide levels in protein fractions from hydrated and dry leaves of resurrection plants and related desiccation-sensitive species. *Annals of Botany*, **45**, 163–171.

Daniel V. and Gaff D.F. (1980b) Desiccation-induced changes in the protein complement of soluble extracts from leaves of resurrection plants and related desiccation-sensitive species. *Annals of Botany* **45**, 173–181.

Eickmeier W.G., Casper C. and Osmond C.B. (1993) Chlorophyll fluorescence in the resurrection plant *Selaginella lepidophylla* (Hook. and Grev.) Spring during high-light and desiccation stress, and evidence for zeaxanthin-associated photoprotection. *Planta*, **189**, 30–38.

Flowers D.J and Ludlow M.M. (1986) Contributions of osmotic adjustment to the dehydration tolerance of water-stressed pigeonpea (*Cajanus cajan* (L.) Millsp.) leaves. *Plant, Cell and Environment*, **9**, 33–40.

Gaff D.F. (1966) The sulphydryl-disulphide hypothesis in relation to desiccation injury of cabbage leaves. *Australian Journal of Biological Sciences*, **19**, 291–299.

Gaff D.F. (1977) Desiccation tolerant vascular plants of southern Africa. *Oecologia*, **31**, 95–105.

Gaff D.F. (1980) Protoplasmic tolerance of extreme stress. In *Adaptation of Plants to Water and High Temperature Stress*, edited by N.C.Turner and P.J. Kramer, pp.207–230. New York: John Wiley & Sons.

Gaff D.F. (1989) Desiccation tolerant resurrection grasses for dryland areas. In *The French Grasslands Society, Proceedings of XVI International Grasslands Congress, Nice*, **2**, 1537–1538.

Gaff D.F., Bartels D., Gaff J.L. and Schneider K. (1993) Gene expression at low RWC in two hardy tropical grasses. *Transactions of the Malaysian Society of Plant Physiology*, **3**, 238–240.

Gaff D.F. and Churchill D.M. (1976) *Borya nitida* Labill. — an Australian species in the Liliaceae with desiccation-tolerant leaves. *Australian Journal of Botany*, **24**, 209–224.

Gaff, D.F. and Latz, P.K. (1978) The occurrence of resurrection plants in the Australian flora. *Australian Journal of Botany*, **26**, 485–492.

Gaff D.F. and Loveys B.R. (1984) Abscisic acid content and effects during dehydration of detached leaves of desiccation tolerant plants. *Journal of Experimental Botany*, **35**, 1350–1358.

Gaff D.F. and Loveys B.R. (1993) Abscisic acid levels in drying plants of a resurrection grass. *Transactions of the Malaysian Society of Plant Physiology*, **3**, 286–287.

Gaff D.F. and McGregor G.R. (1979) The effect of dehydration and rehydration on the nitrogen content of various fractions from resurrection plants. *Biologia Plantarum*, **21**, 92–99.

Gaff D.F., Zee S.Y. and O'Brien T.P. (1976) The fine structure of dehydrated and reviving leaves of *Borya nitida* Labill. — a desiccation-tolerant plant. *Australian Journal of Botany*, **24**, 225–236.

Gaff D.F. and Ziegler H. (1989) ATP and ADP contents in leaves of drying and rehydrating desiccation-tolerant plants. *Oecologia*, **78**, 407–410.

GHasempour-Mahidashti H.R. (1995) Studies of plant growth regulators and sugars in relation to induction of desiccation tolerance in resurrection plants, particularly the grass *Sporobolus stapfianus*. Ph.D. Thesis, Monash University, Clayton, Australia.

Grieve, B.J. (1961) Negative turgor pressures in sclerophyll plants. *Australian Journal of Science*, **23**, 375–377.

Hallam N.D. and Gaff D.F. (1978) Reorganization of fine structure during rehydration of desiccated leaves of *Xerophyta villosa*. *New Phytologist*, **81**, 349–355.

Hallam N.D. and Luff S.E. (1980) Fine structural changes in the mesophyll tissue of leaves of *Xerophyta villosa* during desiccation. *Botanical Gazette*, **141**, 173–179.

Harrison J.G. and Perry D.A. (1976) Studies on the mechanisms of barley seed deterioration. *Annals of Applied Biology*, **84**, 57–62.

Harten J.B. and Eickmeier W.G. (1986) Enzyme dynamics of the resurrection plant *Selaginella lepidophylla* (Hook. & Grev.) Spring during rehydration. *Plant Physiology*, **82**, 61–64.

Heber U. and Santarius K.A. (1964) Loss of adenosine triphosphate synthesis caused by freezing and its relationship to frost hardiness problem. *Plant Physiology*, **39**, 712–719.

Hetherington S.F., Hallam N.D. and Smillie R.M. (1982) Ultrastructural and compositional changes in chloroplast thylakoids of leaves of *Borya nitida* during humidity-sensitive degreening. *Australian Journal of Plant Physiology*, **9**, 601–609.

Hsiao R.C. (1970) Rapid changes in levels of polyribosomes in *Zea mays* in response to water stress. *Plant Physiology*, **46**, 281–285.

Iljin W.A. (1931) Austrocknungresistenz des Farnes *Notochlaena marantae* R.Br. *Protoplasma*, **13**, 322–330.

Kaiser K., Gaff D.F. and Outlaw Jr W.H. (1985) Sugar content of leaves of desiccation-sensitive and desiccation-tolerant plants. *Naturwissenschaften*, **72**, 608–609.

Koostra P.(1973) Changes in seed ultrastructure during senescence. *Seed Science and Technology*, **1**, 417–425.

Kuang J., Gaff D.F., Gianello R.D., Blomstedt C.K., Neale A.D. and Hamill J.D. (1995) Changes in *in vivo* protein complements in drying leaves of the desiccation-tolerant grass *Sporobolus stapfianus* and the desiccation sensitive grass *Sporobolus pyramidalis*. *Australian Journal of Plant Physiology*, **22**, 1027–1034.

Lebkuecher J.G. and Eickmeier W.G. (1991) Reduced photoinhibition with stem curling in the resurrection plant *Selaginella lepidophylla*. *Oecologia*,**88**, 597–604.

Lee-Stadelmann O.Y. and Stadelmann E.J. (1979) Drought tolerance and protoplasmic qualities in mesophytic higher plants. In *Proceedings of the International Arid Land Conference on Plant Resources*, edited by J.R. Gudding and D.K. Northington. Lubbock: Texas Technical University Press.

Levitt, J. (1962) A sulphydryl-disulfide hypothesis of frost injury and resistance in plants. *Journal of Theoretical Biology*, **3**, 355-391.

Levitt J. (1980) *Responses of Plants to Environmental Stresses*. Vol.2, 2nd edition. New York: Academic Press.

Levitt J., Sullivan C.Y. and Krull E. (1960) Some problems in drought resistance. *Bulletin of Research Council of Israel*, **8**D, 177–179.

Maas E.V. and Hoffman M. (1977) Crop tolerance — current assessment. American Society of Civil Engineers Proceedings. *Journal of Irrigation Drainage Division*, **103**, 115–134.

Marr K. (1990) Effects of hormones on desiccation tolerance. B.Sc. (Hons) Report, Monash University, Clayton, Australia.

Maranville J.W. and Paulsen B.W. (1972) Alterations of current protein composition of corn (*Zea mays L.*) seedlings during water stress. *Crop Science*, **12**, 660–663.

Matsumoto H., Wakiuchi N and Takahashi E.(1968) Changes of sugar levels in *Cucumber* leaves during ammonium toxicity. *Physiologia Plantarum*, **21**, 1210–1216.

Michel D., Furini A., Salamini F. and Bartels D. (1994) Structure and regulation of ABA- and desiccation-responsive gene from the resurrection plant *Craterostigma plantagineum*. *Plant Molecular Biology*, **24**, 549–560.

Nir I., Klein S. and Poljakoff-Mayber A. (1969) Effects of moisture stress on submicroscopic structure of maize roots. *Australian Journal of Biological Sciences*, **22**, 17–33.

Oliver M.J. (1991) Influence of protoplasmic water loss on the control of protein synthesis in the desiccation- tolerant moss *Tortula ruralis*. Ramifications for a repair-based mechanism of desiccation tolerance. *Plant Physiology*, **97**, 1501–1511.

Oliver M.J. and Bewley J.D. (1984) Plant desiccation and protein synthesis. VI. Changes in protein synthesis elicited by desiccation of the moss *Tortula ruralis* are effected at the translational level. *Plant Physiology*, **74**, 923–927.

Owoseye J.A. and Sanford W.W. (1972) An ecological study of *Vellozia schnitzleinia*, a drought-enduring plant of northern Nigeria. *Journal of Ecology*, **60**, 807–817.

Piatkowski D., Schneider K., Salamini F. and Bartels, D. (1990) Characterization of five abscisic acid-responsive cDNA clones isolated from the desiccation tolerant plant *Craterostigma plantagineum* and their relationship to other water-stress genes. *Plant Physiology*, **94**, 1682–1688.

Powell A.A. and Mathews S. (1977) Changes in pea seeds (*Pisum sativum L.*) stored in humid or dry conditions. *Journal of Experimental Botany*, **28**, 225–234.

Pruzsinszky S. (1960) Ueber Trocken- und Feuchtluftresistanz des Pollens. *S B Wien Akad Wiss Math Naturw Kl, Abt.*, **169**, 43–100.

Rajeswari M.S., Bhandari N.N. and Dakshini K.M.M. (1993) Reversible structural changes in the mesophyll and bundle sheath cells of the resurrection grass *Oropetium thomaeum* (Linn.f.) Trin. (Poaceae). *Phytomorphology*, **43**, 9–17.

Roberts B.E. and Osborne D.J. (1973) Protein synthesis and viability in rye grains. In *Seed Ecology*, edited by W Heydecker, p 99. London: Butterworth.

Savitskaya N.N. (1965) Free amino acids in barley plants under conditions of soil water deficiency. *Soviet Plant Physiology*, **12**, 298–300.

Schnepf E. (1961) Ueber Veraenderungen der plasmatischen Feinstrukturen wahrend des Welkens. *Planta*, **57**, 156–175.

Schwab K.B. and Gaff D.F. (1986) Sugar and ion contents of drought tolerant plants under water stress. *Journal of Plant Physiology*, **125**, 257–265.

Schwab K.B. and Gaff D.F. (1990) Influence of compatible solutes on soluble enzymes from desiccation tolerant *Sporobolus stapfianus* and desiccation sensitive *Sporobolus pyramidalis*. *Journal of Plant Physiology* **137**, 208–215.

Schwab K.B. and Heber U. (1984) Thyalkoid membranes stability in drought-tolerant and drought-sensitive plants. *Planta*, **161**, 37–45.

Sgherri C.L.M., Loggini B., Bochicchio A. and Navari-Izzo F. (1994a) Antioxidant system in *Boea hygroscopica*: changes in response to desiccation and rehydration. *Phytochemistry*, **37**, 377–381.

Sgherri C.L.M., Loggini B., Puliga S. and Navari-Izzo F. (1994b) Antioxidant system in *Sporobolus stapfianus*: changes in response to desiccation and rehydration. *Phytochemistry*, **35**, 561–565.

Simon E.W. (1974) Phospholipids and plant membrane permeability. *New Phytologist*, **73**, 377–420.

Tymms M.J. (1979) Protein synthesis in a drought tolerant flowering plant *Xerophyta villosa*. Ph.D. Thesis, Monash University, Clayton, Australia.

Tymms M.J., Gaff D.F. and Hallam N.D. (1982) Protein synthesis in the desiccation tolerant angiosperm *Xerophyta villosa*. *Journal of Experimental Botany*, **33**, 332–343.

Vieweg G.H. and Ziegler H. (1969) Zur Physiologie von *Myrothamnus flabellifolia*. *Berichte Deutscher Botanischen Gesellschaft*, **82**, 29–36.

Visser T. (1955) Germination and storage of pollen. *Meded Landbouwhogeschool Wageningen*, **55**, 1–68.

Williams R.J. and Leopold A.C. (1989) The glassy state in corn embryos. *Plant Physiology*, **89**, 977–981.

Wolfe J. and Steponkus P.L. (1983) Tension in the plasma membrane during osmotic contraction. *Cryo-Letters*, **4**, 315–322.

Wood J.N. and Gaff D.F. (1989) Salinity studies with drought resistant *Sporobolus* species. *Oecologia*, **78**, 559–564 .

Wright S.T.C. and Hiron R.W.P. (1969) Abscisic acid, the growth inhibitor induced in detached wheat leaves by a period of wilting. *Nature*, **224**, 719–720.

Zholkevich V.N. and Rogacheva A.Ya. (1967) Effect of respiratory inhibitors on the uptake of oxygen by wilting plants. *Soviet Plant Physiology*, **14**, 424–428.

3. MECHANISMS OF PLANT RESISTANCE TO WATERLOGGING

BINGRU HUANG

Department of Horticulture, Forestry and Recreational Resources, Kansas State University, Manhattan, KS66506, USA.

INTRODUCTION

A prolonged period of rainfall or over-irrigation combined with poor internal soil drainage often causes soil waterlogging. Waterlogging is a world-wide occurring event that has a major impact on both the natural vegetation and agricultural crops. In waterlogged soils, air spaces filled with water delay diffusion of gases between atmosphere and the rhizosphere and roots (Grable, 1966; Jackson, 1985); dissolved oxygen is depleted from the bulk soil solution by respiration of roots and soil microorganisms (Drew and Lynch, 1980).

The depletion of dissolved oxygen in waterlogged soils leads to hypoxia or anoxia within hours to days, depending on the temperature, respiration activity of plant and microorganisms, and the frequency and duration of soil saturation (Drew and Stolzy, 1991). Oxygen deficiency is a major limitation to plant growth and productivity. Survival of waterlogging stress therefore requires the plant to adapt primarily to hypoxic conditions. Waterlogging-resistant plants develop a variety of mechanisms, including an ability of detoxification of reduced ions by roots, an avoidance of oxygen deficiency by internal transport of oxygen from the leaves to the roots and an ability to sustain metabolism, at least at the level of maintenance processes, by anaerobic pathways (Chen *et al.*, 1980; Crawford, 1982; Barclay and Crawford, 1982; Drew and Stolzy, 1991).

Research on plant responses to waterlogging or hypoxia has been conducted in a wide range of areas, including molecular, biochemical, physiological, anatomical and morphological adaptation. A number of reviews have been written on the topic (Drew and Stolzy, 1991; Konings and Lambers, 1991; Kennedy *et al.*, 1992, Voesenek *et al.*, 1992; Perata and Alpi, 1993; Vartapetian, 1993; Drew *et al.*, 1994; Jackson *et al.*, 1994; Pezeshki, 1994; Ricards *et al.*, 1994). The fundamental question raised is the mechanism(s) of plant adaptation to the oxygen stress. Most of the reviews concern adaptation to waterlogging for the above ground component or whole plants. This chapter reviews the literature concerning the mechanisms of metabolic, morphological, anatomical, and physiological adaptations of plants to waterlogging or oxygen stress, particularly for roots.

METABOLIC ADAPTATION

Metabolic responses to hypoxia and anoxia have been studied extensively in many plant species, but a common metabolic mechanism conferring tolerance to these conditions has not been identified. Several theories have been proposed with regard to metabolic adaptation. The first hypothesis which attempts to explain waterlogging tolerance in comprehensive metabolic terms was proposed by McManmon and Crawford (1971). They proposed that the resistance of roots for flood-tolerant wetland plants is accounted for by peculiar features of their metabolism and flood-tolerance in a given plant is dependent on the ability to avoid ethanol accumulation in the glycolytic pathway, thus maintaining a low level of ethanol biosynthesis. In roots of flood-intolerant plants under hypoxia, glucose is metabolized according to the classical scheme of alcoholic fermentation and plant damage in an oxygen deficit medium is a consequence of self-poisoning by end-products of fermentation, mainly ethanol, which accumulates in plant tissues in the course of anaerobiosis; in roots of flood-tolerant plants, the accumulation of toxic ethanol is thought to be avoided by switching glucose breakdown in the direction of malate formation which is less toxic. This theory proposes that in flood-sensitive species, glycolysis will be promoted in response to oxygen stress resulting in ethanol production at faster rates than in flood-tolerant species.

Several major fermentative enzymes are involved in glycolysis, including pyruvate decarboxylase, lactate dehydrogenase, and alcohol dehydrogenase. Alcohol dehydrogenase (ADH) is the enzyme that has been studied most thoroughly in relation to anaerobiosis. In response to anaerobic stress, ADH activity increases in most plants. The extent of ADH induction varies with organs in one species. On a fresh weight basis, induction of ADH in coleoptile and leaves is much less dramatic than in the root tips (Drew *et al.*, 1994). Although ADH activity is apparently not required for growth of maize in air, ADH null mutants demonstrate that ADH activity is essential for extended survival of this species during flooding (Schwartz, 1969, Lemke-Keyes and Sachs, 1989). Maize seedlings survive only a few hours, whereas normal seedlings survive 3 d of anoxia (Kennedy *et al.*, 1992). However, research with *Echinochioa phyllopogon* indicates that no correlation is found between flooding-tolerance and the total level of activity or the number of ADH isozymes (Kennedy *et al.*, 1987). There is also no correlation between ADH activity and the rate of alcohol production under anoxia of *Urtica* roots (Smith *et al.*, 1986). Induction of ADH activity alone is not the factor that leads to improved survival of anoxia in wild type of maize root tips (Roberts *et al.*, 1989; Johnson *et al.*, 1989).

There may be a relationship between ADH activity and its location in plants (Kennedy *et al.*, 1992). Flood-tolerant species such as rice and *E. phyllopogon* exposed to anoxic conditions show a greater increase in ADH activity in leaves than in roots, whereas in flood-intolerant species such as maize and pea more increases in ADH activity occur in roots than in leaves (Cobb and Kennedy, 1987). In wild rice, ADH activity is greater in roots than in shoots after 2 d of induction, suggesting that this relationship does not hold true in flood-tolerant species (Muench *et al.*, 1993). In 5 to 21-d-old soybean seedlings, the induction of ADH is organ specific and is also related to the age of the plants (Newman and Van Toai, 1991). High ADH activity may be important in hypoxia for germinating seeds in wild rice (Muench *et al.*, 1993). The physiological significance of high ADH activities in roots is thus unclear.

Lactate dehydrogenase (LDH) is also induced anaerobically in roots and seeds of several species (Hook and Crawford, 1978; Hoffman *et al.*, 1986; Good and Crosby, 1989). Lactate fermentation often occurs in anaerobic roots and germinating seeds and is thought to have transient role in lactate production (Kenndey *et al.*, 1992). Induced LDH activity supports continued lactate fermentation, maintaining the redox balance without the carbon loss associated with ethanol fermentation. The induction of LDH activity in roots occurs in the absence of any increase in lactate concentration in hypoxic seedlings, suggesting that the physiological advantage of increasing LDH activity in root tissue during long-term hypoxia remains unclear (Muench *et al.*, 1993). Pyruvate decarboxylase (PDC)is perhaps at the most important branch point in anaerobic metabolism and catalyzes the decarboxylation of pyruvate, yielding CO_2 and acetaldehyde, precursor of ethanol. In maize, PDC activity increases 5– to 9–fold during anoxia (Kelly, 1989). However, no relationship is found between ethanol level and the induction of PDC synthesis in maize and rice (Morrell and Greenway, 1989).

To evaluate the importance of glycolytic processes in oxygen stress tolerance, at least several enzymes of the glycolytic pathways need to be determined. An experiment with wheat supports the hypothesis that high rate of alcohol fermentation is one important aspect of tolerance of plant tissue to anoxia since wheat roots are sensitive to anoxia and in these roots the rates of alcohol fermentation and, therefore, glycolysis would usually be limited by low levels of PDC and ADH (Waters *et al.*, 1991). Yet whether increases in ADH or PDC are the most important factors involved in the increased tolerance to anoxia is difficult to establish.

The hypothesis of a fermentation pathway leading to accumulation of ethanol and consequently for flooding injury, however, has been questioned (Rumpho and Kennedy, 1981, 1983; Jackson *et al.*, 1982). Anaerobic conditions result in ethanol production in both flood-tolerant species such as rice (Bertani *et al.*, 1980) and intolerant species such as pea (Smith and Rees, 1979). Flood-tolerant species can contain even more ethanol than intolerant ones (Hook *et al.*, 1971; Phung and Knipling, 1976) or exhibit higher rather than lower rates of glycolysis (Carpenter and Mitchell, 1980). Research with pea indicates that ethanol content is unlikely to have significant effects on plant growth or survival because supplying ethanol in aerobic or anaerobic conditions at similar or 100 times the concentrations to those found in the xylem sap of flooded pea plant causes no or little injury to the plants (Jackson *et al.*, 1982). Anaerobic damage of plant cells by the end-products of fermentation is also in doubt for other reasons. The maintenance and even the stimulation of fermentation by glucose feeding rather facilitates plant cell survival under hypoxia (Vartapetian, 1993). It has been shown that root ultrastructure damage under oxygen-deficit conditions could be prevented if they were fed exogenous glucose as in experiments with rice, pumpkin, and pea roots, and rice coleoptiles and leaves (Vartapetian *et al.*, 1976; Webb and Armstrong, 1983). The biochemical and enzymological relationship between ethanol biosynthesis and flooding injury therefore seems as improbable as that between flooding injury and ethanol content. Since ethanol accumulation is widespread, it would appear that other metabolic responses must be important in conferring long-term tolerance of hypoxia.

Another theory in metabolic adaptation proposed by Davies (1980) tries to explain the role of lactate as an important factor in the induction of other hypoxic proteins by reducing the pH of cells in short-term flood tolerance in some plants. According to the second theory, the cause of plant injury under hypoxia is acidosis of the cytoplasm. Under this hypothesis, the relative rate of synthesis of lactate versus ethanol depends upon the cytoplasmic pH. Under anaerobic conditions, pyruvate is initially converted to lactate, but as the cytosolic pH decreases, LDH activity is inhibited, PDC activity is stimulated, and ethanol synthesis predominates. This hypothesis has recently been demonstrated with the experiments carried out using ^{31}P and ^{13}C nuclear magnetic resonance technique (Roberts *et al.*, 1984; Menegus *et al.*, 1989; Xia and Roberts, 1994). During an extended period of anoxia, the signal for cell death of maize or pea root tips appears to be the distinct lowering of the cytoplasmic pH despite continuous fermentation, suggesting that fermentation is inadequate to maintain energy metabolism for long periods (Roberts *et al.*, 1984). Cytoplasmic acidosis is a determinant of flooding-intolerance in plants and resistance to anoxia is inversely correlated with the extent of cytoplasmic acidification.

Cytoplasmic pH in plants is determined by the combined effects of many reactions, such as lactate excretion (Xia and Roberts, 1994), fermentation to lactate, H^+ transport, carboxylation/decarboxylation (Kurkdjian and Guern, 1989), and fermentation to alanine (Roberts *et al.*, 1992). Further work is necessary to assess the importance of fermentation to lactate and lactate extrusion in regulation of cytoplasmic pH in anoxic root tips. The mechanisms of cytoplasmic injury due to low pH is not known (Roberts *et al.*, 1984; Drew, 1990), but at the electron microscope level, hypoxia causes prominent changes in the fine structure of meristematic cells, especially to mitochondria, Golgi, and chromatin (Aldrich *et al.*, 1985). Although shifts in anaerobic metabolism involving pH changes have been shown to occur in several plant species (Roberts *et al.*, 1984; Menegus *et al.*, 1989), this mechanism does not appear to have a critical role in the long-term survival under hypoxia and is inconsistent for different organs, such as roots and seeds and different flood tolerant species, such as rice and *E. phyllopogon* (Muench *et al.*, 1993).

Under natural conditions, oxygen concentrations in waterlogged soils decline over hours or days so that roots experience an appreciable period of hypoxia before anoxia. Experiments with maize and wheat indicate that the ability of root tips to tolerate anoxia is gradually improved after first subjecting the roots in hypoxic conditions for a period of time (Waters *et al.*, 1991; Hole *et al.*, 1992). The mechanisms underlying the role of hypoxic acclimation in improving oxygen stress resistance is not well-understood (Drew *et al.*, 1994). One possibility is that hypoxia allows synthesis of a transport mechanism to remove lactate from cells when they become anoxic (Xia and Saglio, 1992), thereby delaying cytoplasmic acidosis. An additional explanation is that hypoxic acclimation could improve energy supply, because experiments show that acclimated root tips produce more ATP and total adenylates than unacclimated roots during continuous anoxia (Johnson *et al.*, 1989; Drew *et al.*, 1994; Zhang and Greenway, 1994). Root tips of intact seedlings previously acclimated to hypoxia also have a higher level of ADH activity throughout a subsequent continuous anoxia (Johnson *et al.*, 1989).

From the above discussion, alternatives to the theories of ethanol self-poisoning and cytoplasmic pH lowering are therefore needed. It seems that the theories hold for only short-term waterlogging tolerance, since increases in ADH activity of roots diminishes as root aeration increases (Burdick and Mendelssohn, 1990).

ANATOMICAL AND MORPHOLOGICAL ADAPTATION

The anatomical and morphological adaptations to waterlogging may be important long-term adaptations that contribute to waterlogging tolerance. Waterlogging-tolerant plants develop specific morphological and anatomical characteristics to survive and function under waterlogged conditions, such as formation of aerenchyma (intercellular gas-filled spaces) in the root cortex and development of adventitious roots (Jackson and Drew, 1984; Justin and Armstrong, 1987; Laan *et al.*, 1989).

The presence of aerenchyma in roots is clearly conducive to the survival of the plants under waterlogged conditions and has been considered as an adaptive response (Drew *et al.*, 1979; Kawase, 1981). The functional significance of aerenchyma can be deduced from several aspects of aerenchymatous root characteristics (Konings and Lambers, 1991), including a) a greater oxygen transport. In many species, aerenchyma may provide most, if not all, of the oxygen requirements of the roots, as well as some of the requirements of the surrounding rhizosphere (Thomson *et al.*, 1990; Engelaar *et al.*, 1993). Increases in root porosity significantly enhance the internal oxygen transport in *Rumex* species (Laan *et al.*, 1990) and shoot growth (Ridge, 1987). Even though a close functional relationship between shoot growth and aerenchyma development has been noted, very little research has concentrated on the correlation between these two phenomena; b) the continued growth of the roots in anaerobic conditions. The elongation rate of roots with large root porosity is greater than for roots with less aerenchyma (Thomson *et al.*, 1992); c) the demonstration that aerenchymatous roots in maize have higher values of ATP content, adenylate energy charge and ATP/ADP ratios than non-aerenchymatous roots, when transferred to anoxic conditions (Drew *et al.*,1985); and d) the fact that in aerenchymatous *Senecio* roots, the activity of the cytochrome path is not inhibited upon transferring the roots to an anoxic solution (Lambers *et al.*, 1978).

Aerenchyma forms in roots under hypoxia in many plant species (Jackson and Drew, 1984). The extent of aerenchyma in roots varies among species and genotypes within a species. Extensive aerenchyma forms in wetland species. Aerenchyma is also found in many herbaceous non-wetland species, including wheat and maize. Huang *et al.* (1994 a,b, 1995 a,b) studied several wheat cultivars and found that waterlogging-tolerant cultivars had significantly higher root porosities under waterlogged conditions than the sensitive ones. For the 91 plant species surveyed by Justin and Armstrong (1987), the extent of aerenchyma development was related to waterlogging tolerance. Topa and McLeod (1986) also suggested that the waterlogging tolerance of pond pine (*Pinus serotina*) was positively correlated with formation of aerenchyma.

The mechanisms leading to increased porosity or other adaptive characteristics such as development of adventitious roots of waterlogging-tolerant plants are not completely understood. Roots of wetland plants such as rice form aerenchyma even when well-aerated, indicating that its induction is genetically controlled. In non-wetland plants, however, aerenchyma formation is strongly influenced by the external conditions, particularly by oxygen-deficient conditions. Some evidence exists with respect to the physiological processes leading to cell lysis in maize roots in response to hypoxia. Enhanced endogenous ethylene levels under hypoxia have been shown to induce the breakdown of cortical cell walls and thus promote formation of lysigeneous type of aerenchyma in maize (Drew *et al.*, 1979; Konings, 1982). Indeed, hypoxic roots produce a larger amount of endogenous ethylene compared to the aerated roots, to a greater extent for waterlogging-tolerant genotypes of wheat (Figure 3.1). An increase of endogenous ethylene concentration of approximately 20% above that normally found in aerated roots is sufficient to induce formation of aerenchyma in the cortex of maize roots (Konings, 1982). In addition, exogenous ethylene treatment of roots also stimulates aerenchyma formation, as found in different genotypes of wheat (Figure 3.2). Ethylene gas applied to maize roots in well-aerated solutions promotes aerenchyma formation even at concentrations as low as 0.1 $\mu l\ l^{-1}$. Similarly, the precursor of ethylene, 1-aminocyclopropane-1-carboxylic acid (ACC), promotes aerenchyma formation, when applied in aerated solution (Konings and Lambers, 1991). Inhibition of ethylene production by silver nitrate and cobalt also inhibits aerenchyma formation, further confirming that endogenous ethylene controls the induction of aerenchyma formation in maize roots (Drew *et al.*, 1981; Konings, 1982). The role of ethylene in the lysigeneous breakdown of cortical cells in rice, however, seems to be cultivar dependent. For example, Jackson *et al.* (1985) found no effect of ethylene in one rice cultivar;

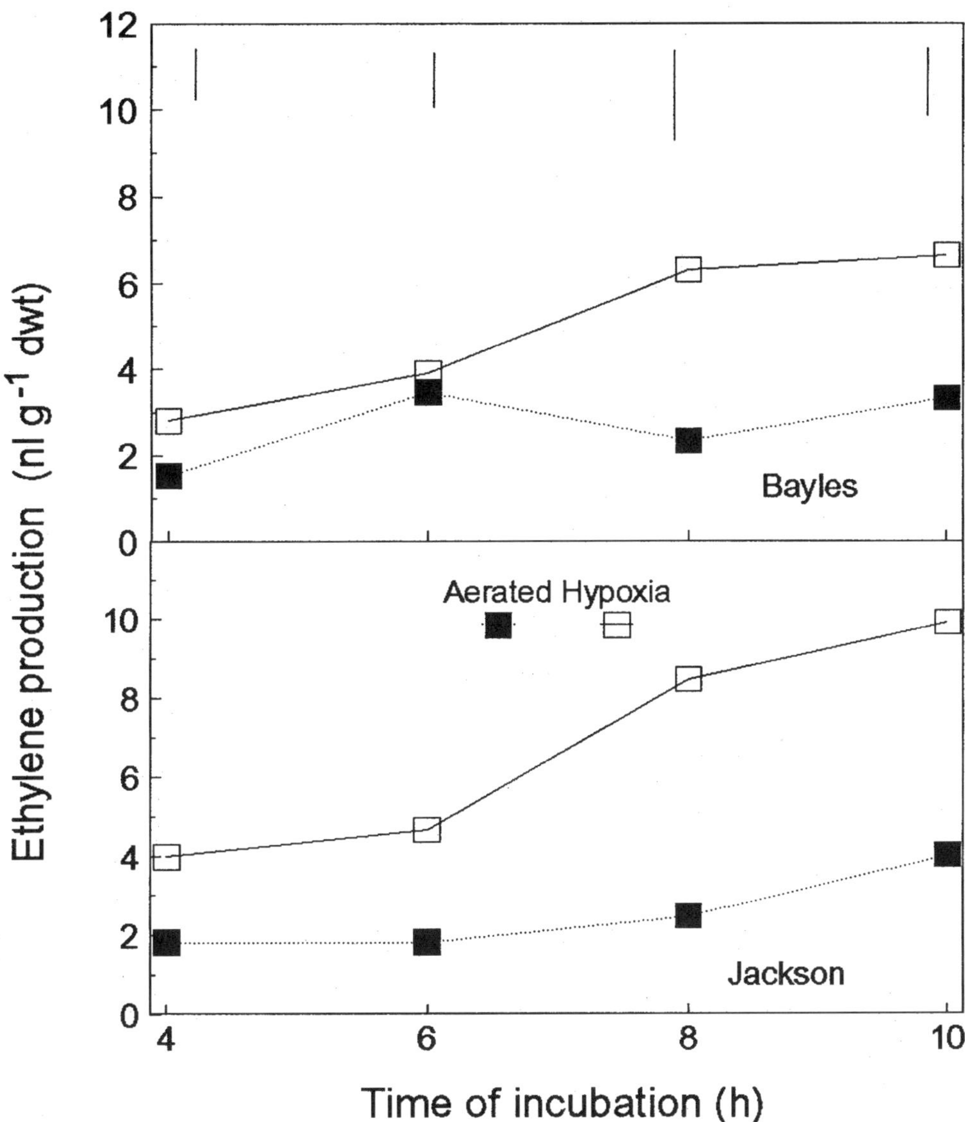

Figure. 3.1. Ethylene production of adventitious (crown) roots for waterlogging-sensitive wheat genotype, Bayles, and waterlogging-tolerant genotype, Savannah, in response to hypoxia. Plants were grown in full-strength Hoagland's solution aerated with ambient air (aerated) or bubbled with 3% O_2 balanced N_2 (hypoxia). Bars indicate LSD for treatment comparisons at P=0.05 level.

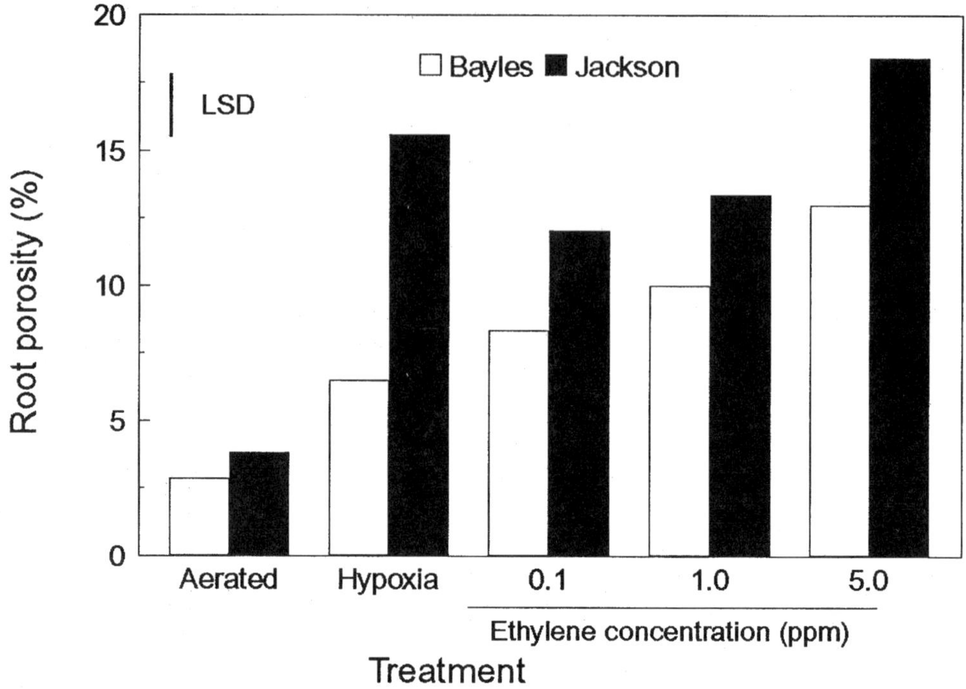

Figure. 3.2. Porosity of adventitious (crown) roots for waterlogging-sensitive wheat genotype, Bayles, and waterlogging-tolerant genotype, Savannah, as affected by hypoxia and exogeneous ethylene treatment. Plants were grown in full-strength Hoagland's solution aerated with ambient air (aerated), bubbled with 3% O_2 balanced N_2 (hypoxia) or 0.1, 1.0, 5.0 ppm ethylene. Bars indicate LSD for treatment comparisons at P=0.05 level.

whereas, a small promotion of aerenchyma formation was demonstrated by Justin and Armstrong (1991) in another cultivar.

Little is known of the biochemistry and enzymology of cell wall breakdown that takes place during lysis (Jackson *et al.*, 1994). The formation of aerenchyma is caused either by the separation of cells during development (schizogeny) or by cell death and dissolution (lysigeny) (Drew and Stolzy, 1991). Lysigenous aerenchyma can develop in already existing roots by the lysis of cortical cells (Huang, unpublished). In some species such as *Rumex*, air channels in the roots develop in a "schizonegous" pattern (Laan *et al.*, 1989). Cell walls of cortical cells separate from each other in such a way that cells remain intact and honey-comb shaped intracellular spaces are formed. The schizonegous aerenchyma formation takes place immediately behind the meristematic zone in the root apex, and appears to be constitutive (at least in adventitious roots), since neither reaeration nor inhibitors of ethylene biosynthesis are able to reduce the root porosity (Visser, 1995).

Ethylene may be involved in lysigenous aerenchyma development in such a way that high ethylene concentration in waterlogged plants triggers an increase of cellulase activity leading to aerenchyma development as demonstrated for stems of sunflower (Kawase, 1981) and roots of maize (He *et al.*, 1994). Cellulase activity in roots increases significantly under hypoxia and with exogenous ethylene treatment, especially for the waterlogging-tolerant genotype of wheat (Figure 3.3). The process of cell lysis in maize roots occurs primarily in cortical tissue

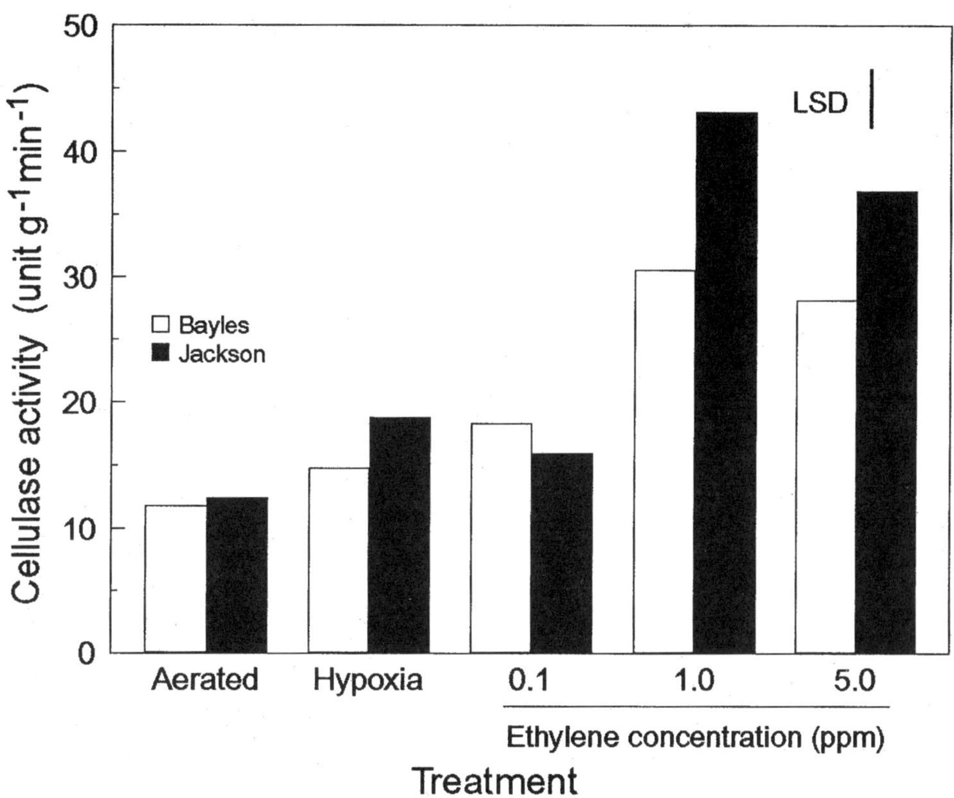

Figure. 3.3. Cellulase activity of adventitious (crown) roots for waterlogging-sensitive wheat genotype, Bayles, and waterlogging-tolerant genotype, Savannah, as affected by hypoxia and exogeneous ethylene treatment. Plants were grown in full-strength Hoagland's solution aerated with ambient air (aerated), bubbled with 3% O_2 balanced N_2 (hypoxia) or 0.1, 1.0, 5.0 ppm ethylene. Bars indicate LSD for treatment comparisons at P=0.05 level.

(Drew *et al.*, 1979; Campbell and Drew, 1983) and not in surrounding cells, suggesting site specific actions for ethylene (Drew, 1990). Whether cellulase synthesis or activity is strictly confined to the midcortical cells that are destined to lyse is worth investigation (Drew *et al.*, 1994). Another aspect with regard to the role of ethylene in aerenchyma formation deserving an examination is the nature of the signal transduction pathway between ethylene and release of a hydrolytic enzyme or aerenchyma formation. He *et al.* (1993) suggested a close involvement of Ca^{2+} in signaling cell lysis, for aerenchyma formation can be greatly modified by manipulation of endogenous Ca^{2+} using Ca^{2+} chelators or a calcium pump blocker.

In some cases with maize roots, the cells collapse without apparent breakdown of cell walls and the rupture of the tonoplast seems to be the earliest event in the process of aerenchyma formation (Konings and Lambers, 1991). Further, Suttle and Kende (1980) found that ethylene hastened the increase of membrane, possibly tonoplast, permeability in senescing flowers. The increased permeability of the tonoplast in maize roots may be preceded by the decomposition of phospholipids, as found in other tissues under the influence of ethylene (Thompson *et al.*, 1982), low oxygen (Vartapetian *et al.*, 1987), and low nutrient supply (Kuiper and Kuiper, 1979). It appears that disfunctioning of the tonoplast is probably the first step in the chain of events leading to cell collapse and aerenchyma formation (Konings, 1982).

Root nutrient status also affects aerenchyma formation. The presence of nitrate, ammonium or phosphate in aerated solution strongly reduces aerenchyma formation (Konings and Verschuren, 1980; Drew and Saker, 1983; Konings and Lambers, 1991; He *et al.*, 1994). The reduced aerenchyma formation could be because of nitrate or ammonium supporting the protein synthesis relevant to the integrity of the tonoplast membrane so that rupture of this membrane and cell collapse is prevented. Nitrate may also positively affect the phospholipid composition of membranes, thus influencing aerenchyma formation. The inhibitory effect of phosphate on aerenchyma formation is lesser than that of nitrate (Drew and Saker, 1983). Phosphate may promote membrane integrity by positively affecting phospholipid metabolism (Ratnayake *et al.*, 1978).

Aerenchyma formation also varies with age or length of roots. A positive linear relationship between root porosity and root length is found over the root length from 20 to 300 mm for rice seedlings (Justin and Armstrong, 1991). Root tips seem to play an important role in the formation of aerenchyma in the more distal sections of the root. Excision of root tip suppresses the lysis of cells proximal to the root apex (Atwell *et al.*, 1988; Konings and Lambers, 1991), suggesting that for maximum aerenchyma development a growing, actively metabolizing root apex is required under anaerobic conditions. One possible role for the apex is to supply the more proximal tissues with ACC or ethylene. The root tip is the site of highest ACC synthesis (Atwell *et al.*, 1988) and transport of ACC from the anaerobic root system to the shoot has been demonstrated (Bradford and Yang, 1980). Therefore, an examination of age or regional differences of roots in aerenchyma formation and in ethylene production or sensitivity would greatly strengthen our current understanding of mechanisms in waterlogging tolerance.

In response to waterlogging, other structural changes also occur in addition to formation of aerenchyma. Suberization and lignification of cell walls of hypodemis are enhanced under oxygen deficient conditions (Clark and Harris, 1981; Erdmann and Wiedenroth, 1988; Armstrong and Armstrong, 1988) which impede the radial, outward diffusion of O_2 in the more proximal mature root zones. Greater quantities of O_2 are conserved internally and remain available for the meristematic zones where O_2 consumption is greatest due to the suberization and lignification. However, radial oxygen loss from roots could be beneficial for plants whose roots develop aerenchyma since oxygen loss helps reducing soil toxins and affects soil microflora (Armstrong, 1971).

Formation of adventitious roots at surface soil under waterlogged conditions is an important morphological characteristic that may contribute to waterlogging tolerance. The development of adventitious surface roots could facilitate oxygen uptake from the aerobic root/soil interface and may replace the losses due to damaged functioning of original roots which occurs under flooded conditions. The development of adventitious roots in close proximity to the aerobic soil surface in flood-tolerant species or tolerant genotypes within a species has been well documented (Drew *et al.*, 1979; Wenkert *et al.*, 1981; Laan *et al.*, 1989; Naidoo and Naidoo, 1992; Huang *et al.*, 1995b). Stimulation of development of adventitious roots under waterlogged conditions has been related to accumulation of ethylene (Reid and Bradford, 1984) and auxin (Visser, 1995). With wheat plants grown under well-aerated conditions, increases in concentrations of ethylene applied to rooting medium restrict root elongation, particularly for waterlogging sensitive genotypes, but enhance production of adventitious roots for the tolerant genotype (Figure 3.4). Waterlogging has major effects on the carbohydrate status of both shoot and roots (Kozlowski, 1992; Huang and Johnson, 1995), which may be of great importance for the initiation of roots, as low concentration of endogenous sugars can negatively influence the rooting performance of cuttings.

Apart from root porosity and production of adventitious roots, the diameter and elongation rate of the adaptive adventitious roots may also be important determinants in waterlogging resistance of a plant species. The increased diameter of adventitious roots and primary roots for two wetland species, *Remux palustris* and *Remux hydrolapathum*, indicates a high resistance of these roots to soil anoxia (Visser, 1995). An increased diameter of the root decreases the relative radial oxygen loss to the rhizosphere and therefore enhances oxygen diffusion to the root tip (Armstrong, 1979). In addition, fast-growing adventitious roots could enhance waterlogging tolerance by rapidly replacing low-porosity primary tap root systems in waterlogged soil. The tap root is probably a significant barrier for the diffusion of oxygen from the shoot to the roots as the woody tap root has a rather low porosity (Visser, 1995).

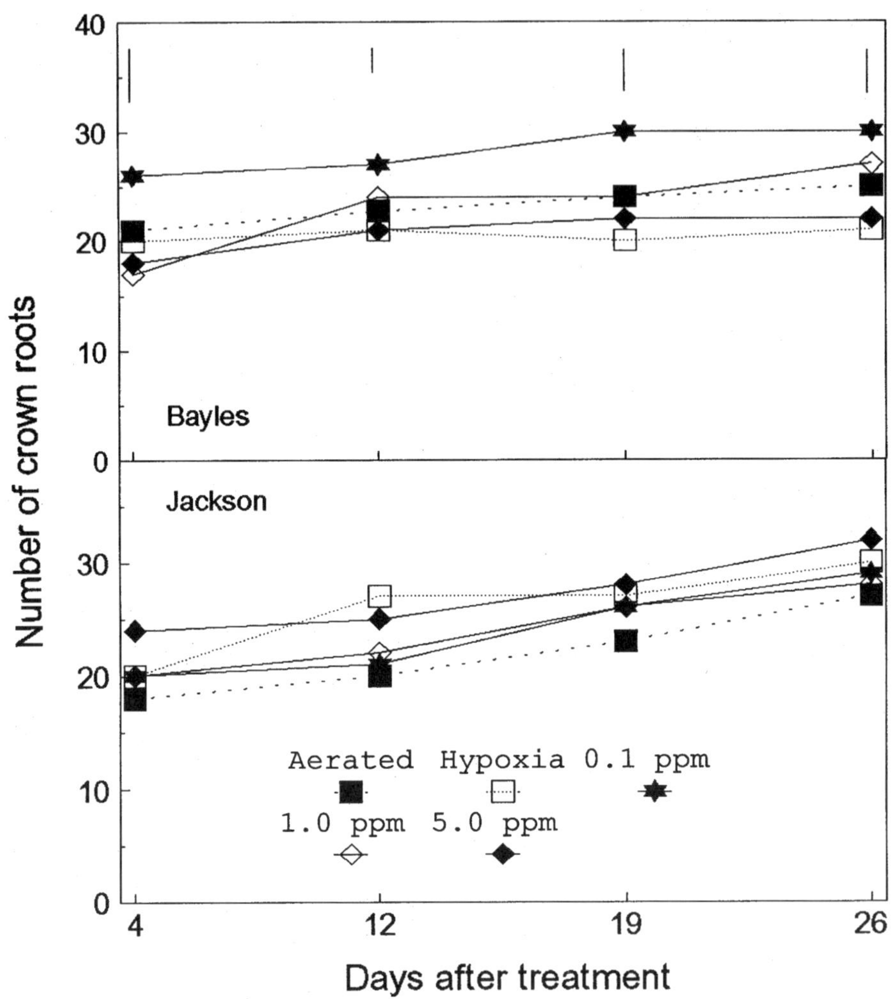

Figure. 3.4. Adventitious (crown) root production for waterlogging-sensitive wheat genotype, Bayles, and waterlogging-tolerant genotype, Savannah, as affected by hypoxia and exogeneous ethylene treatment. Plants were grown in full-strength Hoagland's solution aerated with ambient air (aerated), bubbled with 3% O_2 balanced N_2 (hypoxia) or 0.1, 1.0, 5.0 ppm ethylene. Bars indicate LSD for treatment comparisons at P=0.05 level.

PHYSIOLOGICAL ADAPTATION

The major physiological mechanisms in plant adaptation to waterlogging will be described in three aspects: detoxification of reduced ions, carbohydrate accumulation, and hormone relations.

Detoxification

In waterlogged soils, some micronutrient elements such as ferric form of Fe and manganic form of Mn are converted to the more reduced and soluble ferrous and manganous forms which can cause toxicity to plant roots (Jackson and Drew, 1984). The well-advanced development of aerenchyma improves radial oxygen leakage to the rhizosphere for waterlogging-tolerant plants, such as wetland species (Armstrong, 1971), which could help oxidize reduced forms of Fe and Mn by depositing these elements at the root surface or in cortical cell walls and thus reduce their availability (Henry and Brocklebank, 1985; McKevlin *et al.*, 1987; Fisher and Stone, 1991).

Under waterlogged conditions, concentrations of Fe and Mn are increased in shoots for the waterlogging-sensitive genotype of wheat whereas large amount of Fe and Mn are kept in roots for waterlogging-tolerant genotype (Huang *et al.*, 1995a). Hook *et al.* (1983) observed that flooded loblolly pine seedlings accumulated relatively large amounts of Fe in the shoot, whereas swamp tupelo, a hydrophyte, did not, because swamp tupelo is found to be capable of rhizosphere oxidation. Deposits containing Fe and Mn are present on the roots of streamside *Spartina alterniflora* (Mendelssohn and Postek, 1982), loblolly pine (McKevlin *et al.*, 1987) and rice (Chen *et al.*, 1980). In rice, grain yield and mineral coating is positively correlated (Chen *et al.*, 1980). However, in wheat, a negative relation between grain yield and root mineral coating is observed (Ding and Musgrave, 1995). The mechanisms by which tolerant plant species inhibit Fe translocation to the shoot and formation of mineral coating under waterlogged or anoxic conditions are not well understood.

Carbohydrate Accumulation and Allocation

Tolerance to waterlogging of roots may be related to carbohydrate or sugar accumulation in the root tips or to phloem transport of sugar from shoots or endosperm to the roots. High sugar concentration has been regarded as an adaptive change, facilitating ion uptake via a supply of energy. An adequate supply of carbohydrate can be critical to the survival of plant tissue at low oxygen concentration (Setter *et al.*, 1987; Xia and Saglio, 1992). Sugar accumulation increases in roots of waterlogging-tolerant wheat plants under hypoxia, while it is unchanged in roots of flood-sensitive plants (Huang and Johnson, 1995a). However, Limpinuntana and Greenway (1979) suggest that the sugar accumulation is a consequence of reduced root growth rather than root adaptation to hypoxia. The importance of carbohydrate status of roots in waterlogging tolerance has been further confirmed by experiments with supplying of exogenous sugar to hypoxic or anoxic roots. Glucose supply prolongs the retention of root elongation potential under anoxia (Waters *et al.*, 1991). The addition of glucose to the rooting medium improves the anoxia tolerance of 2-d-old seedlings as treatment with glucose increases the ATP content of these tips (VanToai *et al.*, 1995). Supplies of carbohydrates to root cells strongly influence the ability to survive anoxia.

Adventitious roots of rice survive only 4 h, but this is extended to 44 h with exogenous glucose (Webb and Armstrong, 1983). Root ultrastructural injury under anoxia is prevented after feeding the roots with glucose, demonstrating that there is a positive effect of both exogeneous sugar and endogenous reserves of carbohydrates in survival of plants under anaerobic conditions (Vartapetian, 1993).

Carbon reallocation from below-ground to above-ground components during waterlogging as indicated by reduced root to shoot dry matter ratio is probably also an effective physiological strategy in decreasing the amount of respiring root tissue relative to volume and therefore the O_2 demand of the root system for aerobic respiration (Naidoo and Naidoo, 1992). Resource allocation from below-ground to above-ground occurs under waterlogged conditions in both wetland species (Mendelssohn and Postek, 1982) and non-wetland plants (Huang *et al.*, 1994a, b). The ratio of soluble sugar concentration in roots to sugar concentration in shoots increases for the waterlogging-tolerant genotype of wheat under hypoxia (Huang and Johnson, 1995). The relatively large amount of soluble sugar transported to roots could facilitate energy supply for root respiration and ion uptake.

Hormone Relations

Although one is aware of the important role of hormones in adaptive responses of plants upon flooding, flooding tolerance is related more to aerenchyma development (Armstrong, 1979; Justin and Armstrong, 1987; Laan *et al.*, 1989) and to the terminal products of anaerobic respiration (McManmon and Crawford, 1971; Roberts *et al.*, 1984) than to the role of hormones, such as ethylene, abscisic acid (ABA), cytokinins or indoleacetic acid (IAA). When roots are stressed, the amounts of hormone and or precursor in shoots and roots can be altered, which may affect some physiological, morphological or anatomical processes of plants under flooded conditions.

Ethylene seems to play an important role in improving the ability of roots and thus the whole plant to survive the aerobic conditions of prolonged waterlogging (Jackson, 1985). Since the induction of adventitious roots and aerenchyma and stimulation of root extension are known to be adaptive traits under waterlogged conditions, an enhancement of ethylene production by oxygen deficiency to a certain level should be favorable to these growth responses. The biochemical basis of the stimulation of ethylene biosynthesis by a limited oxygen supply is not clear and several possibilities have been put forward (Jackson *et al.*, 1995). The effect is associated with increased concentrations of 1-aminocyclopropane-1-carboxylate (ACC) in hypoxic roots (Atwell *et al.*, 1988) and the extra ethylene is derived from this ACC (Jackson *et al.*, 1985). Low oxygen supply stimulates ACC synthase production in roots and leaves in flooded tomato plants (Wang and Arteca, 1992). The conversion of ACC to ethylene is limited by low O_2 level and ACC is apparently transported from roots to the leaves where it is converted to ethylene. Also, delivery of ACC from roots into shoot is enhanced by soil flooding (Else *et al.*, 1995). Although partial oxygen shortage stimulates ethylene production in plant tissues (Jackson *et al.*, 1985; Atwell *et al.*, 1988), little is known whether there is a differential response in root endogenous ethylene concentration or production to waterlogging for plant species or genotypes differing in waterlogging-tolerance. Research with wheat has shown that the stimulation of ethylene production for the waterlogging-tolerant genotype occurs to a greater extent than for the waterlogging-tolerant genotype (Figure 3.1).

Root ethylene concentration or production has mostly been calculated or measured for excised root tips or single roots incubated in sealed vials by gas chromatography. This is usually necessary to accumulate sufficient gas for detection by gas chromatograph equipped with flame ionization detectors (FID) or photoionization detectors (PID) which are only sensitive to 10 mm^3 m^{-3} ethylene or more. Unfortunately, this method can induce artifacts related to physical perturbation, changes in gas composition, and escape of ethylene. Additionally, real-time monitoring to follow fast changes in ethylene production is seldom included in studies on ethylene production rates in plant tissues under waterlogged conditions (Harren *et al.*, 1990). A recently developed laser-driven trace gas analyzer can be employed to measure continuous ethylene production, which has a sensitivity that exceeds by several orders of magnitude that of the conventional techniques such as a gas chromatography (Harren *et al.*, 1990; Jackson *et al.*, 1995). The higher sensitivity of laser trace gas detection of ethylene compared to gas chromatographic techniques allows us measure ethylene concentrations in a gas stream flowing slowly over roots of intact seedlings in a continuous flow system without collection traps and destructive sampling. Ethylene can be measured directly in the effluent of the flooding chamber. The principle of laser trace gas analyzer allows measurements of fast and short-lasting changes in ethylene production.

Carbon dioxide can accumulate in high concentrations up to 20% in waterlogged soils or in stagnant solutions (Grable, 1966) and in roots of flooded plants, where it may influence ethylene production, metabolism and action and thus aerenchyma formation (Pearce *et al.*, 1992), However, little is known about the influence of elevated CO_2 concentrations on root ethylene production at low oxygen supply. Brailsford *et al.* (1993) working with one cultivar of maize, found no effect of carbon dioxide on ethylene production rate under aerated or hypoxic conditions. Carbon dioxide level of 10%, however, reduces ethylene production rates in kiwi fruit (Rothan and Nicolas, 1994).

Phytohormones other than ethylene, could also be of physiological importance (Abeles, 1973; Wright, 1980), since the endogenous concentrations of IAA, cytokinin, gibberellin (GA), and ABA in roots have been reported to alter upon flooding (Phillips, 1964; Burrows and Carr, 1969; Reid and Crozier, 1971; Zhang and Davies, 1987). The physiological significance and sources of these hormones however are still unclear. Application of exogenous GA partially relieved the negative effects of waterlogging on leaf area, stomatal conductance and photosynthesis (Bishnoi and Krishnamoorthy, 1992). In some cases, exogenously applied cytokinins overcome the inhibition of shoot growth and stomatal conductance caused by perturbations of the root system (Selman and Sandanam, 1972; Blackman and Davies, 1983). However, it did not alleviate the reduction in growth and stomatal conductance caused by root hypoxia for *Phaseolus vulgaris* and *Populus trichocarpa* x *P. deltoides* (Neuman *et al.*, 1990).

Treatment with exogenous ABA before anoxic stress increases the seedling survival rate for 3-d-old maize seedlings but not for 2-d-old seedlings, because of improved ATP content in the root tips of 3-d-old plants (VanToai *et al.*, 1995). The increases in ABA concentration in either leaves or roots also explain the reason for the decline in stomatal conductance and leaf growth under flooding conditions (Bradford and Hsiao, 1982; Zhang and Davies, 1987). Zhang and Davies (1987) have suggested that root-sourced ABA is responsible for some early stomatal closure in flooded plants although its prolonged effect is doubtful since most roots collapse rapidly and die within the first few days of flooding. Recent research with stem-girdling experiments, however, has shown that ABA accumulation in leaves could not explain the changes in shoot physiology on some trees under flooding and a positive message in the xylem occurs (Smit *et al.*, 1990; Reece and Riha, 1991).

Oxygen deficiency inhibits synthesis of IAA, GA, and cytokinins by roots (Reid and Bradford, 1984) and reduces cytokinin fluxes in xylem sap of roots (Neuman *et al.*, 1990), while production of ABA increases in roots (Zhang and Davies, 1987) and leaves (Jackson and Hall, 1987) and the delivery rate of ABA from roots into shoot is inhibited (Else *et al.*, 1995). Neuman and Smit (1991) reported an increase of ABA in the xylem of *Phaseolus vulgaris* in the early hours of flooding and suggested that extra ABA is produced in flooded roots only in the early days of flooding. Most recent research has suggested that ABA in the xylem may come from roots or from old leaves (Zhang and Zhang, 1994). Jackson *et al.* (1988) have shown that roots could not be the source of ABA in flooded pea plants. Jackson (1985) and Jackson and Hall (1987) have proposed that ABA accumulated in leaves of flooded plants is the ABA which would normally be transported to roots, that flooding of the root environment disturbs this transportation and results in its accumulation in leaves. It would be interesting to examine whether roots under flooded conditions are subjected to water stress since water deficit is the usual prerequisite for ABA accumulation in leaves (Davies *et al.*, 1980; Zhang and Zhang, 1994). Aerenchymatous roots could have limited water potential because aerenchyma in the cortical cells is filled with gases which could result in low water status of the aerenchymatous cortical tissue and xylem vessels by inhibiting water movement across the cortex into the xylem vessels.

Phytohormones other than ethylene also affect formation of aerenchyma in roots. However, phytohormones (with the exception of GA) do not promote or inhibit aerenchyma formation via altered ethylene levels in the roots (Konings and de Wolf, 1984; Konings and Lambers, 1991). They may, however, affect the responsiveness of cortical cells to ethylene. IAA as well as naphthylacetic acid (NAA), prevent aerenchyma formation under the inductive conditions of either non-aeration or absence of nitrate, although they strongly promote ethylene production and thus enhance the endogenous ethylene level in the roots (Konings and Lambers, 1991). ABA also prevents aerenchyma formation, but without affecting the ethylene level, while GA promotes both aerenchyma formation and ethylene production, under aerated and non-aerated conditions. The synthetic cytokinin, kinetin, promotes aerenchyma under aerated and non-aerated conditions, but increases the ethylene level only in aerated solutions and reduces it slightly in non-aerated solutions. The influence of the phytohormones could possibly also be understood by their influence on cell membrane rather than ethylene synthesis (Konings and Lambers, 1991).

CONCLUSION

In this chapter, metabolic, anatomical/morphological, and physiological resistance mechanisms of plants, particularly of roots, to waterlogging or oxygen stress have been discussed. Depending upon plant species, survival of roots in oxygen deficient conditions appears to involve several mechanisms. There is, however, not a single unique mechanism of plant adaptation to waterlogging or oxygen stress.

The two metabolic mechanisms are important for short-term tolerance to waterlogging, hypoxic or anoxic conditions. The presence of aerenchyma in both wetland and non-wetland plants has been consistently shown to have significant long-term effect on survival of roots to anaerobic conditions by supporting aerobic root respiration and plant growth in waterlogged soils. The role of plant hormones in waterlogging tolerance is still unclear. Our understanding of waterlogging or oxygen stress effects and the mechanisms involved has been advanced in the past few years by applying new molecular techniques. However, as discussed here, some of the mechanisms for waterlogging or anaerobic tolerance are still under speculation and need further confirmation. In summary of the existing views, the ability to maintain oxygen supply and energy balance are paramount in the survival of roots and thus the whole plant to waterlogging or oxygen stress.

ACKNOWLEDGEMENT

The author thanks Drs. J.E. Box and D.S. NeSmith for valuable comments on the manuscript.

REFERENCES

Abeles F.B. (1973) *Ethylene in Plant Biology*. New York: Academic Press.

Aldrich H.C., Ferl R.J., Hils M.H and Akin D.E. (1985) Ultrastructural correlates of anaerobic stress in corn roots. *Tissue and Cell*, **17**, 341.

Armstrong J. and Armstrong W. (1988) *Phragmites australis*: A preliminary study of soil-oxidizing sites and internal gas transport pathways. *New Phytologist*, **108**, 373–382.

Armstrong W. (1971) Radial oxygen losses from intact rice roots as affected by distance from the apex, respiration and waterlogging. *Physiologia Plantarum*, **25**, 192–197.

Armsstrong W. (1978) Rootaeration in wetland condition. In *Plant Life in Anaerobic Environments*, edited by D.D. Hook and R.M.M. Crawford, pp. 269-297. Ann Arbor: Ann Arbor Science Publishers.

Armstrong W. (1979) Aeration in higher plants. In *Advances in Botanical Research. Vol. 7*, edited by H.W. Woolhouse, pp. 225–332. London: Academic Press.

Atwell B.J., Drew M.C. and Jackson M.B. (1988) The influence of oxygen deficiency on ethylene synthesis, 1-aminocyclopropane-1-carboxylic acid levels and aerenchyma formation in roots of *Zea mays*. *Physiologia Plantarum*, **72**, 15–22.

Barclay H.M. and Crawford R.M.M. (1982) Plant growth and survival under strict anaerobiosis. *Journal of Experimental Botany*, **22**, 541–549.

Bertani A., Brambilla I. and Menegus F. (1980) Effect of anaerobiosis on rice seedlings: growth, metabolic rate, and fate of fermentation ptoructs. *Journal of Experimental Botany*, **31**, 325–331.

Bishnoi N.R. and Krishnamoorthy H.N. (1992) Effect of waterlogging and gibberellic acid on leaf gas exchange in peanut (*Arachis hypogaea* L.). *Plant Physiology*, **139**, 503–505.

Blackman P.G. and Davies W.J. (1983) The effects of cytokinin and ABA on stomatal behavior of maize and *Commelina*. *Journal of Experimental Botany*, **34**, 1619–1626.

Braford K.J. and Hsiao T.C. (1982) Stomatal behavior and water relations of waterlogged tomato plants. *Plant Physiology*, **70**, 1508–1513.

Bradford K.J. and Yang S.F. (1980) Xylem transport of 1-aminocyclopropane-1-carboxylic acid, an ethylene precursor, in waterlogged tomato plants. *Plant Physiology*, **65**, 322–326.

Brailsford R.W., Voesenek L.A.C.J., Blom C.W.P.M., Smith A.R., Hall M.A. and Jackson M.B. (1993) Ethylene production by primary roots of *Zea mays* L. in response to sub-ambient partial pressures of oxygen. *Plant, Cell and Environment*, **16**, 1071–1080.

Burdick D.M. and Mendelssohn I.A. (1990) Relationship between anatomical and metabolic responses to soil waterlogging in the coastal grass *Spartina patens*. *Journal of Experimental Botany*, **41**, 223–228.

Burrows W.J. and Carr D.J. (1969) Effects of flooding the root system of sunflower plants on the cytokinin content of the xylem sap. *Physiologia Plantarum*, **22**, 1105–1112.

Campbell R. and Drew M.C. (1983) Electron microscopy of gas space (aerenchyma) formation in adventitious roots of *Zea mays* L. subjected to oxygen shortage. *Planta*, **157**, 350–357.

Carpenter J.R. and Mitchell C.A. (1980) Root respiration characteristics of flood-tolerant and intolerant tree species. *Journal of the American Society for Horticultural Science*, **105**, 684–687.

Chen C.C, Dixon J.B. and Turner F.T. (1980) Iron coatings on rice roots: mineralogy and quantity influencing factors. *Journal of the Soil Science Society of America*, **44**, 635–639.

Clark L.H. and Harris W.M. (1981) Observation on the root anatomy of rice. *American Journal of Botany*, **68**, 154–161.

Cobb B.G and Kennedy R.A. (1987) Distribution of alcohol dehydrogenase in roots and shoots of rice (*Oryza sativa*) and *Echinochloa seedlings*. *Plant, Cell and Environment*, **10**, 633–638.

Crawford R.M.M. (1982) Physiological responses to flooding. In *Encyclopedia of Plant Physiology*, edited by O.L. Lange, P.S. Nobel, C.B. Osmond and H. Ziegler, pp. 453–477. Berlin: Springer-Verlag.

Davies D.D. (1980) Anaerobic metabolism and production of organic acids. In *The Biochemistry of Plants: A Comprehensive Treatise, Vol 2*, edited by P.K. Stumpf and E.E. Conn, pp. 581–611. New York: Academic Press.

Davies W.J., Mansfield T.A. and Wellburn A.R. (1980) A role of abscisic acid in drought endurance and drought avoidance by plants. In *Proceedings of the 10th International Conference on Plant Growth Substances*, edited by F. Skoog, pp. 242–253. New York: Springer-Verlag.

Ding N. And Musgrave M.E. (1995) Relationship between mineral coating on roots and yield performance of wheat under waterlogging stress. *Journal of Experimental Botany*, **46**, 939–945.

Drew M.C. (1990) Sensing soil oxygen. *Plant, Cell and Environment*, **13**, 681–693.

Drew M.C., Cobb B.G., Johnson J.R., Andrews D., Morgan P.W., Jordan W. *et al*. (1994) Metabolic acclimation of root tips to oxygen deficiency. *Annals of Botany*, **74**, 281–286.

Drew M.C., Jackson M.B. and Giffard S. (1979) Ethylene-promoted adventitious rooting and development of cortical air spaces (aerenchyma) in roots may be adaptive responses to flooding in *Zea mays* L. *Planta*, **147**, 83–88.

Drew M.C., Jackson M.B., Giffard S.C. and Campbell R. (1981) Inhibition by silver ions of gas space (aerenchyma) formation in adventitious roots of *Zea mays* L. subjected to exogenous ethylene or to oxygen deficiency. *Planta*, **153**, 217–224.

Drew M.C. and Lynch J.M. (1980) Soil anaerobiosis, microorganisms and root function. *Annual Review of Phytopathology*, **18**, 37–66.

Drew M.C., Saglio P.H. and Pradet A. (1985) Larger adenylate energy charge and ATP/ADP ratios in aerenchymatous roots of *Zea mays* in anaerobic media as a consequence of improved internal oxygen transport. *Planta*, **165**, 51–58.

Drew M.C. and Saker L.R. (1983) Induction of aerenchyma formation by nutrient deficiency in well-aerated maize roots. *ARC Letcombe Laboratory Annual Report*, 1982, pp. 41–42.

Drew M.C. and Stolzy L.H. (1991) Growth under oxygen stress. In *Plant Roots: The Hidden Half*, edited by Y. Waisel, A. Eshel and U. Kafkafi, pp. 331–350. New York:Marcel Dekker.

Else M.A., Hall K.C., Arnold G.M., Davies W.J. and Jackson M.B. (1995) Export of abscisic acid, 1-aminocyclopropane-1-carboxylic acid, phosphate, and nitrate from roots to shoots of flooded tomato plants. *Plant Physiology*, **107**, 377–384.

Engelaar W.M.H.G., Van Bruggen, M.W., Van Den Hoek W.P.M., Huyser M.A.H. and Blom C.W.P.M. (1993) Root porosities and radial oxygen losses of *Rumex* and *Plantago* species as influenced by soil pore diameter and soil aeration. *New Phytologist*, **125**, 565–574.

Erdmann B. And Wiedenroth E.M. (1988) Changes in the root system of wheat seedlings following root anaerobiosis: 3. Oxygen concentration in the roots. *Annals of Botany*, **62**, 277–286.

Fisher H.M. and Stone E.L. (1991) Iron oxidation at the surfaces of slash pine roots from saturated soils. *Journal of the Soil Science Society of America*, **55**, 1123–1129.

Good A.G and Crosby W.L. (1989) Induction of alcohol dehydrogenase and lactate dehydrogenase in hypoxically induced barley. *Plant Physiology*, **90**, 860–866.

Grable A.R. (1966) Soil aeration and plant growth. *Advances in Agronomy*, **18**, 57–106.

Harren F.J.M., Bijnen F.G.C., Reuss J., Voesenek L.A.C.J. and Blom C.W.P.M. (1990) Sensitive intracavity photoacoustic measurements with a CO_2 waveguide laser. *Applied Physical Biology*, **50**, 137–144.

He C.J., Morgan P.W. and Drew M.C. (1994) Induction of enzymes associated with lysigenous aerenchyma formation in roots of *Zea mays* during hypoxia or nitrogen starvation. *Plant Physiology*, **105**, 861–865.

He C.J., Morgan P.W., Jordan W.R. and Drew M.C. (1993) The role of calcium in aerenchyma development in roots of maize. *Plant Physiology*, 102s, 159.

Hendry G.A.F. and Brocklebank K.J. (1985) Iron-induced oxygen radical metabolism in waterlogged plants. *New Phytologist*, **101**, 199–206.

Hoffman N.E., Bent A.F. and Hanson A.D. (1986) Induction of lactate dehydrogenase isozymes by oxygen deficit in barley root tissue. *Plant Physiology*, **82**, 658–663.

Hole D.G, Cobb B.G., Hole P.S. and Drew M.C. (1992) Enhancement of anaerobic respiration in root tips of *Zea mays* following low-oxygen (hypoxic) acclimation. *Plant Physiology*, **99**, 213–218.

Hook D.D., Brown C.L. and Kormanik P.P. (1971) Inductive flood tolerance in swamp tupelo (*Nyssa sylvatica* var. Biflora (Walt). Sarg.). *Journal of Experimental Botany*, **22**, 78–89.

Hook D.D. and Crawford R.M.M. (1978) *Plant Life in Anaerobic Environments*. Ann Arbor: Ann Arbor Science Publishers.

Hook D.D., Debell D.S., McKee W.H.Jr. and Askew J.J. (1983) Responses of loblolly pine (mesophyte) and swamp tupelo (hydrophyte) seedlings to soil flooding and phosphorus. *Plant and Soil*, **71**, 387–394.

Huang B. and Johnson J.W. (1995) Root respiration and carbohydrate status of two wheat genotypes in response to hypoxia. *Annals of Botany*, **75**, 427–432.

Huang B., Johnson J.W., NeSmith D.S. and Bridges D.C. (1994a) Growth, physiological and anatomical responses of two wheat genotypes to waterlogging and nutrient supply. *Journal of Experimental Botany*, **45**, 193–202.

Huang B., Johnson J.W., NeSmith D.S. and Bridges D.C. (1994b) Root and shoot growth of wheat genotypes in response to hypoxia and subsequent resumption of aeration. *Crop Science*, **34**, 1538–1544.

Huang B., Johnson J.W., NeSmith D.S. and Bridges D.C. (1995a) Nutrient accumulation and distribution of wheat genotypes in response to waterlogging and nutrient supply. *Plant and Soil*, **173**, 47–54.

Huang B., NeSmith D.S., Bridges D.C. and Johnson J.W. (1995b). Responses of squash to salinity, waterlogging, and subsequent drainage: II Root and shoot growth. *Journal of Plant Nutrition*, **18**, 141–152.

Jackson M.B. (1985) Ethylene and the responses of plants to excess water in their environment - A review. In *Ethylene and Plant Development*, edited by J.A. Roberts and G.A. Tucker, pp. 241–265. London: Butterworths.

Jackson M.B., Attwood P.A., Braisford R.W., Coupland D., Else M.A., English P.J. *et al.* (1994) Hormones and root-shoot relationships in flooded plants - an analysis of methods and results. *Plant and Soil*, **167**, 99–107.

Jackson M.B. and Drew M.C. (1984) Effect of flooding on the growth and metabolism of herbaceous plants. In *Flooding and Plant Growth*, edited by T.T. Kozlowski, pp. 47–128. New York: Academic Press.

Jackson M.B., Fenning T.M. and Jenkins W. (1985) Aerenchyma (gas space) formation in adventitious roots of rice (*Oryza sativa L.*) is not controlled by ethylene or small partial pressures of oxygen. *Journal of Experimental Botany*, **36**, 1566–1572.

Jackson M.B. and Hall K.C. (1987) Early stomatal closure in waterlogged pea plants is mediated by abscisic acid in the absence of foliar water deficits. *Plant, Cell and Environment*, **10**, 121–130.

Jackson M.B., Herman B. and Goodenough A. (1982) An examination of the importance of ethanol in causing injury to flooded plants. *Plant, Cell and Environment*, **8**, 163–172.

Jackson M.B., Young S. and Hall K.C. (1988) Are roots a source of abscisic acid for the shoots of flooded pea plants? *Journal of Experimental Botany*, **42**, 1499–1506.

Johnson J., Cobb B.G. and Drew M.C. (1989) Hypoxic induction of anoxia tolerance in roots of *Zea mays*. *Plant Physiology*, **91**, 837–841.

Justin S.H.F. and Armstrong W. (1987) The anatomical characteristics of roots and plant response to soil flooding. *New Phytologist*, **106**, 465–495.

Justin S.H.F. and Armstrong W. (1991) Evidence for the involvement of ethylene in aerenchyma formation in adventitious roots of rice (*Oryza sativa* L.) *New Phytologist*, **118**, 49–62.

Kawase M. (1981) Effect of ethylene on aerenchyma development. *American Journal of Botany*, **68**, 651–658.

Kelly P.M. (1989) Maize pyruvate decarboxylase mRNA is induced anaerobically. *Plant Molecular Biology*, **13**, 213–222.

Kennedy R.A., Rumpho M.E. and Fox T.C. (1987) Germination physiology of rice and rice weeds:metabolic adaptation to anoxia. In *Plant Life in Aquatic and Amphibious Habitats*, edited by R.M.M. Crawford, pp. 193–203. Oxford: Blackwell Press.

Kennedy R.A., Rumpho M.E. and Fox T.C. (1992) Anaerobic metabolism in plants. *Plant Physiology*, **100**, 1–6.

Konings H. (1982) Ethylene-promoted formation of aerenchyma in seedlings roots of *Zea mays* L. under aerated and non-aerated conditions. *Physiologia Plantarum*, **54**, 119–124.

Konings H. and de Wolf A. (1984) Promotion and inhibition by plant growth regulators of aerenchyma formation in seedling roots of *Zea mays*. *Physiologia Plantarum*, **60**, 309–314.

Konings H. and Lambers H. 1991. Respiratory metabolism, oxygen transport and the induction of aerenchyma in roots. In *Plant Life Under Oxygen Deprivation*, edited by M.B. Jackson, D.D. Davies and H. Lambers, pp. 247–265. The Hague: Academic Publishing bv.

Konings H. and Verschuren G. (1980) Formation of aerenchyma in roots of *Zea mays* in aerated solutions and its relation to nutrient supply. *Physiologia Plantarum*, **49**, 265–270.

Kozlowski T.T (1992) Carbohydrate sources and sinks in woody plants. *Botanica Helvetica*, **58**, 107–223.

Kuiper D. and Kuiper P.J.C. (1979) Comparison of *Plantago* species from nutrient-rich and nutrient-poor conditions: Growth response, ATPase and lipids of the roots, as affected by the level of mineral nutrition. *Physiologia Plantarum*, **45**, 489–491.

Kurkdjian A. and Guern J. (1989) Intercellular pH: measurement and importance in cell activity. *Annual Review of Plant Physiology and Plant Molecular Biology*, **40**, 271–303.

Laan P., Smolders A., Blom C.W.P.M. and Armstrong A. (1989) The relative roles of internal aeration, radial oxygen losses, iron exclusion and nutrient balances in flood-tolerance of *Rumex* species. *Acta Botanica Neerlandica*, **38**, 131–145.

Laan P., Tosserams M., Blom C.W.P.M. and Veen B.W. (1990) Internal oxygen transport in *Rumex* species and its significance for respiration under hypoxic conditions. *Plant and Soil*, **122**, 39–46.

Lambers H., Steingrover E. and Amakman G. (1978) The significance of oxygen transport and metabolic adaptations in flood-tolerance in *Senecio* species. *Physiologia Plantarum*, **58**, 148–154.

Lemke-Keyes C.A. and Sachs M.M. (1989) Anaerobic tolerant null: a mutant that allows adhl nulls to survive anaerobic treatment. *Journal of Heredity*, **80**, 316–319.

Limpinuntana V. and Greenway H. (1979) Sugar accumulation in barley and rice grown in solutions with low concentrations of oxygen. *Annals of Botany*, **43**, 373–381.

McKevlin M.R., Hook D.D., McKee W.H., Wallace S.U. and Woodruff J.R. (1987) Loblolly pine seedling root anatomy and iron accumulation as affected by soil waterlogging. *Canadian Journal of Forestry Research*, **17**, 1257–1264.

McManmon M. and Crawford R.M.M. (1971) A metabolic theory of flooding tolerance: the significance of enzyme distribution and behavior. *New Phytologist*, **70**, 299–306.

Mendelssohn I.A. and Postek M.T. (1982) Elemental analysis of deposits on the roots of *Spartina alterniflora* Loisel. *American Journal of Botany*, **69**, 904–912.

Menegus F., Cattaruzza L., Chersi A. and Fronza G. (1989) Differences in the anaerobic lactate-succinate production and in the changes of cell sap pH for plants with high and low resistance to anoxia. *Plant Physiology*, **90**, 29–32.

Morrell S. And Greenway H. (1989) Evidence does not support ethylene as a cue for synthesis of alcohol dehydrogenase and pyruvate decarboxylase during exposure to hypoxia. *Australian Journal of Plant Physiology*, **18**, 469–475.

Muench D.G., Archibold O.W. and Good A.G. (1993) Hypoxic metabolism in wild rice (*Zizania palustris*): enzyme induction and metabolite production. *Physiologia Plantarum*, **89**, 165–171.

Naidoo L. and Naidoo S. (1992) Waterlogging responses of *Sporobolus virginicus* (L.) Kunth. *Oecologia*, **90**, 445–450.

Neuman D.B., Rood S.B. and Smit B.A. (1990) Does cytokinin transport from root–to-shoot in the xylem sap regulate leaf responses to root hypoxia. *Journal of Experimental Botany*, **41**, 1325–1333.

Neuman D.B. and Smit B.A. (1991) The influence of leaf water status and ABA on leaf growth and stomata of *Phaseolus* seedlings with hypoxic roots. *Journal of Experimental Botany*, **42**, 1499–1506.

Newman K.D. and VanToai T.T. (1991) Developmental regulation and organ-specific expression of soybean alcohol dehydrogenase. *Crop Science*, **31**, 1253–1257.

Pearce D.M.E., Hall K.C. and Jackson M.B. (1992) The effects of oxygen, carbon dioxide and ethylene on ethylene biosynthesis in relation to shoot extension in seedlings of rice (*Oryza sativa*) and barnyard grass (*Echinochloa oryzoides*). *Annals of Botany*, **69**, 441–447.

Perata P. and Alpi A. (1993) Plant responses to anaerobiosis. *Plant Science*, **93**, 1–17.

Pezeshki S.R. (1994) Plant response to flooding. In *Plant-Environment Interactions*, edited by R.E. Wilkinson, pp.289–321. New York: Marcel Dekker.

Phillips I.D.J. (1964) Root-shoot hormone relations II. Changes in endogenous auxin concentration produced by flooding of the root system in *Helianthus annus*. *Annals of Botany*, **28**, 37–45.

Phung H.T. and Knipling E.B. (1976) Photosynthesis and transpiration of citrus seedlings under flooded conditions. *HortScience*, **11**, 131–133.

Ratnayake M., Leonard R.T. and Menge J.A. (1978) Root exudation in relation to supply of phosphorus and its possible relevance to mycorrhizal formation. *New Phytologist*, **81**, 543–552.

Reece C.F. and Riha S.J. (1991) Role of root systems of eastern larch and white spruce in response to flooding. *Plant, Cell and Environment*, **14**, 29–34.

Reid D.M. and Bradford K.J. (1984) Effects of flooding on hormone relations. In *Flooding and Plant Growth*, edited by T.T. Kozlowski, pp. 195–219. Orlando: Academic Press.

Reid D.M. and Crozier A. (1971) Effects of waterlogging on the gibberellin content and growth of tomato plants. *Journal of Experimental Botany*, **22**, 39–48.

Ricards B., Couee I., Raymond P., Saglio P., Saint-Ges V. and Pradet A. (1994) Plant metabolism under hypoxia and anoxia. *Plant Physiology and Biochemistry*, **32**, 1–10.

Ridge I. (1987) Ethylene and growth control in amphibious plants. In *Plant Life in Aquatic and Amphibious Habitats*, edited by R.M.M. Crawford, pp. 53–77. Oxford: Blackwell Scientific Publications.

Roberts J.K.M., Callis J., Jardetsky D., Walbot V. and Freeling M. (1984) Cytoplasmic acidosis as a determinant of flooding intolerance in plants. *Proceedings of the National Academy of Sciences*, USA, **81**, 6029–6033.

Roberts J.K.M., Chang K., Webster C., Callis J. and Walbot V. (1989) Dependence of ethanolic fermentation, cytoplasmic pH regulation, and viability on the activity of alcohol dehydrogenase in hypoxic maize root tips. *Plant Physiology*, **89**, 1275–1278.

Roberts J.K.M., Hooks M.A., Miaullis A.P., Edwards S. and Webster C. (1992) Contribution of malate and amino acid metabolism to cytoplasmic pH regulation in hypoxic maize root tips studied using nuclear magnetic resonance spectroscopy. *Plant Physiology*, **98**, 480–487.

Rothan C. and Nicolas J. (1994) High CO_2 levels reduce ethylene production in kiwifruit. *Physiologia Plantarum*, **92**, 1–8.

Rumpho M.E. and Kennedy R.A. (1981) Anaerobic metabolism in germinating seeds of *Echinochloa crus-gali* (barnyard grass). Metabolite and enzyme studies. *Plant Physiology*, **68**, 165–168.

Rumpho M.E. and Kennedy R.A. (1983) Activity of the pentose phosphate and glycolytic pathways during anaerobic germination of *Echinochloa crus-gali* (barnyard grass) seeds. *Journal of Experimental Botany*, **34**, 893–902.

Schwartz D. (1969) An example of gene fixation resulting from selective advantage in suboptimal conditions. *American Naturalist*, **103**, 479–481.

Selman I.W. and Sandanam S. (1972) Growth responses of tomato plants in non-aerated water culture to foliar sprays of gibberellic acid and benzyladenine. *Annals of Botany*, **36**, 837–848.

Setter T.L., Waters I., Atwell B.J., Kupkanchanakul T. and Greenway H. (1987) Carbohydrate status of terrestrial plants during flooding. In *Plant Life in Aquatic and Amphibious Habitats*, edited by R.M.M. Crawford, pp. 411–443. Oxford: Blackwell Scientific.

Smit B.A., Neuman D.S. and Stachowiak M.L. (1990) Root hypoxia reduces leaf growth. Role of factors in the transpiration stream. *Plant Physiology*, **92**, 1021–1028.

Smith A.M. and Rees T. (1979) Effects of anaerobiosis on carbohydrate oxidation by roots of *Pisum sativum*. *Phytochemistry*, **18**, 1453–1458.

Smith A.M., Hylton C.M. and Woolhouse H.W. (1986) Alcohol dehydrogenase activity in the roots of marsh plants in naturally waterlogged soils. *Planta*, **168**, 130–138.

Suttle J. and Kende H. (1980) Ethylene action and loss of membrane integrity during petal senescence in *Tradescantia*. *Plant Physiology*, **65**, 1067–1072.

Thompson J.E., Mayak S., Shinitsky M. and Halevy A.H. (1982) Acceleration of membrane senescence in cut carnation flowers by treatment with ethylene. *Plant Physiology*, **69**, 859–863.

Thomson C.J, Armstrong W., Waters I. and Greeway H. (1990) Aerenchyma formation and associated oxygen movement in seminal and nodal roots of wheat. *Plant, Cell and Environment*, **13**, 395–405.

Thomson C.J., Colmer T.D., Watkin E.L.J. and Greenway H. (1992) Tolerance of wheat (*Triticum aestivum* cvs. Gamenya and Kite) and triticale (*Triticosecale* cv. Muir) to waterlogging. *New Phytologist*, **120**, 335–344.

Topa M.A. and McLeod K.W. (1986) Aerenchyma and lenticle formation in pine seedlings: a possible avoidance mechanism to anaerobic growth conditions. *Physiologia Plantarum*, **68**, 540–550.

VanToai T.T., Saglio P., Ricards B. and Pradet A. (1995) Developmental regulation of anoxic stress tolerance in maize. *Plant Cell and Environment*, **18**, 937–942.

Vartapetian B.B. 1993. Plant physiological responses to anoxia. In *International Crop Science Proceedings*, pp. 721–726. Madison: Crop Science Society of America.
Vartapetian B.B., Andreeva I.N. and Kozlova G.I. (1976) The resistance to anoxia and the mitochondrial fine structure of rice seedlings. *Protoplasma*, **88**, 215–224.
Vartapetian B.B., Mazliak P. and Lance C. (1987) Lipid biosynthesis in rice coleoptiles grown in the presence or in the absence of oxygen. *Plant Science Letters*, **13**, 321–328.
Visser E.J.W. (1995) *Adventitious root formation in flooded plants.* Dissertation, University of Nijmegen, The Netherlands.
Voesenek L.A.C.J., van der Sman A.J.M., Harren F.J.M. and Blom C.W.P.M. (1992) An amalgamation between hormone physiology and plant ecology: A review on flooding resistance and ethylene. *Journal of Plant Growth Regulation*, **11**, 171–188.
Wang T.W. and Arteca J.M. (1992) Effects of low O_2 root stress on ethylene biosynthesis in tomato plants (*Lycopersicon esculentum* Mill. Cv Heinz 1350). *Plant Physiology*, **135**, 631–634.
Waters I., Kuiper P.J.C., Watkin E. and Greenway H. (1991) Effects of anoxia on wheat seedlings. I. Interactions between anoxia and other environmental factors. *Journal of Experimental Botany*, **42**, 1427–1435.
Waters I, Morrell S., Greenway H. and Colmer T.D. (1991) Effects of anoxia on wheat seedlings. 2. Influence of O_2 supply prior to anoxia on tolerance to anoxia, alcohol fermentation, and sugar levels. *Journal of Experimental Botany*, **42**, 1437–1447.
Webb T. and Armstrong W. (1983) The effects of anoxia and carbohydrates on the growth viability of rice, pea, and pumpkin roots. *Journal of Experimental Botany*, **34**, 579–603.
Wenkert W., Fausey N.R. and Waters H.D. (1981) Flooding responses in *Zea mays* L. *Plant and Soil*, **62**, 351–366.
Wright S.T.C. (1980) The effect of plant growth regulator treatments on the levels of ethylene emanating from excised turgid and wilted wheat leaves. *Planta*, **148**, 381–388.
Xia J. and Roberts J.K.M. (1994) Improved cytoplasmic pH regulation, increased lactate efflux, and reduced cytoplasmic lactate levels are biochemical traits expressed in root tips of whole maize seedlings acclimated to a low-oxygen environment. *Plant Physiology*, **105**, 651–657.
Xia J.H. and Saglio P.H. (1992) Lactic acid efflux as a mechanism of hypoxic acclimation of maize root tips to anoxia. *Plant Physiology*, **100**, 40–46.
Zhang J. and Davies W.J. (1987) ABA in roots and leaves of flooded pea plants. *Journal of Experimental Botany*, **189**, 649–659.
Zhang J. and Zhang X. (1994) Can early wilting of old leaves account for much of the ABA accumulation in flooded pea plants? *Journal of Experimental Botany*, **45**, 1335–1342.
Zhang Q. and Greenway H. (1994) Anoxia tolerance and anaerobic catabolism of aged beetroot storage tissues. *Journal of Experimental Botany*, **45**, 567–575.

4. MECHANISMS OF SALT TOLERANCE IN HIGHER PLANTS

EDWARD P. GLENN, J. JED BROWN and MOHAMMAD JAMAL KHAN

The University of Arizona
Environmental Research Laboratory,
2601 E. Airport Drive, Tucson, AZ 85706 USA.

INTRODUCTION: NATURE OF THE SALINITY PROBLEM FOR HIGHER PLANTS

Components of Salt Stress

Most salinity problems for higher plants are caused by excess NaCl which is widely distributed in coastal and arid zone soils and water supplies. The separate roles of Na and Cl in causing salt stress in plants have not often been explored; most research has concentrated on the effects of Na (Cheeseman, 1988). In this review, we sometimes refer to Na and Cl effects as if they were equivalent, but this is merely a convention not well supported by experimental evidence. High NaCl salinity imposes at least three types of problems for higher plants: 1) the osmotic pressure in the external solution can exceed the osmotic pressure in the plant cells, requiring an osmotic adjustment by the cells to avoid desiccation; 2) the uptake and transport of nutritional ions such as K and Ca is disrupted by excess Na; 3) Na and Cl can have direct toxic effects on membranes and enzyme systems at high levels (see reviews on various aspects of salt tolerance by Greenway and Munns, 1980; Munns and Termaat, 1986; Maas, 1987; Rengel, 1992; Munns, 1993; Ashraf, 1994). The osmotic problem is nearly identical to the problem faced by plants under water stress, leading nearly 100 years ago to the still useful observation that salt stress is a form of physiological drought (Schimper, 1898; further elaborated in Strogonov, 1964).

Higher plant cells normally come in direct contact with the external soil solution only at the point of contact between plant and soil: the epidermal and cortical root cell plasma membranes (Intertidal halophytes can be covered with seawater at each high tide but their cuticles prevent direct contact of leaf and stem cells with seawater). Plant cell membranes can be directly damaged by high Na levels (Leopold and Willing, 1984), and salt tolerant plants may have special adaptations in terms of fatty acid composition of the lipid bilayer to withstand salt damage to the plasma membrane (Kuiper, 1984). However, internal cells and tissues of plants are in general exposed to much lower NaCl levels than occurs in the external solution, and nutritional and osmotic effects are more common than direct salt damage for a plant growing within its salt tolerance range. When a plant's salt tolerance limit is exceeded, NaCl may flood into tissues in the transpiration stream, damaging roots, shoots and leaves by direct toxic effects.

Osmotic Adjustment of Higher Plants Vs. Lower Organisms

Halophilic bacteria accumulate higher levels of salts in their protoplasm than occurs in the external solution, through active uptake of ions, so that the water potential inside the cell remains negative relative to the outside solution; their enzyme systems have evolved to operate under the resulting high protoplasmic salt levels — in some species in excess of 2 M NaCl (Brown, 1983). Less tolerant bacteria do not have adapted enzyme systems and use organic solutes in the protoplasm to regulate osmotic pressure (Le Rudulier, 1993). Eukaryotic cells, as well, do not have adapted enzyme systems. Marine algae accumulate taxon-specific organic solutes for osmotic adjustment and maintain low cytoplasmic levels of Na (e.g. Balnokin *et al.*, 1993, and references cited therein).

Similar to other eukaryotes, cytoplasmic levels of NaCl in higher plants do not generally exceed 0.05–0.15 M, even in halophytes, regardless of the external salinity (Jeschke, 1984; Cheeseman, 1988). As in algae, cytoplasmic osmotic adjustment is achieved through the accumulation of various organic solutes such as glycine-betaine, proline, sugars and other "compatible osmotica" which raise the cytoplasmic osmotic pressure (Albert and Popp, 1977; Storey *et al.*, 1977; Wyn-Jones *et al.*, 1977, 1984). However, higher plants face a special challenge regarding osmotic adjustment due to the large central vacuole of the plant cell, which regulates turgor and cell expansion and has at least 20 times greater fluid volume than the cytoplasm (Jeschke, 1984).

Synthesizing organic solutes is an energy-intensive process for the cell, and if accumulation of organic solutes were the only means of osmotic adjustment of the vacuole as well as the cytoplasm, much of the plant's production of photosynthate would necessarily be involved in osmotic adjustment (see calculations in O'Leary, 1979; Wyn-Jones, 1981; and further discussion in Wyn-Jones *et al.*, 1984). Instead, it is generally agreed that most higher plants accumulate Na and K salts in the vacuole for osmotic adjustment (Jeschke, 1984; Pitman, 1984), although some grasses may utilize organic solutes in the vacuoles as well (Albert and Popp, 1977).

Even though ATPase's are involved in the uptake and sequestration of salts (e.g. Braun *et al.*, 1986), far less energy per mole of solute is theoretically needed to accumulate salts by active uptake than to elaborate organic osmotica.

Potassium is an essential plant nutrient, required in the cytoplasm at approximately 130 mM to maintain membrane integrity and enzyme activity (Jeschke, 1984; Pitman, 1984). It can be in short supply in the external medium and its uptake by root cells can be reduced by competitive effects of Na under saline conditions. Nevertheless, many plants, especially the less salt tolerant species, maintain high K selectivity even at elevated salinities and preferentially accumulate K over Na in vacuoles at low to moderate salinities (up to 200 mM NaCl)

(Jeschke and Wolf, 1993). Halophytes utilize mainly Na (balanced by Cl) for osmotic adjustment of the vacuole, especially at salinities above 200 mM. Since the vacuole is surrounded by cytoplasm, these plants have the ability to transport NaCl across the cytoplasm and into the vacuole at a sufficient rate to support

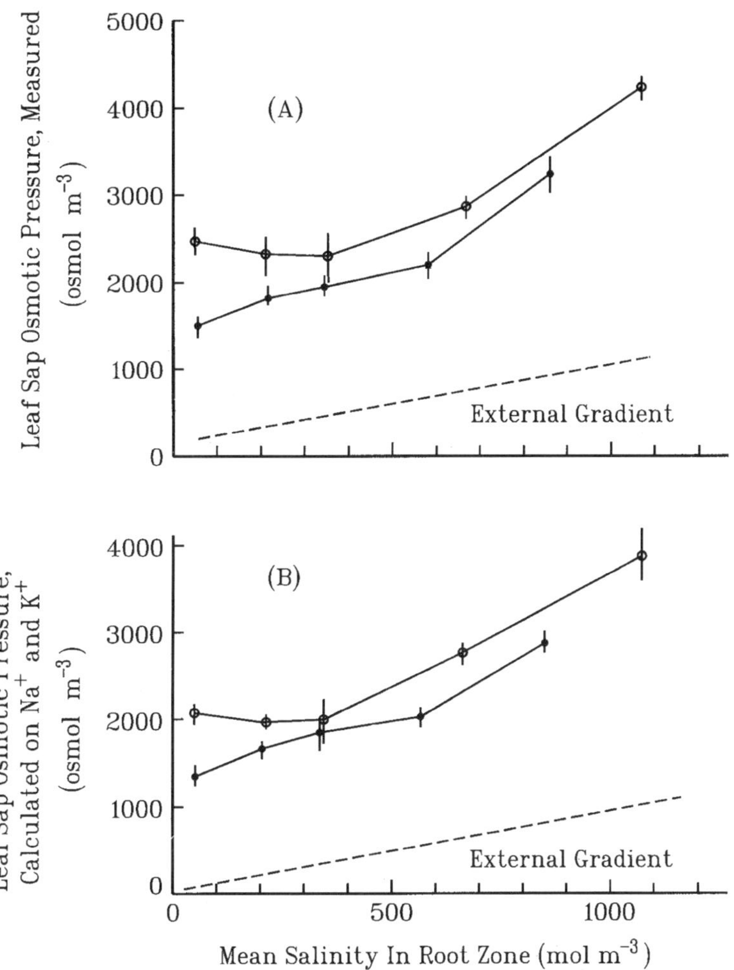

Figure 4.1. Relationship between leaf sap osmotic pressure and salinity of the external solution for two subspecies of the xerohalophyte, *Atriplex canescens*, differing in salt tolerance. Solid line = ssp. *linearis* (more salt tolerant), dashed line = ssp. *canescens* (less salt tolerant). (a) shows measured leaf sap osmotic pressure and (b) shows values calculated based on Na and K in leaves. Both ssp. had similar osmotic pressures at a given salinity and Na + K accounted for greater than 95% of osmotic adjustment of each to the external gradient (+Glenn *et al.*, 1994).

osmotic adjustment of expanding and dividing cells, but without causing a build up of salts in the cytoplasm to toxic levels. Vacuole NaCl concentrations in higher plants can range from 4–600 mM depending upon species and external salinity (Jeschke, 1984); the most tolerant higher plants are osmoconformers, maintaining osmotic pressure in the leaf sap 2–3 times higher than the external salinity through accumulation of Na and to a lesser extent K salts (Figure 4.1) (Glenn and O'Leary, 1984; Glenn, 1987; Glenn *et al.*, 1992b, 1994).

Contrast Between Emergent and Submerged Organisms

A problem faced by higher plants but not by submerged organisms is the need of the former to maintain high transpiration rates for carbon acquisition. Higher plants typically transpire 200–500 g H_2O per g dry matter production, depending upon whether they are C_3, C_4 or CAM plants. This high water flux requires a higher plant to have tight control over the influx of ions in the water stream. For example, consider a terrestrial salt marsh plant such as the C_4 grass *Spartina* growing on seawater (3.2% NaCl). If NaCl entered freely in the transpiration stream and if 200 g H_2O entered the plant for each g dry matter production, the resulting new growth would contain (.032 × 200) 6.4 g NaCl for each g of organic matter. This level of salt entry would certainly exceed the plant's ability to sequester it in vacuoles or excrete it through the leaves.

In actuality, *Spartin*a plants are only 1–2% NaCl by dry weight (Glenn, 1987). Obviously, *Spartina* and all higher plants must have the ability to limit salt influx into the plant tissues while admitting water freely. Higher plants, therefore, face two seemingly opposing challenges in regard to salinity: 1) they must acquire sufficient salts including NaCl in the vacuoles for osmotic adjustment of new and expanding cells, while maintaining low NaCl levels in the cytoplasm, which is along the pathway to the vacuole; and, 2) they must rigorously exclude excess NaCl from entering in the transpiration stream, to prevent it accumulating to damaging levels in the plant's cells, tissues and organs.

Salt entry must particularly be controlled in the terminal organs, the leaves, where photosynthate is produced and where water exits the plant by evaporation. Any salt in the water entering the leaf in the xylem must remain in or on the leaf or be reexported to the roots in the phloem, which as we will see is not a very effective method of controlling NaCl levels in plants. Furthermore, the volume of water in the leaf apoplast is very small relative to the volume in the cells, so that any imbalance in the rate of salt entry vs. salt uptake into cells or excretion onto the leaf surface can result in a rapid increase in apoplast salinity (Flowers *et al.*, 1991). When the apoplast osmotic pressure exceeds the osmotic pressure of the cell, water flows out of the cell into the apoplast, desiccating the cell. In contrast to higher plants, most submerged organisms are bathed in a more or less constant solution and they do not face the same regulatory challenges as higher plants with regard to salt uptake and compartmentation.

Variability in Salt Tolerance Among Higher Plants

Salt tolerance is considered to be a secondary adaptation of angiosperms, since salt tolerance is rare or absent among the lower orders of land plants from which higher plants are thought to have evolved. Higher plants exhibit a wide variety of adaptations to salinity and they vary in salt tolerance from sensitive plants such as rice which is harmed by 0.05 M NaCl in the irrigation supply to halophytes from the euryhaline zone of desert estuaries which grow and complete their life cycles in root zone salinities of 1 M NaCl. Extreme salt tolerance, or halophytism, probably evolved separately in several higher plant families growing in coastal or estuarine locations, then spread to inland habitats (O'Leary and Glenn, 1994). Approximately 2–3,000 species of halophytes, displaying variability in degree of salt tolerance, dominate the world's salt marsh and salt desert floras. Approximately one-third of angiosperm plant families contain halophytic genera.

Table 4.1. Varability of salt tolerance and diversity of tolerance mechanisms among higher plants. Tolerance level is the NaCl salinity that causes 50% reduction in dry matter production. Mechanisms of tolerance are based on literature reports.

Plant	*Tolerance Level (mM)*	*Tolerance Mechansims*	*References*
Bean	40	Na exclusion Na retransport leaf to root	Mass, 1987; Läuchli, 1984
Rice	40	Na exclusion Na storage in older leaves	Mass, 1987 Yeo & Flowers, 1986b
Maize	60	Na exclusion?	Maas, 1987; Cramer *et al.*, 1994
Wheat	140	Partial Na exclusion, storage in leaf vacuoles	Mass, 1987; Schachtman & Munns, 1992
Barley	200	Controlled Na uptake, storage in leaf vacuoles	Mass, 1987; Jeschke & Wolfe, 1993
Diplachne fusca	313	" " + salt glands	Warwick & Halloran, 1994
Atriplex canescens ssp. canescens	350	" " + salt bladders	Glenn *et al.*, 1994
Distichlis palmeri	600	" " + salt glands	Miyamoto *et al.*, 1995 Glenn, 1987
A. canescens ssp. *linearis*	700	Higher Na uptake than ssp. *canescens*	Glenn *et al.*, 1994
Salicornia bigelovii	>720	High Na uptake succulence	Glenn & O'Leary, 1984 Glenn *et al.*, 1991

Despite their differences, higher plants share some underlying similarities in their mechanisms of salt tolerance, even though some species have developed these mechanisms to operate with far greater efficiency than other species, and are therefore much more salt tolerant (representative species along the salt tolerance spectrum and references to their specific mechanisms of tolerance are in Table 4.1). The following sections will document some of the mechanisms of salt tolerance found among higher plants. In most cases, the underlying genetic and molecular mechanisms of adaptation are not yet known even though the physiological processes can be described. For example, efficient transmembrane compartmentation of ions is the hallmark of tolerant plants, yet membrane-bound ion translocases are only starting to be characterized and compared among species. Breeding programs to introduce specific physiological traits associated with salt tolerance into sensitive varieties are only beginning to produce results.

UPTAKE AND COMPARTMENTATION OF IONS

Control of Na, K and Ca Uptake into Plants

The first line of defense against excess Na entry into the plant, as mentioned, is the root cell plasmalemma, which has a low permeability for Na in all species studied; by contrast, root cells exhibit preferential uptake of K which can be accumulated against a concentration gradient (Pitman, 1981, 1984; Pitman *et al.*, 1981). Under the two-carrier hypothesis for univalent cation uptake (Epstein *et al.*, 1963), Carrier I predominates at low external levels of cations (up to 1 mM) and is highly specific for K over Na, whereas Carrier II operates at higher external cation levels and is less discriminatory for K over Na than Carrier I (Na inhibits K uptake). Carrier II has a higher Vmax than Carrier I and dominates cation uptake when external salt levels are high. Hence, under high external salinity, membranes become "leakier" for Na and K uptake may become inhibited.

Poorly adapted plants often exhibit dramatic drops in K uptake at high external Na concentrations as well as enhanced uptake of Na into the shoots (Rains, 1969). These ion distribution patterns are generally interpreted as symptoms of salt damage rather than adaptations to salinity, since they are accompanied by low growth rates and signs of salt damage to shoots (Greenway and Munns, 1980). These observations tend to support the concept of competition between Na and K for common carrier sites. On the other hand, K uptake by halophytes can be level across an external NaCl gradient from 20 to 1000 mM even while Na content increases (illustrated for two subspecies of the xerohalophyte, *Atriplex canescens,* in Figure 4.2), showing that at least some plants have overcome this problem. Jeschke and Wolf (1993) concluded that lupin, castor bean and barley were all able to maintain adequate cytoplasmic K levels in root cells (ca. 130 mM) despite being grown in NaCl solutions without added K; the plants were able to remobilize leaf K for utilization in the roots and shoot growing points. Pitman (1984), as well, has questioned whether K deficiency actually limits plants under salt stress even though Na inhibits K uptake. Adding additional K to the external solution does not enhance salt tolerance to NaCl.

On the other hand, there is direct evidence that Ca is required to maintain plant membrane integrity, and that Na displaces Ca from membranes, leading to disruption of membrane function (Cramer *et al.*, 1989; Rengel, 1992). It has been demonstrated many times that low levels of Ca can partially restore growth of salt-inhibited glycophytes, and that this effect is correlated with a lowering of Na uptake into the plant (illustrative data for wheat are in Figure 4.3). Unfortunately, the economic impact of alleviating salt stress by Ca application is rather limited, since most soils and saline water supplies already have sufficient Ca to

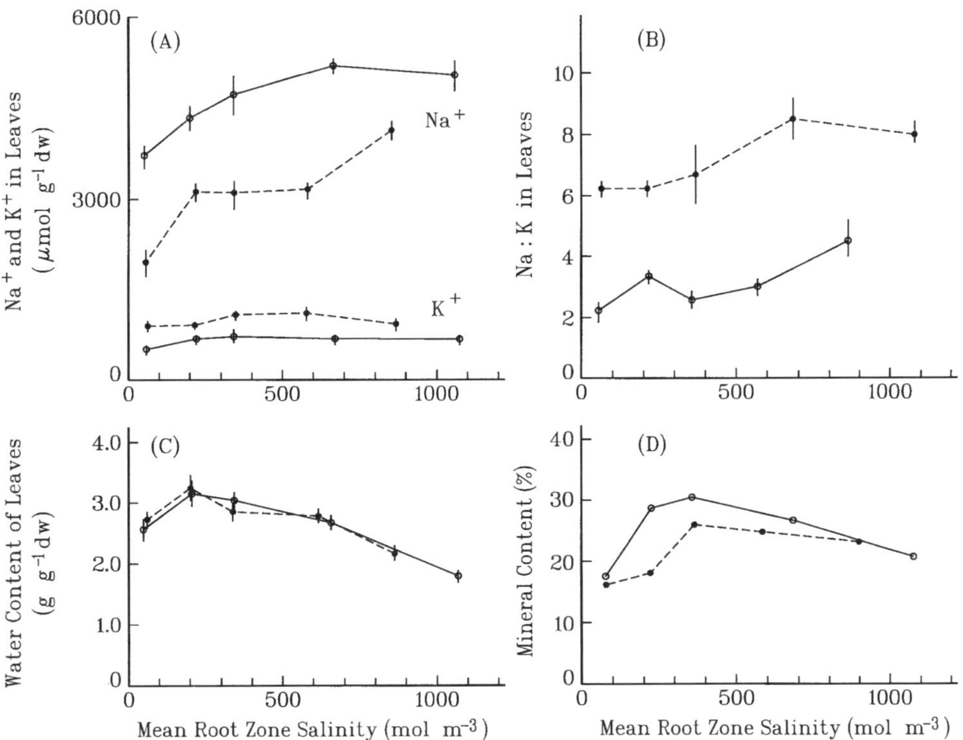

Figure 4.2. Na and K uptake (a), Na:K ratios (b), water content (c) and total mineral content (d) in leaves of two subspecies of *A. canescens* differing in salt tolerance along a salinity gradient (see Figure 4.1). Solid line is ssp. *linearis* (more tolerant); dashed line is ssp. *canescens* (less tolerant). Note greater uptake of Na and higher Na:K ratios in leaves of ssp. *linearis* (Glenn *et al.*, 1994).

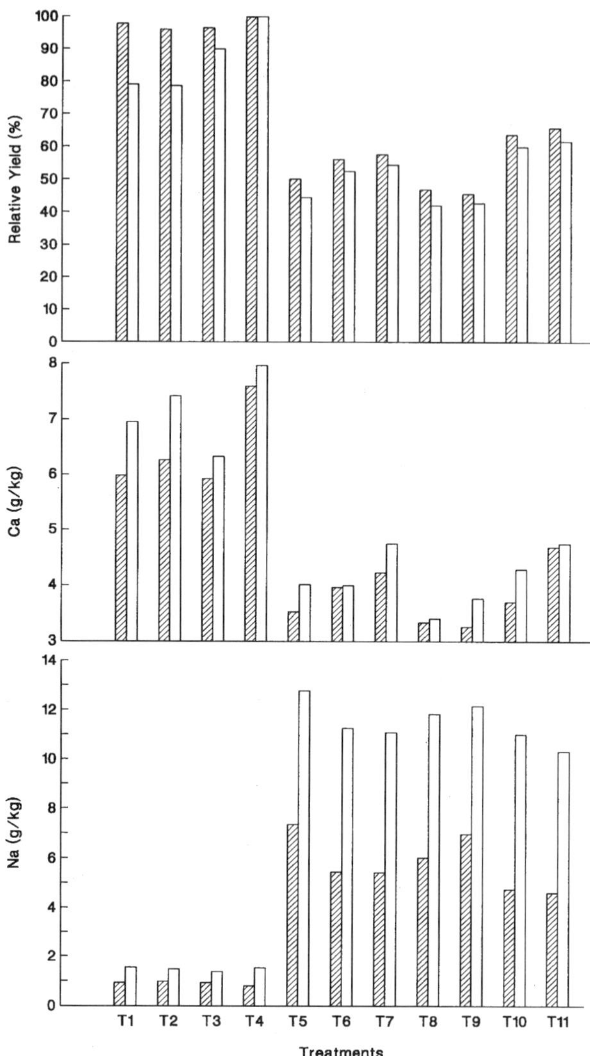

Figure 4.3. Effect of Ca, Mg and NH_4 additions on salt tolerance of a salt-tolerant and an unselected variety of bread wheat. Top graph shows relative yield under different treatments; middle and bottom graphs show leaf Ca and Na contents, respectively. For each treatment, first bar is the salt-tolerant line and the second bar is the unselected line. Treatments 1-11 are: 1) control Hoagland solution (no salt addition); 2) control + 4 mM $CaSO_4$; 3) control + 4 mM $MgSO_4$; 4) control + 4 mM NH_4; 5) control + 100 mM NaCl; 6) control + 100 mM NaCl + 4 mM $CaSO_4$; 7) control + 100 mM NaCl + 8 mM $CaSO_4$; 8) control + 100 mM NaCl + 4 mM $MgSO_4$; 9) control + 100 mM NaCl + 8 mM $MgSO_4$; 10) control + 100 mM NaCl + 4 mM NH_4; 11) control + 100 mM NaCl + 8 mM NH_4. Note that Ca and NH_4 but not Mg partially relieved the negative effects of NaCl on yield and that NH_4 was more effective than Ca.

gain the small restoration of growth that Ca can contribute. Figure 4.3 shows that NH_4 has an even greater protective effect than Ca.

Under even moderate external salinity levels (below 50 mM), Na and Cl enter root cells down an electrostatic gradient at the point of contact with the external solution (Pitman, 1984). Although plant membranes are less permeable to Na than K (permeability ratio of approximately 0.3 in barley roots), root cells exposed to high external Na levels must maintain an active efflux pump to reexport Na from the cytoplasm, as well as an efficient Na pump at the tonoplast, to concentrate Na in the vacuole fluid against the cytoplasmic gradient. Leaf cells of tolerant plants must have the ability to effectively withdraw Na from the apoplast solution and transport it across the tonoplast for sequestration in vacuoles against a concentration gradient.

Role of Ion Translocases in Salt Tolerance

The relative ability of plants to carry out ion translocations across cell and tonoplast membranes must be at the heart of differences in relative salt tolerance among species. Studies of these membrane enzymes are revealing differences among species (Mandala and Taiz, 1985; Braun *et al.*, 1986, 1988; Matoh *et al.*, 1989; Hassidim *et al.*, 1990; Reuveni *et al.*, 1990; Schachtman *et al.*, 1991; Maathuis *et al.*, 1992; Rea *et al.*, 1992; Niu *et al.*, 1993a,b; Martinez-Cortina and Sanz, 1994; Perez-Prat *et al.*, 1994). A diagram showing the possible ion translocases involved in cellular uptake and compartmentation of Na and K is in Figure 4.4.

Braun *et al.* (1986, 1988) and Hassidim *et al. (1990) have investigated ion translocases in membrane fractions of the halophyte Atriplex nummularia* and compared the results to those obtained from a glycophyte (cotton). Root membrane vesicles from plasmalemma and tonoplasts of *A. nummularia* and cotton exhibited Na/H and K/H antiporter activity but activity was much higher in the halophyte than the glycophyte. Na and K added simultaneously to the incubation mix had an additive effect on enzyme activity, leading to the conclusion that there are separate antiporters for each cation in *A. nummularia*. The halophyte had markedly higher levels of Na/H antiporter than K/H antiporter whereas in the glycophyte K/H anitporter activity appeared to be higher. Na/H antiporter activity was present even in *A. nummularia* plants grown without salt but was higher in salt-grown plants.

Thus, high Na/H antiporter activity was associated with salt tolerance in *A. nummularia*. These antiporters presumably evacuate excess Na from the cytoplasm to the apoplast and sequester Na from the cytoplasm into the vacuole for osmotic adjustment. Cl is thought to move passively across membranes in response to the electrical potential difference created by the movement of cations.

In addition to Na/H and K/H antiporters to translocate ions across membranes, H-ATPases (Niu *et al.*, 1993a,b) and (in the tonoplast) H-pyrophosphatases (H-PPases) (Rea *et al.*, 1992; Colombo and Cerana, 1993) are needed

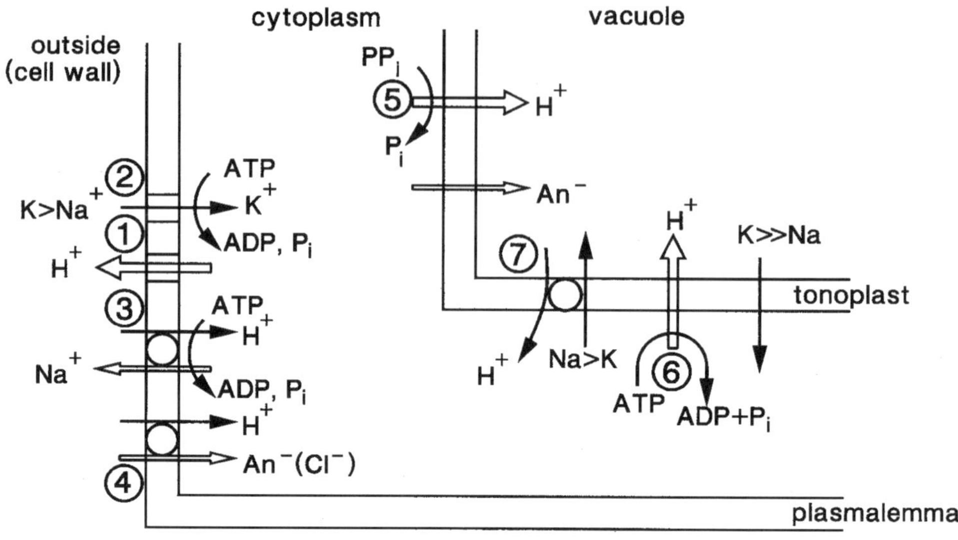

Figure 4.4. Model of the proton-mediated exchange systems at the plasmalemma and tonoplast of higher plants. 1 = proton pump; 2 = K uniport; 3 = H/Na antiport; 4 = anion symport; 5 = H/PPase; 6 = H/ATPase; 7 = H/Na antiport.

to generate H-electrochemical gradients across membranes for ion exchange. NaCl was found to regulate gene expression of plasma membrane H-ATPase in both tobacco and *A. nummularia* plants but, again, the halophyte enzyme activity was much more responsive to NaCl induction than the glycophyte. In *A. nummularia,* mRNA for H-ATPase was induced by NaCl in elongating roots and in expanded leaves (but not expanding leaves or stems) (Niu *et al.*, 1993a). Plasma membrane H-ATPase activity was also regulated by NaCl in *A. nummularia* cell cultures (Niu *et al.*, 1993b). The mRNA response to NaCl was a transcription or post-transcription, multigenic effect exhibited only at a particular point in the cell growth cycle. After initial adaptation to a salt shock, cells did not continue to maintain high levels of H-ATPase mRNA. Once adjusted, cells apparently did not have to maintain high H-ATPase activity to maintain ion homeostasis.

The amount of metabolic energy needed for ion translocation under saline conditions is considerable and may in part explain the growth reduction in plants on salinities beyond 50-200 mM NaCl. In *Suaeda maritima* grown on 200 mM NaCl, 25–30% of ATP-dependant, vacuolar, proton pump capacity is involved in osmotic adjustment to salinity and maintaining compartmentation in mature leaf cells, and the percentage is likely to be even higher in expanding leaves contributing to new growth (Maathuis *et al.*, 1992). At even higher salinities, proportionally greater net transport and sequestration of Na and K salts is needed

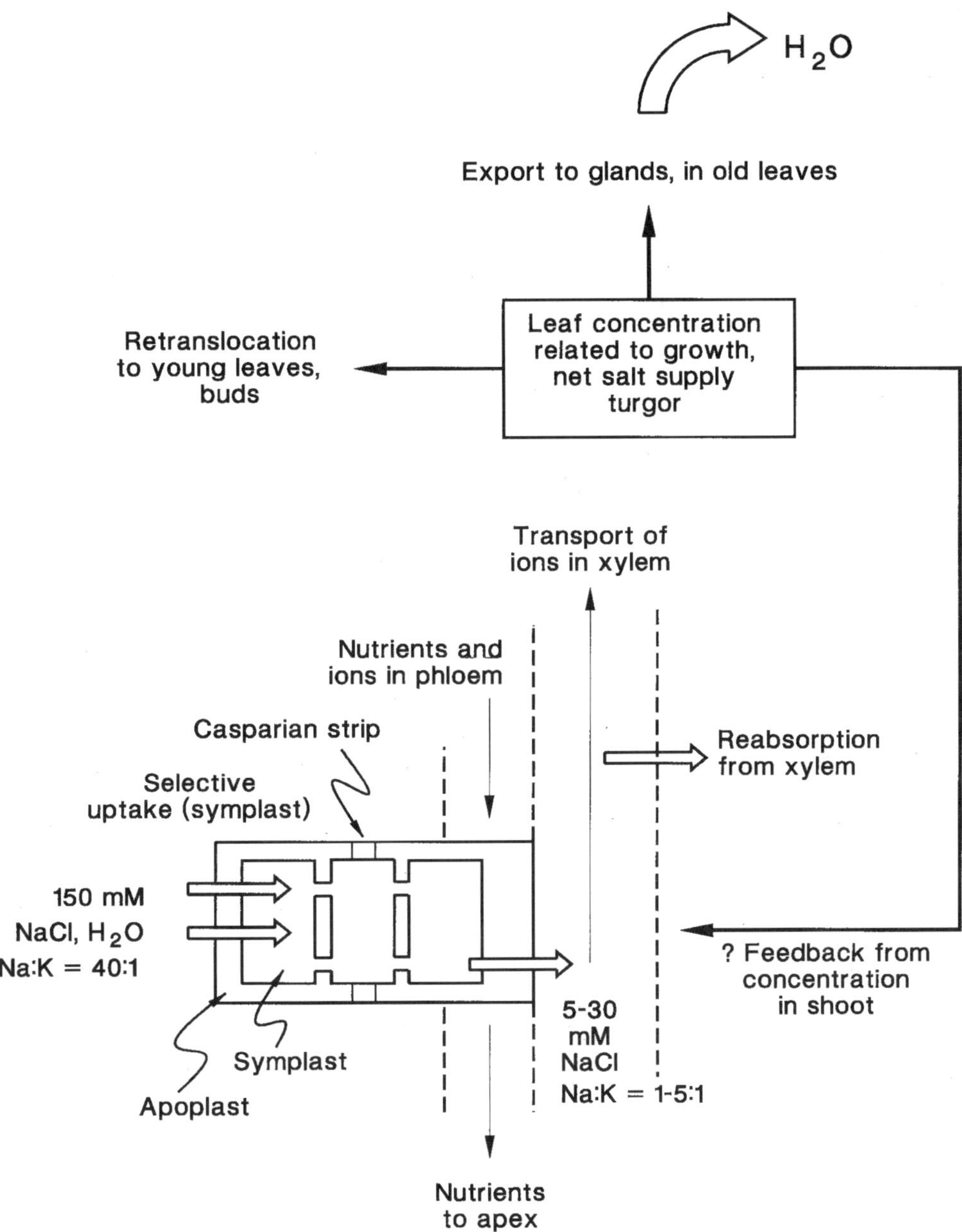

Figure 4.5. Idealized pathways of solute and water movement in a plant exposed to 150 mM NaCl in the external solution. Note that the Casparian strip acts as a barrier to solute entry into the xylem and phloem via the apoplast pathway; entry is via the symplasm where the ion content of the bulk solution is altered.

per unit growth than at lower salinities, so the percentage of metabolic energy diverted to osmotic adjustment increases.

CONTROL OF SALT LEVELS IN TISSUES AND ORGANS

Pathways of Water and Salt Movement Through a Plant

A generalized pathway of water and salt flow through a higher plant is illustrated schematically in Figure 4.5 (based on drawings and discussion in Pitman, 1984; Jeschke and Wolf, 1993). The hypothetical plant in Figure 4.5 is assumed to be growing in saline medium containing 150 mM $NaCl^-$ at the upper end of the tolerance range of many glycophytes but near the optimal salinity for many halophytes. In this section, we will follow the pathway of water and salts through the whole plant, noting control mechanisms along the way. The discussion is qualitative but numerous, quantitative studies of ion uptake and transport have been conducted on glycophytes and halophytes over a period of several decades.

Pathways of Movement through the Roots

Water and salts can follow one of two paths into the root: through the symplast or the apoplast. The symplast is the interconnected cytoplasmic compartments of root cells leading from the epidermis inward to the stele, interconnected by plasmodesmata between adjacent cells; the apoplast is the interconnected cell wall channels of the same cells. Water entering the symplast must cross the plasmalemma of an epidermal or cortical cell, at which point the ionic composition may be altered by active transport of solutes. Entry into the symplast is the most fundamental control point for salt entry into the plant. Water entering in the apoplast is much more representative of the external solution than water in the symplast, although the salt concentration of apoplast water can be modified by uptake into cells along the apoplast pathway and by the ion exchange capacity of the cell walls (Jeschke, 1984; Pitman, 1984).

The plant in Figure 4.5 has a suberized Casparian strip at the endodermis which blocks the apoplast pathway into the stele, where water and solute loading into the xylem vessel elements occurs. The Casparian strip is thought to be a critical control element for the flux of salts into the shoot. If salts could enter the xylem vessel elements through the apoplast pathway alone, the shoots could be flooded with salts; however, the Casparian strip requires that salts and water cross the endodermis by entering the symplasm. It was early recognized that halophytes often have thick layers of suberin or double layers of suberized cells at the endodermis, compared to mesophytes which often have thin Casperian strips (Anderson, 1974; Poljakoff-Mayber, 1975; Kramer, 1984). Nevertheless, there is still disagreement on the percentage of water and solutes that normally enter the transpiration stream via the apoplast and symplast pathways, so the role of the endodermis in controlling salt entry, while logically important, has not been completely resolved and probably differs among species (Jeschke,

1984). Ion distribution patterns have been measured across roots of halophytes and glycophytes using ion microscopy, and invariably there *is* a steep drop in Na content of cells across the endodermis (Pitman *et al.*, 1981; Jeschke, 1984; Kramer, 1984; Sacher and Staples, 1984).

Reabsorption of Na from the Transpiration Stream

Once in the stele, water and solutes in the symplast flow through xylem parenchyma cells into the xylem vessel elements for transport to the shoot. The xylem parenchyma cells surrounding the vessel elements offer an additional membrane-level control point for salt entry into the transpiration stream. Not only can the xylem parenchyma retain salts against a concentration gradient while permitting water flow into xylem vessel elements, they can also reabsorb salts from the transpiration stream during its passage from the root to the shoot. In some species, especially legumes, some of the xylem parenchyma cells around the vessel elements differentiate into special transfer cells that may function in the reabsorption of Na from the transpiration stream (Kramer *et al.*, 1977; Winter, 1982; Läuchli, 1984; Jeschke and Wolf, 1993). In these plants, a large amount of Na may be stored in the roots under saline growth conditions but only a small percentage is transported to the shoots. Maize xylem parenchyma cells also reabsorb Na (Yeo *et al.*, 1977; Cramer *et al.*, 1994).

Na reabsorbed from the xylem may be stored in the xylem parenchyma cells or retranslocated downward. The common reed (*Phragmites communis*) exhibits apparent reabsorption of Na from the xylem vessels directly into the phloem rather than into xylem parenchyma cells (Matsushita and Matoh, 1991, 1992). Active transport of Na from the vessels into the phloem at the shoot base results in a several-fold reduction in Na content of the transpiration stream. This conclusion was based on experiments in which shoot bases were girdled to disrupt the phloem but leave the xylem intact, followed by administration of ^{22}Na; the girdled shoot bases transmitted 284% more radioactive label to the leaves than the ungirdled stems. Staining of sections of vascular bundles for ATPase activity revealed that active uptake was likely to occur in the phloem rather than the xylem parenchyma as in other plants. Presumably, reabsorbed salts are transported back to the root tips in the phloem.

The efficacy of these reabsorption mechanisms has been questioned (Läuchli, 1984). They occur mainly in plants that are at the low end of the salinity tolerance gradient, and may represent adaptations to conditions of temporary high salinity or responses to excessive leakage of NaCl into roots, rather than mechanisms which allow a plant to grow and complete its life cycle on permanently saline substrate. The reabsorbed Na must be handled by the plant. In time this leads to Na excess in the roots and impairment of the translocation ability of the plant. There is also evidence from bean plants that high Na content in the phloem dehydrates the cells, and that reabsorption and translocation of Na from leaves to roots, though observed, does not appear to be a very effective control mechanism for Na exclusion (Läuchli, 1984).

In an informative review on the effectiveness of different salt tolerance mechanisms, Jeschke and Wolf (1993) described the uptake patterns of Na, K and Cl in three glycophyte species differing in salt tolerance: white lupin, *Lupinus albus*, (salt-sensitive); castor bean, *Ricinus communis*, (moderately tolerant); and barley, *Hordeum vulgare*, (tolerant). Summarizing a great deal of detailed evidence, they concluded that lupin was ineffective at controlling Na uptake into the shoot including the leaves. Even under low external NaCl concentration (10 mM) half the total Na entering the plant was transported to the shoots, mainly into leaf laminae, while half remained in the roots. Lupin was also the species with the highest rate of retransport of Na from the shoot back to the root in the phloem. Of the half of the total Na that was transported to the shoot, 75% was retransported back to the root, leaving only 12% of total uptake in the leaf laminae.

By contrast, the more tolerant castor bean was more effective in limiting uptake into the roots and also prevented most transport of Na from root to shoot from occurring. Under 128 mM NaCl, total Na uptake was comparable to lupin on 10 mM NaCl, and 80% of the Na that entered the plant was retained in the root. Retransport from leaf laminae to roots in phloem was less than 10% of total uptake but the laminae levels of Na were low, containing only 2% of total plant Na.

Barley, the most tolerant species, exhibited yet a third pattern with respect to ion transport. Under 100 mM external salinity, total uptake was higher than the other species and approximately half the total Na uptake was transported to the shoot and half remained in the root, as in white lupin. As in castor bean, retransport from leaf laminae to the roots in the phloem was less than 10% of total uptake. Unlike the other species, however, 37% of total Na was stored in the leaf laminae rather than retained in or retransported to the roots. This pattern is similar to the typical ion transport pattern of halophytes, which transport a large proportion of the Na entering the roots into the leaves, where it is sequestered in vacuoles to provide osmotic adjustment.

Control of Salt Levels in Leaves

As shown in Figure 4.6, the salt content in the transpiration stream is low compared to the external solution, even for salt-accumulating halophytes. As examples, barley grown under 150 mM NaCl has approximately 5 mM NaCl in the transpiration stream (Pitman, 1984) and maximum transport rates are reached at an external concentration of just 40 mM NaCl (Delane *et al.*, 1982); *Atriplex canescens* subspecies carry no greater than 20–30 mM NaCl in the transpiration stream even when the external solution is 1 M NaCl and their transport capacity is saturated at an external level of 250–400 mM NaCl (Glenn *et al.*, 1994) (Figure 4.6). Clearly, the roots of most plants are effective in limiting salt transport to the shoots. Nevertheless, even low levels of salts in the transpiration stream are problematic, given the high flux of water through the plant. Salt damage to leaves is one of the earliest symptoms of toxicity. In sensitive plants like rice, even low external levels (50 mM NaCl) can lead to very high and damaging levels (500 mM) in the leaf apoplast due to poor control of NaCl entry in the transpiration stream (Flowers *et al.*, 1991). Rice plants can have nearly the same rate of NaCl transport from root to shoot as halophytes, but unlike halophytes they lack the ability to adequately handle the salts arriving in the leaves (Yeo and Flowers, 1984, 1986b). Therefore, they suffer massive damage and are at the low end of the salt tolerance scale while halophytes are at the upper end.

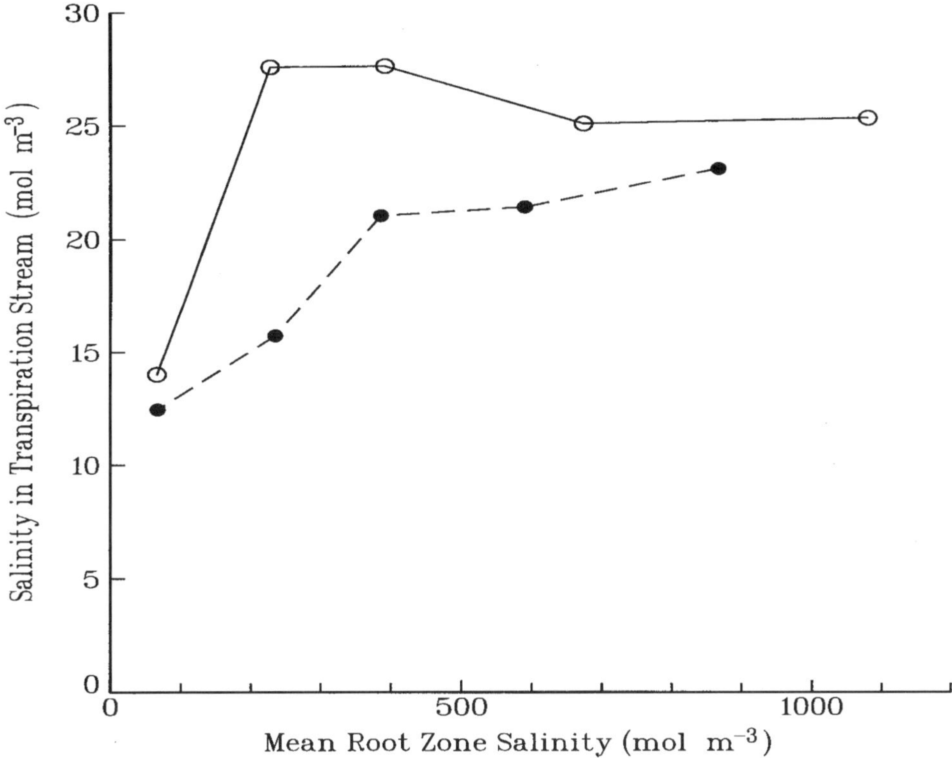

Figure 4.6. Concentration of NaCl in the transpiration stream of two subspecies of *Atriplex canescens* differing in salt tolerance across a salinity gradient. Open circles are ssp. *linearis* (more tolerant) and closed circles are ssp. *canescens* (less tolerant). Note that ssp. *canescens* had lower levels of NaCl in the transpiration stream than ssp. *linearis* (Glenn *et al.*, 1994).

Control of Na Levels by Growth

Rice, wheat and barely have two adaptations to handle high salt levels arriving in the shoots: dilution of salts by growth, and partitioning of salts into older leaves (Greenway and Munns, 1980; Yeo and Flowers, 1984, 1986b; Munns and Termaat, 1986; Wolf *et al.*, 1990; Schachtman and Munns, 1992). Screening trials with rice varieties have shown that selections with inherently higher growth rates often exhibit more salt tolerance than those with lower growth rates (Yeo and Flowers, 1984, 1986b). The varieties may have similar rates of NaCl uptake into shoots, but those with the higher growth rates have lower levels of salt per gram dry weight of shoots, since they have greater shoot biomass production per unit salt uptake compared to slower-growing varieties. Similarly, more tolerant varieties often show greater accumulation of NaCl in older leaves compared to young, expanding leaves. The older leaves are damaged by the salt, but the plant continues to produce new leaves to support photosynthesis and growth by controlling salt transport to the growing points (Greenway and Munns, 1980). In wheat, Na may be preferentially absorbed into the sheath rather than the leaf (Schachtman and Munns, 1992). Whether these are specific adaptations to salt or merely a consequence of different transpiration rates of leaves with different growth rates is not completely resolved. The halophytic grass, *Diplachne fusca*, takes up more NaCl than glycophyte grains but partitions the salts equally into young and old leaves alike (Warwick and Halloran, 1992), a pattern similar to dicotyledonous halophytes such as *Suaeda maritima* (Yeo and Flowers, 1986a).

Salt Glands and Salt Bladders

More tolerant plants often have more targeted methods for handling salt in leaves. Salt glands and bladders for depositing salts onto the outer leaf surfaces are examples of such mechanisms. Salt glands are found in at least eleven plant families (10 dicotyledons and 1 monocotyledon, the Gramineae) (Liphschitz and Waisel, 1982) and are especially widespread among grasses, occurring in 33 genera. Salt bladders, anatomically distinct from salt glands, occur in the Chenopodeacae, especially in *Atriplex* whose 200 species apparently all possess salt bladders (Schirmer and Breckle, 1982; Freitas and Breckle, 1992).

Salt glands in the Gramineae consist of two cells, a basal and a cap cell (Liphschitz and Waisel, 1982). Saline solution collects in the basal cell and is expelled from the cap cell. Both cells have dense cytoplasm with numerous mitochondria, lack central vacuoles and have cell walls covered with suberin or cutin through which channels must exist for the exit of brine from the cap cell. In some genera the basal cell is sunken into the epidermis while in others the salt gland is a trichome-like structure above the epidermis. There appears to be a correlation between salt gland anatomy and salt tolerance: genera with sunken glands (e.g. *Spartina*) have greater tolerance than those with semi-sunken glands (e.g. *Paspalum*) which are more tolerant than those with surface glands (e.g. *Panicum*) (Liphschitz and Waisel, 1982). Grass genera with salt glands include important crop and forage plants: *Saccharum* (sugarcane); *Cynodon* (Bermuda grass); *Sorghum*; and *Chloris* (buffalo grass). However, not all grasses with salt glands exhibit unusually high salt tolerance, and the effectiveness of the glands in excreting salt entering the leaf varies from just a few percent of the total to 50% or more for halophytic grasses such as *Spartina* (Bradley and Morris, 1991) and *Diplachnea* (Warwick and Halloran, 1992). The glands concentrate NaCl against a concentration gradient, with the excreted brine having higher osmotic pressure than the leaf cell sap, and a higher Na/K ratio than in the cells. Salt excretion increases with external salinity only up to a point; beyond an external concentration of approximately 150 mM NaCl, excretion rates actually decrease.

The salt glands of dicot species are more variable in morphology than grasses, varying from two-cell glands to complex structures of up to 16 cells (members of the Plumbaginaceae, e.g. *Statice*) (Liphschitz and Waisel, 1982; Batanouny, 1993). Pores in the suberized cell cap are sometimes visible. Some mangrove species, for example *Avicenia marina*, are highly efficient salt excretors while others with apparently the same degree of salt tolerance, for example *Rhizophora mangle*, do not have salt glands at all (Popp *et al.*, 1993).

The specialized salt bladders of the *Chenopodiaceae* are highly modified trichomes or vesicles consisting of a long, narrow stalk cell (or cells) and a balloon-shaped bladder cell projecting above the epidermis (Schirmer and Breckle, 1982). Different species of *Atriplex* have characteristic shaped bladders which can aid in species identification. Salt solution is transferred against a gradient from the leaf mesophyll cells to the bladder cell via the stalk cell. Brine accumulates in the central vacuole of the bladder cell, which eventually bursts, depositing salt on the leaf surface. The accumulated salt may act to reduce transpiration by increasing the reflectance of the leaf. As much as 80% of the NaCl entering *Atriplex* leaves may be excreted via the bladder cells (Freitas and Breckle, 1992).

Leaf Succulence to Regulate Osmotic Pressure in the Leaves

As mentioned, not all tolerant plants have the ability to excrete salts. Many halophytes and salt tolerant glycophytes handle a temporary excess of salts in the apoplast by increasing the water content of the mesophyll cells (presumably the vacuole water content), thereby diluting the salts and increasing their capacity to further absorb salts from the apoplast solution (Kramer, 1984). Succulent halophytes can also lose water, to concentrate salts in the vacuole, when the rate of arrival of salts is not sufficient to support osmotic adjustment of expanding cells at maximum water content (Glenn and O'Leary, 1984).

Some halophytes exhibit a diurnal variation in succulence as measured by leaf thickness, in which succulence is greatest at night and lowest during the day (Rozema, 1991). In extreme succulent species, such as *Suaeda maritima*, the rate of salt uptake must be very high to support expansion growth and cell division, and under normal conditions growth is limited by ion availability in the leaves (Yeo and Flowers, 1986a). Hence, problems of ion excess would not normally occur in extremely succulent leaves or stems.

Control of Stomatal Opening by Apoplastic Na Levels

Potassium fluxes in and out of guard cells play a central role in the stomatal movements of most plants, but the coastal halophyte, *Aster tripolium*, shows additional control by Na (Perera *et al.*, 1994). This plant lacks salt glands, hence control of Na uptake into leaves is likely to be of great importance. Na tends to be transported into shoots in proportion to the transpiration rate (Pitman, 1981, 1984), hence, uptake into the leaves could be regulated by controlling the transpiration rate through partial stomatal closing. *Aster* leaf epidermal strips were floated on solutions containing NaCl and KCl along with control strips from a non-halophyte, *Commelina communis*, and stomatal apertures were measured.

As expected, *Aster* and *Commelina* both responded to KCl by opening their stomata. Also as expected, *Commelina* stomata opened as well in response to NaCl, with little difference in response to Na or K in the range 0–100 mM, since Na can substitute for K in most plants. However, unlike *Commelina*, *Aster* stomata closed when exposed to 25 mM NaCl, and the effect could not be reversed by KCl. Abscisic acid did not appear to be the mediator of the NaCl response in *Aster* nor could it be reversed by auxin. The authors concluded that *Aster tripolium* has an additional control mechanism for stomatal closure by which regulation of Na intake into the leaves is accomplished. When the leaf cells' ability to sequester Na in vacuoles is exceeded by the rate of import into the leaf, Na builds up in the apoplast and stimulates partial stomatal closure.

These novel findings have not yet been repeated in another halophyte; only a small number of plants yield clean epidermal peels which can be used to study stomatal response to bathing solutions (Perera *et al.*, 1994). If stomatal control by Na is found to be widespread, it would be an important finding on how halophytes regulate Na uptake into leaves.

Coordination of Leaf Water Content and Salt Uptake to Regulate Osmotic Potential of Halophytes

As mentioned, many halophytes are osmoconformers, increasing the osmotic pressure in their leaf tissues in proportion to the external salinity (Glenn and O'Leary, 1984; Glenn, 1987; Glenn *et al.*, 1992b, 1994). The coordinated roles of NaCl uptake and tissue water content in maintaining osmotic pressure of leaf cells is illustrated by two survey studies of halophytes grown on a salinity gradient.

In a screening of 20 dicotyledonous halophytes and putatively salt-tolerant glycophytes from 9 plant families, 9 of the 20 had shoot Na/K ratios < 1 on non-saline control solution but 19 of the 20 had ratios ranging from 2.5 to 15.1 on salinities of 180 mM or higher (Glenn and O'Leary, 1984). Only the non-halophyte, sunflower (*Helianthus annuus*) in that screening maintained a Na/K ratio less than 1 on 180 mM NaCl and it grew poorly at that salinity and did not survive at higher salinities. Only 2 species showed a slight increase in K content with salinity (*H. annuus* and the related *H. venetus*) whereas all the rest had lower K contents with increased salinity. Of the 9 species that survived to 720 mM

NaCl, all showed large increases in Na compared to controls, and their leaf osmotic pressures increased in conformity to the external solution. The 9 plants that survived to 720 mM were stimulated by 180 mM relative to growth rates on control solution. These plants came from such salt tolerant genera as *Salicornia*, *Allenrolfia*, *Batis* and *Atriplex*.

Water contents of the dicot shoots also responded to salinity. Plants grown on control solution had the lowest water content; plants at 180 mM, the growth optimum for some species, had the highest water contents; and 5 of the 9 species that survived to 720 mM showed significantly lower water contents at the highest salinity. Control of water content appeared to be part of the adaptive process for at least some species in maintaining osmotic adjustment, since ion content and water content together determined the total osmotic pressure, which, as noted, conformed closely to the external salinity.

The same basic trends were noted among monocot species (27 from the Poaceae, 1 from the Cyperaceae), though they differed quantitatively from the dicots in water content and Na:K ratios (Glenn, 1987; see also Gorham *et al.*, 1980; Rozema, 1991; for discussion of halophyte physiotypes). The species varied in salt tolerance from those that survived only to 180 mol m^{-3} and are normally considered glycophytes (e.g. salt-tolerant strains of *Triticum*, *Hordeum* and *Cynodon*) to species from salt-marsh genera such as *Distichlis* and *Spartina* that survived on full-strength seawater (540 mol m^{-3}). Na:K ratios were < 0.5 for all species on non-saline control solution, but ranged from 0.6-5.8 for all but one species on 180 mol m^{-3} and from 1.2-5.2 for the 14 species that survived on 540 mol m^{-3}. The only species that appeared to be an "excluder" was the common reed, *Phragmites australis*, which maintained a low Na:K ratio (0.13) even on 180 mol m^{-3} but did not survive on 540 mol m^{-3} (see also Matsushita and Matoh, 1991). All of the species increased their shoot Na^+ content by 1.5-15 times in going from 0 to 180 mol m^{-3}, and all showed a reverse trend with respect to K^+. All but three had lower water contents on 180 mol m^{-3} than on control solution, and all 14 that survived on 540 mol m^{-3} had lower water contents than control plants, in some cases dramatically lower (e.g., *Aleuropis littoralis*, the top-performing species on 540 mol m^{-3} in terms of growth rate, had 5.8 g H_2O g^{-1} dw on 0 mol m^{-3} but only 1.6 g g^{-1} on 540 mol m^{-3}; *P. australis*, as well, depended mainly on water loss for osmotic adjustment). Like the dicots, the monocots were osmoconformers. In a longer-term experiment in the same study (4-6 months to ensure complete adjustment to salinity) *Sporobolus virginicus*, *Pucinellia maritima* and *Paspalum vaginatum*, all increased the osmotic pressure of their leaf sap in proportion to the external salinity gradient, and Na+K balanced by Cl^- could account for 80% or more of the measured osmotic pressure across salinity levels. Hence, organic solutes could have played little part in the overall response (though they were undoubtedly important cytoplasmic osmotica). Since K^+ actually decreased, the osmotic adjustment was due to the same two factors important in dicots: enhanced Na^+ uptake and lower water content.

The overall pattern of halophyte salt uptake for osmotic adjustment is also illustrated in the work by Bradley and Morris (1991) with *Spartina alterniflora* grown on a salinity gradient. *Spartina* is often regarded as an "excluder" with other salt marsh grasses (e.g. Rozema, 1991; Rozema and Van Diggelen, 1991). True enough, 91-97% of salts were excluded based on the theoretical maximum that could have entered in the transpiration stream. Further, K^+ was selectively taken up over Na^+ based on ratios in the external solution compared to the tissues; half the salt taken up was excreted onto the leaves and the excreted salt was almost all NaCl while K^+ was retained. Yet, in terms of osmotic adjustment, Na^+ was the only cation to increase in the tissues when the external salinity was increased from 180 to 720 mol m^{-3} NaCl in the external salinity, approximately doubling in content while the others stayed the same or decreased. Even in so-called "excluder" halophytes, Na^+ invariably turns out to be the cation that is responsible for the plants' osmotic response to an increase in external salinity.

GENETIC MECHANISMS OF SALT TOLERANCE

Overview of Genetics of Salt Tolerance

Progress in selective breeding for salt tolerance and identification of salt-tolerance genes has been reviewed (Hurkman, 1992; Ashraf, 1994). Most of this work does not start from a hypothesis about the mechanism of salt tolerance. For example, breeding programs typically select more tolerant cultivars from mass screenings of seedlings grown along a salinity gradient without regard to the underlying mechanisms which confer greater tolerance. At the molecular level, mRNA's and polypeptides induced by salt stress have been characterized but not yet tied to specific mechanisms of tolerance (see also Winicov, 1994).

For many crop plants, heritable variation in salt tolerance is quite high (Ashraf, 1994). Narrow-sense heritability (the extent to which phenotypes are determined by genes with additive effects on the phenotype transmitted by the parents) is expressed by breeders as a proportion, V_A/V_P, where V_A = additive genetic component and V_P = total phenotypic variance of a trait. A $V_A/V_P > 0.3$ for a trait usually produces a rapid response to selective breeding. V_A/V_P values of 0.19-0.98 were found for a group of 11 range grass and legume species (Ashraf *et al.*, 1986), with only 3 species having values less than 0.3. This and similar studies with crop plants confirm that heritability of salt tolerance is high and amenable to manipulation by breeders.

The number of genes controlling a trait conferring salt tolerance can be relatively small within a species, especially in glycophytes. For example soybean cultivar Lee is salt tolerant whereas cv. Jackson is sensitive, and the difference is apparently due to a single gene restricting Cl^- uptake in cv. Lee (Abel and Mackenzie, 1964). Breeding experiments with rice have indicated that three pairs of genes are involved in resistance to male sterility as a consequence of exposure to salinity (Akbar and Yabuno, 1975).

Transfer of Enhanced K/Na Discrimination Among Wheat Cultivars

Perhaps the most thorough series of experiments connected with defining and transferring a salt tolerance mechanism among plants has been the attempt to transfer enhanced K/Na discrimination among wheat varieties (Kingsbury *et al.*, 1984; Wyn-Jones *et al.*, 1984; Kingsbury and Epstein, 1986; Richards *et al.*, 1987; Rawson *et al.*, 1988; Gorham *et al.*, 1990; Schachtman *et al.*, 1991; Schachtman *et al.*, 1992; Schachtman and Munns, 1992). The progenitor of all cultivated wheats is tetraploid wild emmer wheat, *Triticum dicoccoides*, which has been found to have significant variability in salt tolerance among accessions (Nevo *et al.*, 1992). It has variability as well the ability to maintain a high K/Na ratio in the leaves under salt stress (the enhanced Na exclusion trait), which has appeared to be associated with salt tolerance in this species. Bread wheat (*T. aestivum*) is more salt tolerant than durum wheat (*T. turginsum*) (Schachtman and Munns, 1992). The genome formula for bread wheat is ($2n = 6x = 42$, AABBDD), with the D genome having been added to durum wheat from *T. tauschii*, a wheat progenitor, during the domestication process.

Through conventional crossing experiments, the location of the enhanced K/Na discrimination gene was placed at the Kna1 locus on the 4D chromosome. Two types of experiments demonstrated that, in concept, salt tolerance could be conferred by transfer of the Kna1 trait: Dvorak *et al.* (1994) produced durum wheat strains in which chromosome 4D was recombined with chromosome 4B, so that some of the offspring had the Kna1 locus expressed in a background of durum wheat, whereas others did not receive the Kna1 locus (all recombinants had other parts of the 4D chromosome expressed as well). The +Kna1 families had greater K/Na discrimination and better salt tolerance than the –Kna1 families though the results were provisional. The +Kna1 families did not have significantly lower Na uptake than –Kna1 families but moderately higher K uptake leading to a higher K/Na ratio in the leaves. Grain production was not significantly higher in +Kna1 families than — Kna1 families under two levels of salt stress but biomass production was improved by 15-20%. Schachtman et al. (1992) produced synthetic hexaploids from 5 *T. tauschii* accessions and 2 salt-sensitive accessions of durum wheat. The *T. tauschii* accessions included some that were moderately tolerant to salinity and had a low rate of Na incorporation into shoots and others that were sensitive to salinity and had a higher rate of Na uptake. Parent lines and offspring were grown at 1 and 150 mM NaCl and compared for yield and Na uptake. The hexaploid offspring exhibited greater salt tolerance than durum parents and lower Na uptake, and the degree of salt tolerance exhibited by the offspring varied according to the source of D genome. Relative grain yields ranged from < 10% of controls for durum to 45% for the best hexaploid offspring. Rubio *et al.* (1995) demonstrated that point mutations in the wheat K transporter gene increased plant salt tolerance.

Should Breeders Select for Na Exclusion or Inclusion?

The strategy followed by breeders in the above example, and similar studies with other plants, has been to enhance the ability of crop plants to exclude Na. This strategy is in apparent contradiction to the strategy evolved by halophytes, described above, which invariably accumulate Na in the leaves for osmotic adjustment. The rate at which plants can assimilate and compartmentalize Na+K for osmotic adjustment has been hypothesized to be rate-limiting for the growth of the succulent halophyte *Suaeda maritima* (Yeo and Flowers, 1986a) as well as the leafy xerohalophyte, *Atriplex canescens* at salinities that are supraoptimal for growth (Glenn *et al.*, 1994).

Not all studies with glycophytes find a correlation between Na exclusion and salt tolerance. Albercio and Cramer (1993) and Cramer *et al.*, (1994), working with 7 maize hybrids that differed in Na uptake into shoots and relative salt tolerance, concluded that salt tolerance was not associated with Na exclusion in this series. Barley lacks the Na exclusion trait of wheat yet is more salt tolerant (Rawson *et al.*, 1988; Wolf *et al.*, 1990). Similar to the conclusions concerning halophytes, the rate-limiting step for barley leaf elongation has been attributed to rate of ion transport into the leaves for osmotic adjustment (Delane *et al.*, 1982; Munns *et al.*, 1982). Even in wheat, where an association between Na exclusion and tolerance has been demonstrated, it has also been found that cultivars with the ability to efficiently compartmentalize Na into leaf cell vacuoles can have the same degree of salt tolerance as Na-excluding cultivars despite high uptake rates (Schachtman and Munns, 1992).

Halophytes as Potential Crop Plants

In order to achieve osmotic adjustment during growth, plants need to acquire ions. It may not be sufficient simply to breed crop plants with the ability to exclude Na if high yield under extreme salinity is the goal. Rather, selection for plants that can assimilate and sequester Na at rates that can support high growth rates will be needed. Considerable attention has been devoted to domesticating halophytes as potential sources of animal forage (Watson *et al.*, 1987, 1994; Glenn *et al.*, 1992a; Le Houerou, 1993; Watson and O'Leary, 1993; Squires and Ayoub, 1994) and oilseeds (Glenn *et al.*, 1991) since they are pre-adapted for salt tolerance. Field trials in coastal desert locations using 720 mM seawater as the irrigation source have shown that yields of biomass (10–20 t/ha) and oilseed (2 t/ha) are equivalent to conventional agronomic crops irrigated with fresh water (Glenn and O'Leary, 1984; Glenn *et al.*, 1991).

Domestication of halophytes will require the application of conventional breeding methods to improve uniformity of stand, disease resistance, and other agronomic traits, building upon pre-existing salt tolerance. This approach may yield a true salt tolerant crop more quickly than attempting to introduce salt tolerance mechanisms into plants that lack tolerance (Glenn *et al.*, 1991). On the other hand, attempts to incrementaly improve the salt resistance of conventional crops by building upon their exclusion mechanisms is also a valid approach to dealing with encroaching salinity problems in arid zone irrigation districts around the world. Agronomic methods to use high-salinity water are being developed concurrently with the development of halophyte and conventional crops that can thrive under brackish water irrigation (Miyamoto *et al.*, 1994, 1995).

CONCLUSIONS

Our present knowledge of the mechanisms of salt tolerance in higher plants is incomplete. The fundamental ion transport properties of plant membranes are only starting to be studied, and it is at this level that differences among species with respect to mechanisms of tolerance may become obvious. It is clear from physiological studies that regulation of ion uptake and compartmentation within organelles, cells and tissues forms the basis for salt tolerance of higher plants. Higher plants face a predicament with respect to NaCl uptake: they need to acquire NaCl or other ions for osmotic adjustment of their cells yet they are intolerant of NaCl in their cytoplasm. The spectrum of responses plants exhibit when challenged with salt represent the methods they have evolved to cope with this predicament, some more successfully than other.

Glycophytes and halophytes have in common the ability to exclude most of the NaCl in the external solution from entering the plant while admitting water; otherwise, the plant would be flooded with salts. Morphological adaptations such as thick Casparian strips aid in preventing bulk solution from entering the transpiration stream.

The most salt-sensitive plants apparently have difficulty controlling Na entry into the stele and they have adaptations for the reabsorption and storage of Na in xylem parenchyma cells or for retransport of Na from the shoot back to the root in the phloem. These plants may also lack the ability to efficiently compartmentalize NaCl in leaf cell vacuoles, limiting its role in osmotic adjustment (Schachtman and Munns, 1992). They may rely mainly on K and organic compounds for osmotic adjustment, at considerable expenditure of metabolic energy (Wyn-Jones *et al.*, 1984).

More tolerant plants, such as barley, are able to more efficiently compartmentalize NaCl within the leaves and leaf cells and they utilize Na uptake as a mechanism of osmotic adjustment, allowing growth to proceed at higher salinities than occurs for strict Na-excluders. Halophytes represent the extreme of NaCl uptake, admitting large quantities of ions into the photosynthetic tissues and exhibiting positive growth rates even on 1 M NaCl. Studies of Na/H antiports and H/ATPases are just beginning to reveal differences between glycophytes and halophytes in the activity levels and control mechanisms of ion translocation enzymes. Fundamentally, halophytes must be efficient at transporting Na across the tonoplast against a steep gradient from the cytoplasm to the vacuole. Salt excretion, succulence and feedback control of transpiration through Na effects on stomata are specialized secondary traits that allow halophytes to utilize NaCl as a cheap osmoticum. The success of their adaptations is seen by the fact that halophytes irrigated with seawater can yield as well as conventional crops on fresh water, and by their successful colonization of marine estuarine environments and salt deserts where the soil solution salinity can exceed seawater salinity.

REFERENCES

Abel G.H. and McKenzie A.J. (1964) Salt tolerance of soybean varieties (*Glycine max* L. Merrill) during germination and later growth. *Crop Science*, **4**, 157–161.

Akbar M. and Yabuno T. (1975) Breeding for saline-resistant varieties of rice. III. Responses of F_1 hybrid to salinity in reciprocal crosses between Johna 349 and Magnolia. *Japanese Journal of Breeding*, **25**, 216.

Alberico G.J. and Cramer G.R. (1993) Is the salt tolerance of maize related to sodium exclusion. I. Preliminary screening of seven cultivars. *Journal of Plant Nutrition*, **16**, 2289–2303.

Albert R. and Popp M. (1977) Chemical composition of halophytes from the Neusieder Lake region in Austria. *Oecologia*, **27**, 157–170.

Anderson C.E. (1974) A review of structure in several North Carolina salt marsh plants. In *Ecology of Halophytes*, edited by R.J. Reimold and W.H. Queen, pp. 307–344. New York: Academic Press.

Ashraf, M. (1994) Breeding for salinity tolerance in plants. *CRC Critical Reviews in Plant Sciences*, **13**, 17–42.

Ashraf M., McNeilly T. and Bradshaw A.D. (1986) The potential for evolution of salt (NaC1) tolerance in seven grass species. *New Phytologist*, **103**, 299–309.

Balnokin Yrii, Popova L. and Myasoedov N.A. (1993) Plasma membrane ATPase of marine unicellular alga *Platymonas viridis*. *Plant Physiology and Biochemistry*, **31**, 159–168.

Batanouny K.H. (1993) Adaptation of plants to saline conditions in arid lands. In *Towards the Rational Use of High Salinity Tolerant Plants*, edited by H. Lieth and A. A. Al Masoom, pp. 217–224. Dordrecht: Kluwer Academic Publishers.

Bradley P.M. and Morris J.T. (1991) Relative importance of ion exclusion, secretion and accumulation in *Spartina alterniflora* Loisel. *Journal of Experimental Botany*, **42**, 1525–1532.

Braun Y., Hassidim M., Lerner H. and Reinhold L. (1986) Studies on H^+-translocating ATPases in plants of varying resistance to salinity. *Plant Physiology*, **81**, 1051–1056.

Braun Y., Hassidim M., Lerner H. and Reinhold L. (1988) Evidence for a Na^+/H^+ antiporter in membrane vesicles isolated from roots of the halophyte *Atriplex nummularia*. *Plant Physiology*, **87**, 104–108.

Brown A.D. (1983) Halophilic prokaryotes. In *Encyclopedia of Plant Physiology, Vol.12C*, edited by C.B. Osmond and H. Ziegler, pp.137–163. New York: Springer Verlag.

Cheeseman J.M. (1988) Mechanisms of salinity tolerance in plants. *Plant Physiology*, **87**, 547–550.

Colombo R. and Cerana R. (1993) Enhanced activity of tonoplast pyrophosphatase in NaCl-grown cells of *Daucus carota*. *Journal of Plant Physiology*, **142**, 226–229.

Cramer G.R., Albercio G.J. and Schmidt C. (1994) Salt tolerance is not associated with the sodium accumulation of two maize hybrids. *Australian Journal of Plant Physiology*, **21**, 675–692.

Cramer G.R., Epstein E. and Läuchli A. (1989) Na-Ca interactions in barley seedlings: relationship to ion transport and growth. *Plant, Cell and Environment*, **12**, 551–558.

Delane R., Greenweay H., Munns R. and Gibbs J. (1982) Ion concentration and carbohydrate status of the elongating leaf tissue of *Hordeum vulgare* growing at high external NaCl. I. Relationship between solute concentration and growth. *Journal of Experimental Botany*, **33**, 557–573.

Dvorak J., Noaman M.M., Goyal S. and Gorham J. (1994) Enhancement of the salt tolerance of *Triticum turgidum* L. by the Knal locus transferred from the *Triticum aestivum* L. chromosome 4D by homoeologous recombination. *Theoretical and Applied Genetics*, **87**, 872–877.

Epstein E., Rains D.W. and Elzam O. (1963) Resolution of dual mechanisms of potassium absorption by barley roots. *Proceedings of the National Academy of Sciences USA,* **49**, 684–692.

Flowers T.J., Hajibagheri M.A. and Yeo A.R. (1991) Ion accumulation in the cell walls of rice plants growing under saline conditions: evidence for the Oertli hypothesis. *Plant, Cell and Environment*, **14**, 319–325.

Freitas H. and Breckle S.W. (1992) Importance of bladder hairs for salt tolerance of field-grown *Atriplex* species from a Portuguese salt marsh. *Gustav Fischer Verlag Jena*, **187**, 283–297.

Glenn E.P. (1987) Relationship between cation accumulation and water content of salt-tolerant grasses and a sedge. *Plant, Cell and Environment*, **10**, 205–212.

Glenn E.P. and O'Leary J.W. (1984) Relationship between salt accumulation and water content of dicotyledonous halophytes. *Plant, Cell and Environment*, **7**, 253–261.

Glenn E.P., O'Leary J.W., Watson M.C., Thompson T.L. and Kuehl R.O. (1991) *Salicornia bigelovii* Torr.: an oilseed halophyte for seawater irrigation. *Science*, **251**, 1065–1067.

Glenn E.P., Olsen M.W., Frye R.J. and Moore D.W. (1994) How much sodium accumulation is necessary for salt tolerance of subspecies of the halophyte, *Atriplex canescens*? *Plant, Cell and Environment,* **17**, 711–719.

Glenn E.P., Coates W.E., Riley J.J., Kuehl R.O. and Swingle R.S. (1992a) *Salicornia bigelovii* Torr: a seawater-irrigated forage for goats. *Animal, Feed Science and Technology*, **40**, 21–30.

Glenn E.P., Watson M.C., O'Leary J.W. and Axelson R.D. (1992b) Comparison of salt tolerance and osmotic adjustment of low-sodium and high-sodium subspecies of C_4 halophyte, *Atriplex canescens. Plant, Cell and Environment*, **15**, 711–718.

Gorham J., Bristol A., Young E.M., Wyn-Jones R.G. and Kashour G. (1990) Salt tolerance in the triticeae: K/Na discrimination in barley. *Journal of Experimental Botany*, **41**, 1095–1101.

Gorham J., Hughes L.L. and Wyn-Jones R.G. (1980) Chemical composition of salt-marsh plants from Ynys Mon (Anglesey): the concept of physiotypes. *Plant, Cell and Environment*, **3**, 309–318.

Greenway H. and Munns R. (1980) Mechanisms of salt tolerance in non halophytes. *Annual Review of Plant Physiology*, **31**, 149–190.

Hassidim M., Braun Y., Lerner H.R. and Reinhold L. (1990) Na^+/H^+ and K^+/H^+ antiport in root membrane vesicles isolated from the halophyte *Atriplex* and the glycophyte cotton. *Plant Physiology*, **94**, 1795–1801.

Hurkman W.J. (1992) Effect of salt stress on plant gene expression: a review. *Plant and Soil*, **146**, 145–151.

Jeschke D. (1984) K^+–$N\alpha^+$ exchange at cellular membranes, intracellular compartmentaton of cations and salt tolerance. In *Salinity Tolerance in Plants: Strategies for Crop Improvement*, edited by C. Staples and G.H. Toenniessen, pp. 37–66. New York: John Wiley.

Jeschke D. and Wolf O. (1993) Importance of mineral nutrient cycling for salinity tolerance of plants. In *Towards the Rational Use of High Salinity Tolerant Plants*, edited by H. Lieth and A. Al Masoom, pp. 265–278. London: Kluwer Academic Publishers.

Kingsbury R.W. and Epstein E. (1986) Salt sensitivity in wheat. A case for specific ion toxicity. *Plant Physiology*, **80**, 651–654.

Kingsbury R.W., Epstein E. and Pearcy R.W. (1984) Physiological responses to salinity in selected lines of wheat. *Plant Physiology*, **74**, 417–423.

Kramer D. (1984) Cytological aspects of salt tolerance in higher plants. In *Salinity Tolerance in Plants: Strategies for Crop Improvement*, edited by R.C. Staples and G.H. Toenniessen, pp. 3–15. New York: John Wiley.

Kramer D., Läuchli A., Yeo A.R. and Gullasch J. (1977) Transfer cells in roots of *Phaseolus coccineus*: ultrastructure and possible function in exclusion of sodium from the shoot. *Annals of Botany*, **41**, 1031–1040.

Kuiper P.C. (1984) Functioning of plant cell membranes under saline conditions: membrane lipid composition and ATPases. In *Salinity Tolerance in Plants: Strategies for Crop Improvement*, edited by R.C. Staples and G.H. Toenniessen, pp. 77–91. New York: John Wiley.

Läuchli A. (1984) Salt exclusion: an adaptation of legumes for crops and pastures under saline conditions. In *Salinity Tolerance in Plants: Strategies for Crop Improvement*, edited by R.C. Staples and G.H. Toenniessen, pp. 171–187. New York: John Wiley.

Le Houerou H.N. (1993) Salt-tolerant plants for the arid regions of the Mediterranean isoclimatic zone. In *Towards the Rational Use of High Salinity Tolerant Plants*, edited by H. Lieth and A. Al Masoom, pp. 403–422. Dordrecht: Kluwer Academic Publishers.

Leopold A.C. and Willing R.P. (1984) Evidence for toxicity effects of salt on membranes. In *Salinity Tolerance in Plants: Strategies for Crop Improvement*, edited by R.C. Staples and G.H. Toenniessen, pp. 67–76. New York: John Wiley.

Le Rudulier D. (1993) Elucidation of the role of osmoprotective compounds and smoregulatory genes: the key role of bacteria. In *Towards the Rational Use of High Salinity Tolerant Plants*, edited by H. Lieth and A. Al Masoom, pp. 313–322. Dordrecht: Kluwer Academic Publishers.

Liphschitz N. and Waisel Y. (1982) Adaptation of plants to saline environments: salt excretion and glandular structure. In *Tasks for Vegetation Science*, edited by D.N. Sen and K.S. Rajpurohit, pp. 197–214. The Hague: Dr. W. Junk Publishers.

Maas E.V. (1987) Salt tolerance of plants. In *CRC Handbook of Plant Science in Agriculture*, edited by B.R. Christie, pp. 57–75. Boca Raton, Florida: CRC Press.

Maathuis F.J.M., Flowers T.J. and Yeo A.R. (1992) Sodium chloride compartmentation in leaf vacuoles of the halophyte *Suaeda maritima* (L.) Dum. and its relation to tonoplast permeability. *Journal of Experimental Botany*, **43**, 1219–1223.

Mandala S. and Taiz L. (1985) Partial purification of a tonoplast ATPase from corn coleoptiles. *Plant Physiology*, **78**, 327–333.

Martinez-Cortina C. and Sanz A. (1994) Effect of hormones on sucrose uptake and on ATPase activity of *Citrus sinensis* L. Osbeck leaves. *Annals of Botany*, **73**, 331–335.

Matoh T., Ishikawa T. and Takahashi E. (1989) Collapse of ATP-induced pH gradient by sodium ions in microsomal membrane vesicles prepared from *Atriplex gmelini* leaves. *Plant Physiology*, **89**, 180–183.

Matsushita N. and Matoh T. (1991) Characterization of Na^+ exclusion mechanisms of salt-tolerant reed plants in comparison with salt-sensitive rice plants. *Physiologia Plantarum*, **83**, 170–176.

Matsushita N. and Matoh T. (1992) Function of the shoot base of salt-tolerant reed (*Phragmites communis* Trinius) plants from Na^+ exclusion from the shoots. *Soil Science and Plant Nutrition*, **38**, 565–571.

Miyamoto S., Glenn E.P. and Olsen M.W. (1995) Growth, water use and salt uptake of four halophytes irrigated with highly saline water. *Journal of Arid Environments*.

Miyamoto S., Glenn E.P. and Singh N.T. (1994) Utilization of halophytic plants for fodder production with brackish water in subtropic deserts. In *Halophytes as a Resource for Livestock and for Rehabilitation of Degraded Lands*, edited by V.R. Squires and A.T. Ayoub, pp. 43–75. The Netherlands: Kluwer Academic Publishers.

Munns R. (1993) Physiological processes limiting plant growth in saline soils: some dogmas and hypotheses. *Plant, Cell and Environment*, **16**, 15–24.

Munns R., Greenway H., Deland R. and Gibbs J. (1982) Ion concentration and carbohydrate status of the elongating leaf tissue of *Hordeum vulgare* growing at high external NaCl. II. Cause of the growth reduction. *Journal of Experimental Botany*, **33**, 574–583.

Munns R. and Termaat A. (1986) Whole plant response to salinity. *Australian Journal of Plant Physiology*, **13**, 143–160.

Nevo E., Gorham J. and Beiles A. (1992) Variation for ^{22}Na uptake in wild emmer wheat, *Triticum dicoccoides* in Israel: salt tolerance resources for wheat improvement. *Journal of Experimental Botany*, **43**, 511–518.

Niu X., Narasimhan M., Salzman R., Bressan R. and Hawegawa P. (1993a) NaCl regulation of plasma membrane H^+-ATPase gene expression in a glycophyte and a halophyte. *Plant Physiology*, **103**, 713–718.

Niu X., Zhu K., Narasimhan M., Bressan R. and Hawegawa P. (1993b) Plasma-membrane H^+-ATPase gene expression is regulated by NaCl in cells of the halophyte *Atriplex nummularia* L. *Planta*, **190**, 433–438.

O'Leary J.W. (1979) Yield potential of halophytes and xerophytes. In *Arid Land Plant Resources*, edited by J.R. Goodin and D.K. Northington, pp. 574–581. Lubbock, Texas: Texas Tech University.

O'Leary J. W. and Glenn E.P. (1994) Global distribution and potential for halophytes. In *Halophytes as a Resource for Livestock and for Rehabilitation of Degraded Land, Tasks in Vegetation Science Series*, edited by V.R. Squires and A. Ayoub, pp. 7–18. The Netherlands: Kluwer Academic Publishers.

Perera L.K.R., Mansfield T.A. and Malloch A.J.C. (1994) Stomatal responses to sodium ions in *Aster tripolium*: a new hypothesis to explain salinity regulation in above-ground tissues. *Plant, Cell and Environment*, **17**, 335–340.

Perez-Prat E., Narasimhan M.L., Niu X., Botella M.A., Bressan R., Valpuesta V. *et al.* (1994) Growth cycle stage-dependent NaCl induction of plasma membrane H^+-ATPase mRNA accumulation in de-adapted tobacco cells. *Plant, Cell and Environment*, **17**, 327–333.

Pitman M.G. (1981) Ion uptake. In *The Physiology and Biochemistry of Drought Resistance in Plants*, edited by L.G. Paleg and D. Aspinall, pp. 71–96. Sydney: Academic Press.

Pitman M.G. (1984) Transport across the root and shoot/root interactions. In S*alinity Tolerance in Plants: Strategies for Crop Improvement*, edited by R.C. Staples and G.H. Toennissen, pp. 93–123. New York: John Wiley.

Pitman M.G., Laüchli A. and Stelzer R. (1981) Ion distribution in roots of barley seedlings measured by electron probe x-ray microanalysis. *Plant Physiology*, **68**, 673–679.

Poljakoff-Mayber A. (1975) *The Germination of Seeds*. New York: Pergamon press.

Popp M., Polania J. and Weiper M. (1993) Physiological adaptations to different salinity levels in mangrove. In *Towards the Rational Use of High Salinity Tolerant Plants*, edited by H.Lieth and A.A. AI Masoom, pp. 217–224. Dordrecht: Kluwer Academic Publishers.

Rains D.W. (1969) Sodium and potassium absorption by stem tissue of beans and cotton. *Plant Physiology*, **44**, 547–554.

Rawson H.M., Long M.J. and Munns R. (1988) Growth and development in NaCl treated plants. I: Leaf Na^+, and Cl^- concentration do not determine gas exchange of leaf blades in barley. *Australian Journal of Plant Physiology*, **15**, 519–527.

Rea P.A., Kim Y., Sarafian V., Poole R.J., Davies J.M. and Sanders D. (1992) Vacuolar H^+-translocating pyrophosphatases: a new category of ion translocase. *Trends in Biochemical Science*, **17**, 348–353.

Rengel Z. (1992) The role of calcium in salt toxicity. *Plant, Cell and Environment*, **15**, 625–632.

Reuveni M., Bennett A.B., Bressen R.A. and Hasegawa P.M. (1990) Enhanced H^+ transport capacity and ATP hyrolysis activity of the tonoplast H^+-ATPase after NaCl adaptation. *Plant Physiology*, **94**, 524–530.

Richards R.A., Dennett C.W., Qualset C.O., Epstein E., Norlyn J.D. and Winslow M.D. (1987) Variation in yield of grain and biomass in wheat, barley and triticale in a salt affected soil. *Field Crops Research*, **15**, 227–287.

Rozema J. (1991) Growth, water and ion relationships of halophytic monocotyledonae and dicotyledonae: a unified concept. *Aquatic Botany*, **39**, 17–33.

Rozeman J. and Van Diggelen J. (1991) A comparative study of growth and photosynthesis of four halophytes in response to salinity. *Acta Oecologica*, **12**, 673–681.

Rubio F., Gassmann W. and Schroeder J.I. (1995) Sodium-driven potassium uptake by the plant potassium transporter HKT1 and mutations conferring salt tolerance. *Science*, **270**, 1660–1663.

Sacher R.F. and Staples R.C. (1984) Chemical microscopy for study of plants in saline environments. In *Salinity Tolerance in Plants: Strategies for Crop Improvement*, edited by R.C. Staples and G. H. Toenniessen, pp. 17–36. New York: John Wiley.

Schachtman D.P., Lagudah E.S. and Munns R. (1992) The expression of salt tolerance from *Triticum tauschii* in hexaploid wheat. *Theoretical and Applied Genetics*, **84**, 714–719.

Schachtman D.P. and Munns R. (1992) Sodium accumulation in leaves of *Triticum* species that differ in salt tolerance. *Australian Journal of Plant Physiology*, **19**, 331–340.

Schachtman D.P., Tyerman S.D. and Terry B.R. (1991) The K^+/Na^+ selectivity of a cation channel in the plasma membrane of root cells does not differ in salt tolerance and salt sensitive wheat species. *Plant Physiology,* **97**, 598–605.

Schimper A.F.W. (1898) *Planzengeographie auf physiologischer Grundlage*. Gustav Fischer Verlag Jena.

Schirmer U. and Breckle S.W. (1982) The role of bladders for salt removal in some Chenopodiaceae (mainly *Atriplex* species). In *Tasks for Vegetation Science*, edited by D.N. Sen and K.S. Rajpurohit, pp. 215–232. The Hague: Dr. W. Junk Publishers.

Squires V.R. and Ayoub A.T. (1994) *Halophytes as a Resource for Livestock and for Rehabilitation of Degraded Lands*. Dordrecht: Kluwer Academic Publishers.

Storey R., Ahmad N. and Wyn-Jones R.G. (1977) Taxonomic and ecological aspects of the distribution of glycinebetaine and related compounds in plants. *Oecologia*, **27**, 319–332.

Strogonov B.P. (1964) *Physiological Basis of Salt Tolerance of Plants*. Jerusalem: I.P.S.T.

Warwick N.W. and Halloran G.M. (1992) Accumulation and excretion of sodium, potassium and chloride from leaves of two accessions of *Diplachne fusca* (L.) Beauv. *New Phytologist*, **121**, 53–61.

Watson M.C., Banuelos G.S., O'Leary J.W. and Riley J.J. (1994) Trace element composition of *Atriplex* grown with saline drainage water. *Agriculture, Ecosystems and Environment*, **48**, 157–162.

Watson M.C. and O'Leary J.W. (1993) Performance of *Atriplex* species in the San Joaquin Valley, California, under irrigation and with mechanical harvest. *Agriculture, Ecosystems and Environment*, **43**, 255–266.

Watson M.C., O'Leary J.W. and Glenn E.P. (1987) Evaluation of *Atriplex lentiformis* (Torr.) s. Wats. and *Atriplex nummularia* Lindl. as irrigated forage crops. *Journal of Arid Environments*, **13**, 292–303.

Winicov I. (1994) Gene expression in relation to salt tolerance. In *Stress-induced Gene Expression in Plants*, edited by A.S. Basra, pp. 61–85. Chur: Harwood Academic Publishers.

Winter E.(1982) Salt tolerance of *Trifolium alexandrium*. III. Effects of salts o ultrastructure of phloem and xylem transfer cells in petioles and leaves. *Australian Journal of Plant Physiology*, **9**, 239–250.

Winter E. and Preston J. (1982) Salt tolerance of *Trifolium alexandrinum*. IV. Ion measurement by X-ray microanalysis in unified, frozen hydrated leaf cells at various stages of salt treatment. *Australian Journal of Plant Physiology*, **9**, 251.

Wolf O., Munns R., Tonnet M.L. and Jeschke W.D. (1990) Concentrations and transport of solutes in xylem and phloem along the leaf axis of NaCl-treated *Hordeum vulgare*. *Journal of Experimental Botany*, **41**, 1133–1141.

Wyn-Jones R.G. (1981) Salt tolerance. In *Physiological Processes Limiting Plant Productivity*, edited by C.B. Johnson, pp. 271–292. London: Butterworths.

Wyn-Jones R.G., Gorham J. and McDonnell E. (1984) Organic and inorganic solute content as selective criteria for salt tolerance in the Triticeae. In *Salinity Tolerance in Plants: Strategies for Crop Improvement*, edited by R.C. Staples and G.H. Toenniessen, pp. 189–212. New York: John Wiley.

Wyn-Jones R.G., Storey R., Leigh R.A., Ahmad N. and Pollard A. (1977) A hypothesis on cytoplasmic osmoregulation. In *Regulation of Cell Membrane Activities in Plants*, edited by E. Marré and O. Ciferri, p. 112. Amsterdam: Elsevier.

Yeo A.R. and Flowers T.J. (1984) Mechanisms of salinity resistance in rice and their role as physiological criteria in plant breeding. In *Salinity Tolerance in Plants: Strategies for Crop Improvement*, edited by R.C. Staples and G.H. Toenniessen, pp. 151–170. New York: John Wiley.

Yeo A.R. and Flowers T.J. (1986a) Ion transport in *Suaeda maritima*: its relation to growth and implications for the pathway of radial transport of ions across the root. *Journal of Experimental Botany*, **37**, 143–159.

Yeo A.R. and Flowers T.J. (1986b) Salinity resistance in rice (*Oryza sativa* L.) and a pyramiding approach to breeding varieties for saline soils. *Australian Journal of Plant Physiology*, **13**, 161–173.

Yeo A.R, Kramer D., Läuchli A. and Gullasch J. (1977) Ion Distribution in salt stressed mature *Zea mays* roots in relation to ultrastructure and retention of sodium. *Journal of Experimental Botany*, **28**, 17–29.

5. MECHANISMS OF CHILLING RESISTANCE IN PLANTS

JOHN M. WILSON

School of Biological Sciences, University College of North Wales
Bangor, Gwynedd LL57 2UW, U.K.

INTRODUCTION

Many species, especially those of tropical and sub-tropical origin, suffer injury when exposed to temperatures above the freezing point of tisue but below approximately 15°C; this is called chilling injury to distinguish it from freezing injury. Many economically important crops, such as cotton, soybean, maize, rice and many tropical and sub-tropical fruits are classified as chilling-sensitive (Lyons, 1973). Chilling-induced damage at susceptible phases in the growth and development of plants is a serious cause of reduced crop yields world-wide (Wang, 1990; Bedi and Basra, 1993).

Some physiological processes such as flowering in rice are extremely sensitive to low temperatures and damage may occur at temperatures up to 20°C. Common visible symptoms of chilling injury to the leaves include wilting, bleaching due to photo-oxidation of pigments, waterlogging of the intercellular spaces, browning and eventually leaf necrosis and plant death. Chilling injury to fruits is characterized by the development of necrotic pits on the surface of the fruit, softening of the tissue and rapid browning of the skin (e.g. bananas). Why do these tropical species die at temperatures which are frequently encountered by plants growing in temperate climates such as Britain? In this respect, it is worth noting that the number of plants that have evolved to grown in extremes of cold are relatively few. It has been calculated that there are approximately 360,000 species of angiosperms on earth and in Britain only 2,000 species while further towards the poles even fewer species exist. We know that the earth was once much warmer than at present and therefore we must assume that those plants which could evolve tolerance to chilling temperatures have already done so over the millennia and those that could not have been restricted in their distribution to the tropical regions of the world. Clearly, the acquisition of thermal adaptation to low temperatures by a species is not easily achieved, probably because chilling affects so many different aspects of cellular metabolism.

The complex ways in which lowered temperatures affect plant metabolism, growth and survival has made the elucidation of the mechanisms of injury extremely difficult. At the present time, there are few satisfactory answers to questions such as "what is the primary temperature sensor in plants? What determines the critical temperature at which cells will survive? And how do some species harden or acclimatize to withstand low temperatures?" Because our understanding of the primary mechanisms of chilling injury is so poor, it is difficult to be precise about mechanisms of chilling resistance. In spite of these problems, several theories have been put forward to explain the mechanisms of chilling injury and these will now be discussed in relation to how chilling injury can be prevented.

PREVENTION OF MEMBRANE LIPID PHASE TRANSITIONS

Without doubt the most actively researched hypothesis of chilling injury in the last two decades has been the Lyons-Raison membrane lipid phase change hypothesis (Lyons and Raison, 1970; Lyons, 1973). This hypothesis received considerable impetus in the last 20 years from the development of new techniques for membrane lipid and fatty acid analyses and from techniques such as electron spin resonance for measuring membrane lipid fluidity at different temperatures (Raison *et al.*, 1979). Briefly, the hypothesis stated that the primary temperature sensor in the cell was the degree of fluidity of the membrane lipids and that the primary event on exposure of a chill-sensitive cell to a chilling temperature was a phase transition or change in the degree of molecular ordering of the bulk of the membrane lipids. This phase change was instantly reversible on return of the plant to the warmth as long as the time of chilling was not sufficient to cause injury to the membranes. The temperature at which this phase change occurred was considered to be determined by the degree of unsaturation of the membrane fatty acids. Chill-resistant plants generally had higher degree of unsaturation of their membrane fatty acids than chill-sensitive plants and therefore did not experience phase changes until well below 0°C and injury was therefore prevented.

The thermotropic membrane lipid phase transition hypothesis found early support in the presence of break points and discontinuities in the Arrhenius plots of parameters such as succinic dehydrogenase activity and spin label motion in chilling-sensitive but not in chilling-resistant plants at chilling temperatures. These break points in enzyme activity and associated increases in the activation energy (Ea) of membrane bound enzymes occurred at approximately the same temperatures at which physical changes in the lipids were detected by electron spin resonance and this correlation provided much support for the membrane lipid hypothesis. However, it was not long before other investigators (Miller *et al.*, 1974; Marx and Brinckman, 1979) were able to show similar discontinuities in both these parameters in chill-resistant species such as wheat and broad bean at chilling temperatures.

Many researchers (Martin, 1986) now believe that breaks and discontinuities in Arrhenius plots of physiological and physical properties of membranes are not caused solely by lipid phase transitions in the membrane and there are now examples of Arrhenius breaks in enzyme activity in chill-sensitive plants that are totally unrelated to the lipid state at the chilling temperature (see next section of this chapter). We now know that bulk lipid transitions that were first put forward by Lyons and Raison do not occur in higher plants at chilling temperatures.

Only minor modifications may occur in small domains of lipid (perhaps less than 6% of the total lipid) and appear to be controlled by the amount and fatty acid composition of the lipid phosphatidylglycerol (PG) which is found primarily in the thylakoid membranes of leaf cells (Raison and Orr, 1990). PG is the only lipid in plants with a sufficiently low degree of unsaturation of the fatty acids to give a phase transition in the temperature range 0 to 30°C. High-melting point molecular species of PG that contain only 16:0, 18:0 and 16:1 trans fatty acids undergo the transition from liquid crystalline phase to gel phase at temperatures well above 20°C. All the other cellular lipids have much higher degrees of unsaturation resulting in phase transitions well below 0°C.

Murata (1983) and Roughan (1983) have analyzed the fatty acid composition of PG of many different species and shown that a fairly good correlation exists between the degree of unsaturation of the PG and the known chill-sensitivity of many species. More recently, the molecular species distribution of PG in tobacco and *Arabidopsis* plants has been altered by molecular genetic techniques (Murata *et al.*, 1992: Wolter *et al.*, 1992), supporting their central role in determining chilling sensitivity. However, in some species, this correlation was not found and it is therefore, not possible to categorize a plant as chill-sensitive or chill-resistant solely on the basis of the fatty acid composition of its PG. In the *fatty acid biosynthesis 1 (fab 1)* mutant of *Arabidopsis*, leaf PG contained 43% high-melting-point molecular species—a higher percentage that is found in many chilling-sensitive plants (Wu and Browse, 1995). Nevertheless, the mutant was completely unaffected (when compared with wild-type controls) by a range of low-temperature treatments that quickly led to the death of cucumber and other chilling-sensitive plants. These results clearly demonstrate that high melting-point PGs do not mediate classic chilling damage. However, growth of *fab1* plants was compromised by long-term (>2 weeks) exposure to 2°C. Furthermore, in some species it is thought that other as yet unknown constituents of the membrane may play a similar role to PG.

Raison and Wright (1983) showed that the introduction of disaturated PG into lipids from chill-resistant wheat plants could result in the detection of lipid phase transitions where none existed before. In addition, Wilson and Crawford (1974) have shown that chill hardening of *Phaseolus vulgaris* L. plants at 12°C for 4 days increased the degree of unsaturation of PG and other phospholipids. However, it was shown that these lipid changes were not responsible for preventing the primary cause of chilling injury to the species, namely leaf dehydration which occurs during the early stages of chilling injury (see section on prevention of water loss in this chapter).

In conclusion, although the Lyons-Raison hypothesis has generated much interest in chilling injury research the hypothesis has many difficulties and suggests that chilling resistance is simply a property of the degree of unsaturation of PG. It would be astonishing if such a simple mechanism could explain all the different types of chilling injury to leaves, fruit and roots, especially considering the great diversity of plants that exist in the tropics. Therefore, the prevention of lipid phase transitions can only be considered to be part of the mechanism of chilling resistance. This suggestion is supported by the results of plant breeding experiments indicating that improvements in chilling tolerance are typically multigenic and may be specific to particular stages in the plant life cycle (see Wu and Browse, 1995). A second is that any particular chilling-associated trait may not be found in all chilling-sensitive species because each trait may have arisen more than once during evolution.

PREVENTION OF PROTEIN DENATURATION AND DISSOCIATION

It is known that low temperatures weaken the hydrophobic bonds that are important in determining the tertiary structure of proteins (Brandts, 1969) and this could cause an alteration in the kinetics and/or substrate specificity of a key regulatory enzyme which may lead to chilling injury. These enzyme changes could be independent of any lipid phase change and could therefore be considered as a primary event in the development of chilling injury.

Pyruvate orthophosphate dikinase (PPDK) from chill-sensitive maize (*Zea mays* L.) was the first plant enzyme investigated which showed an abrupt discontinuity in the Arrhenius plot and increase in activation energy at approximately 12°C (Sugiyama and Boka, 1976). In addition, phosphoenol-pyruvate carboxylase (PEPC) activity from tomato plants (chill-sensitive) showed discontinuities in Arrhenius plots at 12°C which were independent of any lipid phase change (Graham *et al.*, 1979). It is reported that this change in enzyme activity and loss of substrate affinity (KM value) at chilling temperatures in tomatoes and other species is caused by the weakening of the hydrophobic bonds which stabilize multimeric aggregates such as PEPC leading to its dissociation into four sub-units at chilling temperatures and an inability to reform into the active enzyme on return to the warmth. However, PPDK, a key enzyme of C_4 plants, splits reversibly in the cold from a tetramer to dimers (Sugiyama *et al.*, 1979). In contrast, when the key photosynthetic enzyme ribulose 1,5-bisphosphate carboxylase (RUBISCO) of C_3 plants is inactivated by cold it does not split into sub-units and the changes in the conformation of the protein are reversible. During the cold hardening of rye (*Secale cereale*), the cold sensitivity of RUBISCO is reduced and this may be due to the fact that the hardened protein has fewer SH groups on its surface or the stabilization the protein by S-S bonding (Levitt, 1980). Therefore, chilling resistance could be associated with the development of enzymes that are much less susceptible to protein denaturation and dissociation at low temperatures. So far, very few enzymes and plant species have been investigated due to problems of enzyme isolation and purification. Additional problems with the protein denaturation hypothesis are that it has been difficult to show variation in the cold lability of PPDK in maize cultivars which differed in their sensitivity to chilling and the PEPC activity of some chill-tolerant *Passiflora* L. species showed properties characteristic of chill-sensitive species (Raison and Orr, 1990). Nevertheless the modification of enzyme sensitivity to low temperature denaturation and dissociation cannot be overlooked in the development of chilling resistance in plants.

PREVENTION OF WATER LOSS

In several crop species (e.g. cotton and *Phaseolus vulgaris*), severe wilting of the leaves occurs within the first few hours of exposure to chilling temperature (Wilson, 1976). This quickly leads to the development of necrotic patches over the leaf surface and the drying of the leaf edges. The Lyons-Raison hypothesis of chilling injury speculated that this rapid water loss and ion leakage may be the result of phase transitions in the membrane lipids of the plasmalemma causing increased permeability of the membranes at chilling temperatures. However, Murata and Tatsumi (1979) have shown that chilling can cause an increase in the permeability of chill-sensitive and chill-resistant tissues to ions at chilling temperatures. Furthermore, in the leaves of *Phaseolus vulgaris*, chilling injury at 5°C can be prevented for up to 9 days by putting the plant inside a polythene bag (thus maintaining a saturated, 100% R.H. atmosphere) before exposure to the chilling temperature (Wilson, 1976). Chilling injury of this type can also be prevented by drought hardening by the withholding of water from the roots so that the leaves undergo 1 or 2 wilt cycles before chilling. Chill hardening at 12°C for 4 days also prevents water loss on transfer to 5°C as the leaves wilt temporarily on first exposure to 12°C and after a few hours at this temperature regain turgor. Spraying the leaves with abscisic acid or an antitranspirant such as Protec 1000 before chilling can also prevent this type of chilling injury.

The most important mechanism of chilling injury to *Phaseolus vulgaris* has been shown to be the abnormal behavior of the stomata at chilling temperatures (Wilson, 1976). If the stomata are open at the start of chilling they remain open ("locked open") in spite of the fact that the leaf is severely wilted and that the water lost from the leaf cannot be replaced from the roots due to the very low permeability of the roots of tropical species to water at chilling temperatures. Furthermore, Eamus and Wilson (1983) showed that endogenous abscisic acid can have the reverse effect in chilled tropical plants and promote stomatal opening thus facilitating increased water loss.

Chill hardening at 12°C prevents chilling injury in *Phaseolus vulgaris* and other species because at this intermediate temperature only mild water stress is induced due to the intermediate locking open of the stomata and lowered permeability of the roots to water at 12°C. After a few hours at 12°C, the leaves can regain turgor due to the closing of the stomata and increased root permeability to water and the stomata remain closed on subsequent chilling at 5°C. Similarly, drought hardening and spraying with abscisic acid can prevent chilling injury to the leaves for varying periods of time by inducing stomatal closure before transfer to the chilling temperature. Anti-transpirants and putting the plants in a polythene bag before chilling simply protect against this type of chilling injury by reducing evapotranspiration rates from the leaves to very low levels.

Raison (personal communication) suggested that the abnormal behavior of the stomata of chill-sensitive plants may be caused by phase changes in the membrane lipids of the guard cells which could alter membrane permeability or the receptors for abscisic acid. Whatever the explanation for this 'locking open' phenomenon, it is clear that the prevention of water loss during the early stages of chilling is very important in development of chilling tolerance in some species. In extremely chill-sensitive species, for example, the ornamental plant, *Episcia reptans* Mart., the speed of chilling injury to the leaves cannot be reduced by any of the hardening and prevention mechanisms outlined above (Wilson, 1976). Clearly, metabolic changes rather than water loss are of primary important in determining the speed of chilling-injury to *Episcia reptans* and to many tropical fruits. Exposure of seedlings of a chilling-sensitive variety of rice (*Oryza sativa* L. cv. Wasetoittu) to water stress (0.5M mannitol, 30 min) induced a degree of chilling resistance (Takahashi *et al.*, 1994). It was suggested that the genes induced by short- term water stress, but not those induced by ABA, were related to acquired chilling resistance in this chilling-sensitive variety.

PREVENTION OF PHOTO-OXIDATION OF PLANT PIGMENTS AND LIPIDS

In cotton, cucumber and other important crop species, bleaching of the upper leaves occurs at chilling temperature and is particularly rapid at high light intensities. Van Hasselt (1972) reported a 50% decrease in the chlorophyll content of cucumber leaves chilled for 24 hours at 1°C and a light intensity of 20,000 lux. The light energy absorbed by pigments at low temperatures cannot be used in photosynthetic reactions, as in most tropical species, little photosynthesis can occur below 15°C due to its temperature sensitivity and photoinhibition. This absorbed light energy is dissipated in the chilled leaf in a range of very damaging photo-oxidative reactions leading to the loss of pigments, lipids and fatty acids especially from the thylakoid membranes (Van Hasselt, 1974). If chilling is prolonged, this type of damage cannot be repaired on return to the warmth and the leaf dies.

Photo-oxidative damage almost invariably means that chilling in the light results in a faster rate of damage to leaves than chilling in the dark. Chill hardening may help prevent this type of injury by increasing the levels of antioxidants such as tocopherol in the leaf or by increasing the levels of other less important pigments such as carotenoids which could be preferentially oxidized at chilling temperatures and thus help to reduce losses of chlorophylls a+b (Van Hasselt, 1972). An increase in the levels of superoxide dismutase and catalse in chill-hardened plants could also help prevent photodynamic damage as the levels of superoxide and its oxidized product singlet molecular oxygen $^1O_2$ may increase in chilled tissue. Superoxide is a powerful oxidant and can peroxidize lipids and unsaturated fatty acids leading to rapid and irreversible membrane damage. Superoxide dismutase could protect plants by converting the superoxide ion to hydrogen peroxide which is then decomposed by catalase. It is known that in chilled plants the levels of these two enzymes decrease rapidly and this would accelerate injury (Taylor *et al.*,1974; Keninga *et al.*, 1979).

Chilling at 4°C of maize seedlings, grown in the dark at 27°C, caused a 4-fold increase in H_2O_2 and led to death (Prasad *et al.*,1994). Tolerance to low temperature was inducible by pre-exposure to 14°C. Interestingly, preadaptation could also be induced by treatment with H_2O_2, suggesting that a transient increase in H_2O_2 might act as a signal to activate genes associated with chilling tolerance.

Chlorophyll synthesis is one of the most temperature-sensitive processes and has been used as a quantitative technique for measuring the chill sensitivity of different species and varieties of the same species (McWilliam *et al.*, 1979). The poor growth of maize in the cold, wet springs of the United Kingdom is caused by the slow rate of chlorophyll synthesis and this can easily be seen by the slowness of the young plants to change from yellow to green. Chilling causes ultrastructural changes to the chloroplasts such as disorganization of the thylakoid membranes (Ilker *et al.*, 1979). The primary site of chilling stress appears to be photosystem II and this leads to a loss of the variable fluorescence of photosystem II and chill inactivation of photosynthesis. This inhibition of photosystem II has made it possible for Greaves and Wilson (1987) to use fluorescence kinetics to classify different species and varieties according to their chill sensitivity and to assess the effectiveness of different hardening treatments in the prevention of chilling injury.

PREVENTION OF A RISE IN CYTOPLASMIC CALCIUM

Minorsky (1985) proposed that increases of up to two fold in cytoplasmic calcium in chill-sensitive plants could be very important in the rapid development of chilling injury as calcium plays an important role in many cellular processes. Sudden increases in cytoplasmic calcium at chilling temperatures could be the result of phase transitions in the membrane lipids of the mitochondria and other organelles which would lead to changes in active transport and leakage of calcium through the membranes. In support of the theory that changes in cytoplasmic calcium may be a primary physiological transducer of chilling injury, the changes in the cytoplasmic strands and actin filaments of the cytoskeleton observed during chilling (see next section of this chapter) are similar to those observed when the concentration of calcium in the cytoplasm was increased (Woods *et al.*, 1984). This type of injury could be prevented in cells during hardening by either lowering the levels of calcium in the cells or by mechanisms which would prevent membrane lipid phase changes which lead to increased membrane permeability.

PREVENTION OF CHANGES TO THE CYTOSKELETON AND PROTOPLASMIC STREAMING

The disruption of protoplasmic streaming in chilled tropical plant cells was one of the earliest discoveries of chilling injury research and yet we still lack a satisfactory explanation for this. Not only does the rate of protoplasmic streaming decrease more rapidly in chilled tropical plants than in temperate species but also the cytoplasmic strands which span the vacuole often disappear within minutes of the start of chilling. This often coincides with the appearance of many small vesicles in the cytoplasm. If the cells are re-warmed, the cytoplasmic strands can reform and protoplasmic streaming resume as long as the period of chilling is not too long. Due to the high speed at which these events take place, they could qualify as primary events and may be caused by changes in the actin filaments and microtubules at low temperatures. The disappearance of microtubules during the chilling of tomatoes has been observed (Ilker *et al.*,1979), and is thought to be due to the dissociation of the microtubules into tubulin sub-units. It is not known whether cold-resistant plants have microtubules that are more stable at low temperatures but the destabilization of microtubules using colchicine increased the severity of chilling injury in plants (Rikin *et al.*, 1980).

Decreased rates of protoplasmic streaming in chilled tropical plants could be due to a decline in the ATP supply for this process. However, measurements of ATP levels in the chilled tissues of some tropical species (Wilson, 1978) showed increases in ATP probably due to very low anabolic rates during chilling. However, on prolonged chilling ATP levels decreased as cellular metabolism became increasingly disorganized, compartmentation was lost and leaves became dehydrated (Wilson, 1978).

PREVENTION OF RESPIRATORY UPSET AND TOXIN ACCUMULATION

The Lyons-Raison hypothesis of chilling injury proposed that a greater effect of chilling on the lipid associated enzymes of mitochondria than on the cytoplasmic enzymes of glycolysis would lead to a metabolic imbalance between glycolysis and the tricarboxylic acid cycle and that this would lead to the accumulation of ethanol, acetaldehyde and alanine as the end products of glycolysis. This proposal assumes that the plant has no homeostatic mechanisms to compensate for the differences in rates of the two processes during chilling. The increases in metabolites such as ethanol would slowly lead to cell death as a result of damage to membrane systems. However, analyses of chilled tropical tissues for ethanol and acetaldehyde have provided little support for this hypothesis. Increases in ethanol and acetaldehyde have been reported in chilled bananas (Murata, 1969) and in lemons (Eaks, 1980) but these increases only occurred after rewarming and may therefore have been the result of tissue damage rather than the cause. Alanine has been reported to increase in the leaves of a chill-sensitive *Passiflora edulis*

L. genotype but not in a chill-resistant genotype. However, ethanol did not accumu late in the chill-sensitive genotype at low temperature (Patterson *et al.*, 1981).

During the initial stages of chilling, respiration often increases and this is then followed by a decline. There may be a further increase during the onset of severe injury. In the extremely chill-sensitive plant *Piscia reptans*, the initial respiratory increase in oxygen uptake is very large over the first few hours of chilling but is associated with a much smaller increase in carbon dioxide evolution. This suggests that massive oxidative reactions are occurring within the cell such as the peroxidation of membrane lipid components and oxidation of phenols in the vacuole. On rewarming, many species show the phenomenon called respiratory overshoot' and have respiration rates much higher than normal for the rewarming temperature. The magnitude of the respiratory rise on rewarming has been used to quantify the degree of chilling injury to some species (Amin, 1969). The cause of respiratory overshoot is not known. It is speculated that it may represent the "burning off "of metabolites such as sugars which may have accumulated during chilling. Alternatively, it may represent the extent to which repair mechanisms are taking place in the cell.

The amount of phenolics may increase in chilled tissues as indicated by the rise in phenylalanine ammonia lyase which catalyzes the first step in the synthesis of chlorogenic acid from phenylalanine (Tanaka and Uritani, 1977). A four to five-fold increase occurs in the chlorogenic acid content of chilled tobacco plants (Levitt, 1980). The prevention of chilling injury by intermittent rewarming and recooling in some species (e.g. plums) may be due to a reduction in toxic levels during rewarming.

PREVENTION OF VARIOUS SECONDARY CHANGES DURING CHILLING

Primary events such as lipid phase changes or protein denaturaiton can lead to a multitude of so called secondary events which may lead directly or indirectly to injury. For example, in some species, chilling causes an increase in the amount of ethylene (especially on rewarming) which may in turn increase the respiratory rate of the tissue (Wang and Adams, 1980) and lead to the accumulation of toxins. If chilling interferes with ATP production or its utilization, then the ion pumps that utilize ATP may not be able to maintain ionic gradients and cell compartmentation would eventually be lost and the tissues leak electrolytes. Phloem transport is severely reduced in many species at chilling temperatures (Giaquinta and Geiger, 1973) and it is speculated that if chilling is prolonged that this could lead to the death of non-photosynthetic parts of the plant (e.g. roots) by starvation. The development of chilling injury to the roots has received little attention in comparison to leaves and fruits as the visual changes are not as dramatic in roots. Usually chilled roots turn brown, show stellar lesions and become flaccid.

However, chilling injury to the roots may be more important to plant survival than chilling injury to the leaves as they live in a 'buffered' environment (the soil) and are not subject to the larger temperature fluctuations experienced by the shoot.

CONCLUSION

A wide variety of changes occur in the behavior of the membranes, lipids, enzymes, pigments and cytoskeleton of chilled tropical plants. Some of these changes can be regarded as primary events which lead to metabolic imbalances, water loss, ion leakage, loss of cellular compartmentation and eventually cell death. The mechanisms by which chill-resistant plants prevent these changes are poorly understood and much more research is required before we can understand how membranes and proteins can be modified to stabilize them at low temperature. The great genetic diversity of tropical plant species and the different symptoms of chilling injury to the various plant organs argues against finding a simple primary event which occurs in all cells and which can explain every type of injury. Chilling resistance is multigenic and therefore plant breeders face a daunting task in trying to transfer chilling resistance to chill-sensitive plants.

REFERENCES

Amin J.V. (1969) Some aspects of respiration and respiration inhibitors in low temperature effects on the cotton plant. *Physiologia Plantaraum*, **22**, 1184–1190.

Bedi S. and Basra A.S. (1993) Chilling injury in germinating seeds: basic mechanisms and agricultural implications. *Seed Science Research*, **3**, 219–229.

Brandts J.F.(1969) Confirmational transitions of proteins in water and aqueous mixtures. In *Structure and Stability of Biological Macromolecules*, edited by S.N. Tunasheff and G.D. Fasman, pp.213–290. New York: Marcel Dekker.

Eaks I.L. (1980) Effect of chilling on respiration and volatiles of California lemon fruit. *Journal of the American Society of Horticultural Science*, **105**, 865–869.

Eamus D and Wilson J.M. (1983) ABA levels and effects in chilled and hardened *Phaseolus vulgaris*. *Journal of Experimental Botany*, **34**, 1000–1006.

Giaquinta R.T. and Geiger D.R. (1973) Inhibition of translocation by chilling temperatures. *Plant Physiology*, **51**, 372– 375.

Graham D., Hockley D.G. and Patterson B.D.(1979) Temperature effects on phosphoenolpyruvate carboxylase from chilling sensitive and chilling resistant plants. In *Low Temperature Stress in Crop Plants: The Role of Membrane*, edited by J.M. Lyons, D. Graham and J.K. Raison, pp.453–461. New York: Academic Press.

Greaves J.A. and Wilson J.M. (1987) Chlorophyll fluorescence analysis—an aid to plant breeders. *Institute of Biology*, **34**, 209–214.

Ilker R., Breidenbach R.W. and Lyons J.M. (1979) Sequence of ultrastructural changes in tomato cotyledons during short periods of chilling. In *Low Temperature Stress in Crop Plants: The Role of Membrane*, edited by J.M. Lyons, D. Graham and J.K. Raison, pp.97–114. New York: Academic Press.

Kaniuga Z., Zabek J. and Michalski W.P. (1979) Photosynthetic apparatus in chilling sensitive plants. VI. Cold and dark induced changes in superoxide dismutase activity in relation to loosely bound manganese content. *Planta*, **145**, 145–150.

Levitt J. (1980) *Response of Plants to Environmental Stresses. Volume 1. Chilling, Freezing and High Temperature Stress*. New York: Academic Press.

Lyons J.M. (1973) Chilling injury in plants. *Annual Review of Plant Physiology*, **4**, 445–466.

Lyons J.M. and Raison J.K. (1990) Oxidative activity of mitochondria isolated from plant tissues sensitive and resistant to chilling injury. *Plant Physiology*, **45**, 386–389.

Martin B (1986) Arrhenius plots and the involvement of thermotropic phase transitions of the thylakoid membrane in chilling impairment of photosynthesis in thermophilic higher plants. *Plant, Cell and Environment*, **9**, 323–331.

Marx R. and Brinkmann K.(1979) Effect of temperature on pathways of NADH oxidation in broad bean mitochondria. *Planta*, **144**, 359–365.

McWilliam W., Manokaran W. and Kipnis T. (1979) Adaptation to chilling stress in *Sorghum*. In *Low Temperature Stress in Crop Plants: The Role of Membrane*, edited by J.M. Lyons, D. Graham and J.K. Raison, pp.491–505. New York: Academic Press.

Miller R.W., Roche I. de la and Pomeroy M.K.(1974) Structural and functional responses of wheat mitochondrial membranes to growth at low temperatures. *Plant Physiology*, **53**, 426–433.

Minorsky P.V. (1985) An heuristic hypothesis of chilling injury in plants: a role for calcium as the primary physiological transducer of injury. *Plant, Cell and Environment*, **8**, 75–94.

Murata N. (1983) Molecular species composition of phosphatidyl glycerols from chilling sensitive and chilling resistant plants. *Plant and Cell Physiology*, **24**, 81–86.

Murata N., Ishizaki-Nishizawa O., Higashi S., Hayashi H., Taska Y. and Mishida I. (1992) Genetically engineered alteration in the chilling sensitivity of plants. *Nature*, **356**, 313–326.

Murata T. (1969) Physiological and biochemical studies of chilling injury in bananas. *Physiologia Plantarum*, **22**, 401–411.

Murata T. and Tatsumi Y.(1979) Ion leakage in chilled plant issues. In *Low Temperature Stress in Crop Plants: The Role of Membrane*, edited by J.M. Lyons, D. Graham and J.K. Raison, pp.141–152. New York: Academic Press.

Patterson B.D., Pearson J.A., Payne L.A. and Ferguson I.B.(1981) Metabolic aspects of chilling resistance: Alanine accumulation and glutamate depletion in relation to chilling sensitivity in passion fruit species. *Australian Journal of Plant Physiology*, **8**, 395–403.

Prasad T.K., Anderson M.D., Martin B.A. and Stewart C.R. (1994) Evidence for chilling-induced oxidative stress in maize seedlings and a regulatory role for hydrogen peroxide. *The Plant Cell*, **6**, 65–74.

Raison J.K., Chapman E.A., Wright L.C. and Jacobs W.L. (1979) Membrane lipid transitions: their correlation with the climatic distribution of plants. In *Low Temperature Stress in Crop Plants: The Role of Membrane*, edited by J.M. Lyons, D. Graham and J.K. Raison, pp.177–180. New York: Academic Press.

Raison J.K. and Orr G.R. (1990) Proposals for a better under standing of the molecular basis of chilling injury. In *Chilling injury in Horticultural Crops*, edited by C.Y. Wang, pp.145–164. Boca Raton: CRC Press.

Raison J.K. and Wright L.C. (1983) Thermal phase transitions in the polar lipids of plant membranes: Their induction by disaturated phospholipids and their possible relation to chilling injury. *Biochemica et Biophysica Acta*, **731**, 69–78.

Rikin A., Atsmon D. and Gitler C. (1980) Chilling injury in cotton (*Gossypium hirsutum* L.): effects of antimicrotubular drugs. *Plant and Cell Physiology*, **21**, 829–837.

Roughan P.G. (1985) Phosphatidyl glycerol and chilling sensitivity in plants. *Plant Physiology*, **77**, 740–746.

Sugiyama T. and Boka K. (1976) Differing sensitivity of pyruvate orthophosphate dikinase to low temperatures in maize cultures. *Plant and Cell Physiology*, **17**, 851–854.

Sugiyama T., Schmitt M.R., Ku S.B. and Edwards G.E. (1979) Differences in cold lability of pyruvate orthophosphate dikinase among C_4 species. *Plant and Cell Physiology*, **20**, 965–971.

Takahashi R., Joshee N. and Kitagawa Y. (1994) Induction of chilling resistance by water stress, and cDNA sequence analysis and expression of water-stress regulated genes in rice. *Plant Molecular Biology*, **26**, 339–352.

Tanaka Y. and Uritani I. (1977) Purification and properties of phenylalanine ammonia lyase in cut-injured sweet potato. *Journal of Biochemistry*, **81**, 963–970.

Taylor A.O., Slack C.R. and McPherson H.G. (1976) Plants under climatic stress. VI. Chilling and light effects on photosynthetic enzymes of sorghum and maize. *Plant Physiology*, **54**, 696–701.

Van Hasselt P.R.(1972) Photo-oxidation of leaf pigments in *Cucumis* leaf discs during chilling. *Acta Botanica Neerlandica*, **21**, 539–548.

Van Hasselt P.R. (1974) Photo-oxidation of unsaturated fatty acids in *Cucumis* leaf discs during chilling. *Acta Botanica Neerlandica*, **23**, 156–166.

Wang C.Y.(1990) *Chilling Injury of Horticultural Crops*. Boca Raton: CRC Press.

Wang C.Y. and Adams D.O. (1980) Ethylene production by chilled cucumbers (*Cucumis sativus* L.). *Plant Physiology*, **66**, 841–843.

Wilson J.M. (1976) The mechanism of chill and drought hardening of *Phaseolus vulgaris* leaves. *New Phytologist*, **76**, 257–270.

Wilsons J.M.(1978) Leaf respiration and ATP levels at chilling temperatures. *New Phytologist*, **80**, 325–334.

Wilson J.M. and Crawford R.M.M. (1974) The acclimatization of plants to chilling temperatures in relation to the fatty acid composition of leaf polar lipids. *New Phytologist*, **73**, 806–820.

Wolter F.P., Schmidt R. and Heinz E. (1992) Chilling sensitivity of *Arabidopsis thaliana* with genetically engineered membrane lipids. *The EMBO Journal*, **11**, 4685–4692.

Woods C.M., Polito V.S. and Reid M.S. (1984) Response to chilling stress in plant cells. II. Redistribution of intracellular calcium. *Protoplasma*, **121**, 17–20.

Wu J. and Browse J. (1995) Elevated levels of high-melting-point phosphatidyl glycerols do not induce chilling sensitivity in an *Arabidopsis* mutant. *The Plant Cell*, **7**, 17–27.

6. MECHANISMS OF FREEZING RESISTANCE OF WOOD TISSUES: RECENT ADVANCEMENTS

ZORAN RISTIC[1] and EDWARD N. ASHWORTH[2]

[1]*Department of Biology, University of South Dakota, Vermillion, South Dakota 57069, U.S.A. and* [2]*Department of Horticulture, Purdue University, West Lafayette, Indiana 47907, U.S.A.*

INTRODUCTION

Woody plants of temperate and boreal regions periodically experience freezing temperatures. In these areas, freezing temperatures limit the geographical distribution of woody species (George *et al.*, 1974, 1982; Burke *et al.*, 1976; Becwar *et al.*, 1981), and, in the case of economically important plants, can cause significant losses to horticultural production. The primary problem that woody plants face when exposed to freezing temperatures is ice formation. Ice crystals can cause severe injuries to the living cells that can lead to the death of the plant. Woody plants, however, have developed the mechanisms by which they can survive freezing temperatures. The effects of freezing on woody plants and the mechanisms of freezing resistance have long interested botanists and agriculturalist, both from a theoretical and a practical point of view (Levitt, 1980).

Two groups of wood tissues are recognized on the basis of their freezing behavior. The first group avoids intracellular ice formation by maintaining their cellular water in a deep supercooled state or metastable equilibrium (Burke *et al.*, 1976; George *et al.*, 1982; Ashworth, 1993). This type of freezing behavior has been termed freezing avoidance or supercooling (Burke *et al.*, 1976; Levitt, 1980). The second group avoids intracellular ice formation by losing cellular water to extracellular ice. This type of freezing behavior is known as freezing tolerance or extracellular freezing (Burke *et al.*, 1976; Levitt, 1980).

Many aspects of freezing behavior of plant tissues including woods are discussed in more details elsewhere (Burke *et al.*, 1976; Burke, 1979; Levitt, 1980; George *et al.*, 1982; George, 1983; Steponkus, 1984; Ashworth, 1993). This chapter will only provide the general outlines of the mechanisms of freezing resistance, with an emphasis on recent advancements in the area of wood cold hardiness.

FREEZING AVOIDANCE

Freezing avoidance is a common way that many deciduous hardwoods survive freezing temperatures (Quamme *et al.*, 1972; George *et al.*, 1974; Quamme, 1976; George and Burke, 1977; George *et al.*, 1977; Rajashekar and Burke 1978;

Becwar *et al.*, 1981; Rajashekar *et al.*, 1982a,b; Ashworth *et al.*, 1983; Rajashekar and Reid, 1989). It is generally believed that during a freezing stress, ice crystals first form in vessels. Ice crystallization proceeds along the vessels from a few nucleation points and reaches all parts of the shoots at a relatively high velocity (Levitt, 1980). Once the ice forms in the vessels, it will spread throughout intercellular spaces and surround the living cells (xylem ray parenchyma cells). Ice crystals, however, do not spread into the living cells, because plasma membrane and cell wall provide an effective barrier to ice (Steponkus, 1984). The formation of ice crystals in vessels and extracellular spaces creates a water potential gradient between extracellular ice and intracellular water. Intracellular water remains liquid at freezing temperatures and supercools (Burke, 1979; Ashworth, 1993). As a result, living cells of supercooling wood tissues do not dehydrate and retain their original shape and volume during freezing (Ashworth *et al.*, 1983; Malone and Ashworth., 1991; Ristic and Ashworth, 1993, 1995).

The supercooling of cellular water, however, is limited. During exposure to subzero temperatures, cellular water can only supercool to the homogeneous nucleation temperature (–38° C) before freezing. It has been proposed that cell death within supercooling tissues is a result of intracellular ice formation (Burke *et al.*, 1976; George and Burke, 1977). Wood tissues exhibiting supercooling characteristics are, therefore, moderately hardy, and their survival is limited to temperatures at or above the ice nucleation point of cellular solution (George *et al.*, 1974; Burke *et al.*, 1976; Ashworth, 1993).

Differential thermal analysis (DTA) of tissues exhibiting supercooling has revealed two distinct exotherms during freezing (Quamme *et al.*, 1972, 1973; George *et al.*, 1974; Burke *et al.*, 1976; George and Burke, 1977; Hong and Sucoff, 1980; Hong *et al.*, 1980; Ashworth *et al.*, 1983). The first, or high-temperature exotherm (HTE), appears to correspond to the freezing of water in the xylem vessels and extracellular spaces. Under laboratory conditions, the HTE has commonly been observed at temperatures between –5° C to –15° C. This temperature range, however, is not representative of what probably occurs in the field where freezing is usually initiated at temperatures between –1°C and –4°C C (Ashworth and Davis, 1984, 1986; Ashworth *et al.*, 1985b). This discrepancy is apparently due to rapid cooling rates used in DTA experiments relative to field cooling rates, and the tendency for small samples such as those used in DTA experiments to supercool to much lower temperatures compared to intact plants (Ashworth, 1993).

The second exotherm, commonly known as low-temperature exotherm (LTE), corresponds to the freezing of a fraction of supercooled water. This exotherm is generally observed at or above –38° C and is correlated with injury to the living cells of the wood tissue (Quamme *et al.*, 1972, 1973; George and Burke, 1977; Hong *et al.*, 1980; Ashworth *et al.*, 1983). A linear relationship between the number of dead cells and the proportion of deep supercooled water has been observed in several wood species (Hong and Sucoff, 1980; Hong *et al.*, 1980; Ashworth *et al.*, 1983).

It has been hypothesized that individual xylem ray parenchyma cells behave as isolated water droplets (Ashworth, 1993). Water in each cell would be isolated from ice in adjacent tissues and independent of neighboring cells (Rasmussen and MacKenzie, 1973). This hypothesis was based upon analogies with water in oil emulsions in which independent water droplets would supercool to –38° C (Rasmussen and MacKenzie, 1973).

The hypothesis that freezing injury in supercooling wood tissues is due to the freezing of a fraction of supercooled water was based almost entirely upon the appearance of LTE (Quamme *et al.*, 1972, 1973; George and Burke, 1977; Hong *et al.*, 1980; Ashworth *et al.*, 1983; Ashworth, 1993), and upon analogies with water in oil emulsions (Rasmussen and MacKenzie, 1973). Unfortunately, although this hypothesis may be appropriate, no direct evidence was available to support it.

Recent studies, however, have provided microscopic evidence that support the hypothesis that intracellular ice formation is a source of freezing injury in supercooling wood tissues. Using freeze substitution and transmission electron microscopy (TEM), Ristic and Ashworth (1993, 1995) studied freezing behavior of supercooling wood tissues of flowering dogwood (*Cornus florida* L.) (Ristic and Ashworth, 1993, 1995), apple (*Malus domestica* Borkh.), and peach (*Prunus persica* [L.] Batsch) (Ristic and Ashworth, 1995). In their experiments, Ristic and Ashworth (1993, 1995) exposed wood tissues of these species to a series of laboratory freezing temperatures ranging from 0 to –60° C. They observed disruption of cell ultrastructure and evidence that intracellular ice formed (Figure 6.1). The temperature at which the ice crystals were first observed depended on the season in which the wood tissue was collected. In flowering dogwood, evidence of intracellular ice crystals were observed at –20, –10, and –5° C in tissues collected in winter, spring, and summer, respectively (Ristic and Ashworth, 1993). Ice crystals were also seen in ray parenchyma cells of wood tissue of flowering dogwood exposed to freezing stress (–17° C) in the field (Ristic and Ashworth, 1995).

The formation of ice crystals within the living cells of supercooling wood tissues seems to be not uniform. It has been observed that in some ray parenchyma cells of cold-acclimated wood tissues of flowering dogwood (Ristic and Ashworth, 1993), apple, and peach (Ristic and Ashworth, 1995), ice crystals were formed in vacuoles and not in the cytoplasm (Figure 6.1F). This suggests that internal cell membranes, such as tonoplasts, may affect the spread of ice. It has been hypothesized that compartmentation of ice crystals in xylem ray parenchyma cells could be the result of an unequal distribution of soluble sugars (Ristic and Ashworth, 1993, 1995). Koster and Lynch (1992) investigated the distribution of soluble sugars within the vacuole and cytoplasm of protoplasts from cold-acclimated *Secale cereale* L., and found that increased amounts of soluble sugars were present during cold acclimation in compartments outside the vacuole. Koster and Lynch (1992) proposed that the increased sugar content in the cytoplasm may contribute to cryoprotection. Likewise, it is possible that the vacuoles of xylem ray parenchyma cells of non-supercooling wood tissues may have lower sugar content than the cytoplasm, and this could influence the nucleation and crystallization of intracellular water (Ristic and Ashworth, 1993). This could partly explain

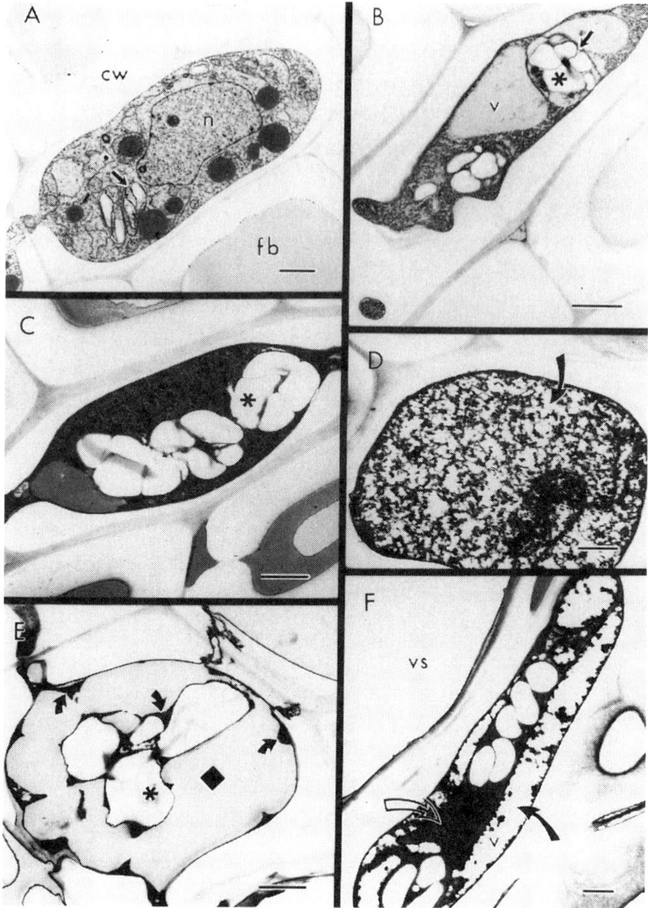

Figure 6.1. The effect of freezing stress on xylem ray parenchyma cells of wood tissue of apple (*Malus domestica* Borkh) collected in winter. Tissue was subjected to laboratory freezing stress at a cooling rate of 5°C h^{-1}, harvested at 0, –5, –15, –30, –45, and –60°C, and prepared for TEM using chemical fixation (A) and the freeze substitution technique (B-F). Two types of injury were evident: 1) intracellular ice (D, F), and 2) fragmented protoplasm (E). A, Xylem ray parenchyma cell in control tissue (tissue harvested at 0 °C); cw, cell wall; fb, fiber, n, nucleus, black arrow, plastid; bar = 2μm. B, Xylem ray parenchyma cell in control tissue; v, vacuole; black arrow, plastid; star, starch grain; bar = 2μm. C, Xylem ray parenchyma cell of tissue freeze stressed at –15°C; note no evidence of large ice crystals; star, starch grain; bar = 2μm. D, Xylem ray parenchyma cell of tissue freeze stressed at –15°C; curved black arrow, large ice crystals; bar = 1μm. E, Xylem ray parenchyma cell of tissue freeze stressed at 15°C; curved black arrows, fragmented protoplasm; black square, empty space between fragmented protoplasm and cell wall; bar = 2μm. F, Xylem ray parenchyma cell of tissue freeze stressed at –60°C; vs, vessel member; curved black arrow, large ice crystals; white curved arrow, cytoplasm without signs of ice crystals; bar = 1μm. (With permission from University of Chicago Press for Ristic & Ashworth, (1995), *International Journal of Plant Sciences*, **156**, 784–792.)

compartmentation of ice crystals in cold-acclimated xylem ray parenchyma cells of flowering dogwood (Ristic and Ashworth, 1993), apple, and peach (Ristic and Ashworth, 1995).

Recent studies have also provided evidence that indicate that, in addition to intracellular ice, another type of freezing injury is apparent in supercooling wood tissues (Ristic and Ashworth, 1993, 1995). When wood tissues of flowering dogwood (Ristic and Ashworth, 1993, 1995), apple, and peach (Ristic and Ashworth, 1995) were exposed to freezing stress, a type of injury manifested by fragmented protoplasm with indistinguishable plasma membrane, damaged cell ultrastructures, and no evidence of intracellular ice formation was observed in ray parenchyma cells (Figure 6.1E). The proportion of cells with fragmented protoplasm, however, differed between cold acclimated and non-acclimated wood tissues (Ristic and Ashworth, 1993). The greater proportion of cells with fragmented protoplasm appeared to be a characteristic of non-acclimated tissue (Ristic and Ashworth, 1993). Ristic and Ashworth (1993) hypothesized that this may be due to seasonal changes in cold hardiness of wood tissue.

It is possible that the type of injury characterized by fragmented protoplasm may be the result of cavitation (Ristic and Ashworth, 1993, 1995). It has been hypothesized that in supercooling wood tissues negative turgor pressure develops during freezing (Burke *et al.*, 1976; George and Burke, 1977; Ashworth, 1993), and cells become exposed to a large dehydrating force (George and Burke, 1977). The rigid cell wall can prevent dehydration as long as cohesion of cell water and its adhesion to the cell components are not exceeded by the dehydrative force (Weiser and Wallner, 1988). When limiting tensions are exceeded, cavitation may occur, resulting in migration of cellular water to extracellular ice and expansion of gas bubbles (Tyree *et al.*, 1984; Tyree and Dixon, 1986). Migration of cell water to extracellular ice and expansion of gas bubbles are likely rapid events, which could cause severe injury to the protoplasm.

The hypothesis that the cavitation may be the source of freezing injury in supercooling wood tissues (Ristic and Ashworth, 1993, 1995), is also supported by the experiment of Weiser and Wallner (1988). Weiser and Wallner (1988) measured the accumulation of acoustic emissions during the freezing of woody plant stems. Experiments with species that exhibited the deep supercooling characteristics noted that acoustic emissions began after the initial HTE and prior to the start of the LTE. Acoustic emissions were not detected below –40 °C. Weiser and Wallner (1988) suggested that cavitation of intracellular water could be a source of acoustic emissions and a possible source of freezing injury.

FREEZING TOLERANCE

Wood tissues exhibiting freezing tolerance are generally cold hardy and can survive temperatures below the homogeneous nucleation point (–38 °C), in some cases as low as –196 °C (Sakai, 1960; Guy *et al.*, 1986). The cold hardiness of these tissues is attributed to their ability to lose cellular water to extracellular ice.

During freezing stress, ice forms in xylem vessels and extracellular spaces (Levitt, 1980). This creates a water potential gradient between extracellular ice and intracellular water. As a result, the cellular water moves from the cytoplasm to extracellular ice (Burke *et al.*, 1976; Levitt, 1980; Ristic and Ashworth, 1994), and, as the tissue temperature declines, living cells of the wood tissue progressively dehydrate.

Studies utilizing nuclear magnetic resonance (NMR) and DTA support the view that the living cells of non-supercooling wood tissues lose water during freezing. Nuclear magnetic resonance studies have observed a continuous decline in the liquid water content with decreasing temperatures (Burke *et al.*, 1974; George and Burke, 1977; Harrison *et al.*, 1978). Likewise, DTA studies have revealed only one exotherm, the HTE during freezing (George *et al.*, 1974, 1982; Burke *et al.*, 1976; Becwar *et al.*, 1981; Malone and Ashworth, 1991). The appearance of HTE has been observed at temperatures slightly below 0 °C (Ashworth and Davis, 1984, 1986; Ashworth *et al.*, 1985a,b). It has been suggested that this exotherm corresponds to the freezing of water in xylem vessels and extracellular spaces (Malone and Ashworth, 1991).

Freeze-induced dehydration creates numerous secondary stresses in the living cells of non-supercooling wood tissues. These include the concentration of cellular components to toxic levels (Lovelock, 1953; Mazur, 1969; Levitt, 1980; Steponkus, 1984), decreased water activity leading to a precipitation of proteins (Levitt, 1980), alterations in hydrophilic and hydrophobic interactions within membranes (Heber *et al.*, 1981; Steponkus, 1984), establishment of a minimum critical cell volume (Meryman, 1971) or surface area (Steponkus and Wiest, 1979; Steponkus *et al.*, 1982; Steponkus, 1984; Dowgert and Steponkus, 1984), and pH changes within the unfrozen cytoplasm (Lovelock, 1953). Presumably, woody tissues that exhibit freezing tolerance have developed mechanisms to tolerate the secondary stresses that accompany extracellular freezing and cellular dehydration. Apparently, the survival of woody tissues exhibiting extracellular freezing is not limited by a particular freezing temperature but rather the survival would be a function of the cell's capacity to withstand the dehydrative stresses that accompany extracellular freezing.

Freeze-induced dehydration also has an enormous impact on the volume and general morphology of the living cells of non-supercooling wood tissues. Because cellular water is being removed from the protoplasm, protoplasm volume must be reduced. This raises an important question. What is the mechanism of the reduction of protoplasm volume? There are two possibilities (Ristic and Ashworth, 1994). One possibility is that during freeze-dehydration both the cell wall and the protoplasm collapse. The alternative would be that cell walls remain rigid and the protoplasm contracts (plasmolyzes). Which one of the two possibilities is more likely to occur during freeze-induced cell dehydration?

It is the traditional view that the reduction in cell liquid water content during freeze-induced dehydration is accompanied by the reduction of cell volume and collapse of the cell wall (Levitt, 1980). Levitt (1980) explains this concept by stating: "since there is no liquid external to the cell, freeze-dehydration cannot plasmolyze the cell in the true sense of the term—a separation of the dehydrated protoplast from the cell wall. Therefore, the cell wall must adhere to the protoplast and both must collapse together as the freeze-dehydration progresses."

Freeze-dehydration-induced collapse of the cell wall has been observed. But, to our knowledge, this has been only seen in tissues of herbaceous plants (Pearce, 1988; Pearce and Ashworth, 1992) and bark tissues of woody plants (Ashworth *et al.*, 1988; Malone and Ashworth, 1991).

Freeze-dehydration-induced collapse of the cell, however, has not been observed in the living cells of wood tissues. Malone and Ashworth (1991) studied freezing response of non-supercooling wood tissues of red osier dogwood (*Cornus sericea* L.), weeping willow (*Salix babylonica* L.), and corkscrew willow (*Salix matsudana* Koidz. f. *tortuosa* Rehd.). In their experiment, wood tissues were exposed to subzero temperatures during either a laboratory freezing test or a natural frost in the field. When appropriate test temperatures were reached, specimens were quench frozen into melted Freon 12 at –150°C. Quench freezing immobilizes water and preserves tissue morphology. Therefore, changes in cell shape, which occurred during freezing-stress, could be observed. Using low temperature scanning electron microscopy (LTSEM) and freeze substitution, Malone and Ashworth (1991) observed that the cell walls of xylem ray parenchyma cells did not collapse in response to a freezing stress. Non-stressed controls and tissues exposed to freezing treatments ranging from 0 to –60oC appeared similar. In addition, cell appearance was similar if either frozen in the laboratory or exposed to a natural freezing-stress in the field. Malone and Ashworth (1991) also observed the same response whether tissues were fixed using freeze substitution, or examined directly in the frozen state using LTSEM. There was no evidence of freeze-induced cell wall contraction within the wood tissues of all three species, red osier dogwood, weeping willow, and corkscrew willow. These observations were unexpected and inconsistent with current concepts of how plant cells respond to freezing.

Using freeze substitution and transmission electron microscopy (TEM), Ristic and Ashworth (1994) examined the status of the plasma membrane and cytoplasm in relation to cell wall in the xylem ray parenchyma cells of red osier dogwood during freezing. Ristic and Ashworth (1994) found that cell walls of ray parenchyma cells remained rigid and the protoplasm contracted during freezing (Figure 6.2). Signs of contraction were seen at –5°C and lower temperatures. Protoplasm contraction always occurred along the radial axes of the parenchyma cells (Figs. 6.2C-F). The cytoplasm of fragmented protoplasm appeared dark and dense, and the internal cell structures such as vacuoles, chloroplasts, and mitochondria appeared normal (Ristic and Ashworth, 1994). Protoplasm contraction and rigid cell wall were apparent under both laboratory (Ristic and Ashworth, 1994) and field freeze conditions (Ristic and Ashworth, 1996).

Interestingly, differences in freeze-induced protoplasm contraction were noticed among wood tissues or red osier dogwood collected at different times of the year (Ristic and Ashworth, 1994). Protoplasm contraction in the winter tissue seemed to occur without significant damage to the protoplasm (Figure 6.2). In

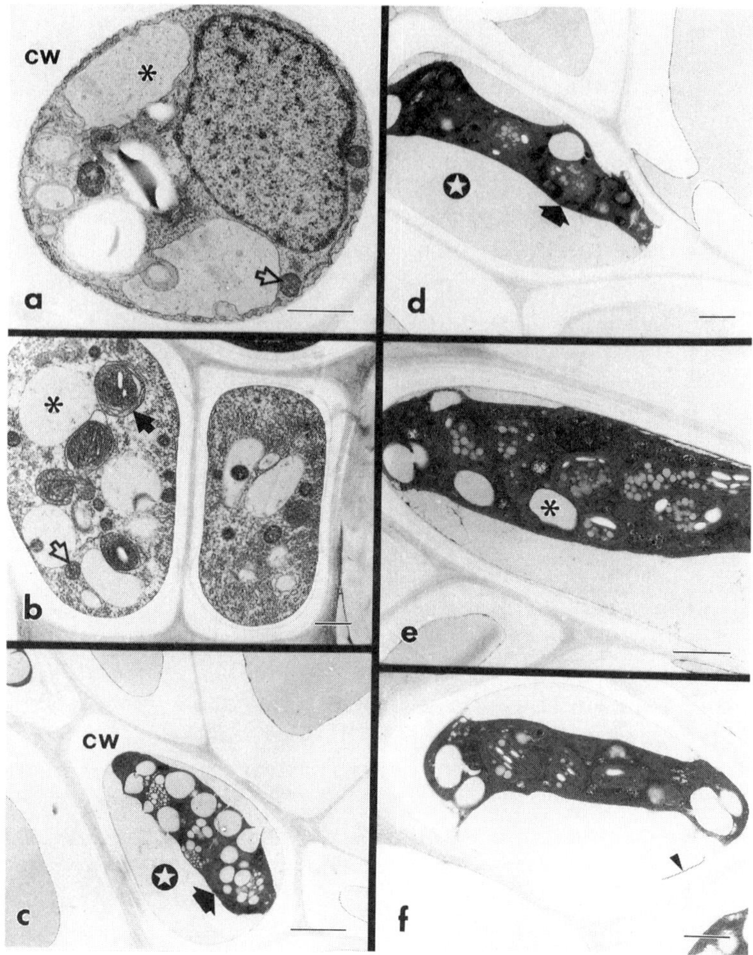

Figure 6.2 The effect of freezing stress on xylem ray parenchyma cells of wood tissue of *Cornus sericea* collected in winter. Tissue was subjected to laboratory freezing stress at a 5°Ch^{-1} cooling rate, harvested at appropriate temperatures, and prepared for TEM using chemical fixation (a) and the freeze substitution technique (b–f). Evidence of protoplasm contraction was noticed at –5°C (c–e) and other freezing temperatures (f). a, Xylem ray parenchyma cell in control tissue. cw, Cell wall; asterisk, vacuole; arrow, mitochondria; bar = 1μm. b, Xylem ray parenchyma cell in control tissue. Asterisk, Vacuole; black arrow, chloroplast; arrow, mitochondria; bar = 1μm. c, Tissue freeze stressed at –5°C. cw, Cell wall; arrow, contracted protoplasm; star, empty space between cell wall and contracted protoplasm; bar = 2μm. d, Tissue freeze stressed at –5°C. Arrow, Contracted protoplasm; star, empty space between cell wall and contracted protoplasm; bar = 1μm. e, Tissue freeze stressed at –5°C. Asterisk, Vacuole; arrow, chloroplast; bar = 1μm. f, Tissue freeze stressed at –30°C. Arrowhead, Broken plasma membrane attached to the cell wall; bar = 1μm. (With permission from American Society of Plant Physiologists, Ristic and Ashworth, 1994)

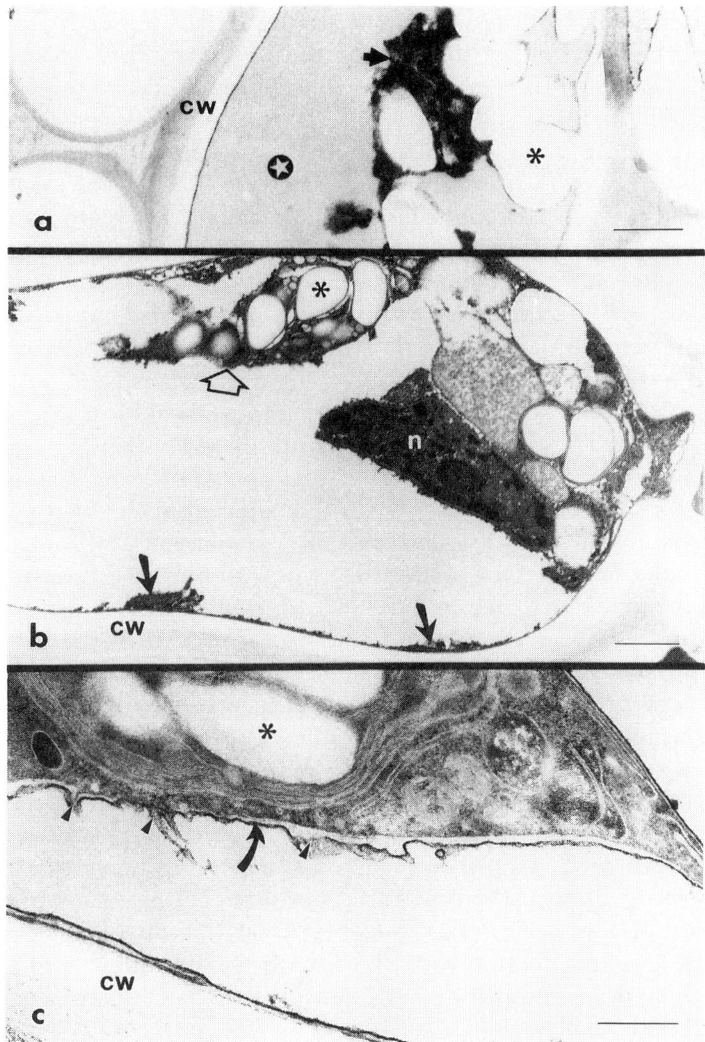

Figure 6.3 The effect of freezing stress on xylem ray parenchyma cells of wood tissue of *Cornus sericea* collected in spring. Tissue was subjected to laboratory freezing stress at a 5°Ch^{-1} cooling rate, harvested at appropriate temperatures, and prepared for TEM using the freeze-substitution technique. Xylem ray parenchyma cells showed two distinct responses: fragmentation of protoplasm (a) and protoplasm contraction (b and c). a, Tissue freeze stressed at –40°C. cw, Cell wall; arrow, piece of fragmented protoplasm; asterisk, starch grain; star, cell lumen; bar = 1μ m. b, Tissue freeze stressed at –40°C. cw, Cell wall; n, nucleus; white arrow, contracted protoplasm; black arrow, pieces of broken protoplasm attached to the cell wall; asterisk, starch grain; bar = 1μm. c, Tissue freeze stressed at –50°C. cw, Cell wall; arrow, plasma membrane; arrowhead, damage on the plasma membrane; asterisk, starch grain; bar = 0.2μm. (With permission from American Society of Plant Physiologists, Ristic and Ashworth, 1994).

contrast, protoplasm contraction in the spring and summer tissues was accompanied by substantial damage to the protoplasm (Figure 6.3). In spring and summer tissues, portions of the plasma membrane and subtending cytoplasm remained attached to the cell wall and cells appeared injured (Figure 6.3B). In some cells, portions of the cytoplasm appeared to stretch between the contracted protoplasm and the cell wall (Figure 6.3C). These observations indicate that adhesion between the cell wall and plasma membrane may change during acclimation and deacclimation (Ristic and Ashworth, 1994). It has been speculated that connections between the plasma membrane and the cell wall may be due to associated adhesion molecules, and that these components change during cold acclimation and deacclimation (Ristic and Ashworth, 1994). More research is needed to investigate this possibility.

Ashworth *et al.* (1993) conducted an experiment simulating the freeze-induced dehydrative stress. They exposed wood tissue of red osier dogwood to a dehydrative stress by incubating specimens at a range of constant relative humidities. These conditions were established by suspending 1 mm thick cross-sections over saturated salt solution within sealed containers (Winston and Bates, 1960). Saturated solutions were selected to establish a range of relative humidities equivalent to the vapor pressure over ice at –20, –40, –60, and –80 °C. Specimens initially lost water rapidly. However, after a few days a constant fresh weight was established. Using LTSEM, Ashworth *et al.* (1993) observed that there was no evidence that the walls of xylem ray parenchyma cells of red osier dogwood collapsed in response to a dehydrative stress. Cell walls of xylem ray parenchyma cells remained rigid and appeared as controls. No evidence of cell wall distortion or separation within middle lamella was apparent in response to the dehydrative stress treatments. Both cold acclimated and non-acclimated specimens appeared similar (Ashworth *et al.*, 1993).

Based on the studies of Malone and Ashworth (1991), Ashworth *et al.* (1993), and Ristic and Ashworth (1994), it becomes clear that current concepts of freezing resistance cannot adequately explain the freezing behavior of living cells of, at least, some non-supercooling wood tissues. Red osier dogwood, weeping willow, and corkscrew willow, for example, are extremely cold hardy (Malone and Ashworth, 1991). The wood tissues of these species do not exhibit LTEs in DTA experiments and survive –60 °C (Malone and Ashworth, 1991); the wood tissue of red osier dogwood can even survive –196 °C (Sakai, 1960; Guy *et al.*, 1986). Based on these criteria, these tissues are expected to undergo extracellular freezing and cell wall collapse in response to freezing (Levitt, 1980). This prediction, however, is proven not correct (Malone and Ashworth, 1991; Ashworth *et al.*, 1993; Ristic and Ashworth, 1994) and an alternative explanation is required.

It has been proposed that during freezing stress cell walls of living cells of non-supercooling wood tissues do not collapse but rather protoplasm contracts (Ristic and Ashworth, 1994). This hypothesis is based on the following evidence. First, as stated earlier, Malone and Ashworth (1991) observed that the ray parenchyma cells of non-supercooling wood tissues of red osier dogwood, weeping willow, and corkscrew willow did not collapse in response to either laboratory or field freezing stress, and their observation was made using LTSEM, a technique which permits observation of tissue in the frozen state. Second, Ashworth *et al.* (1993) noted that in the wood tissue of red osier dogwood that was dehydrated below fiber saturation point (30%), the walls of xylem ray parenchyma cells did not collapse. Third, using freeze substitution and TEM, Ristic and Ashworth (1994) observed no evidence of cell wall collapse but rather protoplasm contraction in freeze-stressed wood tissue of red osier dogwood. Thus, the observations of Malone and Ashworth (1991), Ashworth *et al.* (1993) and Ristic and Ashworth (1994) support the view that in non-supercooling wood tissues, protoplasm contracts and cell walls remain rigid during freezing.

Furthermore, the view that the walls of ray parenchyma cells of wood tissues do not collapse during freezing stress is also supported by mechanical and physical properties of the wood tissue (Ristic and Ashworth, 1994). The thick and lignified cell walls of the wood are rigid and do not have sufficient elasticity to permit significant contraction (U.S. Forest Products Laboratory, 1974). Wood dimensions change very little even after complete removal of cellular water. When wood contracts upon drying, it does not shrink uniformly. Wood shrinks most in the tangential direction (6.1–10.5 %), about one-half as much radially (2.1 – 7.0 %), and only slightly longitudinally (U.S. Forest Products Laboratory, 1974). The combined effects of tangential and radial shrinkage can distort the shape of wood pieces because of difference in shrinkage and the curvature of annual rings (U.S. Forest Products Laboratory, 1974). The behavior of cells within wood tissue would likely be similar.

In the light of physical and mechanical properties of the wood tissue it is, therefore, not surprising that the walls of ray parenchyma cells of non-supercooling wood tissues of several species did not collapse in response to either freezing (Malone and Ashworth, 1991; Ristic and Ashworth, 1994) or dehydrative stress (Ashworth *et al.*, 1993). Ristic and Ashworth (1994) observed no evidence of changes in wood morphology of laboratory freeze-stressed stems of red osier dogwood as would be expected if the dimensions of wood were altered by freezing. Ristic and Ashworth (1996) also noticed no signs of cell wall distortion and/or cell wall separation within the middle lamella of field frozen twigs. Because of tight cell-to-cell connections, it is difficult to imagine that cell walls would collapse without tearing and/or distorting the walls of xylem ray parenchyma cell and those of adjacent cells. Thus, it is likely that the observations using LTSEM (Malone and Ashworth, 1991; Ashworth *et al.*, 1993) and freeze substitution and TEM (Ristic and Ashworth, 1994) provided an accurate representation of cell response during freezing stress, and that the walls of xylem ray parenchyma cells of non-supercooling wood tissues do not collapse but rather protoplasm contracts in response to freezing stress.

Although protoplasm contraction seems to be a likely response of living cells in non-supercooling wood tissues to freezing stress, an important question remains unanswered. What is present between the contracted protoplasm and cell wall? Air is one possibility. Water could be removed from the protoplasm of xylem ray parenchyma cells to the lumen of neighboring fibers and intercellular spaces, either as liquid or vapor (Ristic and Ashworth, 1994). However, at present there are no evidence supporting this hypothesis, and further research is needed.

CONCLUDING REMARKS

Early studies of freezing resistance have anticipated intracellular ice formation as a source of freezing injury in supercooling wood tissues. This hypothesis, however, was based upon the appearance of LTE in DTA experiments and upon analogies with water in model systems, and there was no direct evidence to support the hypothesis. Recent studies have provided microscopic evidence that intracellular ice formation is a source of freezing injury in supercooling wood tissues. Moreover, these studies have also provided evidence that another source of freezing injury, possibly cavitation, may occur in wood tissues as well.

Furthermore, current concepts of freezing resistance do not adequately explain the freezing behavior of non-supercooling wood tissues. Contrary to predictions, the walls of living xylem cells of some non-supercooling tissues do not collapse during freeze-induced cell dehydration. It has been proposed that in non-supercooling wood tissues, protoplasm contracts within the confines of rigid cell walls. The adherence of the protoplasm to the cell wall seems to change during cold acclimation and deacclimation. The ability of the plasma membrane and cell wall to dissociate during freezing might be an important feature in the survival of extremely cold hardy woody plants. Unfortunately, the nature of the connection between plasma membrane and cell wall and the mechanism of their dissociation are unknown. Understanding the mechanism of protoplasm contraction remains an important challenge. Progress in this area will be necessary if we are to understand the extreme cold hardiness of non-supercooling wood tissues and improve cold hardiness in woody plants.

REFERENCES

Ashworth E.N. (1993) Deep supercooling in woody plant tissues. In *Advances in Plant Cold Hardiness*, edited by P.H. Li and L. Christersson, pp. 203–213. Boca Raton: CRC Press.

Ashworth E.N., Anderson J.A. and Davis G.A. (1985a) Properties of ice nuclei associated with peach trees. *Journal of the American Society for Horticultural Science*, **110**, 287–291.

Ashworth E.N., Anderson J.A., Davis G.A. and Lightner G.W. (1985b) Ice formation in *Prunus persica* under field conditions. *Journal of the American Society for Horticultural Science*, **110**, 322–324.

Ashworth E.N. and Davis G.A. (1984) Ice nucleation within peach trees. *Journal of the American Society for Horticultural Science*, **109**, 198–201.

Ashworth E.N. and Davis G.A. (1986) Ice formation in woody plants under field conditions. *HortScience*, **21**, 1233–1234.

Ashworth E.N., Echlin P. Pearce R.S. and Hayes T.L. (1988) Ice formation and tissue response in apple twigs. *Plant, Cell and Environment*, **11**, 703–710.

Ashworth E.N., Malone S.R. and Ristic Z. (1993) Response of woody plant cells to dehydrative stress. *International Journal of Plant Sciences*, **154**, 90–99.

Ashworth E.N., Rowse D.J. and Billmyer L.A. (1983) The freezing of water in woody tissues of apricot and peach and the relationship to freezing injury. *Journal of the American Society for Horticultural Science*, **108**, 299–303.

Becwar M.R., Rajashekar C., Hansen Bristow K.J. and Burke M.J. (1981) Deep undercooling of tissue water and winter hardiness limitations in timberline flora. *Plant Physiology*, **68**, 111–114.

Burke M.J. (1979) Water in plants: the phenomenon of frost survival. In *Comparative Mechanisms of Cold Adaptation*, edited by L.S. Underwood, L.L. Tieszen, A.B. Callahan and G.E. Folk. pp. 259–281. New York: Academic Press.

Burke M.J., Bryant R.G. and Weiser C.J. (1974) Nuclear magnetic resonance of water in cold acclimating red osier dogwood stems. *Plant Physiology*, **54**, 392–398.

Burke M.J. Gusta L.V., Quamme H.A., Weiser C.J. and Li P.H. (1976) Freezing and injury in plants. *Annual Review of Plant Physiology*, **27**, 507–528.

Dowgert M.F. and Steponkus P.L. (1984) Behavior of the plasma membrane of isolated protoplasts during a freeze–thaw cycle. *Plant Physiology*, **75**, 1139–1151.

George M.F. (1983) Freezing avoidance by deep supercooling in woody plant xylem: preliminary data on the importance of cell wall porosity. In *Current Topics in Plant Biochemistry and Physiology*, edited by D.D. Randal, D.G. Blevins, R.L. Larson and B.J. Rapp, pp. 84–95. Columbia: University of Missouri Press.

George M.F., Becwar M.R. and Burke M.J. (1982) Freezing avoidance by deep undercooling of tissue water in winter-hardy plants. *Cryobiology*, **19**, 628–639.

George M.F. and Burke M.J. (1977) Cold hardiness and deep supercooling in xylem of shagbark hickory. *Plant Physiology* **59**, 319–325.

George M.F., Burke M.J., Pellett H.M. and Johnson, A.G. (1974) Low temperature exotherms and woody plant distribution. *HortScience*, **9**, 519–522.

George M.F., Hong S.G. and Burke M.J. (1977) Cold hardiness and deep supercooling of hardwoods: its occurrence in provenance collections of red oak, yellow birch, black walnut, and black cherry. *Ecology*, **58**, 674–680.

Guy C.L., Niemi K.J., Fennell A. and Carter, J.V. (1986) Survival of *Cornus sericea* L. stem cortical cells following immersion in liquid helium. *Plant, Cell and Environment*, **9**, 447–450.

Harrison L.C., Weiser C.J. and Burke M.J. (1978) Freezing of water in red osier dogwood stems in relation to cold hardiness. *Plant Physiology*, **62**, 899–901

Heber U., Schmitt J.M., Krause G., Klosson R.J. and Santarius K.A. (1981) *Effects of Low Temperature on Biological Membranes*, edited by G.J. Morris and A. Clarke. pp. 263–283. London: Academic Press.

Hong S. and Sucoff E. (1980) Units of freezing of deep supercooling water in woody xylem. *Plant Physiology*, **66**, 40–45.

Hong S., Sucoff E. and Lee-Stadelmann O. (1980) Effects of freezing deep supercooled water on the viability of ray cells. *Botanical Gazette*, **141**, 464–468.

Koster K.L. and Lynch D.V. (1992) Solute accumulation and compartmentation during cold acclimation of puma rye. *Plant Physiology*, **98**, 108–113.

Levitt, J. (1980) *Responses of Plants to Environmental Stresses: Chilling, Freezing, and High-Temperature Stresses*, Vol. 1. New York: Academic Press.

Lovelock J. (1953) Hemolysis of human red blood cells by freezing and thawing. *Biochimica et Biophysica Acta*, **10**, 414–426.

Malone S.R. and Ashworth E.N. (1991) Freezing stress response in woody tissues observed using low-temperature scanning electron microscopy and freeze substitution technique. *Plant Physiology*, **95**, 871–881.

Mazur P. (1969) Freezing injury in plants. *Annual Review of Plant Physiology*, **20**, 419–448.

Meryman H.T. (1971) Osmotic stress as a mechanism of freezing injury. *Cryobiology*, **8**, 489–500.

Pearce R.S. (1988) Extracellular ice and cell shape in frost-stressed cereal leaves: a low- temperature scanning electron microscopy. *Planta*, **175**, 313–324.

Pearce R.S. and Ashworth E.N. (1992) Cell shape and localization of ice in leaves of overwintering wheat during frost stress in the field. *Planta*, **188**, 324–331.

Quamme H.A. (1976) Relationship of the low temperature exotherm to apple and pear production in North America. *Canadian Journal of Plant Science,* **56**, 493–500.

Quamme H., Stushnoff C. and Weiser C.J. (1973) The mechanism of freezing injury in xylem of winter apple twigs. *Plant Physiology*, **51**, 273–277.

Quamme H., Weiser C.J. and Stushnoff C. (1972) The relationships of exotherms to cold injury in apple stem tissues. *Journal of the American Society for Horticultural Science*, **97**, 608–613.

Rajashekar C. and Burke M.J. (1978) The occurrence of deep undercooling in the genera *Pyrus*, *Prunus*, and *Rosa*: a preliminary report. In *Plant Cold Hardiness and Freezing Stress*, edited by P.H. Li and A. Sakai. pp. 213–225. New York: Academic Press.

Rajashekar C. and Reid W. (1989) Deep supercooling in stem and bud tissues of pecan. *HortScience,* **24**, 348–350.

Rajashekar C., Pellett H.M. and Burke M.J. (1982a) Deep supercooling in roses. *HortScience,* **17**, 609–611.

Rajashekar C., Westwood M.N. and Burke M.J. (1982b) Deep supercooling and cold hardiness in genus *Pyrus*. *Journal of the American Society for Horticultural Science,* **107**, 968–972.

Rasmussen D.H. and MacKenzie A.P. (1973) Clustering in supercooled water. *Journal of Chemistry and Physics*, **59**, 5003–5013.

Ristic Z. and Ashworth E.N. (1993) Ultrastructural evidence that intracellular ice formation and possibly cavitation are the sources of freezing injury in supercooling wood tissue of *Cornus florida* L. *Plant Physiology*, **103**, 753–761.

Ristic Z. and Ashworth E.N. (1994) Response of xylem ray parenchyma cells of red osier dogwood (*Cornus sericea* L.) to freezing stress: microscopic evidence of protoplasm contraction. *Plant Physiology*, **104**, 737–746.

Ristic Z. and Ashworth E.N. (1995) Response of xylem ray parenchyma cells of supercooling wood tissues to freezing stress: microscopic study. *International Journal of Plant Sciences,* **156**, 784–792

.Ristic Z. and Ashworth E.N. (1996) Response of xylem ray parenchyma cells of red osier dogwood (*Cornus sericea* L.) to field freezing stress, and to freeze-thaw cycle. *Journal of Plant Physiology* (in press).

Sakai A. (1960) Survival of the twig of woody plants at –196 °C. *Nature,* **195**, 393–394.

Steponkus P.L. (1984) Role of the plasma membrane in freezing injury and cold acclimation. *Annual Review of Plant Physiology*, **35**, 543–584.

Steponkus P.L., Dowgert, M.F., Evans R.Y. and Gordon-Kamm, W. (1982) Cryobiology of isolated protoplasts. In *Plant Cold Hardiness and Freezing Stress*, Vol 2, edited by P.H. Li and A. Sakai, pp. 459–474. New York: Academic Press.

Steponkus P.L. and Wiest S.C. (1979) Freeze-thaw induced lesions in the plasma membrane. In *Low Temperature Stress in Crop Plants, The Role of the Membrane*, edited by J.M.Lyons, D. Graham and J.K. Raison. pp. 231–254. New York: Academic Press.

Tyree M.T. and Dixon M.A. (1986) Water stress induced cavitation and embolism in some woody plants. *Physiologia Plantarum*, **66,** 397–*405*.

Tyree M.T., Dixon M.A., Tyree E.L. and Johnson R. (1984) Ultrasonic acoustic emissions from the sapwood of cedar and hemlock. *Plant Physiology*, **75**, 988–992.

U.S. Forest Products Laboratory (1974) *Wood Handbook: Wood as an Engineering Material*. Department of Agriculture Handbook 72 (revised). Washington D.C.: U.S. Department of Agriculture.

Weiser R.L. and Wallner S.J. (1988) Freezing woody plant stems produces acoustic emissions. *Journal of the American Society for Horticultural Science*, **113**, 636–639.

Winston P.W. and Bates D.H. (1960) Saturated solutions for the control of humidity in biological research. *Ecology,* **41**, 232–237.

7. REDUCTION OF HIGH TEMPERATURE STRESS IN PLANTS

JAMES R. MAHAN, B. L. McMICHAEL and D. F. WANJURA

USDA–ARS, Cropping Systems Research Laboratory, Lubbock, Texas 79401,USA.

INTRODUCTION

The negative effects of thermal stress on plants are substantial, pervasive and often result in reduced crop yields worldwide. (Boyer, 1982). Perhaps as important as the losses due to thermal stresses are the geographic limitations placed upon agriculture by adverse thermal environments. Adding to these long-recognized concerns are the recent predictions of global climate change (Houghton, 1989) which suggest that thermal stress may increase as a result of changes in the ambient thermal environment and alterations in patterns of precipitation. In light of current and future effects of thermal stress, continued research into methods for increasing thermal stress resistance is warranted.

While the ability of the plant to survive and/or complete a life cycle under stress is one way to define thermal stress resistance, a more useful definition from an agronomic point of view might be based upon the maintenance of economic value when exposed to thermal stress. The criteria used to assess the maintenance of economic value are highly subjective and depend upon both the crop and climate of interest. Unfortunately, many of the mechanisms that contribute to stress resistance prove to be of little value from an economic standpoint. Stress resistance traits frequently increase plant survival under stress but result in diminished growth and yield under nonstressed conditions (Kramer, 1980). The maintenance of plant performance within high temperature environments is the topic of this chapter.

One of the difficulties in studies of high temperature stress is the definition and quantification of the stress. Numerous definitions of such stress have been proposed (Levitt, 1980; Kramer, 1980; Hale and Orcutt, 1987). Generally, thermal stress is categorized relative to some estimate of an optimal thermal range that is characteristic of each species in question. Hale and Orcutt (1987) postulated a "zero-stress" condition as a necessary starting point in discussions of thermal stress. They defined a zero-stress condition in terms of the optimal conditions for growth and development of the plant. The temperature range over which the plant is in the zero-stress state is its optimal thermal range. The identification of the optimal thermal range provides a means of defining stressful and non-stressful plant temperatures. While the optimal thermal range could be defined in terms of the thermal dependence of a variety of factors, we will focus on estimates of the optimal thermal range that are defined in terms of the thermal dependence of biochemical reactions. It is our thesis that specific knowledge of the optimal thermal range for the metabolism of a plant is an appropriate starting point for the development of methods for reducing the adverse effects of thermal stress. Specific approaches derived from knowledge of the optimal thermal range can be focused on the alteration of the optimal thermal range of the plant and/or the alteration of plant temperature.

IDENTIFICATION OF THE OPTIMAL THERMAL RANGE

The determination of the optimal thermal range is a useful starting point for studies of the plant's ability to resist thermal stress. Numerous investigations on the optimal temperature for various aspects of plant growth and development have been carried out through observations of plants grown under a variety of thermal regimes. Cotton has been the object of numerous investigations of this type (Reddy *et al.*, 1992, 1993; Hodges *et al.*, 1993). This approach though often successful requires long-term studies in controlled environments and is generally quite expensive in terms of time and resources. In these whole plant studies, the changes in plant metabolism that result from thermal variation are observed as changes in the growth and development of the plants.

The direct effect of temperature on the metabolic biochemistry of plants has been investigated in some detail and is well understood in many instances (Somero, 1975; Somero and Low, 1976; Hochachka and Somero, 1984; Patterson and Graham, 1987). The relationship between the normal thermal environment for an organism and thermal dependence of enzymes has been well established (Simon, 1979; Simon *et al.* 1983; Selinioti *et al.*, 1986; Mahan *et al.*, 1991). Teeri and Peet (1978) investigated the thermal dependence of malate dehydrogenase from two populations of *Potentilla glandulosa* Lindl. and reported that the thermal dependencies were indicative of the normal thermal range for the plant. The thermal dependence of the apparent K_m (the substrate concentration that results in a rate equal to one half the maximum velocity) for selected enzymes has been proposed as an indicator of the optimal thermal range of a plant. Mahan *et al.* (1990) used the term Thermal Kinetic Window (TKW) to describe the thermal range over which the apparent K_m is both minimal and stable (<200% of minimum). Burke *et al.* (1988) showed a relationship between the amount of time that the leaf temperatures were within the TKW and the rate of biomass accumulation in cotton and wheat.

Peeler and Naylor (1988) reported differences in the thermal response of the variable fluorescence recovery among plants from different thermal regimes. Burke (1990) and Ferguson and Burke (1991) expanded this work into a method for determining the thermal optimum for the plant. The thermal dependence, determined by this method was found to be: 1) in agreement with that identified by the TKW, 2) stable over the life of the plant and 3) unaltered by water or thermal stress.

The thermal dependence of enzyme function and variable fluorescence recovery provide two methods for the estimation of the optimal thermal range for plants. Optimal thermal ranges defined by both methods are about 10°C wide and are in general agreement with the recognized thermal dependence of the whole plant (compare Burke *et al.*, 1988 with Hodges *et al.*, 1993).

QUANTIFICATION OF THERMAL STRESS IN TERMS OF THE OPTIMAL THERMAL RANGE

An estimate of the optimal thermal range provides a criterion by which the degree of stress associated with a given plant temperature can be determined. Because of the frequent inequality between air and plant temperatures, plant as opposed to air temperatures, should be used in the characterization of the thermal stresses. Thus air temperatures above the optimal thermal range do not necessarily constitute high temperature stress (Figure 7.1). Comparing the

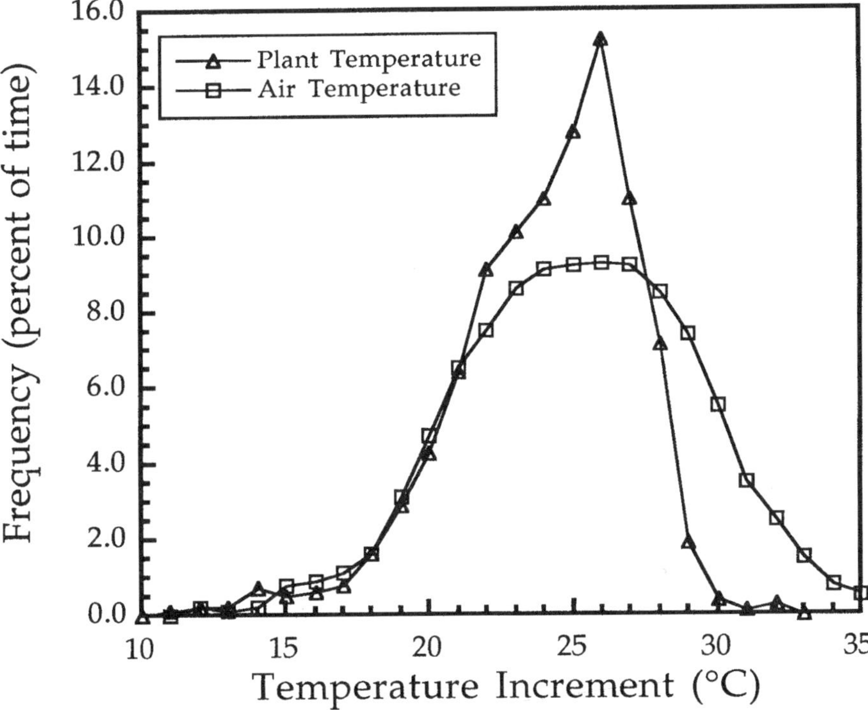

Figure 7.1. Seasonal distribution of air and canopy temperature for cotton during the summer of 1990 in Lubbock, TX. Observations of canopy temperatures every fifteen minutes from day of year 180 to day of year 262 were rounded to the nearest degree centigrade and added to the appropriate temperature increment.

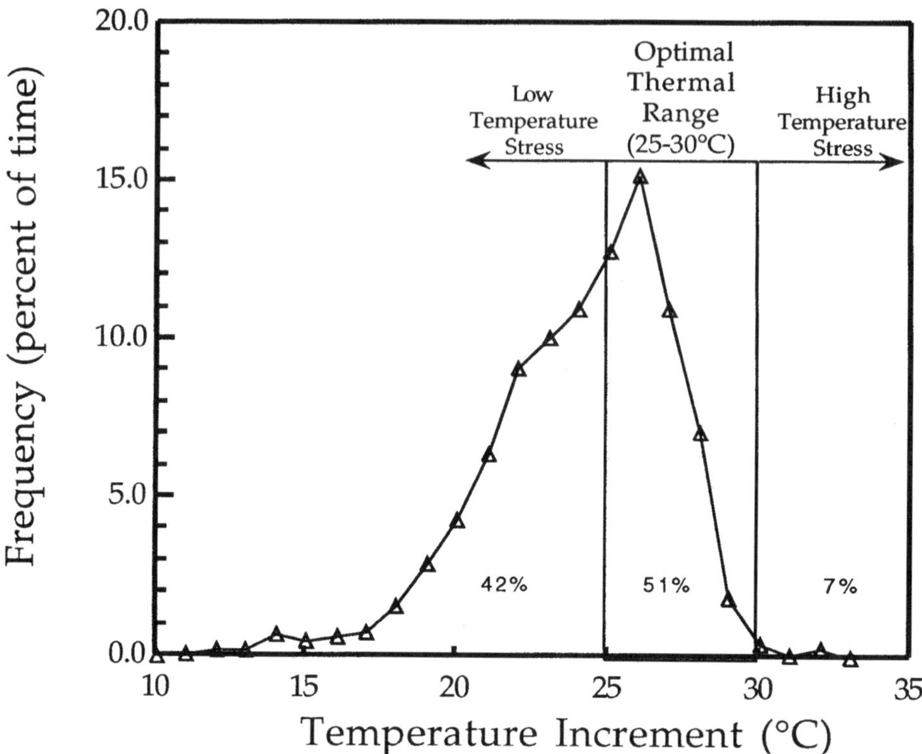

Figure 7.2. Seasonal distribution of canopy temperature for cotton during the summer of 1990 in Lubbock, TX. in comparison with an estimate of the optimal thermal range for cotton. Observations of canopy temperatures every fifteen minutes from day of year 180 to day of year 262 were rounded to the nearest degree centigrade and added to the appropriate temperature increment. Percentages indicate the portion of the season in stress category.

temperature of the plant with the optimal thermal range will help to identify and quantify the thermal stresses experienced by the plant (Figure 7.2). Plant temperatures below and above the optimal thermal range constitute low and high temperature stress, respectively, while those within the optimal thermal range are non-stressful.

METHODS FOR AVOIDING HIGH TEMPERATURE STRESS

If thermal stress is defined as the occurrence of leaf temperatures beyond the limits of the plant's optimal thermal range, then increasing the amount of time that the plant is within its optimal thermal range will reduce the adverse effects of high temperature stress. Within a given thermal environment, the extent of time that the temperature of the plant is within the optimal thermal range might be increased by: 1) altering the plant temperature to increase time within the optimal thermal range and/or 2) altering the optimal thermal range of the plant in such a way that leaf temperatures that previously limited metabolic function no longer impose a limitation.

Optimization of Plant Temperature Relative to the Optimal Thermal Range

The identification of the optimal thermal range for a given plant provides a "target" for modification of the temperature of the plant. Plants have a significant capacity for reducing their temperature relative to that of their environment (Mahan and Upchurch, 1988; Upchurch and Mahan, 1988) and optimization of the energy exchanges between the plant and the environment is one approach to limiting thermal stress. The alteration of plant temperature involves the alteration of the energy exchanges between the plant and its environment. In the following sections approaches to the optimization of plant temperatures will be discussed in terms of water use and canopy architecture.

Optimization of Plant Structure

The energy exchanges within a plant canopy are affected by the size, shape, and orientation of the leaves. The physical factors that define the energy budget and, as a result, the temperature of the plant are well understood (Cambell, 1981; Nobel, 1991). Breeding for specific canopy architecture has the potential to reduce the occurrence of plant temperatures above the optimal thermal range and thus improve the plant's ability to resist high temperature stress. Altering plant temperatures through changes in canopy architecture could serve to reduce thermal stresses with minimal effect on water consumption. Various factors including leaf area, root to leaf ratio, leaf orientation, size and shape, surface characteristics (e.g., pubescence), leaf thickness and the size and distribution of stomata are known to affect transpiration. Nobel (1991) reported that the rotation of leaves relative to the sun could reduce their temperature by up to 6°C. Alterations in leaf size and shape can similarly affect the temperature of the canopy. Generally, small leaf shapes (small leaves, leaflets on compound leaves or dissected leaves) have thinner boundary layers that are more conducive to sensible and latent heat transfers and as a consequence are often cooler than larger leaves in similar environments. The relative importance of the characteristics of individual leaves decreases as the plant canopy becomes more dense. Under these conditions, the aerodynamic characteristics of the canopy play a major role in the energy transfers between the plant and the environment. The effect of canopy closure on the energy balance of plants has been the subject of numerous studies (Ham *et al.*, 1991; Mateos *et al.*, 1991) and in general rapid closure of the canopy will reduce the occurrence of high leaf temperatures. The development of plant varieties with desirable canopy architecture coupled with management practices designed to avoid temperatures beyond the limits of the optimal thermal range may result in improved agronomic performance.

Optimization of Plant Water Use

The transpirational cooling uncouples the temperature of the plant from that of the aerial environment and thus represents a powerful mechanism for limiting the effects of thermal stress. Transpirational cooling depends on the flow of water from the soil through the plant to the atmosphere. Water movement in the plant is influenced by such factors as the morphology and physiology of the root system, the water content of the soil, soil structure, the morphology of the canopy at the plant-atmosphere interface and the dryness of the atmosphere (Kramer, 1983). Optimization of transpirational cooling can be achieved by improving the ability of the plant to extract water from the soil and through irrigation management.

Optimization of Root Systems for Efficient Water Extraction

Transpiration rates can have an effect on the amount of time that leaves in a canopy are within the optimal thermal range for growth and plant function. Transpiration rates are dependent on a number of factors. The hydrostatic gradient between the soil-root interface and the leaf, is the driving force for water movement (Oertli, 1991) and factors that act to increase the gradient tend to increase the canopy temperature as a result of decreased transpiration. The morphology and architecture of the root system at any stage of development influence the hydrostatic gradient. The resistance to water flow in the root system both radially (into the root) and axially (within the xylem stream) may be great enough to significantly increase the gradient, reduce the hydraulic conductivity and increase the canopy temperature. Yamauchi *et al.* (1995) showed that a significant time lag between measurements of the xylem water potential of intact cotton tap roots indicated a large radial resistance to water flow which was considerably higher than axial resistance. Differences in gross morphology, including the degree of branching and rooting depth may decrease the availability of water due to the inability of the root system to explore a larger soil volume and thus increase the gradient. Changes in total root length available to extract water can also impact the gradient to transport water.

It has been shown, for example, that in some species deeper roots were more efficient in extracting water as the soil became drier but that total water uptake was insufficient to maintain plants at their potential transpiration rates (Proffitt *et al.*, 1985). Taylor (1983) suggested that root elongation rate and longevity of branch roots were important factors in the development of root systems that could help to maintain a favorable plant water status. Low root temperatures (12° versus 24°C) resulted in a 27% reduction in transpiration, which resulted in higher leaf temperatures (Nelson, 1967).

Visualization of an "optimum" root system that would maintain sufficient water flow under a wide range of environmental conditions and provide for the maintenance of canopy temperature within the optimal thermal range is a difficult task. It is arguable that an "optimized" root system would possess characteristics such as minimum total resistance to liquid water flow, a high degree of branching, and rapid elongation rates. These characteristics may provide for a more favorable water status for a longer period of time for the plant. However, in a climate where soil water would not be replenished, one might argue that a higher internal root resistance would help to "ration" water use over time.

Since both the environment and the growth of the root system are dynamic processes influenced by a number of factors, there is a need for more detailed experimentation that explores the interactions between roots, shoots, and the environment. Such information would provide insight into the genetic alteration of the development of root systems for maintenance of a more favorable canopy temperature under different management regimes.

Optimization of Canopy Temperature through Breeding

Because stomatal function plays a critical role in thermal behavior of plants, it has been suggested that breeding plants with desirable stomatal characteristics could enhance their ability to resist thermal stress. The potential for the reduction of transpiration under drought stress through alteration of stomatal function has been reviewed (Jones, 1987). Selection of stomatal characteristics to minimize transpiration under water-limited conditions is a strategy for the improvement of drought resistance but perhaps at the cost of plant productivity. Reduced transpiration, while conserving water, may result in an increase in leaf temperature and declines in productivity resulting from increased thermal stress. When water is a limiting factor there is an unavoidable interaction between water conservation and thermal stress. In such cases, the question arises as to which stress, water or temperature, is most detrimental to the plant.

Quisenberry *et al.* (1982), using the transpiration decline technique on individual leaves, found genetic differences in the relative water content of cotton leaves at mean stomatal closure. They reported a significant correlation between the relative water content at mean stomatal closure and the growth rates of the entries when the plants were grown under water-limited conditions (i.e., growth rates were higher when stomata remained open at lower relative water contents). However, the correlation was not significant when the plants were grown under optimal water conditions suggesting that under favorable conditions the ability of the plant to have the stomata remain open at low water contents was of no real advantage.

Extensive variation in stomatal characteristics exists in nature (Meidner and Mansfield, 1968). This variation provides the possibility for identification of desired characteristics for improvement of thermal stress resistance by stomatal function. The development of technology for interspecific gene transfers was suggested as an important factor in the exploitation of interspecific variation in stomatal characteristics.

Optimization of Canopy Temperature through Irrigation Management

The control of irrigation in cotton through the use of canopy temperature thresholds, based upon estimates of the optimal thermal range has been inves-

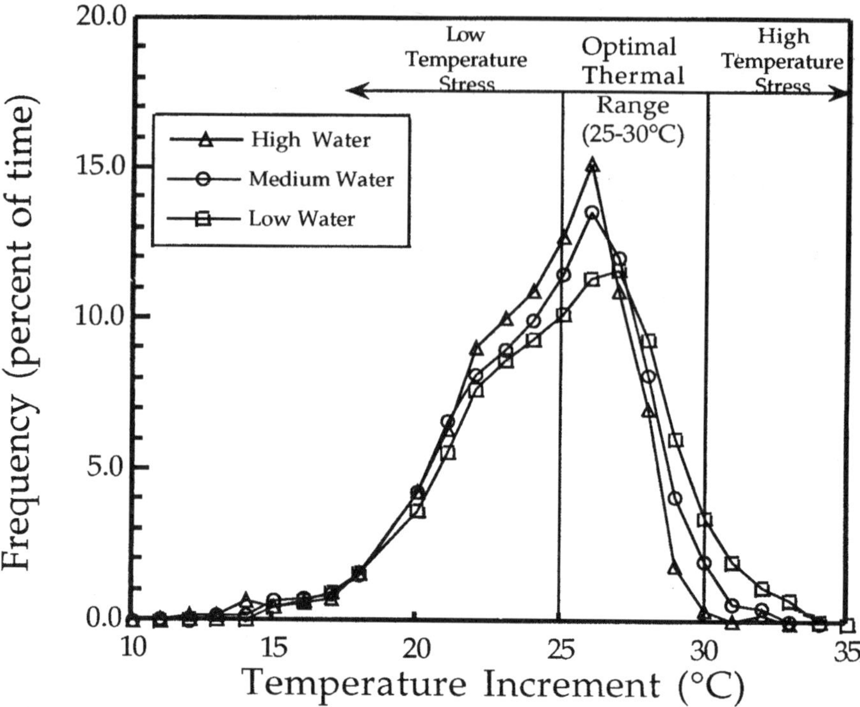

Figure 7.3. Seasonal distribution of canopy temperature for cotton under three levels of irrigation during the summer of 1990 in Lubbock, TX. as compared to an estimate of the optimal thermal range for cotton. Observations of canopy temperatures every fifteen minutes from day of year 180 to day of year 262 were rounded to the nearest degree centigrade and added to the appropriate temperature increment.

tigated (Wanjura *et al.*, 1990, 1992; Wanjura and Mahan, 1994). This approach, which has been referred to as temperature threshold irrigation scheduling, irrigates in such a way as to reduce the occurrence of plant temperatures in excess of the threshold value. Temperature threshold irrigation scheduling is based upon the theory that a biochemically-based optimal thermal range is an appropriate indicator of the thermal dependence of metabolism and that canopy temperatures within that range are not thermally stressful. The advantage of using a biologically-based optimal thermal range for irrigation decision making is that control is directed towards a specific goal; to maximize the amount of time that the plant temperature is within the optimal thermal range through the timely addition of water.

There are several assumptions that have been used in the development of temperature threshold irrigation scheduling: 1) Plant growth and development will increase as time within the optimal thermal range is increased, 2) A thermal kinetic window is an adequate definition of the optimal thermal range, 3) Irrigation management can be used to alter the temperatures experienced by the plant.

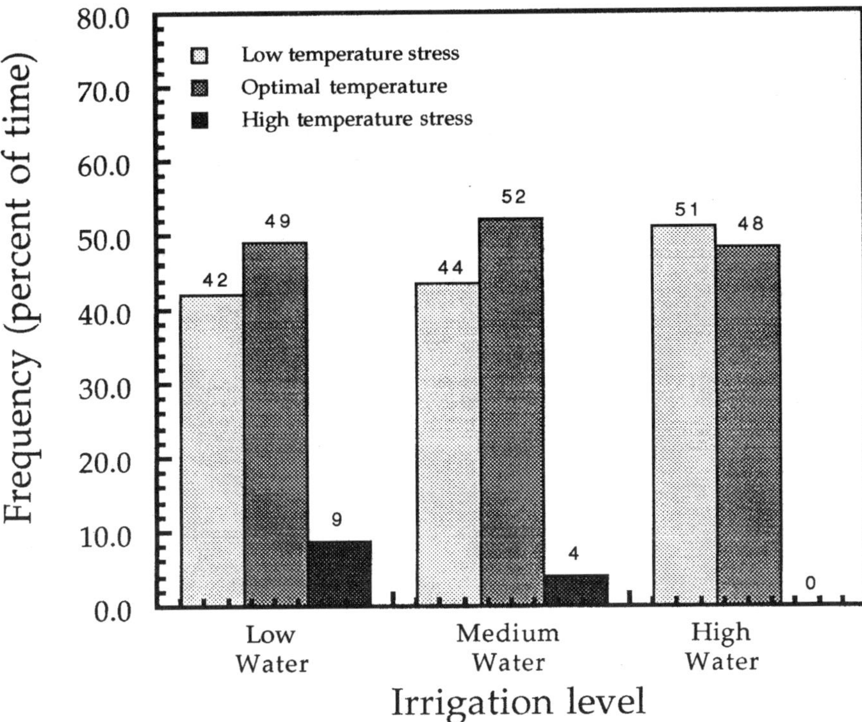

Figure 7.4. Thermal stress experienced by cotton under three levels of irrigation during the summer of 1990 in Lubbock, TX. as determined from comparison between optimal thermal range and canopy temperatures. Percentage of time indicated by parentheses.

Experiments on temperature threshold irrigation scheduling with cotton have included treatments using a variety of temperature thresholds (Wanjura *et al.*, 1992). The use of temperature thresholds for irrigation scheduling results in changes in the seasonal patterns of plant temperature (Figure 7.3). This resulted in alterations in the thermal stresses experienced by the plant (Figure 7.4). The results of these studies suggest that temperature threshold irrigation scheduling is a promising technique for irrigation management. Both yield and biomass production consistently responded to 2°C alterations in the temperature threshold with 28°C apparently superior to 26, 30 or 32°C thresholds. While the thermal stresses experienced by the plant are altered through the use of temperature thresholds, the changes in time at stressful temperatures are small (<5%).

Alteration of the Optimal Thermal Range

The thermal dependence of critical metabolic components of plants is an area in which there is potential for improving high temperature stress resistance. The

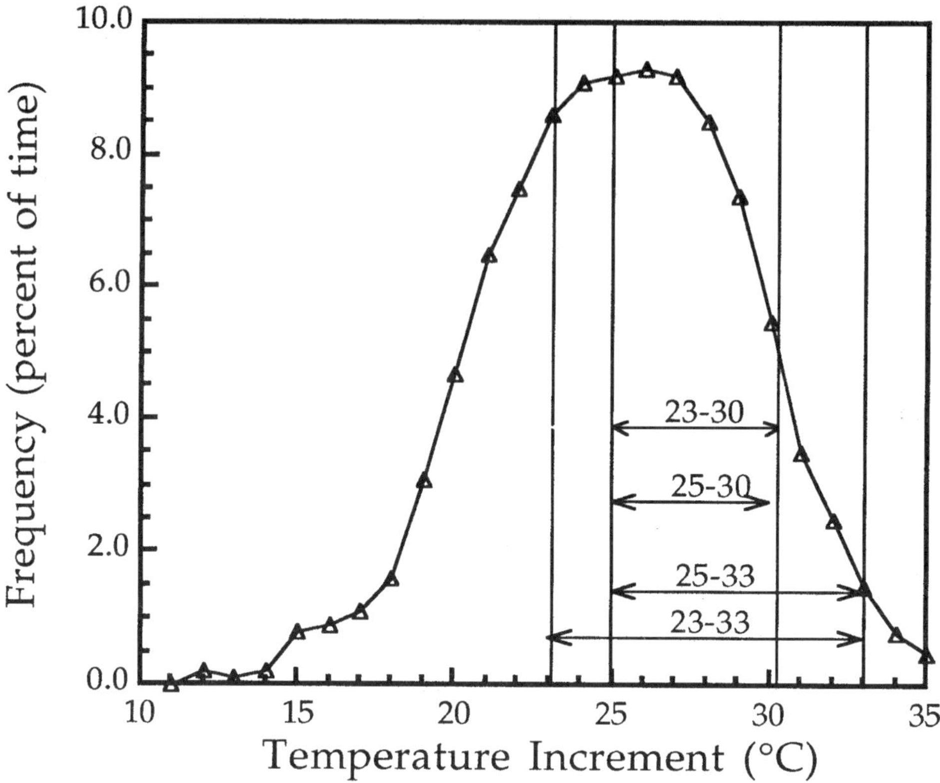

Figure 7.5. Seasonal distribution of canopy temperature for cotton during the summer of 1990 in Lubbock, TX. in comparison with various examples of possible optimal thermal ranges for cotton.

extent to which the optimal thermal range for a plant can be altered by changes in one, or even a few, enzyme(s) is ultimately dependent on the extent to which the enzyme is affected by thermal stress. Enzymes that are thought to have protective functions (e.g., superoxide dismutase, ascorbate peroxidase, glutathione reductase, catalase) might be the best candidates for alteration. The potential for altering the thermal dependence of more complex plant processes such as photosynthesis through the alteration of the thermal dependence of a small number of enzymes is probably much lower.

The ability to transfer genetic material between species via molecular biological techniques suggests that alterations in the optimal thermal range of plants might be accomplished by the incorporation of enzymes with desirable thermal dependencies into plants. The wide variation in the thermal dependence of enzyme function that has been reported in both plants and animals suggests that such transfers may be possible. In addition to the transfer of genes between species, it should be possible in the near future to "engineer" enzymes with desired optimal thermal ranges (Wilks *et al.*, 1988; Kati and Wolfenden, 1989). The "molecular

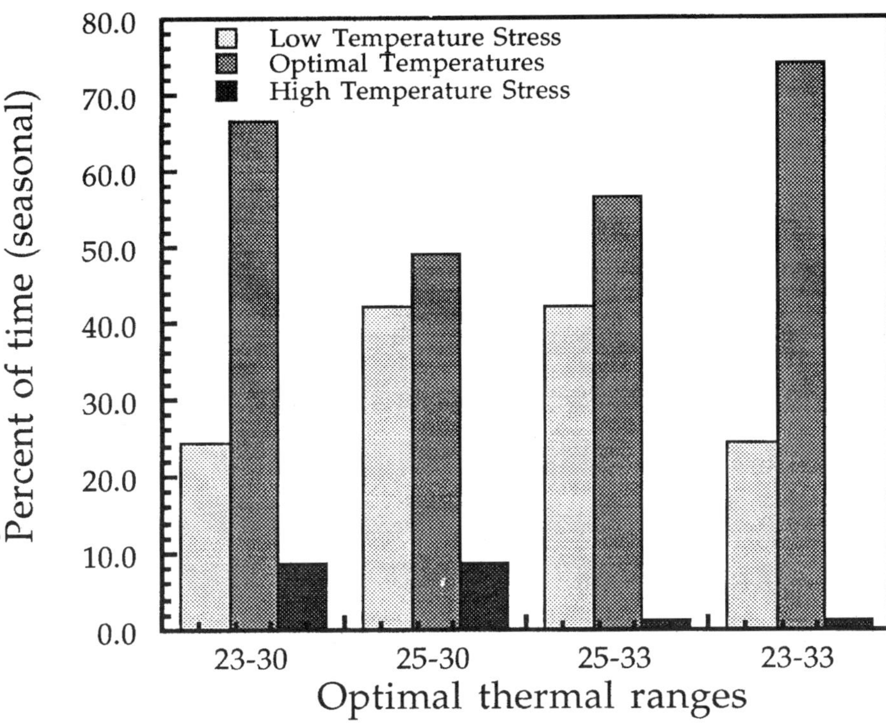

Figure 7.6. Thermal stress experienced by cotton during the summer of 1990 in Lubbock, TX. as determined from comparison between canopy temperatures and four theoretical estimates of the optimal thermal range for cotton.

engineering" of enzymes to alleviate thermal stress requires the identification of the factors that contribute to the observed thermal dependence of the enzyme. A simple model of the thermal dependence of enzyme kinetics has been used to explore the effects of alterations in the previously mentioned factors on thermal dependence of the function of malate dehydrogenase from cotton (Mahan and Wanjura, 1990; Mahan *et al.*, 1991). The results of these modeling efforts suggest that changes in such factors as activation energy, thermal kinetic window, and thermostability can alter the thermal dependence of the function of the enzyme. The incorporation of enzymes with differing thermal dependencies into a plant would serve to widen its optimal thermal range (Figure 7.5) and thus alter the pattern of high temperature stress experienced by the plant (Figure 7.6). In addition to alterations in the thermal dependence of the kinetics of enzyme(s) it is also possible to increase the activity of enzymes through increasing the amount present in the tissue. Alterations in the amount of glutathione reductase through transgenic expression of pea glutathione reductase in the cytosol, chloroplasts and mitochondria of tobacco have shown mixed results (Broadbent *et al.*, 1995). Some transformants showed increased resistance to oxidative stresses resulting from exposure to paraquat and ozone.

SUMMARY AND CONCLUSIONS

The identification of the plant's optimal thermal range provides a means for quantifying thermal stress and a "thermal target" for assessing the effectiveness of stress resistance mechanisms. Knowledge of the optimal thermal range for a plant presents two options for improving thermal stress resistance: 1) alterations of the plant temperature through optimization of energy exchanges between the plant and environment or 2) alterations in the thermal dependence of the plant's metabolism through molecular biological methodology. The implementation of the mechanisms will require information and expertise from a variety of disciplines; biochemistry and plant physiology for the identification of the optimal thermal range, environmental and soil physics for the characterization of the thermal environment and the optimization of energy exchanges between the plant and its environment, molecular biology and breeding for altering the optimal thermal range, and finally agricultural engineering for the implementation of improved management practices. Cooperation among such diverse disciplines, though not commonplace today, can certainly be realized.

REFERENCES

Boyer J.S. (1982) Plant productivity and environment. *Science,* **218**, 443–448.

Broadbent P., Creissen G.P., Kular B., Wellburn A.R., and Mullineaux P.M. (1995) Oxidative stress responses in transgenic tobacco containing altered levels of glutathione reductase activity. *The Plant Journal,* **8**, 247–255.

Burke J.J. (1990) Variation among species in the temperature dependence of the reappearance of variable fluorescence following illumination. *Plant Physiology,* **93**, 652–656.

Burke J.J., Mahan J.R. and Hatfield J.L. (1988) Crop-specific thermal kinetic windows in relation to wheat and cotton biomass production. *Agronomy Journal,* **80**, 553–556.

Campbell G.S. (1981) Fundamentals of radiation and temperature relations. In *Physiological Plant Ecology I,* edited by O.L. Lange, P.S. Nobel, C.B. Osmond and H. Ziegler, pp 11–41. New York:Springer Verlag.

Ferguson D.L. and Burke J.J. (1991) Influence of water and temperature stress on the temperature dependence of the reappearance of variable fluorescence following illumination. *Plant Physiology,* **97**, 188–192.

Hale M.G. and Orcutt D.M. (1987) *The Physiology of Plants Under Stress.* pp 1–4. New York: John Wiley & Sons.

Ham J.M., Heilman J.L. and Lascano R.J. (1991) Soil and canopy balances of a row crop at partial cover. *Agronomy Journal,* **83**, 744–753.

Hochachka P.W. and Somero G.N. (1984) *Biochemical Adaptation.* Princeton, New Jersey:Princeton University Press.

Hodges H.F., Reddy K.R., Mckinion J.M. and Reddy V.R. (1993) Temperature effects on cotton. Mississippi Agricultural and Forestry Experiment Station, Bulletin no: 990.

Houghton R.A. (1989) Global Climate Change. *Scientific American,* **260**, 36.

Jones H.G. (1987) Breeding for stomatal characters. In *Stomatal Function,* edited by E. Zeiger, G.D. Farquhar and I.R. Cowan. Stanford, California:Stanford University Press.

Kati W.M. and Wolfenden R. (1989) Major enhancement of the affinity of an enzyme for a transition-state analog by a single hydroxyl group. *Science,* **243**, 1591–1593.

Kramer P.J. (1980) Drought stress, and the origin of adaptations. In *Adaptation of Plants to Water and High Temperature Stress*, edited by N.C. Turner and P.J. Kramer, pp 7–20. New York: John Wiley and Sons.

Kramer P.J. (1983) *Water Relations of Plants*. Orlando, Florida:Academic Press, Inc.

Levitt J. (1980) *Responses of Plants to Environmental stresses*. Vol. I. *Chilling Freezing and High Temperatures Stresses*, pp 3–17 and 347–348. New York: Academic Press.

Mahan J., Burke J. and Orzech K. (1990) The thermal dependence of the apparent K_m of glutathione reductases from three plant species. *Plant Physiology*, **93**, 822–824.

Mahan J.R. and Upchurch D.R. (1988) Maintenance of constant leaf temperature by plants-I. Hypothesis-Limited Homeothermy. *Environmental and Experimental Botany*, **28**, 351–357.

Mahan J.R. and Wanjura D.F. (1990) An enzyme-based model for predicting leaf temperature. *Winter Meeting ASAE,* Chicago, Illinois. Paper 904–505.

Mahan J.R., Wanjura D.F. and Upchurch D.R. (1991) The thermal dependence of enzyme function in cotton; a seasonal projection. ASA Abstracts, p 130.

Mateos L., Smith R.C.G. and Sides R. (1991) The effect of soil surface on the crop water stress index. *Irrigation Science,* **12**, 73–83.

Meidner H. and Mansfield T.A. (1968) *Physiology of Stomata*. London: McGraw-Hill.

Nelson L.E. (1967). Effect of root temperature variation on growth and transpiration of cotton (*Gossypium hirsutum* L.) seedlings. *Agronomy Journal*, **59**, 391–395.

Nobel P.S. (1991) *Physiochemical and Environmental Plant Physiology*. New York:Academic Press.

Oertli J.J. (1991) Transport of water in the rhizosphere and roots. In *Plant Roots*: *The Hidden Half*, edited by Y. Waisel, A. Eshel and U.Kafkafi, pp 559–588. New York:Marcel Dekker.

Patterson B.D. and Graham D. (1987) Temperature and metabolism. In *The Biochemistry of Plants, A Comprehensive Treatise*, edited by P.K. Stumpf and E.E. Conn, pp 153–199. New York:Academic Press.

Peeler T.C. and Naylor A.W. (1988) The influence of dark adaptation temperature on the reappearance of variable fluorescence following illumination. *Plant Physiology,* **86**, 152–154.

Profitt A.P.B., Berliner P.R. and Oosterhuis D.M. (1985) A comparative study of root distribution and water extraction efficiency by wheat grown under high-and low-frequency irrigation. *Agronomy Journal*, **77**, 655–662.

Quisenberry J.E., Roark B. and McMichael B.L. (1982) Use of transpiration decline curves to identify drought-tolerant cotton germplasm. *Crop Science,* **22**, 918–922.

Reddy K.R., Hodges H.F. and Mckinion J.M. (1993) Temperature effects on Pima cotton leaf growth. *Agronomy Journal*, **85**, 681–686.

Reddy K.R., Hodges H.F., Mckinion J.M. and Wall G.W. (1992) Temperature effects on Pima cotton growth and development. *Agronomy Journal,* **84**, 237–243.

Selinioti E., Manetas Y. and Gavalas N.A. (1986) Cooperative effects of light and temperature on the activity of phosphoenoplyruvate carboxylase from *Amaranthus paniculatus* L. *Plant Physiology,* **82**, 518–522.

Simon J.P. (1979) Adaptation and acclimation of higher plants at the enzyme level: latitudinal variations of thermal properties of NAD malate dehydrogenase in *Lathyrus japonicus* Willd. (Leguminosae). *Oecologia,* **39**, 273–287.

Simon J.P., Potvin C. and Blanchard M. (1983) Thermal adaptation and acclimation of higher plants at the enzyme level: Kinetic properties of NAD malate dehydrogenase and glutamate oxaloacetate transaminase in two genotypes of *Arabidopis thaliana* (Brassicaceae). *Oecologia*, **60,** 143–148.

Somero G.N. (1975) Temperature as a selective factor in protein evolution: The adaptional strategy of "compromise". *Journal of Zoology*, **194**, 175–188.

Somero G.N. and Low P.S. (1976) Temperature: a shaping force in protein evolution. *Biochemistry Society Symposia*, **41**, 33–42.

Taylor H.M.(1983) Managing root systems for efficient water use: an overview. In *Limitations to Efficient Water Use In Crop Production*, edited by H.M. Taylor, W.R. Jordan and T.R. Sinclair, pp 87–113. Madison, Wisconsin:American Society of Agronomy.

Teeri J.A. and Peet M.M. (1978) Adaptation of malate dehydrogenase to environmental temperature variability in two populations of *Potentilla glandulosa* Lindl. *Oecologia*, **34**, 133–141.

Upchurch D.R. and Mahan J.R. (1988) Maintenance of constant leaf temperature by plants. II. Experimental observations in cotton. *Environmental and Experimental Botany*, **28**, 359–366.

Wanjura D.F., Upchurch D.R. and Mahan J.R. (1990) Evaluating decision criteria for irrigation scheduling of cotton. *Trans ASAE,* **33**, 512–518.

Wanjura D.F., Upchurch D.R. and Mahan J.R. (1992) Automated irrigation based on threshold canopy temperature. *Trans ASAE,* **35**, 1411–1417.

Wanjura D.F. and Mahan J.R. (1994) Thermal environment of cotton irrigated using canopy temperature. *Irrigation Science,* **14**, 199–205.

Wilks H.M., Hart K.W., Feeney R., Dunn C.R., Muirhead H., Chia W.N., *et al.*, (1988) A specific, highly active malate dehydrogenase by redesign of a lactate dehydrogenase framework. *Science*, **242**, 1541–1544.

Yamuchi A., Taylor H.M., Upchurch D.R., and Mcmichael B.L.(1995). Axial resistance to water flow of intact cotton taproots. *Agronomy Journal,* **87**, 439–445.

8. RESISTANCE OF PLANTS TO THE EFFECTS OF ULTRAVIOLET RADIATION

TERENCE M. MURPHY

Department of Plant Biology University of California Davis, Davis, CA 95616, USA.

INTRODUCTION

The world is bathed in electromagnetic radiation from the sun, radiation that spans a wide range of wavelengths and corresponding energies (Figure 8.1). In the ultraviolet region of the spectrum, photons have relatively high energy and can excite binding electrons in many inorganic and organic molecules. It is thought that UV radiation provided much of the energy that powered the synthesis of the first bioorganic molecules on the primordial Earth. However, excitation of the electrons of nucleic acids, proteins, and other biochemical molecules can lead to chemical reactions that destroy the biological function of the molecules and damage or kill cells, tissues, and whole organisms. The appearance in the atmosphere of oxygen and ozone, which absorb most of the higher-energy UV photons from the sun, is one of the factors that allows life to exist on this planet. Nevertheless, the UV that still penetrates the atmosphere is sufficiently strong to affect living organisms adversely.

Obligate photosynthetic organisms like green plants are naturally exposed to high fluences of sunlight and to relatively high fluences of the ultraviolet photons that do penetrate the atmosphere. It is interesting to consider how much the ultraviolet radiation stresses these organisms and how they protect themselves from the radiation. This chapter focuses on the effects of UV on cellular functions. As will be seen, UV affects a variety of physiological processes in plants. In some cases, it is possible to identify the target molecules. But in others, it is not only difficult to identify the targets and the chain of events by which physiology is altered, but also to decide whether the observed effects represent the consequences of damage or aspects of a defensive response.

The mechanisms of resistance may take several forms. The most basic is the attenuation of the radiation within cells and tissues by absorption or reflection. Another is the repair of damage to sensitive molecules, especially DNA. A third involves interference with the mechanism by which ultraviolet radiation damages molecules, for instance, by the scavenging of activated oxygen species. In the following sections, I will describe studies on all three mechanisms.

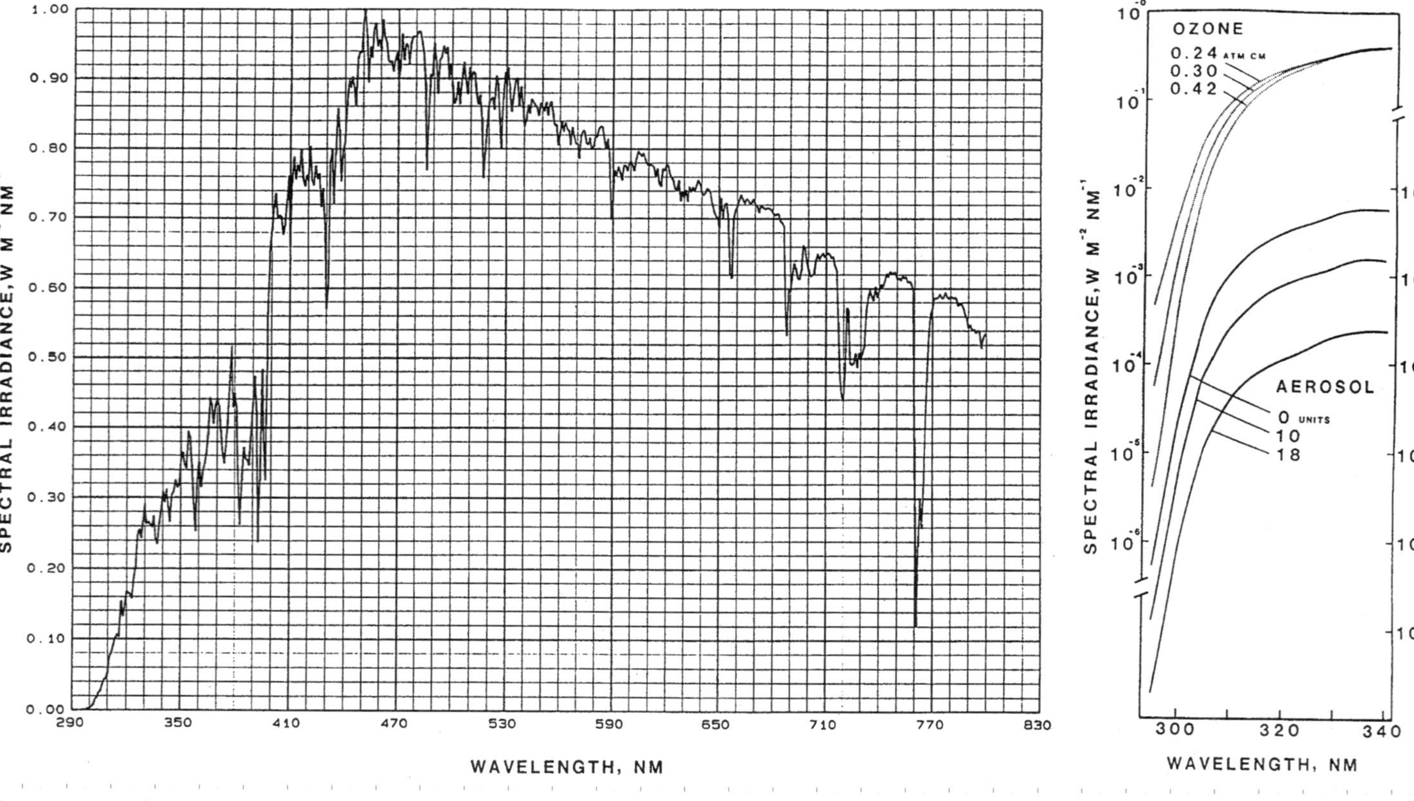

Figure 8.1. The solar spectrum. The left-hand graph shows the UV and visible spectrum of sunlight on March 10, 1990, 1400 h, at Orlando, FL, USA (courtesy of Optronic Laboratories, Inc., Orlando, FL, USA). Values should be multiplied by 0.894; note the linear ordinate. The right-hand graph shows the effect of changes in stratospheric ozone and of relative aerosol concentration on the UV-B portion of the spectrum as predicted by a semi-empirical model (Björn and Murphy, 1985); note the logarithmic ordinate.

THE DAMAGE CAUSED BY UV RADIATION

Spectral Ranges

Ultraviolet radiation is by convention divided into three bands (CIE, 1987), each with a different energy and a different physiological and ecological significance. UV-C (190–280 nm) includes the germicidal radiation of low pressure mercury vapor lamps (primarily at 254 nm) often used for sterilization in microbiology. It is strongly cytotoxic, but UV-C from the sun is absorbed by diatomic oxygen and ozone in the upper atmosphere, so it is not now a factor in the natural environment. Early experiments demonstrated the basic effects of germicidal-lamp UV-C on leaves: a "bronzing" color change ascribed to necrosis of the epidermis of *Phaseolus* primary leaves (Bawden and Kleczkowski, 1952), accelerated chlorosis of *Glycine* leaf disks (Tanada and Hendricks, 1953), and reduction in growth capacity of several organs (Klein, 1967). It was observed that the sensitivity of plants to UV-C varied greatly among species (Cline and Salisbury, 1966).

UV-B (280–315 nm) is the band of lowest wavelength and highest energy that penetrates the ozone layer of the stratosphere. It is absorbed by most biological molecules less efficiently than UV-C, but fluctuations in ozone concentration and in the pathlength through the ozone layer (angle of incidence of sunlight relative to the ground), as well as humidity, aerosols, and other factors (Figure 8.1, Björn and Murphy, 1985) make large differences in the amount of UV-B radiation that reaches the ground. Thus, resistance to radiation in this band is of major concern in considerations of changes in the stratospheric ozone layer. The effects of UV-B on the growth and function of crops and wild plants have been studied intensively over the last two decades. In general, relatively high fluences of UV-B have effects on growth, especially leaf expansion (Teramura and Sullivan, 1987) and anatomy (Bornman and Vogelmann, 1991), photosynthesis, biomass production and allocation, and reproduction, with high variability among species and among cultivars of a single species (Caldwell, 1981; Teramura, 1983; Tevini and Teramura, 1989).

UV-A (315–400 nm) is absorbed by a few highly unsaturated biological compounds, and it can influence some physiological processes. However, it is not highly cytotoxic. Because of this and because it represents a high and fairly constant proportion of the energy in sunlight (with or without an ozone layer), the question of resistance to radiation of this band has not been seriously considered.

Photobiochemical Aspects

The molecules that can be altered by photons of ultraviolet radiation include some of the most essential components of every living cell. Nucleic acids, proteins, and cofactors like flavins and porphyrins absorb ultraviolet radiation, and many other compounds can react with products formed by UV-excited pigments.

All the bases of DNA and RNA absorb ultraviolet radiation, primarily in the UV–C band (peaking at 260 nm) but also to a significant degree in the UV-B band. The major photoproducts detected after irradiation of DNA and RNA *in vitro* are pyrimidine complexes: cyclobutadipyrimidines and (6–4)pyrimidine-pyrimidone adducts. Cyclobutadipyrimidines have been detected in DNA extracted from irradiated plant cells (Soyfer, 1983; Pang and Hays, 1991; Quait *et al.*, 1992). Photoproducts in DNA prevent the DNA from acting as a template for replication and transcription. Thus cells of all types of organisms, including plants, have evolved ways of repairing UV-induced damage to DNA. Photoproducts in RNAs also prevent them from functioning, with mRNA being the most sensitive. There is evidence for repair of UV-damaged viral RNA in plants (first report, Bawden and Kleczkowski, 1959), but no evidence for repair of the plants' own RNA molecules or for reactivation of processes in irradiated plant cells that can be ascribed to repair of RNA (Wright and Murphy, 1975).

The aromatic amino acids of proteins, as well as cysteine, absorb and can be oxidized by UV-C. The sensitivity of tryptophan extends into the UV-B band. In general, the inactivation of enzymes requires higher fluences than does damage to nucleic acids. A plant enzyme that has been demonstrated to be inactivated by relatively low fluences of UV-C and UV-B radiation is the plasma membrane H^+-ATPase (Imbrie and Murphy, 1984a,b,c).

The photochemical reactions stimulated by ultraviolet light in living cells almost always take place in the presence of oxygen, and oxygen can participate in three main ways (Foote, 1991).

In certain organic molecules, for instance tryptophan, that have been excited by ultraviolet photons, the electrons form triplet states with a low but predictable frequency. The energy in an excited triplet-state molecule can be transferred efficiently to O_2, a natural triplet-state molecule. The product, singlet oxygen, is much more reactive than triplet oxygen, and it adds readily to unsaturated carbons in proteins and lipids and to guanine in DNA (photodynamic Type II mechanism). In the case of the plasma membrane H^+-ATPase, tryptophan excitation leads to the production of singlet oxygen and to the oxidation of the tryptophans in the protein. The loss of tryptophan is considered the cause of the inactivation of the enzyme (Imbrie and Murphy, 1984a,b).

Other organic molecules that absorb ultraviolet radiation may oxidize or reduce a substrate molecule. The product radicals may react with oxygen (photodynamic Type I mechanism).

If the excited molecule loses one electron to oxygen, superoxide (O^-_2) ions are formed. The superoxide ions can add to unsaturated carbons, although in practice they are not very reactive. More importantly, superoxide can form hydrogen peroxide by further reduction or by dismutation; then more superoxide can reduce Fe^{3+}, and the resulting Fe^{2+} can react with hydrogen peroxide to form hydroxyl ions, which are extremely reactive, oxidizing DNA, lipids, and other molecules (see Halliwell and Gutteridge, 1989).

It is important to realize that not all the effects of UV radiation on plants need result from the direct or indirect damage described above (Hashimoto *et al.*,

1993). Plants have adapted to respond to light of various wavelengths. Two putative photoreceptors specifically respond to ultraviolet wavelengths: a UV-A and blue light photoreceptor, sometimes called "cryptochrome" (Mohr and Drumm-Herrell, 1983), and a UV-B photoreceptor (Wellman, 1976). Each of these affects plant development in various ways. Thus, a change in growth pattern or in a metabolic pathway need not reflect "stress," as defined by damage or even by a lack of optimum conditions. It may be a naturally programmed response, and it may have (although it need not have) an adaptive role in allowing a plant to survive higher fluences of UV radiation.

Photophysiology

Gene expression

In spite of the fact that nucleic acids are among the biological molecules most sensitive to UV radiation, and in spite of considerable information that UV-induced damage to nucleic acids affects their various functional properties, there have been few studies in plants demonstrating and quantifying the inhibition of transcription or translation by UV.

The addition of UV-B radiation at moderate intensities to the illumination of *Pisum sativum* plants caused a rapid decrease in the concentration of mRNAs for both the small and large subunits of ribulose 1,5 bisphosphate carboxylase/oxygenase (Rubisco) (Jordan *et al.*, 1992). Strid (1993) observed a similar decrease in the concentration of mRNA for chlorophyll a/b binding protein and mRNA for a chloroplast superoxide dismutase. Because the concentrations of mRNAs are affected by several processes, including rates of synthesis and degradation, it is not possible to conclude that the decreases reflected damage to template DNA. Indeed, in the experiments reported by Strid (1993), the mRNAs for chalcone synthase and glutathione reductase increased in concentration.

Some early studies demonstrated the sensitivity of the components of the wheat embryo translational system *in vitro* to UV-C radiation (Murphy *et al.*, 1973): mRNA was by far the most sensitive component. In an extension of this work, it was shown that UV-C inhibited amino acid incorporation by suspension cultured tobacco cells (Murphy *et al.*, 1975). Yet while an incident fluence of 1200 J m^{-2} inhibited *in vivo* amino acid incorporation by approximately 90%, polysomes isolated from such cells retained 40–85% of their *in vitro* amino acid incorporating activity. The cause of the additional loss of protein synthetic capability *in vivo* has never been explained.

If gene expression were sensitive to the levels of UV-B in sunlight, a possibility for which there is little data, a plant's ability to grow and to respond to other stresses could be impaired.

Photosynthesis

The photosynthetic system is particularly sensitive to damage by ultraviolet radiation (see Iwanzik *et al.*, 1983; Tevini and Iwanzik, 1983 for references). The primary sites of damage are in photosystem II; photosystem I is resistant to inactivation by UV-C and UV-B at the fluences that inactivate photosystem II (Iwanzik *et al.*, 1983; Kulandaivelu and Noorudeen, 1983).

Both UV-C and UV-B inhibit oxygen evolution, variable fluorescence, and regeneration of oxidized photosystem I in isolated chloroplasts (Kulandaivelu and Noorudeen, 1983). In this study, the effects of UV-C seemed to be more complex than those of UV-B, as analyzed with different electron donors and receptors. Inhibitory effects of UV-C were inferred at the water-splitting enzyme, the photosystem II reaction center, and plastoquinone. In contrast, UV-B radiation apparently damaged only the reaction center, since systems in which the oxygen-splitting enzyme was bypassed or removed and systems in which plastoquinone was bypassed were inhibited just as much as control systems (Iwanzik *et al.*, 1983; Kulandaivelu and Noorudeen, 1983).

Some further work, however, indicated that UV-B radiation inhibits oxygen evolution by photosystem II membrane fragments from spinach through a destruction of the water-splitting function. Bornman *et al.* (1984) noted a substantial reversal of the effects of 280 nm radiation on fluorescence kinetics of spinach thylakoids by hydroxylamine, which donates electrons to the oxidizing side of photosystem II. Renger *et al.* (1989) demonstrated that inhibition by UV-B was not observed in the presence of diphenylcarbazide, which also substitutes for water as a donor of electrons. Furthermore, UV-B slowed the relaxation kinetics of the flash-induced absorbance change at 830 nm (a measure of P_{680} redox state) in a manner expected for a destruction of the water-splitting site. Although there was no loss of protein from the membrane fragments as assessed by SDS-polyacrylamide gel electrophoresis, UV-B radiation resulted in the loss of atrazine binding sites, and Renger *et al.* (1989) suggested that the D-1/D-2 reaction center proteins sustained the primary damage, perhaps through release of a critical manganese atom (Figure 8.2). Melis *et al.* (1992) obtained somewhat conflicting results, since inhibition was still observed in the presence of diphenylcarbazide; they found inhibition of the bound quinone, Q_A, and the overall reduction of the plastoquinone pool. A coordinated loss of manganese, oxygen evolution, and herbicide-binding capacity has been observed in other contexts (Jursinic and Stemler, 1983). The action spectrum for the inhibition observed by Renger *et al.* (1989) was relatively featureless, rising in efficiency as the wavelength decreased from 350 to 250 nm. This spectrum did not identify a particular chromophore as responsible for the damage, nor did it suggest that the mechanism of damage differed through this wavelength range.

Greenberg *et al.* (1989) demonstrated that UV radiation accelerated the turnover of radioactively labeled 32–kDa photosystem II reaction center protein (D-1 protein) in *Spirodela oligorrhiza* chloroplasts. By reducing the amount of chlorophyll, the authors could distinguish the photoreceptor for UV-induced turnover from the photoreceptor(s) for turnover induced by visible light. Visible-light turnover probably involved chlorophyll and other accessory pigments. The action spectrum for the UV-specific effect extended through the UV-A, B, and C ranges and bore a resemblance to the absorbance spectrum of the plastosemiquinone radical. Filtering UV-B radiation from sunlight reduced the rate of turnover of the D-1 protein, suggesting that UV-B was responsible for a substantial part (ca 33%) of the destruction of the protein under normal conditions. From their work with spinach thylakoids, Melis *et al.* (1992) also suggested that plastoquinone was the photosensitizer in the UV-B-stimulated turnover of photosystem II D-1/32-kDa protein.

In chloroplasts of *Vigna sinensis*, as in those of spinach, the UV-B-dependent inhibition of photosystem II activity, estimated with water or $MnCl_2$ as electron donor, was also alleviated when diphenylcarbazide or hydroxylamine was the electron donor (Nedunchezhian and Kulandaivelu, 1991). However, the changes in chloroplast proteins were more complex, with several polypeptides, including subunits of Rubisco and photosystem II reaction center and antenna complexes, decreasing in amount and rate of synthesis. A few polypeptides, particularly some from isolated photosystem II particles, increased in amount, although the mechanism and significance of the increase is not known.

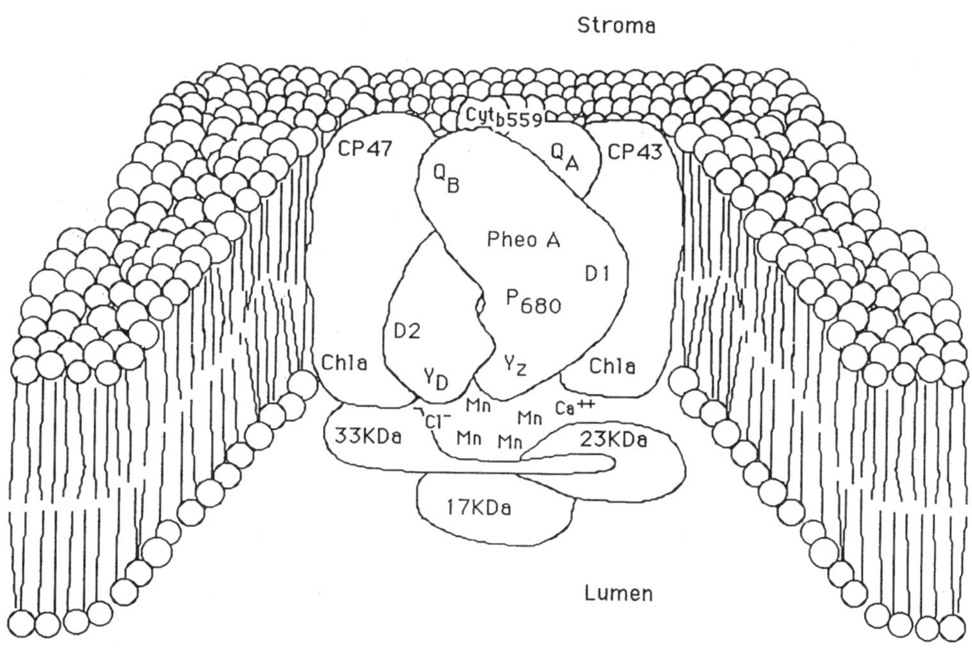

Figure 8.2. Model of photosystem II, adapted from Yocum (1991), showing the central position of the D1 and D2 polypeptide chains. The model demonstrates how damage to D1 could affect water splitting, the reaction center, and the atrazine binding site.

The results obtained with irradiated chloroplasts and thylakoids are reflected in the properties of chloroplasts isolated from UV-B-irradiated plants. Strid *et al.* (1990), who compared pea plants irradiated (or not) between the period 17 to 24 days after germination, found that irradiation reduced photosystem II reaction centers, when measured on a chlorophyll basis, but not photosystem I or cytochrome *f*. They also found a reduction in ATP synthase activity and in the amount of Rubisco in chloroplasts from irradiated plants

In summary, UV radiation damages the water-splitting function and the reaction center of photosystem II. The radiation also accelerates the turnover of the D-1 protein and other associated proteins, perhaps because the break in the smooth flow of electrons stimulates rogue oxidation-reduction reactions.

Membrane structure and function

Numerous lines of evidence suggest that membrane integrity may be compromised by ultraviolet radiation (see Murphy, 1983 for review). For instance, irradiated cells leak solutes through the plasma membrane, tonoplast, or chloroplast envelope. In an ultrastructural study, Bornman *et al.* (1983) found gaps in the images of the envelopes of chloroplasts isolated from UV-B-irradiated sugar beet leaves. Despite a large number of descriptive and analytical studies, there is much to learn about the basics of UV-induced damage to membranes. The damage could occur through the oxidation of unsaturated lipids or membrane proteins. It might represent a direct absorption of photons by membrane components or an indirect reaction with products of a non-membrane-associated primary photochemical reaction. Or oxidizing agents could be formed by metabolic reactions triggered by the UV, rather than by photodynamic effects. A loss of membrane integrity could also represent autolytic processes in a necrotic cell, as seems likely when cells or protoplasts lose the ability to accumulate fluorescein after several hours of irradiation with UV-B radiation (Bornman *et al.*, 1982).

Perturbations of ion transport across the plasma membrane of *Chara corralina* were studied extensively by Doughty and Hope (1973; 1976a,b,c). Irradiation of giant internodal cells of this organism with UV-C or UV-B produced a transient depolarization of the membrane by 10 to 30 mV, with kinetics that depended on the pH of the external medium. The authors attributed the depolarization to a rapid increase in the permeability of the membrane to Cl^- and to a subsequent inactivation of an electrogenic pump. The two effects could be distinguished, because the inactivation of the pump predominated at higher pH (7 to 8), when the cells were hyperpolarized. Also, an increase in the permeability to Cl^- increased the conductivity of the membrane, while the inactivation of the pump decreased the conductivity. A loss of K^+ and Na^+ followed the loss of Cl^-, presumably in response to the depolarization of the membrane. These effects, together with changes in the peak height and duration of the tonoplast action potential, indicate that UV radiation affects several components of membranes.

Similar conclusions were drawn by Stolarek and Karcz (1987), who studied *Nitellopsis obtusa*, a relative of *Chara*. In *N. obtusa*, UV-C radiation produced a more marked depolarization, up to 80 mV, to which depolarization of the plasma membrane and hyperpolarization of the tonoplast both contributed, with a return in potential in approximately 10 min to a steady (but still depolarized) state. In cells hyperpolarized by visible light or by the addition of IAA, UV-C induced an action potential (very rapid depolarization of ca 170 mV with a rapid recovery). The increase in conductivity across both membranes was attributed to open Cl^- channels. The depolarization after recovery to a steady state, nullifying the hyperpolarizing effects of light and IAA, was attributed to inactivation of the plasma membrane proton pump.

Effects at least superficially similar to those seen in the Characeae occur in the cells of higher plants. Irradiation with UV-C induced a depolarization of cells of *Malus domestica* (Karcz and Stolarek, 1981) and *Zea mays* (Karcz and Stolarek, 1988) with kinetics that were like those seen with *Chara* and *Nitellopsis*, that is, a depolarization in 1–2 min followed by a return to a steady, but still depolarized, state in 10–20 min.

Irradiation of cultured, non-photosynthetic plant cells (from tobacco, rose, or tomato, for example) induces an efflux of K^+ (Wright and Murphy, 1978; Murphy and Wilson, 1982), with a total loss of up to 33% of the cells' stores of this ion. The effect begins 10–15 minutes after a short (2 min) exposure to approximately 1000 J m^{-2} of UV-C and proceeds for approximately 60 minutes. While UV-C is most effective, wavelengths in the UV-B range also produce the effect (Murphy *et al.*, 1985). In fact, in the UV-B region, the action spectrum for K^+ efflux is very similar to other spectra identified with the UV-B photoreceptor (Wellman, 1976; Hashimoto *et al.*, 1993).

The K^+ efflux is considered to be a change in the ion transport properties of the plasma membrane, rather than a loss of integrity due to physical or chemical damage, for three main reasons: (a) The efflux is specific. There is little or no influx of mannitol into or efflux of 2-deoxyglucose or fluorescein out of irradiated cells (Wright and Murphy, 1978); there is also little effect on net Cl^- and Ca^{2+} fluxes into or out of the medium and no effect on net nitrate and phosphate fluxes (Murphy and Wilson, 1982). The membrane potential, while reduced, is still present. (b) The efflux requires active metabolism. Treatments that reduce the energy charge of the cells (anoxia, carbonylcyanide m-chlorophenylhydrazone), while they may cause a slow leakage of K^+ themselves, inhibit the more rapid response to UV-C (Murphy and Wilson, 1982; Murphy, 1988a). (c) The efflux is inhibited by treatments that tend to interfere with membrane function. It is blocked at 0°C, even though cold treatments generally increase the permeability of lipid bilayers to ions. It is also inhibited by a one-hour treatment at 40°C given before the radiation (Murphy and Huerta, 1989c). Finally, UV-induced efflux of K^+ is reduced to zero as the concentration of Ca^{2+} is reduced from 10^{-4} M to 10^{-5} M, a change that also depolarizes the plasma membrane and blocks uptake of amino acids (Murphy, 1988b).

The nature of the change in ion transport that is induced by UV is still not fully understood. The rapid changes in efflux of K^+ (accelerating at 15 minutes after irradiation, stopping after 60 minutes) do not correlate with changes in membrane potential (which occurs sooner and remains constant for five hours) nor with external K^+ concentration (Huerta and Murphy, 1989b). K^+ channels are known to open following a depolarization (Köhler *et al.*, 1985). But the lack of a change in membrane potential at the time when K^+ efflux begins and ends suggests that K^+ leaves the cells by a non-electrogenic pathway, that is, closely coupled with the efflux of an anion or the influx of another cation.

It is clear that the efflux of K^+ is just one of a suite of effects induced by UV radiation (Table 8.1).

Table 8.1. Membrane-related responses of cultured rose cells to UV-C irradation.

Response	Time after treatment	Reference
Depolarization	Starting in seconds, partial recovery in 5 min, lasting for hours	Huerta and Murphy, 1989b
pH change	Starting in 5–10 min, lasting 30–120 min	Murphy and Wilson, 1982; Murphy *et al.*, 1983.
Inhibition of ferricyanide reduction	Starting in <10 min	Murphy *et al.*, 1991
K^+ efflux	Starting in 8–15 min, lasting 60–90 min; recovery of medium K^+ in 6–10 h	Murphy and Wilson, 1982; Huerta and Murphy, 1989b
H_2O_2 synthesis	Starting in 8–15 min; lasting 60-90 min	Murphy and Huerta, 1990a
Appearance of oxidized glutathione (GSSG)	Starting in <30 min; continuing at least 6 h	Huerta and Murphy, 1989c

The appearance of K^+ in the medium of UV-irradiated rose cells is matched in part by an appearance of HCO_3^-. In the cultured rose cells, which are not photosynthetic but instead respire rapidly, producing a continual efflux of CO_2, the appearance of HCO_3^- could result from an actual efflux of HCO_3^-, an influx of H^+, or an efflux of OH^- together with CO_2. It has been difficult to determine which process in fact occurs. However, the UV-induced efflux of K^+ occurs as rapidly at pH 9 as at pH 5, suggesting that H^+ influx *per se* is not essential (Huerta and Murphy, 1989a).

The UV-induced ion fluxes in rose cells are accompanied by an acidification of the cytoplasm by 0.2 to 0.3 pH units, as determined by ^{31}P-NMR spectroscopy (Murphy *et al.*, 1983). The source of the protons is most likely carbonic acid, formed by respiratory CO_2, followed by export of HCO_3^-, although as explained above it could involve a different process at the plasma membrane. The change in cytoplasmic pH is not as great as expected, given the amount of HCO_3^- lost by the cell. Some of the protons are transported into the vacuole, where the pH drops by 0.25 units. A large fraction (up to 50%) of the protons may be absorbed by the reactions involved in the reduction of nitrate to γ-aminobutyric acid, a pathway that is stimulated in the irradiated cells (Murphy *et al.*, 1983). In this context, it is interesting that UV-B radiation induces the accumulation of putrescine and spermidine in soybean and cucumber plants (Kramer *et al.*, 1991a,b). The synthesis of these organic amines from nitrate would also remove protons from the cytosol.

The irradiation of rose cells with UV-C also induces a rapid synthesis of H_2O_2 by the cells. Although in principle H_2O_2 could be formed from photodynamically produced O_2^-, in this case it seems to be a product of metabolic reactions, since it does not appear until 8–10 minutes after a 3-minute irradiation and accumulates for approximately 60 minutes (to a maximum of 8–9 μM). Its synthesis may depend in some way on the efflux of K^+, since it appears in the extracellular medium at the same time, and its appearance is inhibited by many of the same environmental factors that inhibit UV-induced K^+ efflux (Murphy and Huerta, 1990a). Gross *et al.* (1977), Halliwell (1978), and others have postulated a mechanism for the extracellular synthesis of H_2O_2 by horseradish roots. This mechanism depends on the flow of reducing equivalents (in their model, malic acid) across the plasma membrane, as well as on cell wall peroxidase, to produce O_2^-. in the apoplast; H_2O_2 is formed from the dismutation of O_2^-. UV-A and blue light inhibit rose cell peroxidase, and they inhibit the synthesis of H_2O_2 by UV-C-irradiated cells. On the other hand, added malate does not stimulate the formation of H_2O_2 by rose cells, so in these cells the supply of reducing equivalents may come from another source (Murphy and Huerta, 1990b). Isolated plasma membranes from cultured rose cells have NAD(P)H oxidase/O_2^- synthase activities. Diphenylene iodonium, which inhibits H_2O_2 synthesis by these cells, blocks the activity of the plasma membrane O_2^- synthases, but has no effect on peroxidase (Auh and Murphy, 1995), suggesting that the synthesis of O_2^- by UV-stimulated cells may occur through trans-plasma membrane electron transport.

The irradiation of cultured rose cells with UV-C also affects the electron transport properties of the plasma membrane (Murphy *et al.*, 1991; Murphy and Auh, 1992), with a moderate inhibition of the ability of the cells to reduce extracellular ferricyanide ion to ferrocyanide. While the authors suggested that changes in the electron transport properties of the plasma membrane may be related to the synthesis of H_2O_2, reflecting a diversion of electrons to enzymes that can reduce O_2 to superoxide, evidence in support of that proposition is still lacking.

Some very basic questions about these UV-stimulated changes in membrane function remain unanswered. One concerns the photoreceptor that is sensitive to the UV radiation. Is it the same as the UV-B receptor described by Wellmann (1976)? Indeed, is it the same for each of the responses described above (depolarization, K^+ efflux, H_2O_2 synthesis, inhibition of ferricyanide reduction)? Two possibilities for a photoreceptor may be mentioned: one is the H^+-ATPase. Two forms of plasma membrane H^+-ATPase were reported by Imbrie and Murphy (1984c), one form 15-fold more sensitive than the other. Even the less sensitive form is more sensitive than most enzymes (37% activity after irradiation *in vivo* with 3500 J m^{-2} UV-C). A direct inhibition of the ATPase would have a depolarizing effect on the plasma membrane. Other responses might be triggered by the depolarization. A second possible photoreceptor is the cytoplasmic translational apparatus: partial inhibition might in some way release a signal that affects membrane function. UV radiation inhibits translation. Furthermore, two inhibitors, cycloheximide and emetine, also partially inhibit translation in rose cells and stimulate K^+ efflux (Murphy, 1988a). Without more definitive evidence, these suggestions remain speculative.

The intermediate steps in the transduction of the signal from UV photoreceptor to response organ are also mysterious. Do some of the responses, such as depolarization, represent such intermediates, as suggested above? Do other, well recognized signal pathways contribute? External calcium at 10^{-4} M or greater is necessary for UV-induction of K^+ efflux, but so far the evidence suggests that the calcium is needed more for membrane function than for induction (Murphy, 1988b): that is, when the external calcium concentration is reduced to 10^{-5} M, the membrane potential drops to almost nothing.

Because the loss of K^+ from cultured plant cells is substantial, the change in cytoplasmic pH is significant, and the H_2O_2 outside the cells reaches toxic levels, the effects of UV radiation may contribute to an impairment of the cell's overall metabolism. However, there is no evidence that this syndrome contributes substantially to the mortality of the plant cells irradiated with ultraviolet light. With low fluences of UV, the syndrome can be reversible, and it can be induced in cultured rose cells by other stresses which do not kill the cells, for instance, low concentrations of cycloheximide (Murphy, 1988a) or fungal cell-wall elicitor (Arnott and Murphy, 1991). The same observations suggest that the plasma membrane responses are not symptoms of progressive necrosis in cells that are slated to die. Instead, they may be aspects of an adaptive response to radiation stress. They may also be related to the mechanism for closing of stomata, which is seen in leaves that are irradiated with UV-C and UV-B.

Stomata

Many investigators have reported that the irradiation of illuminated leaves with UV-C causes previously open stomata to close (Herçík, 1964; Wright and Murphy, 1982; Negash and Björn, 1983). The closing of the stomata correlates with the extensive loss of K^+ (or $^{86}Rb^+$) from guard cells, as seen in tobacco (Wright and Murphy, 1982) and *Vicia faba* (Negash *et al.*, 1987). The effective wavelengths extend into the UV-B; UV-A, in contrast, causes an increased uptake of $^{86}Rb^+$ (Negash and Björn, 1986). An increased fluence of white light or UV-A can effect some reversal of UV-C-induced closure in *Eragrostis tef*, although not in *V. faba* (Negash and Björn, 1986; Negash, 1987).

The closing of stomata could be responsible for part of the inhibition of photosynthetic gas exchange observed in irradiated plants, and in fact gas exchange analysis confirms this in some, although not all, cases (see Teramura, 1983; Day and Vogelmann, 1995). Thimann and Satler (1979a,b) have also correlated the closing of stomata with accelerated senescence. As mentioned above, UV-C has been shown to accelerate senescence of detached leaves (Tanada and Hendricks, 1953). If the connection between stomatal aperature and senescence is real, then reversal of stomatal closure by visible light could provide an alternative explanation for the "photoreactivation" of UV-induced senescence (proposed by Tanada and Hendricks, 1953 to represent repair of nucleic acid damage).

Growth and development

In general, UV radiation inhibits the growth of plant organs. There may be several mechanisms involved, and it is not possible to assume that the inhibition represents simple damage to the tissues.

De Zeeuw and Leopold (1957) and Klein (1967) showed that irradiation of various plant organs with UV-C altered the way in which these organs responded to auxin. Irradiated rice coleoptile segments, for instance, required a higher level of naphthalene acetic acid for maximum elongation (Klein, 1967). The maximum amount of elongation was also reduced by the irradiation, an observation that could be explained by effects on gene expression or plasma membrane proton efflux.

UV-B radiation also inhibits the elongation of cucumber hypocotyls (Ballaré *et al.*, 1991) and of cress hypocotyls and roots (Steinmetz and Wellmann, 1986). Distinctions in the temperature dependence of the inhibition of cucumber hypocotyl growth suggests that there are at least two different mechanisms responsible, one predominating at 280–300 nm, the other at 300–320 nm (Takeuchi *et al.*, 1993).

It is a common observation that plants grown under moderate levels of UV-B radiation produce smaller leaves (for example, cucumber, Takeuchi *et al.*, 1989 and Adamse and Britz, 1992; soybean, Sullivan and Teramura, 1990; rice, Teramura *et al.*, 1991). Often, although not always, the leaves are thicker and have a greater weight per unit area. This is not necessarily an inhibition of growth, but may be a change in the pattern of cell division and elongation, leading to a thicker leaf with more cell layers. It may have an important adaptive effect in contributing to the shielding of sensitive targets in the tissue.

When cotyledons of 5-day-old *Brassica napus* seedlings are irradiated with UV-B for 100 minutes, they respond by curling toward the radiation source 12 to 24 hours later (Wilson and Greenberg, 1993). This growth response probably represents an inhibition of the rate of expansion of the irradiated, adaxial side of the cotyledons relative to the unirradiated, abaxial side. Neither UV-A nor blue light is effective, so it seems to be a true UV-B effect (UV-C was not tested, however). The mechanism was not determined, but a reduction in expansibility of the adaxial cells, perhaps through cross-linking of the wall components, is a possibility.

Motility

Planktonic algae are much more sensitive to UV radiation than are higher plants. This is true both for UV-B (Worrest, 1983) and UV-A (Manabe, 1986). These algae lack extensive tissue structures that scatter and attenuate radiation. Instead, they may adapt to fluctuations in UV-B and UV-A (and visible light) by moving to safer depths. With this in mind, it is particularly significant that UV-B radiation damages the ability of many prokaryotes and protists to move and to control their movements. UV-B irradiation of the gliding cyanobacterium, *Phormidium*, inhibited movement of a substantial fraction of the filaments and independently affected phototaxis and the step-up and step-down photophobic responses (Häder, 1984; Häder *et al.*, 1986). UV-B irradiation of the flagellated protist, *Euglena gracilis*, also inhibited movement and negative (but not positive) phototaxis (Häder, 1985; Häder and Häder, 1988). Similar effects have been found with non-photosynthetic protists (*Dictyostelium discoidium*, Häder, 1983). In each case, the loss of motility and/or control, which occurs at low UV-B fluences, tends to inhibit an organism's ability to find areas with the optimum light intensity, subjecting it either to damagingly intense sunlight or to insufficient radiation for photosynthesis (Häder and Worrest, 1991).

Because the control of cellular motility involves electrical and ion transport properties of the plasma membrane (Machemer and Sugino, 1989), the inhibition by UV-B of movement or of changes in movement may be another aspect of the effect of UV on membrane function.

Summary

The general sensitivity of various bioorganic compounds to ultraviolet radiation means that there are many potential targets of UV in any plant cell.

In practice, irradiation of plants and cultured plant cells with UV-C or UV-B alters several aspects of their physiology. Among the most sensitive functions are those that depend on nucleic acids as templates, photosynthesis, ion transport across the plasma membrane, and motility.

There are no *in vivo* cases in which the full chain of events, from absorption of UV to change in cell function, is known. In most cases, even the actual target of the radiation is unknown.

Some responses of irradiated cells that appear to be "damage" may in fact be aspects of defense against harm.

If many of the targets are biological molecules that are found in every cell, then the variation in sensitivity among different plant species, varieties within a species, organs of one plant, and plants of the same variety grown in different environments indicates that plants differ in their ability to resist the damage.

RESISTANCE BY SHADING

Effectiveness of Pigments

Higher plants contain compounds in their cuticle and epidermis that selectively absorb ultraviolet radiation. Caldwell and his colleagues demonstrated that the transmittance of ultraviolet radiation through epidermal peels varied among species and that it varied according to growing conditions; in particular, transmittance was lower when plants were exposed to UV-B (Caldwell, 1968; Robberecht and Caldwell, 1978, 1983). UV-absorptive compounds (presumably flavonoids and other phenolic compounds) could be extracted from epidermal peels with aqueous/methanolic solvents, and it was calculated that they accounted for 20–60% of the attenuation of radiation (Robberecht and Caldwell, 1978, 1983). The remaining attenuation probably resulted from phenolic compounds covalently connected to the cuticle or the epidermal cell walls.

The penetration of UV-B through the epidermis and into the mesophyll of leaves of various alpine plants has been measured directly in intact organs, using fiber-optic microprobes (Day *et al.*, 1992; DeLucia *et al.*, 1992; Cen and Bornman, 1993). Results vary greatly among species, with the distances of penetration of biologically effective UV-B ranging from ca 2 μm to 170 μm. Interestingly, among alpine plants, there is a strong correlation between the depth of penetration of UV-B radiation and the taxonomic position and/or anatomical form of the plants (Day *et al.*, 1992). Conifers allow very little penetration, absorbing virtually all UV-B within the epidermis; monocots and woody dicots allow an intermediate amount of penetration, with some UV-B reaching the photosynthetic tissues in the mesophyll; herbaceous dicots allow the most penetration, with considerable amounts of UV-B extending into and through the palisade layer. This classification is strongly reminiscent of the results of Cline and Salisbury (1966), who found that the sensitivity of various plants to UV-C varied, with conifers (and some xerophytic monocots and dicots) being least sensitive, several members of Gramineae having an intermediate sensitivity, and herbaceous dicots being most sensitive.

There are several studies that demonstrate the protective effect of tissues that contain UV-absorbing compounds. The ability of water-soluble substances in the epidermis of *Oxalis* leaves to protect the photosynthetic mesophyll was suggested by the correlation between the greater amount of absorptive material extracted from adaxial than from abaxial epidermal peels and the smaller effect on photosynthesis, measured as inhibition of fluorescence rise time, by UV irradiation incident from the adaxial than from the abaxial side of the leaf (Björn *et al.*, 1983). The effectiveness of flavonoids in rye seedlings was demonstrated by Tevini *et al.*(1991), who showed that short wavelength UV-B radiation (295 nm) had less effect on the photosystem II fluorescence of seedlings that contained more flavonoids and had a lower transmittance of UV-B light through their epidermis.

Studies of mutants that are deficient in the synthesis of pigments support the idea that the pigments have a protective function. Wild type seeds of *Arabidopsis thaliana*, irradiated dry with UV-C at 26 W m^{-2} for up to 96 hours (a very high fluence), show no inhibition of germination. When transparent testa mutants (e.g., *tt*-5), which are blocked in the flavonoid pathways, are irradiated under the same conditions for only 4 hours, only 50% of the seeds germinate within one week (A.B. Britt, personal communication). While these fluences are unusually high, in keeping with the relative insensitivity of dry cells and tissues (of various life forms) to UV radiation, the difference in sensitivity between wild-type and mutant seeds demonstrates the potential effectiveness of even thin UV-absorptive tissues. Similar work with seedlings makes the same point: both *tt*-4 and *tt*-5 mutants of *Arabidopsis* are more sensitive to UV-B radiation than wild type plants (Li *et al.*, 1993; Rao *et al.*, 1995). The greater sensitivity of *tt*-5, relative to *tt*-4, was ascribed to its lack of sinapate esters as well as flavonoids (Li *et al.*, 1993). Similarly, the DNA in leaf sheaths from a maize mutant blocked in anthocyanin synthesis was damaged by UV-C and UV-B more than DNA in wild-type tissues (Stapleton and Walbot, 1994), and rye leaves blocked in phenylpropanoid synthesis were more sensitive to UV-B than untreated leaves (Reuber *et al.*, 1993).

However, evidence has not always supported the idea that UV-absorbing compounds are effective in protecting exposed tissues from the stress generated by UV-B. In a survey of rice varieties, chosen as growing at various latitudes and altitudes, Teramura *et al.* (1991) measured the UV-B (300 nm) absorption of acidic methanol-extractable compounds in the leaves and compared these to changes in biomass accumulation caused by the UV-B treatments. There was no clear correlation between the sensitivity of biomass accumulation and either the absolute level of absorptive material or the changes in amount of absorptive material induced by UV-B radiation.

The sensitivity of plant growth and physiological processes to UV-B radiation depends strongly on the conditions under which a plant is grown. Crop plants grown in a greenhouse are more sensitive to UV-B than are plants of the same species grown in the field (Teramura, 1983). *Phaseolus* plants in growth chambers were much less sensitive to UV-B when grown at high (700 mmol m^{-2} s^{-1}) photosynthetically active radiation (PAR) than when grown at low (230 mmol m^{-2} s^{-1}) PAR (Cen and Bornman, 1990). Various wavelength bands have different effects (Panagopoulos *et al.*, 1990; Caldwell *et al.*, 1994). The synthesis of many pigments in plants, including products of the phenolic pathway, is regulated by environmental conditions, especially light. For instance, *Brassica napus* plants grown in chambers with UV-B radiation (8.9 kJ m^{-2} day^{-1}) showed a 20% increase in UV-B-screening epidermal pigments, relative to plants grown without UV-B (Cen and Bornman, 1992). There is also considerable variation among plant species in pigment synthesis and accumulation. Therefore, it seems likely that changes in accumulation of pigments accounts for much of the species differences in sensitivity to UV-C (Cline and Salisbury, 1966) and much of the reduction in sensitivity to UV-B that accompanies growth in high light (Cen and Bornman, 1990).

Inducibility of Pigment Synthesis

The compounds that absorb UV radiation in the epidermis of higher plants are often products of the phenylpropanoid pathway: flavonoids, including anthocyanins, coumarins, lignins, and others. Most of the enzymatic steps of this pathway are well known (Figure 8. 3).

The synthesis of phenolic compounds is, in most plants, regulated by environmental conditions, primarily light. For instance, Tevini *et al.* (1991) showed that flavonoids accumulated in rye seedlings (at least in part in the epidermis) starting 4 hours after irradiation with relatively long wavelength UV-B radiation (305 nm). The stimuli required may or may not be complex, depending on species.

Mohr and Drumm-Herrel (1983), studying the synthesis of anthocyanins in seedlings, argued that the major controlling factor was phytochrome. For some species (mustard), red light was sufficient for induction of the pigments, but some species (milo, tomato) required activation of a blue-UV-A receptor (cryptochrome) and some (wheat) required activation of a UV-B receptor in order to become sensitive to Pfr. Ensminger and Schäfer (1992) discuss evidence that a membrane-bound flavin (sensitive to blue light) interacts with a reduced flavin

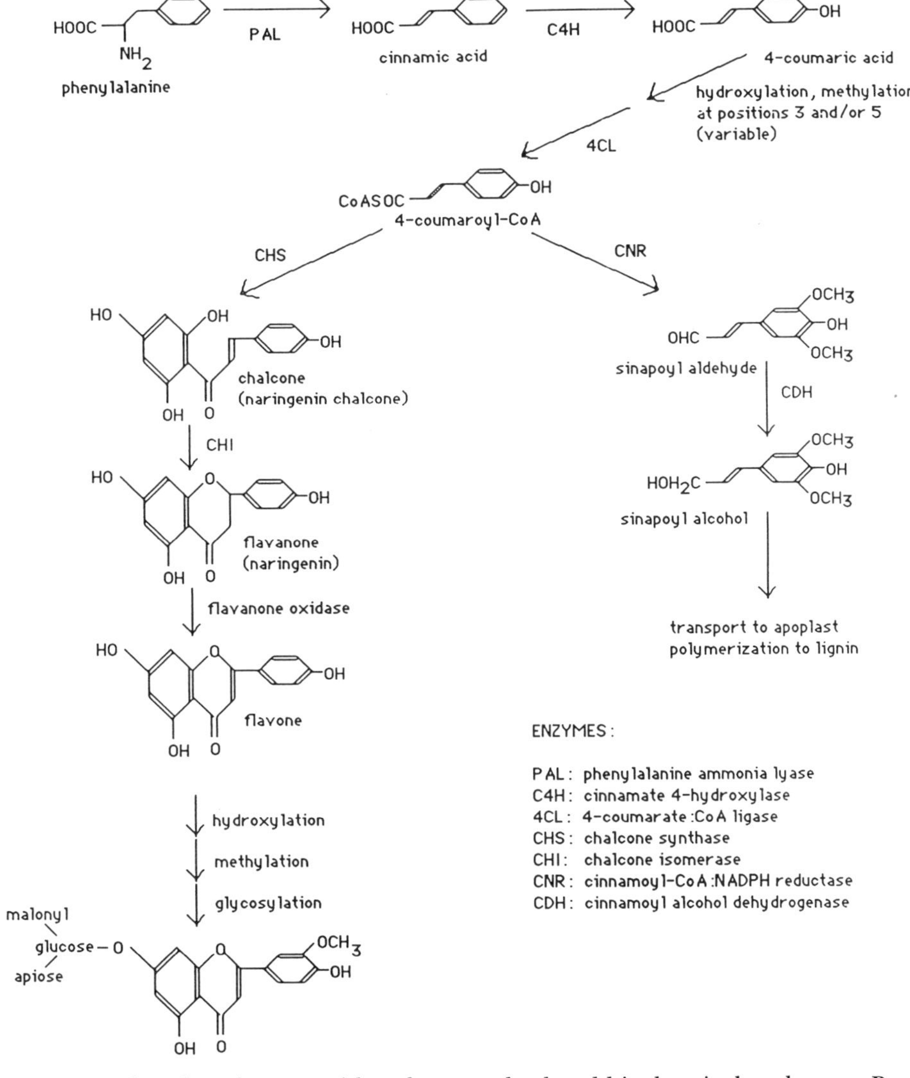

Figure 8.3. The phenylpropanoid pathway and related biochemical pathways. Based on information from Hahlbrock (1977), Hahlbrock and Grisebach (1979), and Hahlbrock and Scheel (1989).

or pterin (sensitive to UV-B) in the light-stimulated production of flavonoids in parsley.

A series of experiments with cultured parsley cells, conducted by Hahlbrock and his colleagues (for review, see Hahlbrock and Scheel, 1989), established that (a) UV-containing white light induces the synthesis of a specific set of phenolic compounds (flavonoids); (b) the synthesis depends on the appearance of enzymes of the phenolic pathway, which themselves are synthesized *de novo* following UV irradiation; (c) the increased rate of synthesis of the enzymes reflects the increased level of mRNA; (d) mRNAs for different enzymes in the pathway show similar patterns of accumulation, suggesting that they are coordinately regulated (there are two groups, an early group, represented by the first enzymes of the pathway, and a late group, consisting of enzymes specifically for flavonoid synthesis); (e) the mRNAs are synthesized more rapidly following UV irradiation.

This work has progressed with the isolation, cloning, and analysis of genes for phenylalanine ammonia-lyase (from common bean, Cramer *et al.*, 1989; parsley, Lois *et al.*, 1989; *Arabidopsis*, Ohl *et al.*, 1990), 4–coumarate:CoA ligase (from parsley, Douglas *et al.*, 1987), and chalcone synthase (from snapdragon, Sommer and Saedler, 1986; parsley, Schulze-Lefert *et al.*, 1989; soybean, Wingender *et al.*, 1989). There are 3–5 genes for phenylalanine ammonia-lyase, which show some variation in inducibility by UV. Parsley 4-coumarate:CoA ligase is coded by two genes, both of which are induced by UV treatment. Parsley has one chalcone synthase gene, which is stimulated by blue light, as does snapdragon; of three soybean chalcone synthase genes tested, only one is stimulated by UV radiation.

The mechanism by which UV radiation stimulates the activity of genes for flavonoid-synthetic enzymes has been the subject of intensive study. As with other regulated genes, the stimulation may be assumed to involve *cis*-regulatory elements and *trans*-regulatory factors. A *cis*-regulatory element is a sequence of bases adjacent to the regulated gene, generally found "upstream" or 5' to the transcription start site. A *trans*-regulatory factor is a soluble molecule that binds to the *cis* element and stimulates or inhibits transcription of the gene by RNA polymerase.

A comparison of the 5' promoter of phenyalanine ammonia-lyase genes from different species identified two consensus sequences, each present one to three times in each promoter: "Box 1", $^{T}_{A}$CTCACCTACC$^{C}_{A}$; "Box 2", $^{C}_{T}$CAA-C$^{A}_{C}$AACCCC (Ohl *et al.*, 1990, choice of consensus mine). In the parsley promoter, these sequences show UV-inducible *in vivo* footprints, which appear when *trans*-regulatory factors have bound to a promoter region protecting it from cleavage by DNAase (Lois *et al.*, 1989). By fusing various segments of the *Arabidopsis* PAL-1 gene to a β-glucuronidase (GUS) gene and inserting these hybrids into *Arabidopsis* plants by *Agrobacterium tumefaciens* transformation, Ohl *et al.* (1990) demonstrated regions that conferred sensitivity to UV, but did not identify these sequences with the consensus sequences.

The regions of the parsley chalcone synthase gene (CHS) that are responsible for the induction of transcription by light were detected by *in vivo* footprinting and confirmed by combining sections of the CHS promoter to a β–glucuronidase gene and measuring transient expression of glucuronidase in parsley protoplasts (Schulze-Lefert *et al.*, 1989). Two regions, each containing two footprints, contributed to light-stimulation of the glucuronidase gene. Each region alone gave a 3 to 10-fold stimulation of activity in the light; together the stimulation was 30-to 40-fold. A common light-regulated motif, CACGTG, was included in one of the regions (Box 2).

The parsley protoplast expression system has also been used to identify the UV-responsive promoter regions in the soybean *chs*1 gene (Wingender *et al.*, 1990) and the snapdragon CHS gene (Lipphardt *et al.*, 1988; Staiger *et al.*, 1990). In soybean, UV induction requires the presence of sequences extending from the region –134 to –175 to downstream from –75. It is interesting that the UV-induction of the soybean gene is successful in parsley protoplasts, implying that the *trans*-regulatory factors of parsley associated with UV-induction interact effectively with the soybean *cis*-element. (Soybean cell cultures and protoplasts do not respond to UV with increased transcription of the *chs*1 gene, presumably because they lack the appropriate *trans*-regulatory factors.) In snapdragon, there are three regulatory regions: one that is UV-responsive (–39 to –197); one that increases the UV-responsiveness of the first, but is not itself UV-responsive (–197 to –357); and one general enhancer (–564 to –661). A further analysis of the snapdragon CHS promoter using tobacco plants transformed with CHS promoter-β-glucuronidase hybrid genes (Fritze *et al.*, 1991) elaborates on this theme, differentiating the effects of four *cis*-regulatory regions and showing that the effects of any one region could be different in different organs (root, stem, leaf, developing seed, and petal). When UV-induction was tested in leaves, promoter sequences extending upstream through the transcription start site to –357 were sufficient for expression of the β-glucuronidase, and the expression was UV-responsive. The general enhancer region (between –564 and –661) intensified the expression and the UV-responsiveness. The region from –39 to –197, previously found to confer UV-responsive expression in parsley protoplasts (Lipphardt *et al.*, 1988), was not functional in leaves in the absence of the regions from –197 to –357. Reversing the –39 to –197 region in the presence of the rest of the promoter sequences (upstream to –1050) greatly reduced promoter activity, but the remaining activity was UV-responsive, suggesting that the binding of this region to light-responsive *trans*-regulatory factors is independent of orientation and/or position, but that further interactions with enhancer factors are not.

A similar analysis of the gene in parsley that codes for 4-coumarate:CoA ligase identified *cis*-acting segments that are responsible for the UV-A-induced transcription of the gene (Douglas *et al.*, 1991). A gene that includes as few as 174 base pairs 5' to the transcribed segment retains UV-A responsiveness, although adding bases from the upstream region increases the level of transcription in the light. Surprisingly, the 5' region is not sufficient to confer UV-A responsiveness — sequences in the exons are also needed–since the ligase promoter by itself does not confer light responsiveness on a β-glucuronidase gene. Likewise, the exonic sequences of the ligase gene are not sufficient for UV-A responsiveness, since they are not induced by UV-A when driven by a CaMV 35S promoter.

Because the effectiveness of flavonoid compounds as UV protective agents depends in part on their localization in the epidermal cells, it is interesting to determine the regulatory factors that are responsible for the tissue specificity of flavonoid synthesis and see how they interact with UV-responsive factors. In general, the 5' promoter regions of the phenylpropanoid genes have the information for tissue specificity (e.g., Ohl *et al.*, 1990). The genetic information (*cis*-acting) for localization of expression of the 4-coumarate:CoA ligase gene is in the 5' region, since the ligase promoter, fused to the β-glucuronidase gene, can be induced (in the epidermis) by wounding (Douglas *et al.*, 1991).

Proteins have also been identified that bind to *cis*-regulatory regions that are important in the UV-induction response. Nuclear factors that bind to the snapdragon CHS promoter, specifically to a region with the CACGTG sequence and to another upstream region, have been isolated (Staiger *et al.*, 1989, 1990). Using as probes DNA sequences from the parsley CHS gene that were shown to be both necessary and sufficient for UV-induced synthesis of chalcone synthase, Weisshaar *et al.* (1991) identified clones from a parsley cDNA library that expressed proteins that bound to the probes. The mRNA for one of those proteins is itself induced by UV-A radiation, being present at the highest levels at the time when the expression of *chs* gene is most rapid. These observations strongly suggest that the clone represents a *trans*-acting factor that contributes to the UV-dependent regulation of the CHS gene.

This situation has an analog in an animal system. Mammals show a "UV response," characterized by the induction of various genes. Using HeLa cells, Stein *et al.* (1992) demonstrated a UV-stimulated induction of the mRNA for *c-jun*. The induction was not inhibited by cycloheximide and thus represents a posttranslational activation of one or more *trans*-regulatory transcription factors. These factors were shown to bind to two AP-1-like *cis*-regulatory elements in the 5' regions of the *c-jun* gene. Since *c-jun* itself produces an AP-1 *trans*-regulatory factor, the Jun protein, its induction by UV can be expected to show positive feedback, as well as to induce other genes in HeLa cells. Further experiments (Devary *et al.*, 1992) have shown that the initial activation of the *c-jun* gene involves the phosphorylation of serines on pre-existing *jun* proteins. The phosphorylation involves activated Raf-1 and Ha-Ras proteins. The initial step may be the activation of Src family tyrosine kinases. c-Src and c-Fyr kinases are rapidly (in 5 min) activated by UV-C radiation, and the induction of *c-jun* is reduced by Src inhibitors. Because most Src and Ha-Ras proteins are associated with the plasma membrane, Devary *et al.* (1992) suggest that the UV reception occurs at that organelle. They further suggest that the signal involves oxidative stress at the membrane, bringing to mind the UV-induced synthesis of H_2O_2 in plant cells (Murphy and Huerta, 1990).

Some transcription factors can also be activated through the reduction of a cysteine residue. The Arp protein from *Arabidopsis*, produced in *Escherichia coli* cells from its cloned cDNA, *in vitro* reduces the human *jun* protein and thus increases its binding to DNA. Arp also is an apurinic/apyrimidinic endonuclease and might be affected by UV damage to DNA, thus suggesting another mechanism to connect UV radiation to the induction of specific genes (Babiychuk *et al.*, 1994).

As mentioned previously, a substantial fraction of the UV-absorptive material in epidermal peels is not extractable with acidic methanol and is probably bound covalently to the cuticle or cell walls. Lignin represents an example of a phenolic material that accumulates in the extracellular space in high amounts. There may be smaller amounts of phenolic material in areas not recognized as being lignified. The polymerization of lignin, and its attachment to other cell wall components, proceeds through free radicals of phenolic molecules (*p*-coumaryl, coniferyl, and sinapyl alcohols), the free radicals being generated in the extracellular space by H_2O_2 and peroxidase (Gross *et al.*, 1977; Halliwell, 1978). It seems possible that the attachment of UV-induced phenolic molecules to the cuticle and cell walls of epidermal cells occurs through a similar mechanism (Kolattukudy, 1980). If true, this would provide a rationale for the burst of synthesis of H_2O_2 by UV-irradiated cells (Murphy and Huerta, 1990). Another possible connection between UV-induced membrane effects (one of which is the acidification of the cytoplasm) and the phenylpropanoid pathway lies in the observation (Hagendoorn *et al.*, 1991) that a variety of effectors that acidify the cytoplasm of *Petunia* cells stimulate the synthesis of phenylalanine ammonia lyase by those cells.

Relationship to Resistance to Other Stresses

It is recognized that many different stresses stimulate the accumulation of phenolic materials in plant cells (Dixon and Paiva, 1995). Considerable attention has been given to wounding and plant pathogens, which often elicit the synthesis of antibiotic materials (phytoalexins) in plants. Phytoalexins are often, although not always, phenolic compounds. Their synthesis generally requires the enzymes of the initial part of the phenolic pathway, phenylalanine ammonia lyase, cinnamate 4-hydroxylase, and *p*-coumarate CoA ligase, as well as enzymes in branch pathways specific to their synthesis and generally different from those involved in flavonoid synthesis. Several different types of "elicitors," products often released from the cell walls of pathogens, have been identified (Darvill and Albersheim, 1984; Davis *et al.*, 1991; Parker *et al.*, 1991).

The connection between UV-radiation-induced and elicitor-induced stimulation of the phenolic pathway and its products has been made in several systems. Hadwiger and Schwochau (1971) demonstrated that UV-C induced the formation of phenylalanine ammonia lyase and pisatin, a phenolic phytoalexin, in peas. Fritzemeier and Kindl (1981) used UV-C radiation to study the induction of phenylalanine ammonia lyase, cinnamate 4-hydroxylase, and stilbene synthase in the leaves of various members of Vitaceae; stilbenes are potent anti-fungal compounds. The addition of fungal elicitors to parsley cell cultures stimulated the formation of furanocoumarins, related to the UV-stimulated flavonoids by the initial steps of their biosynthetic pathway (Scheel *et al.*, 1986). However, the parsley furanocoumarins are secreted, while the flavonoids are accumulated in the vacuole.

Elicitors also affect membrane processes in plant cells (reviewed by Boller, 1989), resulting in membrane depolarization, efflux of K^+, and synthesis of H_2O_2. Apostol *et al.* (1989) suggest that the burst of H_2O_2 is another aspect of the anti-pathogen response, contributing to the killing of infective agents or possibly sealing off an infection by killing the potential host cells. H_2O_2 may also play a role as a signal to convey the perception of stress to nearby cells (Levine *et al.*, 1994). Arnott and Murphy (1991) compared the effects of UV-C radiation and an elicitor preparation from a *Phytophthora* species on K^+ efflux and H_2O_2 synthesis by cultured rose cells. There were similarities in time course and extent of the effects, but various differences (for instance, aged cells remained sensitive to UV but not to elicitor) suggested that the two stimuli acted through different receptors.

The significance of these observations is hard to assess, but it is possible that natural UV radiation helps induce pathogen resistance in plants. A UV treatment applied to tobacco leaves increased their resistance to infection by tobacco mosaic virus (Brederode *et al.*, 1991). Some groups have tested UV-C irradiation as a method of increasing resistance of fruits to postharvest diseases (Lu *et al.*, 1991; Chalutz *et al.*, 1992).

Both the first part of the phenolic pathway and membrane functions of plant cells are sensitive to a number of stresses, including heavy metals (and other reagents that react with sulfhydryl groups) and antibiotic compounds. Does each stress act through a different, specific receptor, or do some of these stresses interact with intermediaries in a single signaling pathway? It is clear that the connection between UV-induced and elicitor-induced responses is sufficiently strong to allow us to learn about one set of responses from studies on the other.

Summary

The surface and epidermis of plant organs can be very effective in shielding the internal tissues from UV radiation. The degree of protection varies greatly among species, and it depends strongly on the growth conditions. Light, often including UV-B radiation, leads to enhanced protection.

Products of the phenylpropanoid pathway, most frequently flavonoids, are among the most common UV-absorbing compounds in the epidermis of plants. The synthesis of these compounds is stimulated by light and other environmental stresses.

The molecular basis for the regulation of the genes of the phenylpropanoid pathway is under intensive study, and much progress has been made in identifying the *cis*-and *trans*-controlling factors. One may expect that, working backward step by step from the activation of *trans*-acting factors, cell biologists will identify the sequence of events by which light (UV) activates this pathway.

DNA REPAIR PROCESSES

Evidence for DNA Repair in Plants

There are several processes, photorepair, excision repair, and recombinational repair, that reverse the effects of ultraviolet radiation on DNA (Sancar and Sancar, 1988; Grossman and Yeung, 1990; Britt, 1995; Figure 8.4). Photorepair involves the formation of a complex of the enzyme DNA photolyase specifically with a pyrimidine dimer. Photolyases are generally considered to be specific for cyclobutadipyrimidines, although Todo *et al.* (1993) and Kim *et al.* (1994) have described a photorepair enzyme from *Drosophila melanogaster* that repairs (6–4)pyrimidine pyrimidone photoproducts and Chen *et al.* (1994) have reported a light-dependent pathway for the repair of these photoproducts in *Arabidopsis*. With the absorption of a photon of UV-A or visible light, the bonds between the pyrimidines are broken, and the enzyme is released. Excision repair involves the removal of photoproducts and adjacent bases from one strand of DNA, followed by resynthesis of that strand using the opposite strand as a template. Recombinational repair refers to processes that allow DNA synthesis to occur without being blocked by UV photoproducts. Excision repair and recombinational repair do not require light (except perhaps as an energy source in photosynthetic organisms), and they are thus sometimes referred to as "dark repair" systems.

Early suggestions of photorepair in plant tissues were made on the basis of reduced damage (necrosis, chlorosis) in UV-C-irradiated tissues that had been illuminated with visible light, relative to irradiated but unilluminated controls (Bawden and Kleczkowski, 1952; Tanada and Hendricks, 1953). However, light has many effects on plants, and apparent photorepair could involve processes other than direct reversal of DNA photoproducts. Supporting evidence for true DNA photorepair in plant cells comes from studies of the loss of photoproducts in DNA from irradiated tobacco and *Ginkgo* (but not *Happlopappus*) cells (Trosko and Mansour, 1968, 1969) and from work with photolyase extracted from plants (Small and Greimann, 1977; Langer and Wellmann, 1990; Pang and Hays, 1991).

Early experiments failed to demonstrate any loss of cyclobutadipyrimidines in UV-irradiated plant cells incubated in the dark (Trosko and Mansour, 1968) or to detect unscheduled (non-S-phase) DNA synthesis (an indication of excision and recombinational repair) in UV-and X-irradiated plant cells (Painter and Wolfe, 1973; Swinton and Hanawalt, 1973). Later, however, several workers demonstrated the operation of excision repair processes in several plant systems (*Daucus carota* cultured cell protoplasts: Howland, 1975; *Chlamydomonas*: Small and Greimann, 1977; *Lathyrus sativus* embryos: Soyfer and Cieminis, 1977, *Petunia* pollen: Jackson and Linskens, 1978, 1979; *Arabidopsis* stems and leaves, Pang and Hays, 1991; for reviews, see Cieminis *et al.*, 1987; Jackson, 1987; McLennan, 1987; Small, 1987; Britt, 1995). The successes probably related, at least in part, to the use of lower UV fluences in concert with more sensitive assays. Howland (1975) showed that the degree of repair depended strongly on the UV fluence used, with higher fluences reducing the fraction of damage repaired.

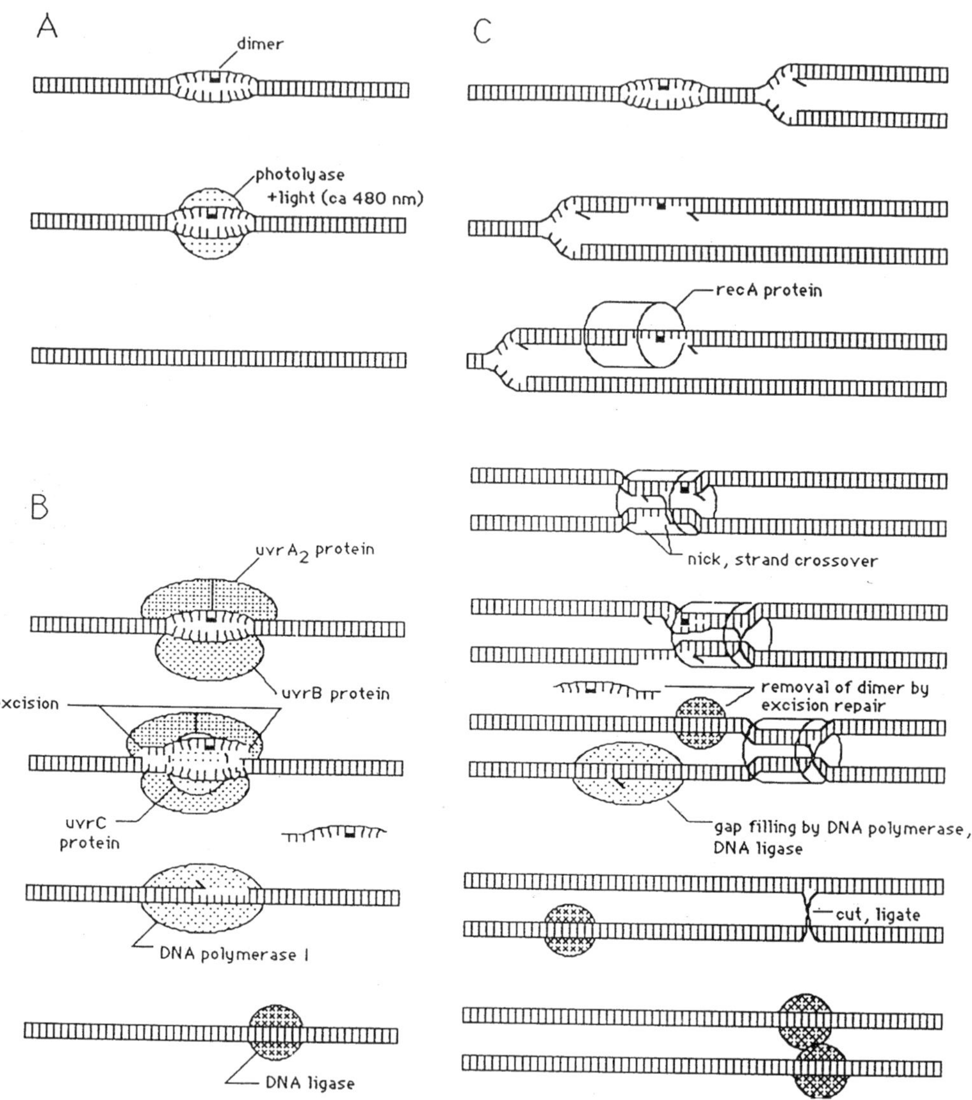

Figure 8.4. Models of DNA repair processes. A. Photorepair of cyclobutadipyrimidines. B. Excision repair of DNA damage in *Escherichia coli.* C. Recombinational repair (bypass) of DNA damage in *E. coli.* Based on information in Jagger (1985) and Sancar and Sancar (1988).

Plant Repair Enzymes

DNA photolyase

DNA photolyase (the photorepair enzyme) has been extracted from *Phaseolus* hypocotyls and assayed by demonstrating a loss of cyclobutadithymines from extract-treated salmon sperm DNA (Langer and Wellmann, 1990). The enzyme has also been extracted from *Arabidopsis* leaves and assayed by its repair of transforming activity of UV-C-irradiated plasmid DNA (Pang and Hays, 1991). The enzyme, which like all other photolyases requires light in its catalytic cycle, was maximally active at 375 to 400 nm, like the enzyme from *Escherichia coli*. The *Arabidopsis* enzyme showed a marked sensitivity to heat, being rapidly inactivated at 50°C *in vitro* (in the absence of UV-irradiated DNA) and apparently at an even lower temperature, 37°C, *in vivo*.

Langer and Wellman (1990) found that the amount of photolyase activity extracted from *Phaseolus* hypocotyls was increased by treating the plants with 5 minutes of red light, 6 hours of far-red light, or 6 hours of blue light. The red-light-induced stimulation was expressed only after 2 hours, and it could be reversed by 5 minutes of far-red light, suggesting that the expression of the photolyase gene was controlled by phytochrome. The induced photolyase synthesis was inhibited by UV-B irradiation, and this inhibition was itself reversed by photorepair (Buchholz *et al.*, 1995).

UV endonuclease

The first step in the excision repair system is the recognition of damaged base(s) and the production of one or two breaks near to the damaged site. There are at least two mechanisms for catalyzing this first step. One is exemplified by the *Escherichia coli* uvrABC excinuclease, which recognizes a wide range of base alterations as substrates, hydrolyzing phosphodiester bonds on either side of the substrate. The other is represented by the *Micrococcus luteus* glycosylase/apurine-apyrimidine endonuclease, which specifically recognizes pyrimidine dimers, cleaving the glycosidic bond between the 5' pyrimidine and its deoxyribose and also a phosphodiester bond adjacent to the apyrimidinic site.

In contrast to the extensive studies of repair endonucleases in bacteria, relatively little is known about the UV endonucleases from eukaryotes. In particular, the enzymes from higher plants have received relatively little attention. Studies on the UV endonuclease from four different plant materials have found quite different characteristics for the activity (Velemínsk *et al.*, 1980; McLennan and Eastwood, 1984; Doetsch *et al.*, 1989; Strickland *et al.*, 1991; Murphy *et al.*, 1993). These workers used different species, different preparation methods, and different assays, and it is not clear which of these differences explains the differences in the enzymes. There are almost certainly several UV-endonucleases in plant cells, and different workers have probably assayed different ones. Three enzymes are compared in Table 8.2.

Table 8.2 Plant endonucleases reported to act on UV-damaged DNA.

Parameter	Murphy et al.[a]	McLennan and Eastwood[b]	Doetsch et al.[c]
Plant	Tobacco leaf	Carrot suspension culture	Spinach leaf
Preparation	nuclear lysate	chromatin pellet, several stages of protein purification	crude extract several stages of protein purification
Assay	supercoil relaxation	$^{32}P\text{-}PO_4$ release	supercoil relaxation
UV treatment	0.024–10 kJ/m^2	20 kJ/m^2, 0.5 mM acetophenone	0.05–10 kJ/m^2
Cation requirement	none	Mg^{2+}	$Zn^{2+}(>Mg^{2+})$
Substrates	(6–4) pyrimidine pyrimidone adducts?	cyclobutadi-pyrimidines, psoralen adducts	single-strand DNA
Molec. mass			
(gel)			30–45 kDA
(gradient)	40 kDa	140 kDa	

[a]Murphy *et al.* (1993)
[b]McLennan and Eastwood (1986)
[c]Doetsch *et al.* (1989)

Velemínsk *et al.* (1980) lysed purified chloroplasts isolated from barley leaves and assayed nucleases by the release of label from ^{3}H-DNA into acid-soluble material. After precipitation of the chloroplast DNA with streptomycin sulfate, the supernatant showed more activity toward UV-irradiated and apurinated DNA than toward native DNA. Two fractions of enzymatic activity showing fairly high specificity toward UV-irradiated DNA could be distinguished by their chromatographic behavior on UV-DNA-cellulose affinity chromatography and by polyacrylamide gel electrophoresis.

McLennan and Eastwood (1986) partially purified a UV endonuclease from chromatin extracted from suspension cultured cells of *Daucus carota*, using as assay the exposure of phosphate residues to phosphatase release. They found a large (Mr 140,000) enzyme that required Mg^{2+} ions for activity. It cleaved native DNA irradiated with UV-B in the presence of acetophenone (containing cyclobutadipyrimidines), trimethylpsoralen-treated DNA, and 2-acetyl aminofluorine-treated DNA, but not heat-denatured DNA or depurinated DNA. The authors pointed out that the characteristics of this enzyme matched those of the enzyme from *E. coli*, except that, unlike the *E. coli* enzyme, the carrot enzyme did not require ATP.

Doetsch *et al.* (1989) partially purified an endonuclease from a crude extract of spinach leaves, using a plasmid supercoil-relaxation assay, and Strickland *et al.* (1991) further purified the enzyme to homogeneity. This enzyme was smaller (Mr 35–40,000) than the carrot enzyme. Like the carrot enzyme, it required divalent cations, but it was stimulated more by Zn^{2+} than by Mg^{2+}. The enzyme cleaved end-labeled DNA at defined sites (generally A bases), but the site specificity and the UV-specificity were lost at low salt concentrations. The authors suggested that the enzyme cleaved single-stranded DNA, which at high salt concentration would be formed by the torsional strains induced by UV photoproducts and at low salt concentrations would be formed by ionic repulsion between phosphates in opposite DNA chains.

Murphy *et al.*'s (1993) preparation was extracted from nuclei isolated from tobacco leaves and detected using a plasmid supercoil-relaxation assay similar to that of Doetsch *et al.* (1989). The enzyme activity was highly specific for UV-C-irradiated DNA. Unlike the enzyme of Doetsch *et al.* (1989), it did not lose UV specificity when the ionic strength of the assay solution was lowered to as little as 16 mM. The tobacco enzyme preparation did not require divalent cations, unlike enzymes reported by McLennan and Eastwood (1986) and by Doetsch *et al.* (1989). Unlike *E. coli* ABC excinuclease but like the carrot enzyme, its activity was not changed by the presence (or absence) of ATP. Unlike the enzyme of McLennan and Eastwood (1986), it did not cleave acetophenone-sensitized DNA (containing cyclobutadipyrimidines); furthermore, the degree of cleavage was not reduced when cyclobutadipyrimidines in UV-C-irradiated DNA were removed using *E. coli* photolyase.

The enzyme activity in Murphy *et al.*'s preparation cleaved the irradiated pUC19 substrate at very specific sites, as determined by electrophoresis of end-labeled defined DNA fragments. These included all sites with the sequence TCA/ and TTA/ ("/" represents the point of cleavage), most sites with the sequence G/CC or A/CC, some sites in other regions containing pairs of pyrimidines, and some sites adjacent to A bases but without adjacent pyrimidine pairs. Some of the specificity reflected a particular sensitivity of certain sites on the DNA to different types of radiation-induced damage: some sites were more prone to form cyclobutadipyrimidines (as determined by sensitivity to T4 endonuclease V); some, (6–4)pyrimidine pyrimidone adducts (as determined by sensitivity to heating in piperidine). Some of the specificity probably reflected the selectivity of the enzyme(s). The observation that some cuts were made 3' to pyrimidine pairs (presumptive cyclobutadipyrimidine or (6-4) photoproducts) and some were 5' suggests that the preparation contained different enzymes with different specificities and mechanisms of action. The fact that the enzyme(s) did not recognize cyclobutadipyrimidines, but cleaved near pyrimidine pairs, suggests that it recognized (6–4)pyrimidine pyrimidone adducts.

Gallagher *et al.* (1993) described an endonuclease from cultured tobacco cells that cleaved UV-irradiated, 3'-end-labeled DNA at cytosines, releasing cytosine hydrates. Oxidized DNA was cleaved at thymines, releasing thymine glycols. The authors suggest that this enzyme is similar to endonuclease III from *Escherichia coli* or human redoxyendonuclease. Babiychuk *et al.* (1994) cloned a gene from *Arabidopsis* that showed sequence similarity to *E. coli* endonuclease III; the protein, when expressed in *E. coli*, had apurinic/apyrimidinic endonuclease activity.

There are many questions concerning DNA repair that remain to be investigated. What photoproducts occur in plant DNA irradiated *in situ*, especially following irradiation with UV-B? Do pyrimidine hydrates, guanine photoproducts, or cross-links to chromatin proteins play a significant role? Do the steps that follow endonuclease cleavage in the repair of UV-damaged plant DNA require repair-specific proteins, perhaps interacting in a complex fashion such as is found in *E. coli* (Grossman and Yeung, 1990)? There is some suggestion that topoisomerases are involved in DNA repair in eukaryotes (Kato, 1974). Topoisomerase II has been isolated from cultured tobacco cells. Irradiation of the cells with fluorescent sunlamps emitting visible and UV-B radiation led to an increase and an increased persistence of the topoisomerase II activity in these cells (S. Heath-Pagliuso, personal communication).

Mutants are becoming available to demonstrate how important DNA repair processes are in plants growing under enhanced ultraviolet radiation or sunlight. Harlow *et al.* (1993, 1994) have reported the isolation of an *Arabidopsis* mutant (*uvh1*), the shoots of which are hypersensitive to UV-C and UV-B radiation. The function lost by the mutation was not identified, although the authors demonstrated that UV-C produced the same degree of DNA damage in *uvh1* and wild-type strains—i.e., the mutant does not differ in shielding. The *uvh1* mutant retains a photorepair ability, measured by reduced necrosis of UV-B-irradiated leaves under cool-white light, relative to gold fluorescent light. Seeds of this mutant are also hypersensitive to γ-radiation from ^{60}Co. These characteristics are consistent with, but do not prove, the hypothesis that the mutant lacks a DNA-dark-repair function.

Starting with the *tt5* flavonoid-deficient strain, Britt *et al.* (1993) identified an *Arabidopsis* mutant (*uvr1*) in which primary root growth is hypersensitive to UV-B radiation, and they showed that this mutant lacks the ability to remove (6–4) photoproducts from DNA in the dark. Rather surprisingly, both the mutant and the parent strains are very slow at removing cyclobutadipyrimidines from their DNA. The mutant retains the ability to photorepair both cyclobutadipyrimidines and (6–4) photoproducts, with both photorepair functions contributing to the capacity for continued root growth in UV-B-irradiated plants (Chen *et al.*, 1994). Preliminary results suggest, however, that the *uvr1* mutant, grown under cool-white fluorescent lamps in my laboratory, is no more sensitive to a single exposure to sunlight (in terms of leaf chlorosis and necrosis) than is its *tt5* parent (T.M.M., unpublished experiments), although both of these strains are considerably more sensitive than wild-type.

Summary

Plants have very effective enzymatic pathways for the repair of UV damage to both nuclear and chloroplastic DNA. These include photorepair and excision repair processes.

A start has been made in isolating and characterizing the repair enzymes in plants, but relative to work with enzymes from *E. coli*, yeast, and man, the work is just beginning. There is evidence for multiple UV-endonucleases in both chloroplasts and nuclei.

In the future, work with repair mutants will contribute a great deal to our understanding of the mechanism and significance of DNA repair processes to plants.

OTHER ASPECTS OF RESISTANCE

The two most studied mechanisms of resistance to UV radiation—absorptive compounds that reduce the radiation that penetrates to the inner sensitive cells and enzymatic processes that repair damage to the most sensitive molecule in cells—probably are not the only mechanisms that help plant cells withstand radiation stress. General repair and synthetic mechanisms that replace the cellular components lost during normal turnover, such as those involved in the replacement of the D-1 protein of photosystem II, are undoubtedly essential. Other mechanisms that might contribute to UV resistance include those that quench the reactions that UV radiation initiates, as described below.

Energy Dissipation in the Chloroplast

The chloroplasts of green and brown algae and of higher plants contain a "xanthophyll" cycle by which different oxygenated forms of carotenoids are interconverted. Under low light conditions, the epoxidated form, violaxanthin, predominates. Under high fluences of photosynthetically active radiation or conditions in which photochemical use of absorbed energy is inhibited (2% O_2, no CO_2), there is a rapid accumulation of the reduced, de-epoxidated form, zeaxanthin. The amount of zeaxanthin correlates negatively with variable fluorescence and with O_2 evolution, suggesting that zeaxanthin dissipates energy from excited chlorophyll in antenna complexes (Demmig *et al.*, 1987; Demmig-Adams, 1990). Furthermore, the absence of the xanthophyll cycle (as in cyanobacteria) correlates positively with a sustained photoinhibition by excess light, thought to result from damage to photoreaction centers (Demmig-Adams *et al.*, 1990a,b). Thus zeaxanthin, by removing excitation energy that cannot be accommodated by normal photosynthetic electron transport, protects the photosynthetic system from rogue oxidative reactions. The xanthophyll cycle represents a method of rapidly changing the amount of zeaxanthin to balance the demand for photosynthetic efficiency (in low light) and protection (in high light)(Demmig-Adams and Adams, 1992).

It is not known whether zeaxanthin or other carotenoids function to protect the photosynthetic system from exposure to UV-B. Most of the damage caused by UV-B is at the reaction center of photosystem II. The absence of substantial damage to the antennae may indicate that energy from the generally low fluences of UV-B can be dissipated without the need for zeaxanthin, or it may mean that zeaxanthin functions very effectively. Adams *et al.* (1990) suggest that while most of the high-light-induced fluorescence quenching involves the antennae, there are other induced quenching reactions at or near the photosystem II reaction center. Under those conditions where UV-B-induced damage to the photosynthetic system can be demonstrated, it may be assumed that these are of limited effectiveness. Fortunately, it is possible to make comparisons on organisms with and without zeaxanthin (e.g., cyanobacteria and green algae: Demmig-Adams *et al.*, 1990a,b). Comparing the effects of UV-B on the components of photosynthesis in these organisms would provide a good start toward learning whether zeaxanthin plays a role in protection against UV-B.

UV-B radiation inhibits de-epoxidation of violaxanthin in isolated chloroplasts (Pfündel *et al.*, 1992). Although high fluences are apparently needed for the same effect on intact leaves, it is possible that UV-B could sensitize photosynthetic cells to photoinactivation by high intensity photosynthetically active (visible) radiation. Synergistic inhibition of photosynthesis by UV-B and visible light has been observed in algae (Cullen *et al.*, 1992).

Protection against Oxygen Radicals

All plant cells produce activated species of oxygen through their metabolic activities, and they contain enzyme systems that help prevent activated oxygen from damaging functional molecules (Elstner, 1982).

Although there is little direct evidence that photodynamic production of activated oxygen is responsible for damage to cells irradiated with UV-B or UV-C *in vivo*, that mechanism remains a possibility. Furthermore, there is a clear stimulation of the metabolic synthesis of H_2O_2 in cultured rose cells that have been irradiated with UV-C. Oxidized lipids, measured as thiobarbituric acid-reactive material, appear after high doses of UV-C (Wright *et al.*, 1981) and UV-B (Kramer *et al.*, 1991b). Thus, UV radiation apparently increases the demand for scavenging systems that reduce the amounts of the various species of activated oxygen.

Superoxide dismutase accelerates the decay of superoxide, and ascorbate and glutathione scavenge this molecule. Catalase and peroxidases reduce the amounts of H_2O_2. In particular, ascorbate peroxidase removes H_2O_2 in chloroplasts. The oxidized ascorbate is re-reduced either directly by the electron transport chain or by glutathione. Reduced glutathione is regenerated by NADPH and glutathione reductase (see Foyer *et al.*, 1994 for review).

Pea plants irradiated with UV-B showed an increase in glutathione reductase mRNA, but a decrease in superoxide dismutase mRNA (Strid, 1993). In cultured rose cells irradiated with UV-C and then incubated for 60 minutes, there was no change in catalase or peroxidase activity; there was a small decrease in superoxide dismutase activity (Murphy and Huerta, 1990a). Three hours after rose cells were irradiated, there was no change in reduced glutathione (GSH), but there was a three-fold increase in oxidized glutathione (GSSG, Huerta and Murphy, 1989c). It is likely that the increase of GSSG reflected oxidation of GSH by activated oxygen species and that feedback mechanisms (presumably involving increased synthesis of GSH) kept the level of GSH constant. GSSG can be inhibitory to protein synthesis (Halliwell and Gutteridge, 1989) and to enzymes, includingihe plasma membrane H^+-ATPase (Qian and Murphy, 1993), so it is not clear to what degree this process protects or endangers the cell.

In a comparison of soybean cultivars, Murali *et al.* (1988) found that one variety, which was more resistant to UV-B (in terms of leaf expansion and dry mass accumulation), had much higher levels of one particular isozyme of peroxidase than did a more UV sensitive cultivar. It is possible that this isozyme played some role in UV resistance, although because there were other differences between the two cultivars, this suggestion remains speculative.

CONCLUSION

The effects of ultraviolet radiation on plant cells are complex. They include damage to central molecules, like the DNA, but they also include many interrelated changes in the metabolism and functions of the cell. It is not known what precipitates the latter effects.

It is clear that plants have at least two important mechanisms for protecting themselves against damage by UV radiation. Shielding internal tissues by erecting reflective or absorptive barriers (the strategy of "avoidance," Caldwell, 1981) has the advantage of protecting all potential targets. The cost may be high, however, in terms of the energy required to synthesize the barriers. Of course, UV-protection may also serve other functions: reflective cuticles may reduce the heat load from infrared radiation and limit excess visible radiation; some UV-absorbing phenolic compounds may also have fungicidal or insecticidal properties.

The high sensitivity of nucleic acids to UV radiation means that some damage may be expected from the UV-B in sunlight, even with shielding from the epidermis. This makes DNA repair systems important. McLennan (1987) has calculated that repair capacity is close to matching the rate of damage in a few systems that have been studied. It would be interesting to learn how much DNA repair capacity varies among different plants as a function of UV exposure and degree of shielding. Establishing, maintaining, and operating the DNA repair systems will also involve a cost to the organism, which suggests that regulating the repair capacity would be worthwhile. However, a repair system will be necessary, even in the absence of UV, since DNA can be damaged by other environmental and endogenous factors, for instance, by metabolically produced activated oxygen species (Halliwell and Gutteridge, 1989).

In general, the responses of plants to supplemental UV-B radiation have been subtle, often overshadowed by environmental variables like visible light, temperature, and availability of water. I believe that this is due, not to the insensitivity of biochemical systems to UV, but to a complex network of adaptations that have evolved in land plants in response to constant stress from solar UV radiation.

ACKNOWLEDGEMENTS

Much of the work in the author's laboratory has been supported by grants from the Exploratory Research Program of the U.S. Environmental Protection Agency.

REFERENCES

Adams W.W., III, Demmig-Adams B. and Winter K. (1990) Relative contributions of zeaxanthin-related and zeaxanthin-unrelated types of "high-energy-state" quenching of chlorophyll fluorescence in spinach leaves exposed to various environmental conditions. *Plant Physiology*, **92**, 302–309.

Adamse P. and Britz S.J. (1992) Amelioration of UV-B damage under high irradiance. I: Role of photosynthesis. *Photochemistry and Photobiology*, **56**, 645–650.

Arnott T. and Murphy T.M. (1991) A comparison of the effects of a fungal elicitor and ultraviolet radiation on ion transport and hydrogen peroxide synthesis by rose cells. *Environmental and Experimental Botany*, **31**, 209–216.

Auh C.K. and Murphy T.M. (1995) Plasma membrane redox enzyme is involved in the synthesis of O_2- and H_2O_2 by *Phytophthora* elicitor-stimulated rose cells. *Plant Physiology*, **107**, 1241–1247.

Babiychuk E., Kushnir S., Van Montagu M. and Inzé D. (1994) The *Arabidopsis thaliana* apurinic endonuclease Arp reduces human transcription factors Fos and Jun. *Proceedings of the National Academy of Sciences USA*, **91**, 3299–3303.

Ballaré C.L., Barnes, P.W. and Kendrick R.E. (1991) Photomorphogenic effects of UV-B radiation on hypocotyl elongation in wild type and stable-phytochrome-deficient mutant seedlings of cucumber. *Physiologia Plantarum*, **83**, 652–658.

Bawden F.C. and Kleczkowski A. (1952) Ultra-violet injury to higher plants counteracted by visible light. *Nature*, **169**, 90–91.

Björn L.O., Bornman J.F. and Olsson E. (1983) Effects of ultraviolet radiation on fluorescence induction kinetics in isolated thylakoids and intact leaves. In *Stratospheric Ozone Reduction, Solar Ultraviolet Radiation and Plant Life*, edited by R.C. Worrest and M.M. Caldwell, pp. 185–197. Berlin: Springer-Verlag.

Björn L.O. and Murphy T.M. (1985) Computer calculation of solar ultraviolet radiation at ground level. *Physiologie Végétalc*, **23** 555–561.

Boller T. (1989) Primary signals and second messengers in the reaction of plants to pathogens. In *Second Messengers in Plant Growth and Development*, edited by W.F. Boss and D.J. Morré, pp 227–255. New York: Alan R. Liss.

Bornman J.F. Björn L.O. and Åkerlund H.E. (1984) Action spectrum for inhibition by ultraviolet radiation of photosystem II activity in spinach thylakoids. *Photobiochemistry and Photobiophysics*, **8**, 305–313.

Bornman J.F., Bornman C.H. and Björn L.O. (1982) Effect of ultraviolet radiation on viability of isolated *Beta vulgaris* and *Hordeum vulgare* protoplasts. *Zeitschrift für Pflanzenphysiologie*, **105**, 297–306.

Bornman J.F., Evert R.F. and Mierzwa R.J. (1983) The effect of UV-B and UV-C radiation on sugar beet leaves. *Protoplasma*, **117**, 7–16.

Bornman J.F. and Vogelmann T.C. (1991) Effect of UV-B radiation on leaf optical properties measured with fibre optics. *Journal of Experimental Botany*, **42**, 547–554.

Brederode F.Th., Linthorst H.J.M. and Bol J.F. (1991) Differential induction of acquired resistance and PR gene expression in tobacco by virus infection, ethephon treatment, UV light and wounding. *Plant Molecular Biology*, **17**, 1117–1125.

Britt A.B. (1995) Repair of DNA damage induced by ultraviolet radiation. *Plant Physiology*, **108**, 891–896.

Britt A.B., Chen J.J., Wykoff D. and Mitchell D. (1993) A UV-sensitive mutant of *Arabidopsis* defective in the repair of pyrimidine-pyrimidinone(6–4) dimers. *Science*, **261**, 1571–1574.

Buchholz G., Ehmann B. and Wellmann E. (1995) Ultraviolet light inhibition of phytochrome-induced flavonoid synthesis and DNA photolyase formation in mustard colyledons (*Sinapis alba* L.). *Plant Physiology*, **108**, 227–234.

Caldwell M.M. (1968) Solar ultraviolet radiation as an ecological factor for alpine plants. *Ecological Monographs*, **38**, 243–268.

Caldwell M.M. (1981) Plant response to solar ultraviolet radiation. *Encyclopedia of Plant Physiology*, **12A**, 169–197.

Caldwell M.M., Flint S.D. and Searles P.S. (1994) Spectral balance and UV-B sensitivity of soybean: a field experiment. *Plant, Cell and Environment*, **17**, 267–276.

Cen Y.P. and Bornman J.F. (1990) The response of bean plants to UV-B radiation under different irradiances of background visible light. *Journal of Experimental Botany*, **41**, 1489–1495.

Cen Y.P. and Bornman J.F. (1993) The effect of exposure to enhanced UV-B radiation on the penetration of monochromatic and polychromatic UV-B radiation in leaves of *Brassica napus*. *Physiologia Plantarum*, **87**, 249–255.

Chalutz E., Droby S., Wilson C.L. and Wisniewski M.E. (1992) UV-induced resistance to postharvest diseases of citrus fruit. *Journal of Photochemistry and Photobiology. B. Biology*, **15**, 367–371.

Chen J.J., Mitchell D. and Britt A.B. (1994) A light-dependent pathway for the elimination of UV-induced pyrimidine (6–4) pyrimidinone photoproducts in *Arabidopsis thaliana*. *The Plant Cell*, **6**, 1311–1317.

Cieminis K.G.K., Rančelienė V.M., Prijalgauskienė A.J., Tiunaitienė N.V., Rudzianskaitė A.M. and Janč ys Z.J. (1987) Chromosome and DNA and their repair in higher plants irradiated with shortwave ultraviolet light. *Mutation Research* **181**, 9–16.

Climatic Impact Assessment Program (CIAP) (1975) *Impacts of Climatic Change on the Biosphere*. Washington D.C.: Dept. of Transportation.

Cline M.M. and Salisbury F.B. (1966) Effects of ultraviolet radiation on the leaves of higher plants. *Radiation Botany*, **6**, 151–163.

Commission Internationale de l'eclairage (CIE) (1987) Vocabulaire international de l'eclairage. *Publication No. 17.4*. Paris: Bureau Central de la CIE.

Cramer C.L., Edwards K., Dron M., Liang X., Dildine S.L., Bolwell G.P. *et al.*(1989) Phenylalanine ammonia-lyase gene organization and structure. *Plant Molecular Biology*, **12**, 367–383.

Cullen J.J., Neale P.J. and Lesser M.P. (1992) Biological weighting function for the inhibition of phytoplankton photosynthesis by ultraviolet radiation. *Science*, **258**, 646–650.

Darvill A.G. and Albersheim P. (1984) Phytoalexins and their elicitors—a defense against microbial infection in plants. *Annual Review of Plant Physiology*, **35**, 243–275.

Davis D.A., Low P.S. and Heinstein P.F. (1991) Isolation and characterization of an elicitor of phytoalexins in cotton isolated from *Verticillium dahliae* culture filtrates. *Plant Physiology*, **96**, S70.

Day T.A. and Vogelmann T.C. (1995) Alterations in photosynthesis and pigment distributions in pea leaves following UV-B exposure. *Physiologia Plantarum*, **94**, 433–440.

Day T.A., Vogelmann T.C. and DeLucia E.H. (1992) Are some plant life forms more effective than others in screening out ultraviolet-B radiation? *Oecologia*, **92**:513–519.

DeLucia E.H., Day T.A. and Vogelmann T.C. (1992) Ultraviolet-B and visible light penetration into needles of two species of subalpine conifers during foliar development. *Plant, Cell and Environment*, **15**, 921–929.

Demmig B., Winter K., Krüger A. and Czygan F.C. (1987) Photoinhibition and zeaxanthin formation in intact leaves: a possible role of the xanthophyll cycle in the dissipation of excess light energy. *Plant Physiology*, **84**, 218–224.

Demmig-Adams B. (1990) Carotenoids and photoprotection in plants: a role for the xanthophyll zeaxanthin. *Biochimica et Biophysica Acta*, **1020**, 1–24.

Demmig-Adams B. and Adams W.W., III (1992) Photoprotection and other responses of plants to high light stress. *Annual Review of Plant Physiology and Plant Molecular Biology*, **43**, 599–626.

Demmig-Adams B., Adams W.W., Czygan F.C., Schreiber U. and Lange O.L. (1990a) Differences in the capacity for radiationless energy dissipation in the photochemical apparatus of green and blue-green algal lichens associaed with differences in carotenoid composition. *Planta*, **180**, 582–589.

Demmig-Adams B., Adams W.W., Green T.G.A., Cygan F.C. and Lange O.L. (1990b) Differences in the susceptibility to light stress in two lichens forming a phycosymbiodeme, one partner possessing and one lacking the xanthophyll cycle. *Oecologia*, **84**, 451–456.

Devary Y., Gottlieb R.A., Smeal T. and Karin M. (1992) The mammalian ultraviolet response is triggered by activation of Src tyrosine kinases. *Cell*, **71**, 1081–1091.

de Zeeuw D. and Leopold A.C. (1957) The prevention of auxin responses by ultraviolet light. *American Journal of Botany*, **44**, 225–228.

Dixon R.A. and Paiva N.L. (1995) Stress-induced phenylpropanoid metabolism. *The Plant Cell*, **7**, 1085–1097.

Doetsch P.W., McCray W.H., Jr. and Valenzuela M.R.L. (1989) Partial purification and characterization of an endonuclease from spinach that cleaves ultraviolet light-damaged duplex DNA. *Biochimica et Biophysica Acta*, **1007**, 309–317.

Doughty C.J. and Hope A.B. (1973) Effects of ultraviolet radiation on the membranes of *Chara corallina*. *Journal of Membrane Biology*, **13**, 185–198.

Doughty C.J. and Hope A.B. (1977a) Effects of ultraviolet radiation on the plasma membranes of *Chara corallina*. I. The hyperpolarized state. *Australian Journal of Plant Physiology*, **3**, 677–685.

Doughty C.J. and Hope A.B. (1977b) Effects of ultraviolet radiation on the plasma membranes of *Chara corallina*. II. The action potential. *Australian Journal of Plant Physiology*, **3**, 687–692.

Doughty C.J. and Hope A.B. (1977c) Effects of ultraviolet radiation on the plasma membranes of *Chara corallina*. III. Action spectra. *Australian Journal of Plant Physiology*, **3**, 693–699.

Douglas C., Hoffman H., Schultz W. and Hahlbrock K. (1987) Structure and elicitor or u.v.-light-stimulated expression of two 4-coumarate:CoA ligase genes in parsley. *The EMBO Journal*, **6**, 1189–1195.

Douglas C.J., Hauffe K.D., Ites-Morales M.E., Ellard M., Paszkowski U., Hahlbrock K. *et al.* (1991) Exonic sequences are required for elicitor and light activation of a plant defense gene, but promoter sequences are sufficient for tissue specific expression. *The EMBO Journal*, **10**, 1767–1775.

Elstner E.F. (1982) Oxygen activation and oxygen toxicity. *Annual Review of Plant Physiology*, **33**, 73–96.

Ensminger P.A. and Schäfer E. (1992) Blue and ultraviolet-B light photoreceptors in parsley cells. *Photochemistry and Photobiology*, **55**, 437–447.

Foyer C.H., Descourvières P. and Kunert K.J. (1994) Protection against oxygen radicals: an important defence mechanism studied in transgenic plants. *Plant, Cell and Environment*, **17**, 507–523.

Fritze K., Staiger D., Czaja I., Walden R., Schell J. and Wing D. (1991) Developmental and UV light regulation of the snapdragon chalcone synthase promoter. *The Plant Cell*, **3**, 893–905.

Fritzemeier K.H. and Kindl H. (1981) Coordinate induction by UV light of stilbene synthase, phenylalanine ammonia-lyase and cinnamate 4-hydroxylase in leaves of Vitaceae. *Planta*, **151**, 48–52.

Foote C.S. (1991) Definition of type I and type II photosensitized oxidation. *Photochemistry and Photobiology*, **54**, 659.

Gallagher P.E., Lenhart J.R. and Weiss R.B. (1993) An endonuclease from tobacco cells which releases ring-saturated pyrimidines from damaged DNA. *Photochemistry and Photobiology*, **57**, 36S.

Greenberg B.M., Gaba V., Canaani O., Malkin S., Mattoo A.K. and Edelman M. (1989) Separate photosensitizers mediate degradation of the 32-kDa photosystem II reaction center protein in the visible and UV spectral regions. *Proceedings of the National Academy of Sciences USA*, **86**, 6617–6620.

Grossman L. and Yeung A.T. (1990) The uvrABC endonuclease of *Escherichia coli*. *Photochemistry and Photobiology*, **51**, 749–755.

Häder D.P. (1983) Inhibition of phototaxis and motility by UV-B irradiation in *Distyostelium discoideum* slugs. *Plant & Cell Physiology*, **24**, 1545–1552.

Häder D.P. (1984) Effects of UV-B on motility and photoorientation in the cyanobacterium, *Phormidium uncinatum*. *Archives of Microbiology*, **140**, 34–39.

Häder D.P. (1985) Effects of UV-B on motility and photobehavior in the green flagellate, *Euglena gracilis*. *Archives of Microbiology*, **141**, 159–163.

Häder D.P. and Häder M. (1988) Ultraviolet-B inhibition of motility in green and dark bleached *Euglena gracilis*. *Current Microbiology*, **17**, 215–220.

Häder D.P. and Worrest R.C. (1991) Effects of enhanced solar ultraviolet radiation on aquatic ecosystems. *Photochemistry and Photobiology*, **53**, 717–725.

Häder D.P., Watanabe M. and Furuya M. (1986) Inhibition of motility in the cyanobacterium, *Phormidium uncinatum*, by solar and monochromatic UV irradiation. *Plant & Cell Physiology*, **27**, 887–894.

Hadwiger L.A. and Schwochau M.E. (1971) Ultraviolet light-induced formation of pisatin and phenylalanine ammonia lyase. *Plant Physiology*, **47**, 588–590.

Hagendoorn M.J.M., Poortinga A.M., Wong Fong Sang H.W., van der Plas L.H.W. and van Walraven H.S. (1991) Effect of elicitors on the plasmamembrane of *Petunia hybrida* cell suspensions: role of Δ pH in signal transduction. *Plant Physiology*, **96**, 1261–1267.

Hahlbrock K. (1977) Regulatory aspects of phenylpropanoid biosynthesis in cell cultures. In *Plant Tissue Culture and Its Bio-technological Application*, edited by W. Barz, E. Reinhard and M.H. Zenk, pp. 99–111. Berlin: Springer-Verlag.

Hahlbrock K. and Grisebach H. (1979) Enzymic controls in the biosynthesis of lignin and flavonoids. *Annual Review of Plant Physiology*, **30**, 105–130.

Hahlbrock K. and Scheel D. (1989) Physiology and molecular biology of phenylpropanoid metabolism. *Annual Review of Plant Physiology and Plant Molecular Biology*, **40**, 347–369.

Halliwell B. and Gutteridge J.M.C. (1989) *Free Radicals in Biology and Medicine, Second Edition*. Oxford: Clarenden Press.

Harlow G.R., Jenkins M.E., Davies C. and Mount D.W. (1993) Isolation of *Arabidopsis thaliana* mutants hypersensitive to UV-B light or ionizing radiation. In *Frontiers of Photobiology: Proceedings of the 11th International Congress on Photobiology, Kyoto, Japan, 7–12 September 1992*, edited by A. Shima, M. Ichahashi, Y. Fujiwara and H. Takebe, pp. 319–321. Amsterdam: Elsevier Science Publishers B.V.

Harlow G.R., Jenkins M.E., Pittalwala T.S. and Mount D.W. (1994) Isolation of *uvh1*, an *Arabidopsis* mutant hypersensitive to ultraviolet light and ionizing radiation. *The Plant Cell*, **6**, 227–235.

Hashimoto T., Kondo N. and Tezuka T. (1993) Harmful and beneficial effects of solar UV light on plant growth. In *Frontiers of Photobiology: Proceedings of the 11th International Congress on Photobiology, Kyoto, Japan, 7–12 September 1992*, edited by A. Shima, M. Ichahashi, Y. Fujiwara and H. Takebe, pp. 551–554. Amsterdam: Elsevier Science Publishers B.V..

Herčík F. (1964) Effect of ultraviolet light on stomatal movement. *Biologia Plantarum*, **6**, 70–72 .

Howland G.P. (1975) Dark-repair of ultraviolet-induced pyrimidine dimers in the DNA of wild carrot protoplasts. *Nature*, **254**, 160-161.

Huerta A.J. and Murphy T.M. (1989a) Effects of extracellular pH on UV-induced K^+ efflux from cultured rose cells. *Plant Physiology*, **90**, 749–753.

Huerta A.J. and Murphy T.M. (1989b) Kinetics of efflux of K^+ from cultured rose cells treated with ultraviolet light. *Physiologia Plantarum*, **77**, 525–530.

Huerta A.J. and Murphy T.M. (1989c) Control of intracellular glutathione and its effect on ultraviolet radiation-induced K^+ efflux in cultured rose cells. *Plant, Cell and Environment*, **12**, 825–830.

Imbrie C.W. and Murphy T.M. (1984a) Photoinactivation of detergent-solubilized plasma membrane ATPase from *Rosa damascena:* action spectra. *Plant Physiology*, **74**, 617–621.

Imbrie C.W. and Murphy T.M. (1984b) Mechanism of photoinactivation of plant plasma membrane ATPase. *Photochemistry and Photobiology*, **40**, 243–248.

Imbrie C.W. and Murphy T.M. (1984c) Proton pumping by vesicles reconstituted from two fractions of solubilized rose-cell plasma membrane ATPase. *Biochimica et Biophysica Acta*, **778**, 229–232.

Iwanzik W., Tevini M., Dohnt G., Voss M., Weiss W., Gräber P. *et al.*, (1983) Action of UV-B radiation on photosynthetic primary reactions in spinach chloroplasts. *Physiologia Plantarum*, **58**, 401–407.

Jackson J.F. (1987) DNA repair in pollen: a review. *Mutation Research*, **181**, 17–30.

Jackson J.F. and Linskens H.F. (1978) Evidence for DNA repair after ultraviolet irradiation of *Petunia hybrida* pollen. *Molecular and General Genetics*, **161**, 117–120.

Jackson J.F. and Linskens H.F. (1979) Pollen DNA repair after treatment with the mutagens 4-nitroquinoline-1-oxide, ultraviolet and near-ultraviolet irradiation, and boron dependence of repair. *Molecular and General Genetics*, **176**, 11–16.

Jagger J. (1985) *Solar UV Actions on Living Cells*. New York: Praeger Publishers.

Jordan B.R., He J., Chow W.S. and Anderson J.M. (1992) Changes in mRNA levels and polypeptide subunits of ribulose 1,5-bisphosphate carboxylase in response to supplementary ultraviolet-B radiation. *Plant, Cell and Environment*, **15**, 91–98.

Jursinic P. and Stemler A. (1983) Changes in $[^{14}C]$ atrazine binding associated with oxidation-reduction state of the secondary quinone acceptor of photosystem II. *Plant Physiology*, **73**, 703–708.

Karcz W. and Stolarek J. (1981) Electric potential changes in free cells of *Malus domestica* induced by ultraviolet radiation. *Proceedings of the Sixth School on Biophysics of Membrane Transport*, Poland, May 4–13, p. 379.

Karcz W. and Stolarek J. (1988) Effects of UV-C radiation on extension growth, H^+-extrusion and transmembrane electric potential in maize coleoptile segments. *Physiologia Plantarum*, **74**, 770–774.

Kato H. (1974) Possible role of DNA synthesis in formation of sister chromatid exchanges. *Nature*, **252**, 739–741.

Kim S.T., Malhotra K., Smith C.A., Taylor J.S. and Sancar A. (1994) Characterization of (6–4) photoproduct DNA photolyase. *Journal of Biological Chemistry*, **269**, 8535–8540.

Klein R.M. (1967) Influence of ultraviolet radiation on auxin-controlled plant growth. *American Journal of Botany*, **54**, 904–914.

Köhler K., Steigner W., Simonis W. and Urbach W. (1985) Potassium channels in *Eremosphaera viridis*. I. Influence of cations and pH on resting membrane potential and on an action-potential-like response. *Planta*, **166**, 490–499.

Kolattukudy P.E. (1980) Biopolyester membranes of plants: cutin and suberin. *Science*, **208**, 990–1000.

Kramer G.F., Krizek D.T. and Mirecki R.M. (1991a) Influence of visible and UV-B radiation on polyamine levels in soybean. *Plant Physiology*, **96**, S114.

Kramer G.F., Norman H.A., Krizek D.T. and Mirecki R.M. (1991b) Influence of UV-B radiation on polyamines lipid peroxidation and membrane lipids in cucumber. *Phytochemistry*, **30**, 2101–2108.

Kulandaivelu G. and Noorudeen A.M. (1983) Comparative study of the action of ultraviolet-C and ultraviolet-B radiation on photosynthetic electron transport. *Physiologia Plantarum*, **58**, 389–394.

Langer B. and Wellman E. (1990) Phytochrome induction of photoreactivating enzyme in *Phaseolus vulgaris* L. seedlings. *Photochemistry and Photobiology*, **52**, 861–864.

Levine A., Tenhaken R., Dixon R. and Lamb C. (1994) H_2O_2 from the oxidative burst orchestrates the plant hypersensitive disease resistance response. *Cell*, **79**, 583–593 .

Li J., Ou-Lee T.M., Raba R., Amundson R.G. and Last R.L. (1993) *Arabidopsis* flavonoid mutants are hypersensitive to UV-B radiation. *The Plant Cell*, **5**, 171–179.

Lipphardt S., Brettschneider R., Kreuzaler F., Schell J. and Dangl J.L. (1988) UV-inducible transient expression in parsley protoplasts identifies regulatory *cis*-elements of a chimeric *Antirrhinum majus* chalcone synthase gene. *The EMBO Journal*, **7**, 4027–4033.

Lois R., Dietrich A., Hahlbrock K. and Schulz W. (1989) A phenylalanine ammonia-lyase gene from parsley: structure, regulation, and identification of elicitor and light-responsive *cis*-acting elements. *The EMBO Journal*, **8**, 1641–1648.

Lu J.Y., Stevens C., Khan V.A., Kabwe M. and Wilson C.L. (1991) The effect of ultraviolet irradiation on shelf-life and ripening of peaches and apples. *Journal of Food Quality*, **14**, 299–305.

Machemer H. and Sugino K. (1989) Electrophysiological control of ciliary beating: a basis of motile behavior in ciliated protozoa. *Comparative Biochemistry and Physiology*, **94A**, 365–374.

Manabe E., Hirosawa T., Tsuzuki M. and Miyachi S. (1986) Effect of solar near ultraviolet radiation on growth of *Chlorella* cells. *Physiologia Plantarum*, **67**, 598–603.

McLennan A.G. (1987) The repair of ultraviolet light-induced DNA damage in plant cells. *Mutation Research*, **181**, 1–8.

McLennan A.G. and Eastwood A.C. (1986) An endonuclease activity from suspension cultures of *Daucus carota* which acts upon pyrimidine dimers. *Plant Science*, **46**, 151–157.

Melis A., Nemson J.A. and Harrison M.A. (1992) Damage to functional components and partial degradation of photosystem II reaction center proteins upon chloroplast exposure to ultraviolet-B radiation. *Biochimica et Biophysica Acta*, **1100**, 312–320.

Mohr H. and Drumm-Herrel H. (1983) Coaction between phytochrome and blue/UV light in anthocyanin synthesis in seedlings. *Physiologia Plantarum*, **58**, 408–414.

Murali N.S., Teramura A.H. and Randall S.K. (1988) Response differences between two soybean cultivars with contrasting UV-B radiation sensitivities. *Photochemistry and Photobiology*, **48**, 653–657.

Murphy T.M. (1983) Membranes as targets of ultraviolet radiation. *Physiologia Plantarum*, **38**, 381–388.

Murphy T.M. (1988a) Induced K^+ efflux from cultured rose cells: effects of protein-synthesis inhibitors. *Plant Physiology*, **86**, 830–835.

Murphy T.M. (1988b) Ca^{2+} dependence and La^{3+} interference of ultraviolet radiation-induced K^+ efflux from rose cells. *Physiologia Plantarum*, **74**, 537–543.

Murphy T.M. and Auh C.K. (1992) *Phytophthora* elicitor inhibits ferricyanide reduction by cultured rose cells. *Environmental and Experimental Botany*, **32**, 487–496.

Murphy T.M., Auh C.K., Schorr R. and Grobe C. (1991) Modification of plasma membrane electron transport in cultured rose cells by UV-C radiation and fungal elicitor. *Plant Physiology*, **96**, S142

Murphy T.M. and Huerta A.J. (1990a) Hydrogen peroxide formation in cultured rose cells in response to UV-C radiation. *Physiologia Plantarum*, **78**, 247–253.

Murphy T.M. and Huerta A.J. (1990b) Effect of broad-band ultraviolet and visible radiation on hydrogen peroxide formation by cultured rose cells. *Physiologia Plantarum*, **80**, 63–68.

Murphy T.M. and Huerta A.J. (1990c) Stress-induced efflux of K^+ ion: environmental control. In *New Trends in Plant Physiology*, edited by K.K. Dhir, I.S. Dua and K.S. Chark, pp. 265–268. New Delhi: Today & Tomorrow Printers & Publishers.

Murphy T.M., Hurrell H.C. and Sasaki T.L. (1985) Wavelength dependence of ultraviolet radiation-induced mortality and K^+ efflux in cultured cells of *Rosa damascena*. *Photochemistry and Photobiology*, **42**, 281–286 .

Murphy T.M., Kuhn D.N. and Murphy J.B. (1973) Ultraviolet irradiation of the components *of the* wheat embryo *in vitro* protein synthesizing system. *Biochemistry*, **12**, 1782–1789.

Murphy T.M., Martin C.P. and Kami J. (1993) Endonuclease activity from tobacco nuclei specific for ultraviolet radiation-damaged DNA. *Physiologia Plantarum*, **87**, 417–425.

Murphy T.M., Matson G.B. and Morrison S.L. (1983) Ultraviolet-stimulated $KHCO_3$ efflux from rose cells: regulation of cytoplasmic pH. *Plant Physiology*, **73**, 20–24.

Murphy T.M. and Wilson C. (1982) The UV-stimulated K^+ efflux from rose cells. *Plant Physiology*, **70**, 709–713.

Murphy T.M., Wright L.A., Jr. and Murphy J.B. (1975) Inhibition of protein synthesis in cultured tobacco cells by ultraviolet radiation. *Photochemistry and Photobiology*, **21**, 219–225.

Nedunchezhian N. and Kulandaivelu G. (1991) Evidence for the ultraviolet-B (280–320 nm) radiation induced structural reorganization and damage of photosystem II polypeptides in isolated chloroplasts. *Physiologia Plantarum*, **81**, 558–562.

Negash L. (1987) Wavelength-dependence of stomatal closure by ultraviolet radiation in attached leaves of *Eragrostis tef*: Action spectra under backgrounds of red and blue light. *Plant Physiology and Biochemistry*, **25**, 753–760.

Negash L. and Björn L.O. (1986) Stomatal closure by ultraviolet radiation. *Physiologia Plantarum*, **66**, 360–364.

Negash L., Jensén P. and Björn L.O. (1987) Effect of ultraviolet radiation on accumulation and leakage of $^{86}Rb^+$ in guard cells of *Vicia faba*. *Physiologia Plantarum*, **69**, 200–204.

Ohl S., Hedrick S.A., Chory J. and Lamb C.J. (1990) Functional properties of a phenylalanine ammonia-lyase promoter from *Arabidopsis*. *The Plant Cell*, **2**, 837–848.

Painter R.B. and Wolff S. (1973) Apparent absence of repair replication in *Vicia faba* after X-irradiation. *Mutation Research*, **19**, 133–137.

Panagopoulos I., Bornman J.F. and Björn L.O. (1990) Effects of ultraviolet radiation and visible light on growth, fluorescence induction, ultraweak luminescence and peroxidase activity in sugar beet plants. *Journal of Photochemistry and Photobiology B. Biology*, **8**, 73–87.

Pang Q. and Hays J.B. (1991) UV-B-inducible and temperature-sensitive photoreactivation of cyclobutane pyrimidine dimers in *Arabidopsis thaliana*. *Plant Physiology*, **95**, 536–543.

Parker J.E., Schulte W., Hahlbrock K. and Scheel D. (1991) An extracellular glycoprotein from *Phytophthora megasperma* f. sp. *glycinae* exicits phytoalexin synthesis in cultured parsley cells and protoplasts. *Molecular Plant-Microbe Interactions*, **4**, 19–27.

Pfündel E.E., Pan R.S. and Dilley R.A. (1992) Inhibition of violoxanthin deepoxidation by ultraviolet-B radiation in isolated chloroplasts and intact leaves. *Plant Physiology*, **98**, 1372–1380.

Qian Y.C. and Murphy T.M. (1993) Regulation of rose cell plasma membrane ATPase activity by glutathione. *Plant Physiology and Biochemistry*, **31**, 901–909.

Quait F.E., Sutherland B.M. and Sutherland J.C. (1992) Action spectrum for DNA damage in alfalfa lowers predicted impact of ozone depletion. *Nature*, **358**, 576–578.

Rao M.V., Paliyath G. and Ormrod D.P. (1995) Differential response of photosynthetic pigments, rubisco activity and rubisco protein of *Arabidopsis thaliana* exposed to UV-B and ozone. *Photochemistry and Photobiology*, **62**, 727–735.

Renger G., Völker M., Eckert H.J., Fromme R., Hohm-Veit S. and Gräber P. (1989) On the mechanism of photosystem II deterioration by UV-B irradiation. *Photochemistry and Photobiology*, **49**, 97–105.

Reuber S., Leitsh J. Krause J.H and Weissenboeck G. (1993) Metabolic reduction of phenyl propanoid compounds in primary leaves of rye *(Secale Cereale* L.) leads to increased UV-B sensitivity of photosynthesis. *Zeitschrift für Naturforschung*, **48**, 749–756.

Robberecht R. and Caldwell M.M. (1978) Leaf epidermal transmittance of ultraviolet radiation and its implications for plant sensitivity to ultraviolet-radiation induced injury. *Oecologia*, **32**, 277–287.

Robberecht R. and Caldwell M.M. (1983) Protective mechanisms and acclimation to solar ultraviolet-B radiation in *Oenothera stricta*. *Plant, Cell and Environment*, **6**, 477–485.

Saito N. and Werbin H. (1969) Evidence for a DNA-photoreactivating enzyme in higher plants. *Photochemistry* and *Photobiology*, **9**, 389–393.

Sancar A. and Sancar G.B. (1988) DNA repair enzymes. *Annual Review of Biochemistry*, **57**, 29–67.

Scheel D., Hauffe K.D., Jahnen W. and Hahlbrock K. (1986) Stimulation of phytoalexin formation in fungus-infected plants and elicitor-treated cell cultures of parsley. In *Recognition in Microbe-Plant Symbiotic and Pathogenic Interaction*, edited by B. Lugtenberg, pp. 325–331. Berlin: Springer-Verlag.

Schulze-Lefert P., Becker-André M., Schulz, W., Hahlbrock K. and Dangl J.L. (1989) Functional architecture of the light-responsive chalcone synthase promoter from parsley. *The Plant Cell*, **1**, 707–714.

Small G.D. (1987) Repair systems for nuclear and chloroplast DNA in *Chlamydomonas reinhardtii. Mutation Research*, **181**, 31–36.

Small G.D. and Greimann C.S. (1977) Photoreactivation and dark repair of ultraviolet light induced pyrimidine dimers in chloroplast DNA. *Nucleic Acids Research*, **4**, 2893–2902

Sommer H. and Saedler H. (1986) Structure of the chalcone synthase gene of *Antirrhinum majus*. *Molecular and General Genetics*, **202**, 429–434.

Soyfer V.N. (1983) Influence of physiological conditions on DNA repair and mutagenesis in higher plants. *Physiologia Plantarum*, **58**, 373–380.

Soyfer V.N. and Cieminis K.K. (1977) Excision of thymine dimers from the DNA of UV-irradiated plant seedlings. *Environmental and Experimental Botany*, **17**, 135–143.

Staiger D., Kaulen H. and Schell J. (1989) A CACGTG motif of the *Antirrhinum majus* chalcone synthase promoter is recognized by an evolutionarily conserved nuclear protein. *Proceedings of the National Academy of Science USA*, **86**, 6930–6934.

Staiger D., Kaulen H. and Schell J. (1990) A nuclear factor recognizing a positive regulatory upstream element of the *Antirrhinum majus* chalcone synthase promoter. *Plant Physiology*, **93**, 1347–1353.

Stapleton A.E. and Walbot V. (1994) Flavonoids can protect maize DNA from the induction of ultraviolet radiation damage. *Plant Physiology*, **105**, 881–889.

Stein B., Angel P., van Dam H., Ponta H., Herrlich P., van der Eb A. *et al.* (1992) Ultraviolet-radiation induced *c-jun* gene transcription; two AP-1 like binding sites mediate the response. *Photochemistry and Photobiology*, **55**, 409–415.

Steinmetz V. and Wellmann E. (1986) The role of solar UV-B in growth regulation of cress *(Lepidium sativum* L.) seedlings. *Photochemistry and Photobiology*, **43**, 189–193.

Stolarek J. and Karcz W. (1987) Effects of UV-C radiation on membrane potential and electric conductance in internodal cells of *Nitellopsis obtusa*. *Physiologia Plantarum*, **70**, 473–748.

Strickland J.A., Marzilli L.G., Puckett J.M., Jr. and Doetsch P.W. (1991) Purification and properties of nuclease SP. *Biochemistry*, **30**, 9749–9756.

Strid Å. (1993) Alteration in expression of defence genes in *Pisum sativum* after exposure to supplementary ultraviolet-B radiation. *Plant & Cell Physiology*, **34**, 949–953.

Strid Å., Chow W.S. and Anderson J.M. (1990) Effects of supplementary ultraviolet-B radiation on photosynthesis in *Pisum sativum*. *Biochimica et Biophysica* Acta, **1020**, 260–268.

Sullivan J.H. and Teramura A.H. (1990) Field study of the interaction between solar ultraviolet-B radiation and drought on photosynthesis and growth in soybean. *Plant Physiology*, **92**, 141–146

Swinton D.C. and Hanawalt P.C. (1973) Absence of ultraviolet light stimulated repair replication in the nuclear and chloroplast genomes of *Chlamydomonas reinhardii*. *Biochimica et Biophysica Acta*, **294**, 385–395.

Takeuchi Y., Akizuki M., Shimizu H., Kondo N. and Sugahara, K. (1989) Effect of UV-B (290–320 nm) irradiation on growth and metabolism of cucumber cotyledons. *Physiologia Plantarum*, **76**, 425–430.

Takeuchi Y., Ikeda S. and Kasahara H. (1993) Dependence of wavelength and temperature of growth inhibition induced by UV-B irradiation. *Plant & Cell Physiology*, **34**, 949–953.

Tanada T. and Hendricks S.B. (1953) Photoreversal of ultraviolet effects in soybean leaves. *American Journal of Botany*, **40**, 634–637.

Teramura A.H. (1983) Effects of ultraviolet-B radiation on the growth and yield of crop plants. *Physiologia Plantarum*, **58**, 415–427.

Teramura A.H. and Sullivan J.H. (1987) Soybean growth responses to enhanced levels of ultraviolet-B radiation under greenhouse conditions. *American Journal of Botany*, **74**, 975–979.

Teramura A.H., Ziska L.H. and Sztein A.E. (1991) Changes in growth and photosynthetic capacity of rice with increased UV-B radiation. *Physiologia Plantarum*, **83**, 373–380.

Tevini M., Braun J. and Fieser G. (1991) The protective function of the epidermal layer of rye seedlings against ultraviolet-B radiation. *Photochemistry and Photobiology*, **53**, 329–333.

Tevini M. and Iwanzik W. (1983) Inhibition of photosynthetic activity by UV-B radiation in radish seedlings. *Physiologia Plantarum*, **58**, 395–400.

Tevini M. and Teramura A.H. (1989) UV-B effects on terrestrial plants. *Photochemistry and Photobiology*, **50**, 479–487.

Todo T., Takemori H., Ryo H. Ihara M., Matsunaga T., Nikaido O., Sato K. and Nomura T. (1993) A new photoreactivating enzyme that specifically repairs ultraviolet light-induced (6–4)photoproducts. *Nature*, **361**, 371–374.

Trosko J.E. and Mansour V.H. (1968) Response of tobacco and *Happlopappus* cells to ultraviolet irradiation after posttreatment with photoreactivating light. *Radiation Research*, **36**, 333–343.

Trosko J.E. and Mansour V.H. (1969) Photoreactivation of ultraviolet light-induced pyrimidine dimers in *Ginkgo* cells grown *in vitro*. *Mutation Research*, **7**, 120–121.

Velemínsk J. and Angelis K.J. (1990) DNA repair in higher plants. *Progress in Clinical and Biological Research*, **340A**, 195–203.

Velemínsk J., Svachulová J. and Satava J. (1980) Endonucleases for UV-irradiated and depurinated DNA in barley chloroplasts. *Nucleic Acids Research*, **8**, 1373–1381.

Watson R.T., Prather M.J. and Kurylo M.J. (and Ozone Trends Panel, Ad Hoc Theory Panel, and NASA Panel for Data Evaluation) (1988) *NASA RP-1208*. Washington, D.C.: NASA Office of Space Science and Applications, NASA.

Weisshaar B., Armstrong G.A., Block A., da Costa e Silva O. and Hahlbrock K. (1991) Light-inducible and constitutively expressed DNA-binding proteins recognizing a plant promoter element with functional relevance in light responsiveness. *The EMBO Journal*, **10**, 1777–1786.

Wellman E. (1976) Specific ultraviolet effects in plant morphogenesis. *Photochemistry and Photobiology*, **24**, 659–660.

Wilson M.I. and Greenberg B.M. (1993) Specificity and photomorphogenic nature of ultraviolet-B-induced cotyledon curling in *Brassica napus* L. *Plant Physiology*, **102**, 671–677.

Wingender R., Röhrig H., Höricke C. and Schell J. (1990) *cis*-Regulatory elements involved in ultraviolet light regulation and plant defense. *The Plant Cell*, **2**, 1019–1026.

Wingender R., Röhrig H., Höricke C., Wing, D. and Schell J. (1989) Differential regulation of soybean chalcone synthase genes in plant defence symbiosis and upon environmental stimuli. *Molecular and General Genetics*, **218**, 315–322.

Worrest R.C. (1983) Impact of solar ultraviolet-B radiation (290–320 nm) upon marine microalgae. *Physiologia Plantarum*, **58**, 428–434.

Wright L.A., Jr. and Murphy T.M. (1975) Photoreactivation of nitrate reductase production in *Nicotiana tabacum var. xanthi*. *Biochimica et Biophysica Acta*, **407**, 338–346.

Wright L.A., Jr. and Murphy T.M. (1978) Ultraviolet radiation-stimulated efflux of ^{86}Rubidium from cultured tobacco cells. *Plant Physiology*, **61**, 434–436.

Wright L.A., Jr. and Murphy T.M. (1982) Shortwave ultraviolet light closes leaf stomata. *American Journal of Botany*, **69**, 1196–1199.

Wright L.A., Jr., Murphy T.M. and Travis R.L (1981) The effect of ultraviolet radiation on wheat root vesicles enriched in plasma membrane. *Photochemistry and Photobiology*, **33**, 343–348

Yocum C.F. (1991) Calcium activation of photosynthetic water oxidation. *Biochimica et Biophysica Acta*, **1059**, 1–15.

9. PLANT STRATEGIES FOR COPING WITH VARIABLE LIGHT REGIMES

ALAN K. KNAPP and PHILIP A. FAY

Division of Biology, Ackert Hall, Kansas State University Manhattan, KS 66506-4901, USA

INTRODUCTION

Plant survival, growth, and yield in natural, agricultural and horticultural settings requires successfully coping with varying availability of critical resources (Chapin *et al.*, 1987). Chronic extremes of water, light, temperature and nutrients are well-known stressors to most basic plant processes (Osmond *et al.*, 1987). However, shorter-term variation in resource availability may also decrease plant performance. Light may be the most temporally variable essential plant resource. There are virtually no ecosystems where light intensity reaching photosynthetic surfaces does not vary, often by an order of magnitude, at time scales from seconds to hours.

Despite the fact that most plants experience varying light availability, most of what we know about plant responses to light is based on steady-state, constant-irradiance analyses from laboratory and field studies (Boardman, 1977; Bunce, 1983; Osmond, 1983; Read, 1985; Turnbull *et al.*, 1993). In fact, there has often been active avoidance of variable light conditions during field studies (Knapp and Smith, 1981; Roessler and Monson, 1985), reflecting the common perspective that fluctuating light levels represent uncontrolled variability rather than an environmental characteristic to be studied.

There has been considerable recent interest in plant responses to short-term, seconds to minutes-long variations in light and the potential for this variability to act as a stress (Chazdon, 1988; Pearcy, 1988; Knapp and Smith, 1990b). Although earlier studies established that photosynthesis and stomata respond to fluctuations in sunlight (Willis and Balasubramanian, 1968; McCree and Loomis, 1969; Pollard, 1970; Woods and Turner, 1971; Davies and Kozlowski, 1975; Lakso and Barnes, 1978; Gross and Chabot, 1979), it was not until the 1980's that laboratory and field instrumentation became capable of measuring rapid transient gas exchange responses to variable light. Thus, our understanding of how plants cope with variable light regimes has progressed rapidly in the last 15 years.

Our goals for this chapter are to (1) review the types of variable light regimes experienced by plants and their causes, (2) examine key plant responses to fluctuations in light independent of one another, and (3) evaluate how the component responses form integrated strategies whereby plants cope with varying light intensities. Strategies used by plants to minimize the negative impacts of variable light can provide unique insights into plant-environment interactions (Knapp and Smith, 1990b). Indeed, strategies for coping with variable light can lead to interpretations of plant physiological adaptations that contrast sharply with interpretations based on more traditional steady-state analyses of plant responses to the environment (Knapp and Smith, 1989).

CAUSES AND PATTERNS OF TEMPORAL VARIABILITY IN SUNLIGHT

Causes of short-term variability in sunlight fall broadly into four categories: cloud cover, overstory canopy shading, within- canopy shading and wind-induced leaf movements. These in turn result in a continuum of light regimes varying in background light intensity and frequency of sun-to-shade transitions. Along this continuum, there are three more-or-less discrete light environments for which plant responses have been studied (Figure 9.1). We will define these as: *sunpatch* (or shadefleck) regimes in which background light levels are high (often full sunlight) but short-term, minutes-long periods of low light (shade) occur due to intermittent clouds or one leaf/plant shading another (Knapp and Smith, 1987, 1988). Plants growing in open environments, beneath open canopies, or those that have dominant canopy positions would most likely experience this light regime. In *Leaf flutter* light regimes, leaves experience rapid, short-term fluctuations between high and low light due to wind-induced leaf movements (Lakso and Barnes, 1978; Roden and Pearcy, 1993a). During calm conditions, background light levels are high because only plants in open sites or in dominant canopy positions can experience this light regime. *Sunfleck* light regimes are characterized by a background of very low light with brief periods of moderate to high light (Figure 9.1; Pearcy and Calkin, 1983; Chazdon, 1988; Pearcy, 1990). Plants growing beneath closed forest canopies, or leaves near the bottom of dense tree or herbaceous canopies (Norman *et al.*, 1971; Baldocchi *et al.*, 1986) are most likely to experience this type of light regime (Smith *et al.*, 1989).

Where do most plants fall along the continuum of light regimes (Figure 9.1)? Sunpatch light regimes are probably the most commonly encountered. Although leaves within an individual plant might experience quite different light regimes simultaneously (Pearcy *et al.*, 1990), the leaf area distribution in most plant communities is skewed towards upper canopy positions where light is most readily available (Stern and Donald, 1962; Hirose and Werger, 1987; Hollinger, 1989). Sunpatch regimes may be especially prevalent in mountainous areas where orographic lift of air masses (Salisbury *et al.*, 1968; Wardle, 1986) causes disproportionate cloud cover. For example, on typical partly-cloudy days in the Rocky Mountains (USA), full sun periods averaging 9.9 minutes alternated with mean shade periods of 3.8 minutes at ca. 450 $\mu mol\ m^{-2}\ s^{-1}$ photon flux density (PFD; Knapp and Smith, 1988). Shade periods occurred on an average 35 times per day; over 100 per day were observed. Cloud-induced sunpatches occur frequently in non-mountainous regions as well (Marotz and Henry, 1978), since cloud cover increases as precipitation increases (Anonymous, 1981). Sunpatch light regimes

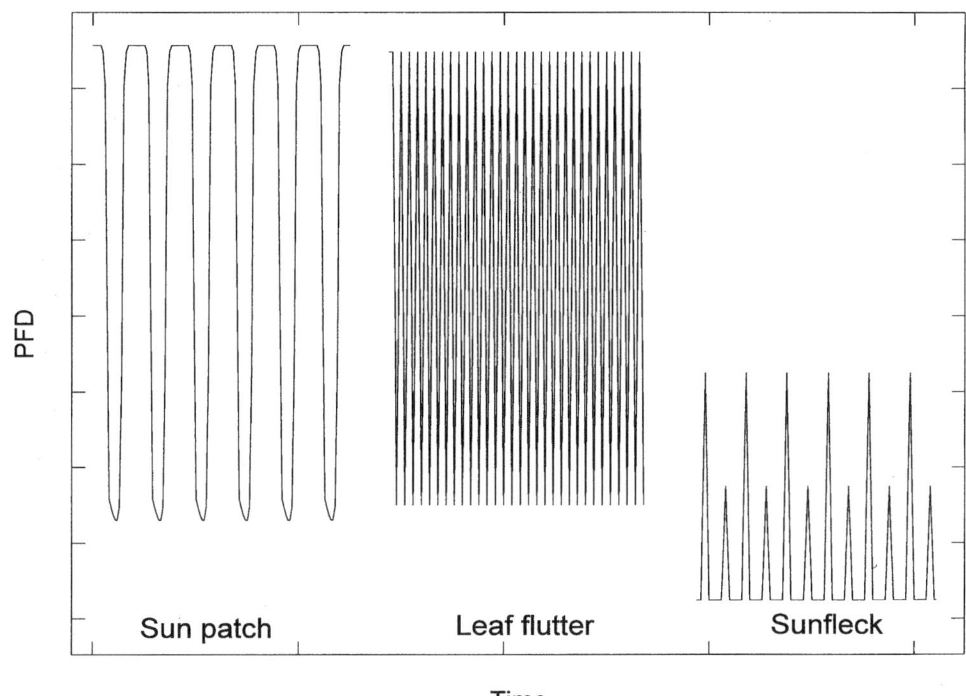

Figure 9.1. Relative light level (PFD) and frequencies of change for typical sunpatch, leaf flutter, and sunfleck light regimes experienced by most plants. See text for description.

are probably rare only in a few situations. In arid regions, cloud cover is rare and plant canopies are sparse (Knapp and Smith, 1991), so that most leaves receive consistent high light most of the time. The seedling stage of row crop agriculture might also typically receive constant high light at the leaf level. However, as the crop canopy develops, sunpatch frequency will increase.

The degree of leaf flutter-induced variation in light depends primarily on microclimatic conditions and species-specific petiole characteristics. In poplars, Roden and Pearcy (1993a) noted that periods of high and low PFD on an average lasted 0.4 s each, with thousands of transitions occurring each day. These values typify most other studies on the light environment of fluttering leaves.

Plants exclusively experiencing extreme sunfleck regimes typically occur in dense forest understories in tropical and temperate regions (Chazdon, 1988; Pearcy, 1988; Pfitsch and Pearcy, 1989). In tropical forest understories, Pearcy (1988) reported that background diffuse light levels were usually only 10-20 $\mu mol\ m^{-2}\ s^{-1}$ PFD and that sunflecks averaged 50 seconds in duration, most with

PFD between 50 and 400 μmol m^{-2} s^{-1}. Sunflecks were extremely important to the daily carbon budgets of these understory species.

PLANT RESPONSES TO SUNLIGHT VARIABILITY

Photosynthetic Responses

The relationship between net photosynthesis (A) and steady-state PFD has been documented for hundreds of plant species, but fewer data exist for responses in A to varying PFD. Photosynthetic responses to rapid PFD decreases from above to below light saturation are, with some exceptions, nearly instantaneous given the resolution of available measurement techniques. (Figure 9.2). Reduced A is often accompanied in C_3 species by a burst of CO_2 (Vines *et al.*, 1982, 1983), indicative of photorespiration (Peterson, 1983). After this transient burst, A rapidly stabilizes.

Upon an abrupt increase in PFD, A increases rapidly in most plants. Exceptions to this rapid increase may occur when stomatal closure during preceding

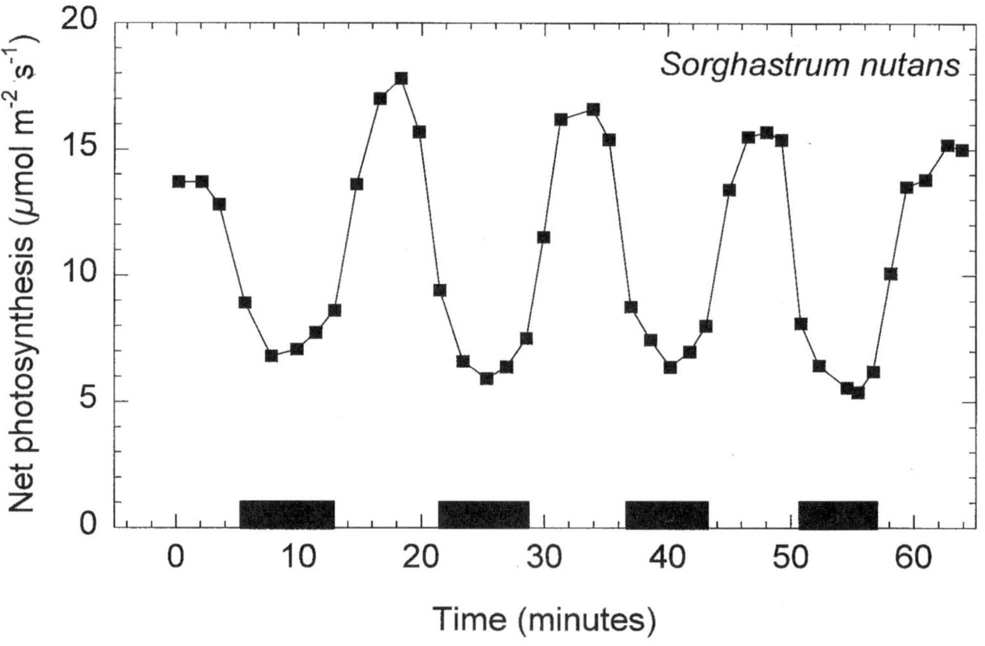

Figure 9.2. Response of net photosynthesis to a sunpatch light regime (300-400 μmol $m^{-2}s^{-1}$ during shade periods, shown by boxes) for the native tallgrass prairie C_4 grass, *Sorghastrum nutans.*

shade limits photosynthetic recovery (see below) and in understory species growing in extreme sunfleck environments. A during sunflecks depends strongly on the induction state of the leaf (Chazdon and Pearcy, 1986; Pearcy 1988, 1989, 1990; Pfitch and Pearcy, 1989). After an extended period of low light, there is an induction requirement in the photosynthetic apparatus such that maximum photosynthetic capacity will not be achieved for up to 20 minutes, even at saturating PFD levels. Such an induction requirement may be related to the light-activation requirement of Rubisco (Pearcy, 1988).

Sunflecks in understory species can lead to CO_2 uptake greater than that predicted from steady-state measurements of A. Light use efficiencies greater than 100% are possible because some of the light energy absorbed during a sunfleck can be stored and used for CO_2 uptake after low PFD has returned to the leaf (post-illumination CO_2 fixation; Pearcy, 1988). Such responses may be particularly important for plants in extremely low PFD environments where PFD in sunflecks may comprise almost 60% of the total daily PFD available. Similar effects occur in leaves near the bottom of dense herbaceous canopies. Where sunfleck regimes are experienced, induction state can be quite variable (Pearcy and Seeman, 1990; Pons and Pearcy, 1992). Under leaf-flutter light regimes, photosynthetic responses to individual PFD transitions are difficult to measure, but integrated CO_2 uptake when sun/shade periods alternated at 0.4 sec intervals indicated that light use efficiencies can exceed 100% (Roden and Pearcy, 1993b).

Stomatal Responses

Steady-state responses in stomatal conductance (g) to PFD typically are similar to responses in A. This is not the case under conditions of sunlight variability. Woods and Turner (1971) were among the first to document dynamic stomatal responses to variable PFD. Knapp and Smith (1987, 1989, 1991) and Fay and Knapp (1993, 1995) have extended these studies to include over 25 different native and agricultural species. Dynamic stomatal response patterns seem to depend on the light environment, the species under study, and plant water status.

For example, in deeply shaded understory environments, g is relatively unresponsive to sunflecks (Hollinger, 1987; Pearcy *et al.*, 1994). In some species, the time required for stomatal responses to sunflecks may be similar to, or longer than, the time required for photosynthetic induction (Kirschbaum and Pearcy, 1988; Pearcy *et al.*, 1994), even though the absolute change in g during the sunfleck is small. In other species, g may remain intermediate between steady-state levels for sun and shade PFD throughout the day (Mooney *et al.*, 1983; Pearcy *et al.*, 1994). Carbon gain during brief sunflecks can then proceed with minimal diffusional limitations.

During sunpatches in open understory environments, the stomata of understory herbs respond rapidly, with g changing up to 4-fold in several minutes (Young and Smith, 1979, 1980; Knapp *et al.*, 1989). In fully open environments, a range of stomatal responses to a standard experimental sunpatch regime (8 min of full sun alternating with 5 min of shade) has been documented (Figure

9.3). Growth form was a good predictor of stomatal responses and sunpatches for subalpine species (Knapp and Smith 1989). Minimal stomatal responses were seen in woody species (see also Whitehead and Teskey, 1995), with rapid responses in herbaceous species. In contrast, in desert herbs that rarely experience sunpatches, stomatal responses to experimental sunpatches were much reduced compared to subalpine herbs (Knapp and Smith, 1991).

In grasslands, stomatal responses to sunpatches have been associated with photosynthetic rate (Knapp, 1992), photosynthetic pathway (Knapp, 1993), and plant water status (Knapp and Smith, 1988; Knapp, 1993; Fay and Knapp, 1996; see below). Overall, when responses from subalpine, grassland, and agricultural species are combined, the best predictor of stomatal response to sunpatches is the photosynthetic rate of the plant. Thus, species with high photosynthetic rates usually have stomata that are very responsive to variable PFD (Figure 9.4).

Stomatal responses in poplars to leaf-flutter light environments are minimal (Roden and Pearcy, 1993a,b). This could be because the light environment changes too rapidly for guard cells to track PFD. However, poplars have long been noted for having stomata that are relatively unresponsive to light level and water stress (Schulte *et al.*, 1987; Ceulemans *et al.*, 1989). Thus, the impact that

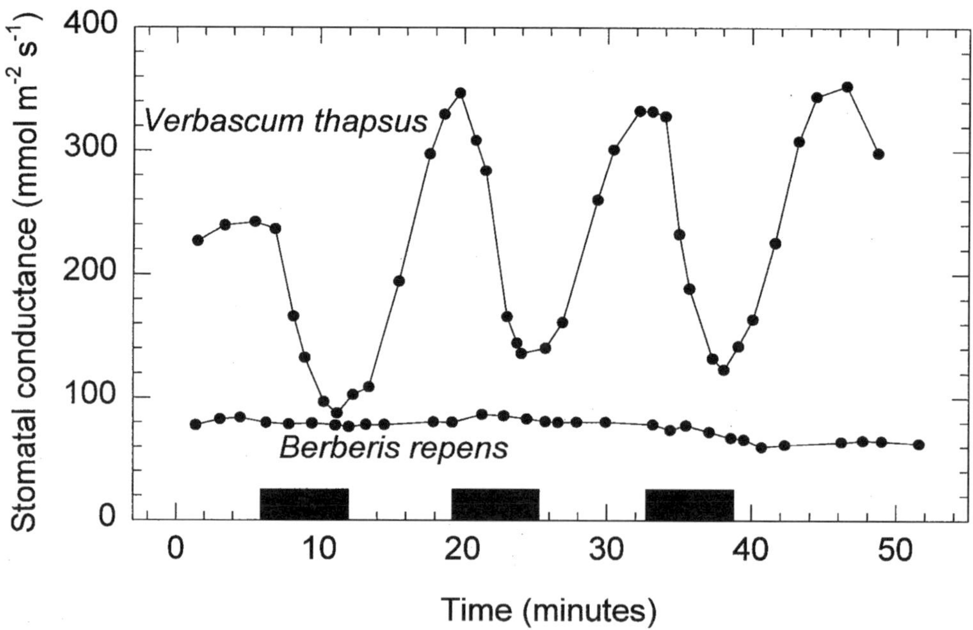

Figure 9.3. Responses of stomatal conductance to a sunpatch light regime (300-400 μmol m^{-2} s^{-1} during shade periods, shown by boxes) for a species with highly responsive stomata, *Verbascum thapsus*, and a species with unresponsive stomata, *Berberis repens*.

rapid changes in PFD may have on other species during leaf flutter requires further study, and will be difficult to separate from other environmental changes associated with leaf fluttering, such as high wind (Grace, 1988).

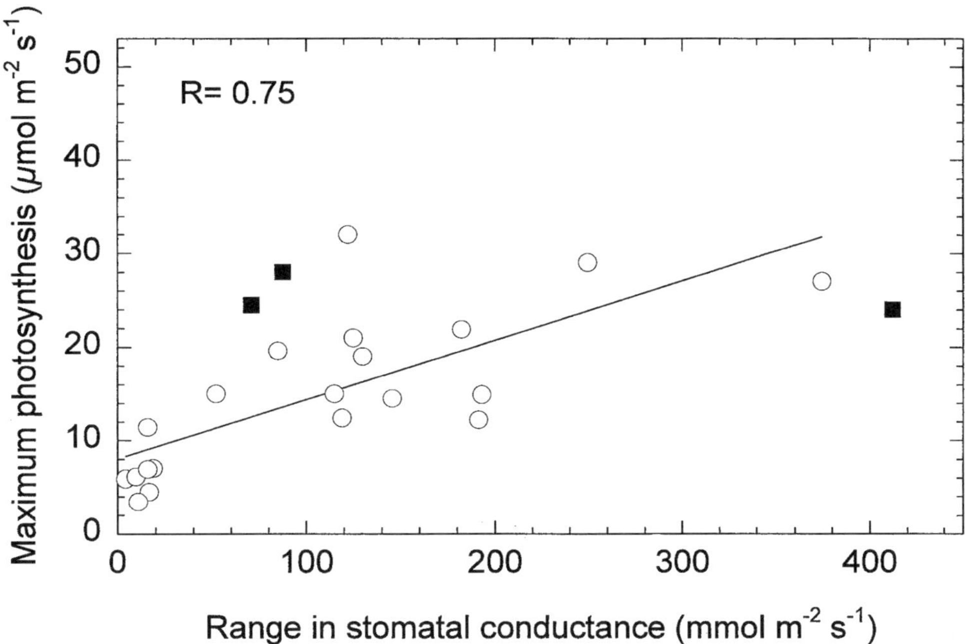

Figure 9.4. Relationship between maximum net photosynthesis and stomatal responsiveness (range in conductance from high to low PFD) to a sunpatch light regime for native (open circles) and crop species (closed squares) measured in the subalpine and in tallgrass prairie.

Responses in Plant Water Status

Plant water status should vary under dynamic PFD conditions, because the resulting variable thermal load will affect transpirational water loss. In sunfleck environments, where PFD levels during sun periods are relatively low, variations in water status have not been reported. But for plants in open understory environments, sunpatches caused large reductions in leaf xylem pressure potential (ψ_{leaf}) and leaf wilting (Young and Smith, 1979, 1980; Knapp and Smith, 1990b). ψ_{leaf} recovered quickly during shade in these understory species. In plants from open environments, fluctuations in ψ_{leaf} during sunpatches can be substantial (Figure 9.5) but the degree of variability may depend on plant growth form. For example, herbs may undergo ψ_{leaf} fluctuations, but trees and other species capable of storing water in trunks or taproots are buffered from large fluctuations in ψ_{leaf} (Knapp and Smith, 1989; Knapp, 1992). No measurable effects of leaf flutter on ψ_{leaf} have been noted (Roden and Pearcy, 1993c).

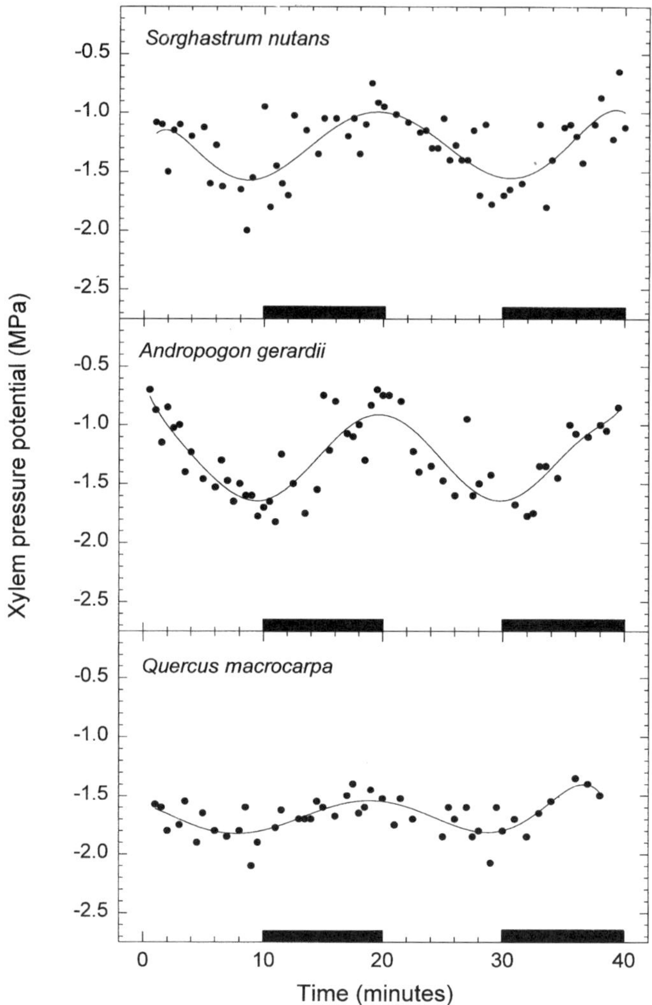

Figure 9.5. Plant water status responses to a sunpatch light regime in the grasses *Sorghastrum nutans* and *Andropogon gerardii*, and the tree *Quercus macrocarpa*. Redrawn from Knapp, 1992. Boxes designate periods of 300-400 μmol m^{-2} s^{-1} shade.

Responses in Leaf Temperature

Responses in leaf temperature (T_l) during periods of variable PFD are linked to stomatal and transpirational responses. In sunfleck environments, PFD levels are so low and increased PFD periods so brief that variability in T_l is small relative to sunpatch environments. Similarly, variability in T_l is small during leaf flutter due to efficient convective heat exchange (Roden and Pearcy, 1993c). During sunpatches, much larger fluctuations in T_l may occur (Figure 9.6), especially in large-leafed plants and under low wind conditions (Young and Smith, 1980; Young 1985). For example, in the large-leafed subalpine herb, *Frasera speciosa*,

7-8°C reductions in T_l may occur in shade relative to sun periods (Knapp and Smith, 1987). In *Cucurbita foetidissima*, a grassland vine with very large leaves, up to 14°C variations in T_l were recorded during sunpatches (Knapp and Fahnestock, 1990). In species with high transpiration rates and unresponsive stomata, T_l may fall well below air temperatures under low wind, shade conditions due to transpirational cooling. Conversely, if g is reduced during short-term shade periods, T_l may initially increase well above air temperatures when full sun returns, until stomata re-open and transpirational cooling is re-established. In either case, environmental changes and physiological responses combine to increase variability in T_l as PFD varies.

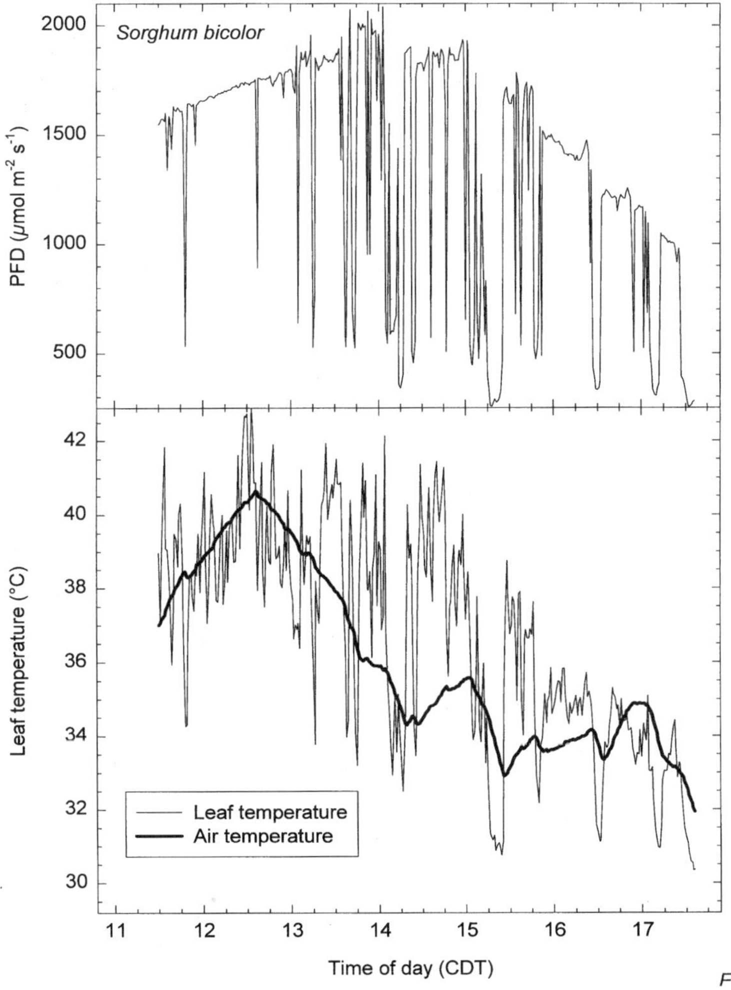

Fi

Figure 9.6. Leaf temperature responses to naturally-varying light intensities for an upper canopy *Sorghum bicolor* leaf. Redrawn from Fay and Knapp, 1995.

INTEGRATED STRATEGIES FOR COPING WITH VARIABLE SUNLIGHT

Here we summarize integrated plant responses to variable PFD. There are a variety of possible interactions among photosynthetic, stomatal, water relations and temperature responses to variable PFD. Looking at the integrated responses of these parameters can provide insights into how plants cope with variable light regimes beyond what may be deduced from responses of the individual parameters. In this section, we assume that periods of variable light act as a stress with the potential to reduce carbon gain and/or water use efficiency. Strategies that minimize these costs might be selected for in both native and agricultural settings.

Sunfleck and Leaf Flutter Strategies

The strategies that minimize C losses during variable PFD are surprisingly similar in plants experiencing sunfleck and leaf- flutter light regimes. The general strategy is to maximize light use efficiency by attaining photosynthetic induction as rapidly as possible and maintaining induction as long as possible through low PFD periods (Chazdon and Pearcy, 1986; Pearcy and Seeman, 1990; Pearcy, 1990; Gildner and Larson, 1992; Küppers and Schneider, 1993; Poorter and Oberbauer, 1993). Evaporative demand is usually low in sunfleck-dominated understory environments and soil water is often non-limiting. Thus, there is little cost associated with water loss for this strategy. Under either light regime, leaf temperatures and leaf water status remain relatively constant, and stomatal conductance is either maintained at high levels during PFD fluctuations (Mooney *et al.*, 1983; Pearcy *et al.*, 1994), or induction state limits C uptake more than diffusion through stomata (Kirschbaum *et al.*, 1983; Kursar and Coley, 1993).

Sunpatch Strategies

Plant responses to sunpatches tend to be more complex and more variable within and among species than sunfleck response strategies, because photosynthetic, stomatal, leaf water status and leaf temperature responses may all occur during sunpatches. As a result, plant and environmental characteristics that influence any of these responses independently (such as water storage organs, soil water availability, leaf size, vapor pressure deficit, etc.) will likely influence the integrated strategy plants use to cope with sunpatche light regimes.

Native Plants

Gas exchange responses to sunpatches have been evaluated for numerous native species in subalpine, forest, grassland and desert environments (Young and Smith, 1980; Knapp and Smith, 1990b, 1991; Knapp, 1993; Whitehead and Teskey, 1995). Responses vary between two extremes (Figure 9.7). In both cases, A decreases rapidly during shade periods, but stomatal responses may vary from essentially no response (Figure 9.7a, a "non-tracking" pattern) to rapid reductions in g during shade and rapid increases in full sun (Figure 9.7b, a "tracking" pattern). When g remains relatively static despite shading, photosynthetic recovery during sun periods is essentially immediate. In contrast, when g tracks shade, photosynthetic recovery may be delayed by reduced stomatal conductance, because turgor mediated changes in g are not as rapid as potential increases in A. Intercellular CO_2 concentrations are also reduced from steady-state levels during the initial portions of sun periods, consistent with stomatal limitations slowing photosynthetic recovery (Figure 9.8; Knapp and Smith, 1988; Fay and Knapp, 1996).

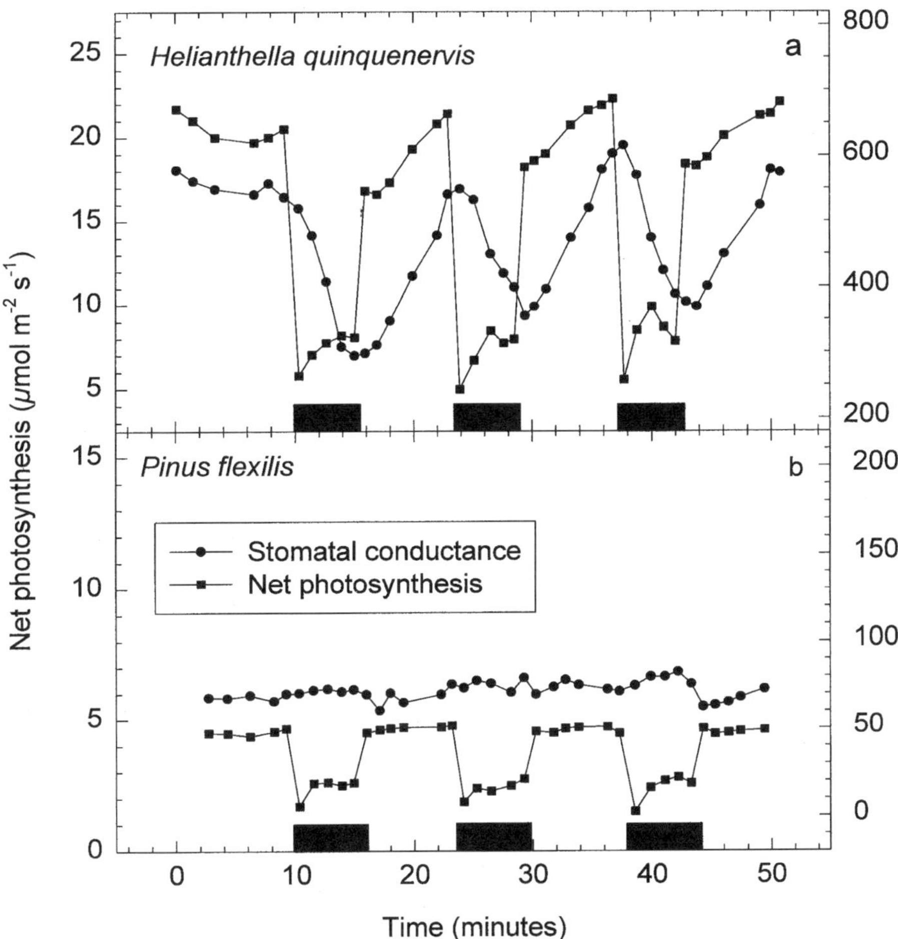

Figure 9.7. Comparison of photosynthetic and stomatal responses to a sunpatch light regime in a "tracking" species, *Helianthella quinquenervis*, and a "non-tracking" species, *Pinus flexilis*. Boxes designate periods of 300-400 μmol $m^{-2}s^{-1}$ shade. Redrawn from Knapp and Smith, 1990b.

Why don't all plants maintain g at high levels during short-term shade periods so that delayed photosynthetic recovery is avoided? The cost of this strategy is a large reduction in water use efficiency (WUE). During shade periods, g and transpiration may be high while A is light-limited. As a result, WUE during shade can be reduced by as much as 60%, relative to steady-state WUE (Figure 9.9; Knapp and Smith, 1989; Whitehead and Teskey, 1995). Hence, species with responsive stomata conserve water and maintain higher WUE, although overall carbon gain may be reduced. In contrast, species with non-responsive stomata maximize carbon gain at a substantial cost to WUE. In general, plants with high steady-state rates of gas exchange typically have low WUE and stomata that track alterations in PFD to minimize further WUE reductions during shade. Con-

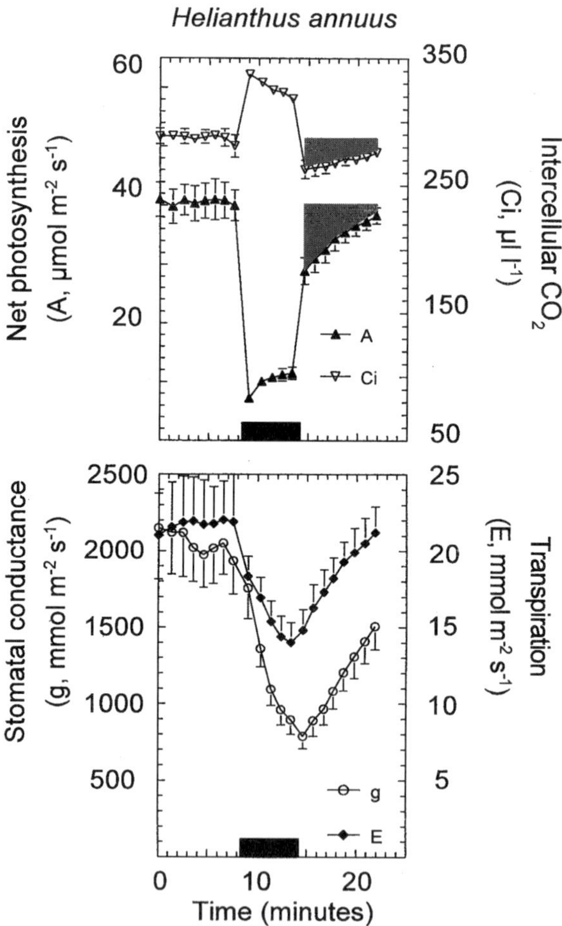

Figure 9.8. Stomatal limitation of photosynthetic recovery (shaded area) after 5 minutes shade (300-400 µmol m^{-2} s^{-1}, boxes) in the native C_3 herb *Helianthus annuus*. Means ± 1 SE. Redrawn from Fay and Knapp, 1996.

versely, species with low photosynthetic and transpiration rates under steady-state conditions typically have higher WUE but experience greater reductions in WUE during shade. In this way, combinations of traits favoring carbon gain or water conservation under constant environmental conditions are reversed when assessed under more typical variable PFD conditions (Knapp and Smith, 1989).

Inter-and intra-specific differences in leaf temperature and water status may mediate gas exchange responses during sunpatches. For example, in large-leafed herbs, large reductions in T_l during shade may reduce transpiration and maintain steady-state WUE even without stomatal adjustments (Knapp and Smith, 1987). In such species, a "tracking" strategy has few benefits. Fast growing riparian herbs with access to abundant water also may have non-tracking stomatal responses during sunpatches (Knapp and Smith, 1990a). In other herbs, stomatal responsiveness changes as water stress increases seasonally; reductions in g become faster, and increases in g become slower (Davies and Kozlowski, 1975; Knapp and Smith, 1988; Jones *et al.*, 1995). This response strategy leads to greater carbon gain when water is available, but greater water conservation as water stress increases. Indeed, as noted earlier, shade periods may increase plant

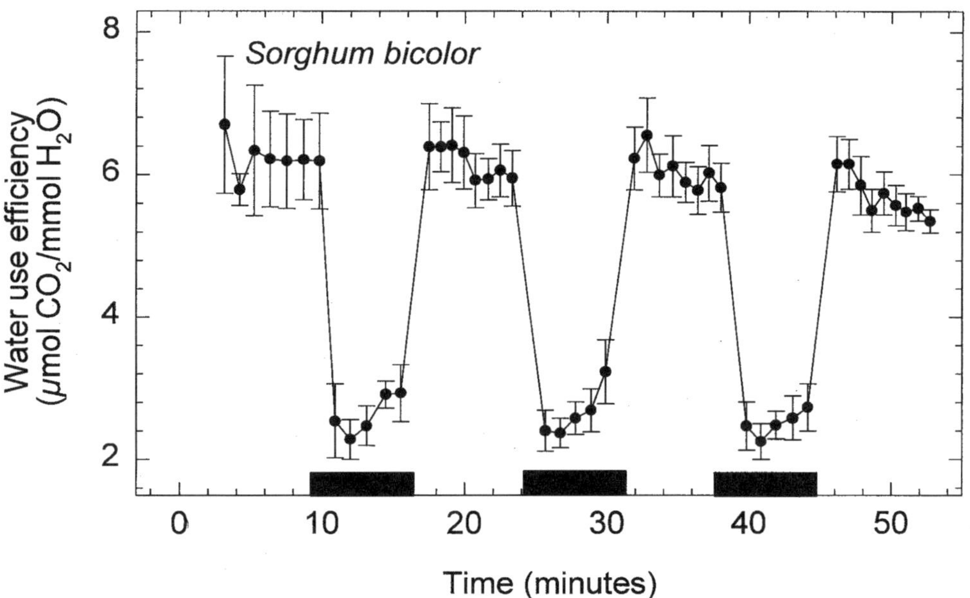

Figure 9.9. Water use efficiency (WUE) responses to shade periods (300-400 μmol m^{-2} s^{-1}, boxes) during a sunpatch light regime in *Sorghum bicolor*. Means ±1 SE. Redrawn from Fay and Knapp, 1995.

water status and enhance carbon gain in subsequent sun periods (Knapp and Smith, 1990b). Knapp and Smith (1989) found that enhanced plant water status during shade was correlated with smaller changes in g. Finally, Tinoco-Ojanguren and Pearcy (1993) reported that stomatal responses to sunpatches were strongly modulated by vapor pressure deficit (VPD) such that stomata respond in a more water conservative fashion at high VPD in some understory species.

Water availability, leaf temperatures and VPD usually vary substantially during a growing season. As a result, sunpatch response strategies might differ between species growing in different portions of the growing season. Fay and Knapp (1996) compared gas exchange responses to variable PFD in two fast-growing herbs, one restricted to early spring growth conditions when evaporative demand is low and water availability is high, vs. one active during the warmest and driest portions of the growing season. They found that the herb with spring phenology displayed a "non-tracking" stomatal response to sunpatches, maximizing carbon gain. In contrast, the summer active herb had a "tracking" stomatal response, minimizing water loss.

Agricultural and Horticultural Species

Domesticated plants artificially selected for rapid growth rates or high WUE might employ nonsteady-state responses to sunpatches differing from native plants. However, most of the agricultural and horticultural species studied conform to generalizations derived from studies of native species. For example, short-term reductions in PFD simulating natural cloud cover did not alter g or result in induction loss in well-watered fruit trees (Andersen, 1991). Well-watered *Chrysanthemum* also had minimal stomatal responses to variable PFD (Stoop *et al.*, 1991). In contrast, in sorghum and millet, water stress increased stomatal responses to shade periods (Jones *et al.*, 1995). *Phaseolus* stomatal responses to light and VPD were similar to responses of native species (Barradas *et al.*, 1994). Field-grown oat, wheat, soybean and sorghum had stomata that tracked fluctuations in PFD just as well as stomata of native grasses (Fay and Knapp, 1993; 1995).

Generalizations About Plant Responses to Variable Light Regimes

When searching for patterns in plant responses to variable PFD, it is useful to adopt the cost/benefit perspective successfully employed in other assessments of plant-environment interactions (Cowan, 1977, 1982; Farquhar and Sharkey, 1982). Assuming that the primary benefits and costs for plants coping with variable PFD regimes are carbon gain and water loss, respectively, the following generalizations can be made. Plants will maximize carbon gain through reduced stomatal responses to variable PFD when growing in environments causing little short-term leaf-level water stress, or when possessing physiological, morphological, or phenological characteristics functionally uncoupling them from short-term soil water depletion or excessive evaporative demand. Thus, most trees and shrubs, with their capacity to store water in their woody tissues, herbs with water storage organs, species whose roots exploit relatively stable water sources, and species usually experiencing low evaporative demand (VPD) display this response strategy (Knapp and Smith, 1989; Andersen, 1991; Whitehead and Teskey, 1995; Fay and Knapp, 1996). Species with low photosynthetic capacity and g should also maximize carbon gain since water loss is low and WUE is high. A unifying characteristic of these "non-tracking" species is that variability in ψ_{leaf} is low during variable PFD.

Conversely, species in which ψ_{leaf} is more tightly coupled to short-term fluctuations in supply and demand of water will respond to variable sunlight with more rapid stomatal alterations, conserving water at the cost of carbon gain. Species with high photosynthetic capacity and g usually employ this strategy since the water-loss costs associated with high g necessitate reducing g during shade periods.

Although some species may "switch" strategies as environmental conditions vary, response patterns may not be plastic in all species. Hamerlynck and Knapp (1994) compared gas exchange responses to sunpatches in *Quercus macrocarpa* leaves differing in photosynthetic capacity and g. Rates of increase and reduction in g during periods of sun and shade were similar in all leaves. Perhaps in this mid-continent grassland oak, occupying an environment with a long-term history of seasonal water stress and severe multi-year droughts (Weaver, 1968), a water-conserving stomatal response to variable PFD has become obligate rather than facultative.

Long-term selective pressure for water conservation may help explain differences in responses of C_3 and C_4 grasses to variable PFD. The higher WUE of C_4 species enables them to dominate hot, seasonally dry grasslands (Terri and Stowe, 1976; Pearcy and Ehleringer, 1984). Both C_3 and C_4 grasses have rapid stomatal responses to sunpatches. C_3 responses were faster than C_4 responses (Knapp, 1993). Though responses of the C_4 grasses were slower, the water loss cost during shade periods was less than for C_3 grasses due to C_4's inherently lower g; smaller changes in g were necessary to restabilize g after a change in PFD. Minimizing transition times from one steady-state g level to another after PFD changes will always reduce costs in both lost potential carbon gain and reduced WUE (Knapp, 1993).

CONCLUSIONS AND FUTURE RESEARCH NEEDS

Are Plants More Stressed Under Variable or Constant Sunlight Regimes?

If the majority of plant species have evolved in environments characterized by variable sunlight regimes, and most leaves on plants experience short-term fluctuations in PFD due to intra-and inter-plant canopy shading, perhaps many physiological and morphological traits adapt plants better to variable sunlight, making constant light (full sun in particular) the greater stress. Certainly, plants adapted to low PFD understory environments can be stressed by exposure to constant high PFD.

But there is also evidence that plants from more open environments may be stressed by extended periods of full sunlight. For example, the subalpine herb, *Arnica cordifolia*, grows in a variety of microsites with a wide range in daily PFD (from relatively low PFD understory sites to open sites). The preferred microsites, based on productivity and reproductive effort, are large gaps with extended periods of full sunlight (Young and Smith, 1982). On cloud-free days, full sunlight typically wilted *Arnica's* leaves (Young and Smith, 1979, 1980) suggesting stressful conditions. No wilting occurred on days with intermittent cloud cover (Knapp and Smith, 1990b). Furthermore, integrated CO_2 uptake (Figure 9.10) and ψ_{leaf} were greater during sunpatches compared to equal periods of full sun. Thus, although *Arnica* prefers gap microsites where PFD is high, the relatively few clear days in their subalpine habitat may be more stressful than more typical partly cloudy days.

Similar results were found during periods of moderate water stress in a subhumid grassland. In the tallgrass prairie of North America, an extended period of variable PFD led to greater carbon gain in C_3 grasses than a similar period of constant full sunlight (Seastedt and Knapp, 1993). Potential losses in photosynthesis during shade were more than compensated for by improvements in plant water status during shade. The recovery of ψ_{leaf} during shade allowed higher

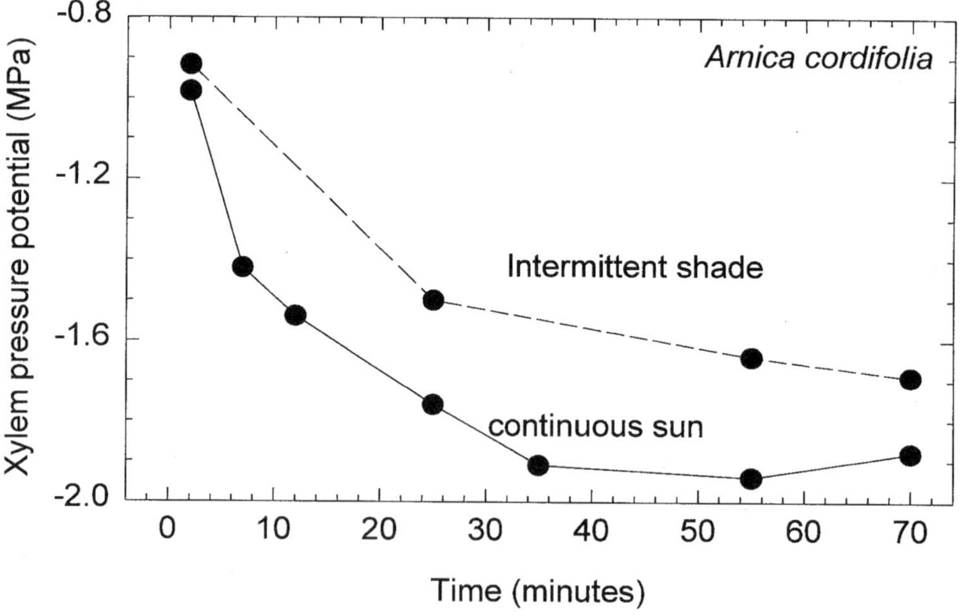

Figure 9.10. Leaf xylem pressure potential time courses for *Arnica cordifolia* under continuous full sun and a sunpatch light regime. Redrawn from Knapp and Smith, 1990b.

rates of CO_2 uptake during subsequent sun periods. Midday wilting occurs in a number of open and open-understory species on cloud-free days (Rawson, 1979; Smith, 1981; Chiariello *et al.*, 1987), as do midday photosynthetic depression and stomatal closure (Cowan, 1982; Tenhunen *et al.*, 1984; Küppers *et al.*, 1986a,b). These observations suggest that extended periods of full sunlight can be stressful to a number of plant species by negatively impacting plant water status. Photoinhibition may also occur during these periods, but the role that PFD variability may play in reducing this damage is unknown.

Future Research Directions

There are two general research directions that require exploration before a more comprehensive evaluation of plant responses to variable sunlight will be possible. More mechanistic studies are clearly needed, and with our growing knowledge of the complex ways in which elevated atmospheric CO_2 levels can affect a variety of plant processes, an assessment of how plants will respond to variable light regimes under future climate scenarios is needed.

From a mechanistic perspective, several topics deserve further study, particularly with regard to stomatal dynamics. For example, what is the mechanistic basis of stomatal responsiveness to light (Zeiger *et al.*, 1987)? Are "trackers" different from "non-trackers" in stomatal responses to blue or red light (Johnsson *et al.*, 1976; Morison and Jarvis, 1983; Assmann, 1988; Assmann and Grantz, 1990)? Root signals and leaf-level hormonal effects have recently been implicated in drought responses via stomatal closure in a variety of plant species (Zhang *et al.*, 1987; Downton *et al.*, 1988; Terashima *et al.*, 1988; Meinzer *et al.*, 1991). Are there interactions between these hormones and plant responses to variable PFD? When a shade period occurs and stomatal conductance is reduced, do all stomata respond uniformly, or is stomatal closure patchy across the leaf (Spence, 1987)? Patchy stomatal closure may cause inaccuracies in calculated gas exchange parameters such as intercellular CO_2 concentrations (Downton *et al.*, 1988). If stomatal responses are patchy, then interpretations of stomatal limitation of photosynthesis require reevaluation. Finally, all studies thus far have focused on entire leaves being shaded or sunlit uniformly. In reality, as light filters through plant canopies, individual leaf surfaces experience a mosaic of sun and shade patches (Norman *et al.*, 1971). Do these leaf segments respond to sunpatches like uniformly-illuminated leaves?

Although climate models may predict warmer, cooler, wetter or drier conditions for specific geographic regions worldwide, two aspects of future climate change scenarios will most directly impact plant responses to variable PFD. First is the prediction that cloud cover will increase in the future. A number of studies have already documented increases in cloud cover worldwide (Henderson-Sellers, 1986a,b). Depending on the region, increased cloud cover could either increase PFD variability or reduce it. Regardless, an understanding of how plants will respond to this change in light regime is critical.

More certain is that atmospheric CO_2 levels will continue to increase. Elevated CO_2 causes general reductions in stomatal conductance, and usually increases in plant and soil water status (Cure and Acock, 1986; Dahlman, 1993; Knapp *et al.*, 1993). What impact will elevated CO_2 have on stomatal responses to variable sunlight? In the only study done thus far, WUE was enhanced in a C_4 grass during periods of variable PFD due to both lower overall g and reduced transition times between sun and shade levels (Knapp *et al.*, 1994). Additional studies should be done to determine if this result can be generalized to other species. The potential for interactions among stomatal responses to elevated atmospheric CO_2, alterations in plant water relations, changes in light variability and possible temperature increases under future climates confirm the need for a more comprehensive understanding of plant responses to variability in sunlight.

ACKNOWLEDGEMENTS

This research was supported by the National Science Foundation Population Biology, Physiological Ecology, and Long-term Ecological Research programs, and the United States Department of Agriculture Plant Responses to the Environment program. Contribution number 96-336-B of the Kansas Agriculture Experiment Station.

REFERENCES

Andersen P.C. (1991) Leaf gas exchange of 11 species of fruit crops with reference to sun-tracking/non-sun-tracking responses. *Canadian Journal of Plant Science*, **71**, 1183–1193.

Anonymous. (1981) *Solar Radiation Energy Resource Atlas of the U.S. Solar Energy Research Institute Publication*. SP-6421037

Assmann S.M. (1988) Enhancement of the stomatal response to blue light by red light, reduced intercellular concentrations of CO_2, and low vapor pressure differences. *Plant Physiology*, **87**, 226–231.

Assmann S.M. and Grantz D.A.(1990) The magnitude of the stomatal response to blue light. *Plant Physiology*, **93**, 701–709.

Baldocchi D., Hutchinson B., Matt D. and McMillen R. (1986) Seasonal variation in the statistics of photosynthetically active radiation penetration in an oak-hickory forest. *Agricultural and Forest Meteorology*, **36**, 343–361.

Barradas V.L., Jones H.G. and Clark J.A.(1994) Stomatal responses to changing irradiance in *Phaseolus vulgaris* L. *Journal of Experimental Botany*, **45**, 931–936.

Boardmann N. (1977) Comparative photosynthesis of sun and shade plants. *Annual Review of Plant Physiology*, **28**, 355–377.

Bunce J.A. (1983) Photosynthetic characteristics of leaves developed at different irradiances and temperatures: an extension of the current hypothesis. *Photosynthesis Research*, **4**, 87–97.

Ceulemans R., Hinckley T.M. and Impens I. (1989) Stomatal response of hybrid poplar to incident light, sudden darkening and leaf excision. *Physiologia Plantarum*, **75**, 174–182.

Chapin F.S., Bloom A.J., Field C.B. and Waring R.H. (1987) Plant responses to multiple environmental factors. *BioScience*, **37**, 49–57.

Chiariello N.R., Field C.B. and Mooney H.A. (1987) Midday wilting in a tropical pioneer tree. *Functional Ecology*, **1**, 3–11.

Chazdon R.L. (1988) Sunflecks and their importance to forest understory plants. *Advances in Ecological Research*, **18**, 1–63.

Chazdon R.L. and Pearcy R.W. (1986) Photosynthetic responses to light variation in rainforest species. I. Induction under constant and fluctuating light conditions. *Oecologia*, **69**, 517–523.

Cowan I.R. (1977) Stomatal behavior and environment. *Advances in Botanical Research*, **4**, 117–228.

Cowan I.R. (1982) Regulation of water use in relation to carbon gain in higher plants. In *Physiological Plant Ecology II, Vol 12B*, edited by O.L. Lange, P.S. Nobel, C.B. Osmond and H. Ziegler, pp 589–613, Berlin: Springer-Verlag.

Cure J.D. and Acock B. (1986) Crop responses to carbon dioxide doubling: a literature survey. *Agricultural and Forest Meteorology*, **38**, 127–145.

Dahlman R.C. (1993) CO_2 and plants: revisited. *Vegetatio*, **104/105**, 339–355.

Davies W.J. and Kozlowski T.T. (1975) Stomatal responses to changes in light intensity as influenced by plant water stress. *Forest Science*, **21**, 129–133.

Downton W.J.S., Loveys B.R. and Grant W.J.R. (1988) Non-uniform stomatal closure induced by water stress causes putative non-stomatal inhibition of photosynthesis. *New Phytologist*, **110**, 503–509.

Farquhar G.D. and Sharkey T.D. (1982) Stomatal conductance and photosynthesis. *Annual Review of Plant Physiology*, **33**, 317–345.

Fay P.A. and Knapp A.K. (1993) Photosynthetic and stomatal responses of *Avena sativa* (Poaceae) to a variable light environment. *American Journal of Botany*, **80**, 1369–1373.

Fay P.A. and Knapp A.K. (1995) Stomatal and photosynthetic responses to variable light in sorghum, soybeans and eastern gamagrass. *Physiologia Plantarum*, **94**, 613–620.

Fay P.A. and Knapp A.K. (1996) Photosynthetic and stomatal responses to variable light in a cool-season and a warm-season prairie forb. *International Journal of Plant Sciences, in press.*

Gildner B.S. and Larson D.W. (1992) Photosynthetic response to sunflecks in the desiccation-tolerant fern *Polypodium virginianum*. *Oecologia*, **89**, 390–396.

Grace J. (1988) Plant response to wind. *Agriculture, Ecosystems and Environment*, **22/23**, 71–88.

Gross L.J. and Chabot B.F. (1979) Time course of photosynthetic response to changes in incident light energy. *Plant Physiology*, **63**, 1033–1038.

Hamerlynck E.P. and Knapp A.K. (1994) Stomatal responses to variable sunlight in Bur Oak (*Quercus macrocarpa Michx.*) leaves with different photosynthetic capacities. *International Journal of Plant Science*, **155**, 583–587.

Henderson-Sellers A. (1986a) Cloud changes in a warmer Europe. *Climatic Change*, **8**, 25–52.

Henderson-Sellers A. (1986b) Increasing clouds in a warming world. *Climatic Change*, **9**, 267–309.

Hirose T. and Werger M.J.A. (1987) Maximizing daily photosynthesis with respect to the leaf nitrogen pattern in the canopy. *Oecologia*, **72**, 520–526.

Hollinger D.Y. (1987) Photosynthesis and stomatal conductance patterns of two fern species from different forest understoreys. *Journal of Ecology*, **75**, 925–935.

Hollinger D.Y. (1989) Canopy organization and foliage photosynthetic capacity in a braod-leaved evergreen montane forest. *Functional Ecology*, **3**, 53–62.

Johnsson M., Issaias S., Brogardh T. and Johnsson A. (1976) Rapid, blue-light-induced transpiration response restricted to plants with grass-like stomata. *Physiologia Plantarum*, **36**, 229–232.

Jones H.G., Hall D.O., Corlett J.E. and Massacci A. (1995) Drought enhances stomatal closure in response to shading in Sorghum (*Sorghum bicolor*) and in Millet (*Pennisetum americanum*). *Australian Journal of Plant Physiology*, **22**, 1–6.

Kirschbaum M.U.F. and Pearcy R.W. (1988) Gas exchange analysis of the relative importance of stomatal and biochemical factors in photosynthetic induction in *Alocasia macrorrhiza. Plant Physiology*, **86**, 782–785.

Knapp A.K. (1992) Leaf gas exchange in *Quercus macrocarpa* (Fagaceae): rapid stomatal responses to variability in sunlight in a tree growth form. *American Journal of Botany*, **79**, 599–604.

Knapp A.K. (1993) Gas exchange dynamics in C_3 and C_4 grasses: consequences of differences in stomatal conductance. *Ecology*, **74**, 113–123.

Knapp A.K. and Fahnestock J.T. (1990) Influence of plant size on the carbon and water relations of *Cucurbita foetidissima*. *Functional Ecology*, **4**, 789–797.

Knapp A.K., Fahnestock J.T. and Owensby C.E. (1994) Elevated CO_2 alters dynamic stomatal responses to sunlight in a C_4 grass. *Plant, Cell and Environment*, **17**, 189–195.

Knapp A.K., Hamerlynck E.P. and Owensby C.E. (1993) Photosynthetic and water relations responses to elevated CO_2 in the C_4 grass *Andropogon gerardii. International Journal of Plant Sciences*, **154**, 459–466.

Knapp A.K. and Smith W.K. (1981) Water relations and succession in subalpine conifers in southeastern Wyoming. *Botanical Gazette*, **142**, 502–511.

Knapp A.K. and Smith W.K. (1987) Stomatal and photosynthetic responses during sun/shade transitions in subalpine species: influence on water use efficiency. *Oecologia*, **74**, 62–67.

Knapp A.K. and Smith W.K. (1988) Effect of water stress on stomatal and photosynthetic responses in subalpine plants to cloud patterns. *American Journal of Botany*, **75**, 851–858.

Knapp A.K. and Smith W.K. (1989) Influence of growth form and water relations on stomatal and photosynthetic responses to variable sunlight in subalpine plants. *Ecology*, **70**, 1069–1082.

Knapp A.K. and Smith W.K. (1990a) Contrasting stomatal responses to variable sunlight in two subalpine herbs. *American Journal of Botany*, **77**, 226–231.

Knapp A.K. and Smith W.K. (1990b) Stomatal and photosynthetic responses to variable sunlight. *Physiologia Plantarum*, **78**, 160–165.

Knapp A.K. and Smith W.K. (1991) Gas exchange responses to variable sunlight in two Sonoran desert herbs: comparison with subalpine species. *Botanical Gazette*, **152**, 269–274.

Knapp A.K., Smith W.K. and Young D.R. (1989) Importance of intermittent shade to the ecophysiology of subalpine herbs. *Functional Ecology*, **3**, 753–758.

Küppers M., Matyssek R. and Schulze E.D. (1986a) Diurnal variations of light-saturated CO_2 assimilation and intercellular carbon dioxide concentration are not related to leaf water potential. *Oecologia*, **69**, 477–480.

Küppers M., Wheeler A.M., Küppers B.I.L., Kirschbaum M.U.F. and Farquhar G.D. (1986b) Carbon fixation in eucalypts in the field. *Oecologia*, **70**, 273–282.

Küppers M. and Schneider H. (1993) Leaf gas exchange of beech (*Fagus sylvatica* L.) seedlings in lightflecks: effects of fleck length and leaf temperature in leaves grown in deep and partial shade. *Trees*, **7**, 160–168.

Kursar T.A. and Coley P.D. (1993) Photosynthetic induction times in shade-tolerant species with long and short-lived leaves. *Oecologia*, **93**, 165–170.

Lakso A.N. and Barnes J.E. (1978) Apple leaf photosynthesis in alternating light. *HortScience*, **13**, 473–474.

Marotz G.A. and Henry J.A. (1978) Satellite-derived cumulus cloud statistics for western Kansas. *Journal of Applied Meteorology*, **17**, 1725–1736.

McCree K.J. and Loomis R.S. (1969) Photosynthesis in fluctuating light. *Ecology*, **50**, 422–428.

Meinzer F.C., Grantz D.A. and Smit B. (1991) Root signals mediate coordination of stomatal and hydraulic conductance in growing sugarcane. *Australian Journal of Plant Physiology*, **18**, 329–338.

Mooney H.A., Field C., Vazquez-Yanes C. and Chu C. (1983) Environmental controls on stomatal conductance in a shrub of the humid tropics. *Proceedings of the National Academy of Sciences USA*, **80**, 1295–1297.

Morison J.I.L. and Jarvis P.G. (1983) Direct and indirect effects of light on stomata. I. Scots pine and Sitka spruce. *Plant, Cell and Environment*, **6**, 95–101.

Norman J.M., Miller E.E. and Tanner C.B. (1971) Light intensity and sunfleck-size distributions in plant canopies. *Agronomy Journal*, **63**, 743–748.

Osmond C.B. (1983) Interactions between irradiance, nitrogen nutrition, and water stress in the sun-shade responses of *Solanum dulcamara*. *Oecologia*, **57**, 316–321.

Osmond C.B., Austin M.P., Berry J.A., Billings W.D., Boyer J.S., Dacey J.W.H., *et al.* (1987) Stress physiology and the distribution of plants. *Bioscience*, **37**, 38–47.

Pearcy R.W. (1988) Photosynthetic utilisation of lightflecks by understory plants. *Australian Journal of Plant Physiology*, **15**, 223–238.

Pearcy R.W. (1989) Regulation of photosynthetic CO_2 assimilation during sunflecks. In *Photosynthesis*, edited by W.R. Briggs, pp 407-424. New York: Alan R. Liss.

Pearcy R.W. (1990) Sunflecks and photosynthesis in plant canopies. *Annual Review of Plant Physiology and Plant Molecular Biology*, **41**, 421–453.

Pearcy R.W. and Calkin H.W. (1983) Carbon dioxide exchange of C_3 and C_4 tree species in the understory of a Hawaiian forest. *Oecologia*, **58**, 26–32.

Pearcy R.W., Chazdon R.L., Gross L.J. and Mott K.A. (1994) Photosynthetic utilization of sunflecks: a temporally patchy resource on a time scale of seconds to minutes. In *Exploitation of Environmental Heterogeneity by Plants*, edited by M.M. Caldwell and R.E. Pearcy, pp 175-208. New York: Academic Press.

Pearcy R.W. and Ehleringer J. (1984) Comparative ecophysiology of C_3 and C_4 plants. *Plant, Cell and Environment*, **7**, 1–13.

Pearcy R.W. and Seemann J.R. (1990) Photosynthetic induction state of leaves in a soybean canopy in relation to light regulation of ribulose-1-5-bisphosphate carboxylase and stomatal conductance. *Plant Physiology*, **94**, 628–633.

Pearcy R.W., Roden J.S. and Gamon J.A. (1990) Sunfleck dynamics in relation to canopy structure in a soybean (*Glycine max* (L.) Merr.) canopy. *Agricultural and Forest Meteorology*, **52**, 359–372.

Pfitsch W.A. and Pearcy R.W. (1989) Daily carbon gain by *Adenocaulon bicolor* (Asteraceae), a redwood forest understory herb, in relation to its light environment. *Oecologia*, **80**, 465–470.

Peterson R.B. (1983) Estimation of photorespiration based on the initial rate of postillumination CO_2 release. I. A nonsteady state model for the measurement of CO_2 exchange transients. *Plant Physiology*, **73**, 978–982.

Pollard D.F.W. (1970) The effect of rapidly changing light on the rate of photosynthesis in largetooth aspen (*Populus grandentata*). *Canadian Journal of Botany*, **48**, 823–829.

Pons T.L. and Pearcy R.W. (1992) Photosynthesis in flashing light in soybean leaves grown in different conditions. II. Lightfleck utilization efficiency. *Plant, Cell and Environment*, **15**, 577–584.

Poorter L. and Oberbauer S.F. (1993) Photosynthetic induction responses of two rainforest tree species in relation to light environment. *Oecologia*, **96**, 193–199.

Rawson H.M. (1979) Vertical wilting and photosynthesis, transpiration, and water use efficiency of sunflower leaves. *Australian Journal of Plant Physiology*, **6**, 109–120.

Read J. (1985) Photosynthetic and growth responses to different light regimes of the major canopy species of Tasmanian cool temperate rainforest. *Australian Journal of Ecology*, **10**, 327–334.

Roden J.S. and Pearcy R.W. (1993a) Effect of leaf flutter on the light environment of poplars. *Oecologia*, **93**, 201–207.

Roden J.S. and Pearcy R.W. (1993b) Photosynthetic gas exchange response of poplars to steady-state and dynamic light environments. *Oecologia*, **93**, 208–214.

Roden J.S. and Pearcy R.W. (1993c) The effect of leaf flutter on the temperature of poplar leaves and its implications for carbon gain. *Plant, Cell and Environment*, **16**, 571–577.

Roessler P.G. and Monson R.K. (1985) Midday depression in net photosynthesis and stomatal conductance in *Yucca glauca*. *Oecologia*, **67**, 380–387.

Salisbury F.B., Spomer G.G., Sobral M. and Ward R.T. (1968) Analysis of an alpine environment. *Botanical Gazette*, **129**, 16-32.

Schulte P.J., Hinckley T.M. and Stettler R.F. (1987) Stomatal responses of *Populus* to leaf water potential. *Canadian Journal of Botany*, **68**, 255–260.

Seastedt T.R. and Knapp A.K. (1993) Consequences of non-equilibrium resource availability across multiple time scales: the transient maxima hypothesis. *American Naturalist*, **141**, 621–633.

Smith W. K. (1981) Temperature and water relation patterns in subalpine understory plants. *Oecologia*, **48**, 353–359.

Smith W.K., Knapp A.K. and Reiners W.A. (1989) Penumbral effects on sunlight penetration in plant communities. *Ecology*, **70**, 1603–1609.

Spence R.D. (1987) The problem of variability in stomatal responses, particularly aperture variance, to environmental and experimental conditions. *New Phytologist*, **107**, 303–315.

Stern W.R. and Donald C.M. (1962) Light relationships in grass- clover swards. *Australian Journal of Agricultural Research*, **13**, 599–614.

Stoop J.M.H., Willits D.H., Peet M.M. and Nelson P.V. (1991) Carbon gain and photosynthetic response of *Chrysanthemum* to photosynthetic photon flux density cycles. *Plant Physiology*, **96**, 229–536.

Tenhunen J.D., Lange, O.L., Gebel, J., Beyschlag, W. and Weber, J.A. (1984) Changes in photosynthetic capacity, carboxylation efficiency, and CO_2 compensation point associated with midday stomatal closure and midday depression of net CO_2 exchange of leaves of *Quercus suber*. *Planta*, **162**, 193–203.

Terashima I., Wong S., Osmond C.B. and Farquhar G.D. (1988) Characterisation on non-uniform photosynthesis induced by abscisic acid in leaves having different mesophyll anatomies. *Plants and Cell Physiology*, **29**, 385–394.

Terri J.A. and Stowe L.G. (1976) Climatic patterns and the distribution of C_4 grasses in North America. *Oecologia*, **23**, 1–12.

Tinoco-Ojanguren C. and Pearcy R.W. (1993) Stomatal dynamics and its importance to carbon gain in two rainforest *Piper* species. *Oecologia*, **94**, 388–394.

Turnbull M.H., Doley D. and Yates D.J. (1993) The dynamics of photosynthetic acclimation to changes in light quantity and quality in three Australian rainforest tree species. *Oecologia*, **94**, 218–228.

Vines H.M., Armitage A. M., Chen S., Tu Z. and Black C.C. Jr. (1982) A transient burst of CO_2 from geranium leaves during illumination at various light intensities as a measure of photorespiration. *Plant Physiology*, **70**, 629–631.

Vines H. M., Tu Z., Armitage A.M., Chen S. and Black C.C. Jr. (1983) Environmental responses of the post-lower illumination CO_2 burst as related to leaf photorespiration. *Plant Physiology*, **73**, 25–30.

Wardle P. (1986) Frequency of cloud cover on New Zealand mountains in relation to subalpine vegetation. *New Zealand Journal of Botany*, **24**, 553–565.

Weaver J.E. (1968) *Prairie Plants and Their Environment*. Lincoln: University of Nebraska Press.

Whitehead D. and Teskey R.O. (1995) Dynamic response of stomata to changing irradiance in loblolly pine (*Pinus taeda L.*). *Tree Physiology*, **15**, 245–251.

Willis A.J. and Balasubramanian S. (1968) Stomatal behaviour in relation to rates of photosynthesis and transpiration in *Pelargonium*. *New Phytologist*, **67**, 265–285.

Woods D.B. and Turner N.C. (1971) Stomatal response to changing light by four tree species of varying shade tolerance. *New Phytologist*, **70**, 77–84.

Young D.R. (1985) Microclimatic effects on water relations, leaf temperatures, and the distribution of *Heracleum lanatum* at high elevations. *American Journal of Botany*, **72**, 357–364.

Young D.R. and Smith W.K. (1979) Influence of sunflecks on the temperature and water relations of two subalpine understory congeners. *Oecologia*, **43**, 195–205.

Young D.R. and Smith W.K. (1980) Influence of sunlight on photosynthesis, water relations, and leaf structure in the understory species *Arnica cordifolia*. *Ecology*, **61**, 1380–1390.

Young D.R. and Smith W.K. (1982) Simulation studies of the influence of understory location on transpiration and photosynthesis of *Arnica cordifolia* on clear days. *Ecology*, **63**, 1761–1770.

Zeiger E., Farquhar G.D. and Cowan I.R. (1987) *Stomatal function*. Stanford: Stanford University Press.

Zhang J., Schurr U. and Davies W.J. (1987) Control of stomatal behaviour by abscisic acid which apparently originates in the roots. *Journal of Experimental Botany*, **38**, 1174–1181.

10. MECHANISMS OF PLANT RESISTANCE TO NUTRIENT DEFICIENCY STRESS

J. NICK PEARSON[1] and ZDENKO RENGEL[2]

[1]*Department of Plant Science, Waite Agricultural Research Institute, The University of Adelaide, Glen Osmond SA 5064, AUSTRALIA*

and

[2]*Soil Science and Plant Nutrition, School of Agriculture, The University of Western Australia, Nedlands WA 6009, AUSTRALIA*

INTRODUCTION

Plants suffer nutrient deficiency stress when the availability of nutrients are below that required for growth. This may be the result of an inherently low nutrient status of the soil, low mobility of nutrients within the soil, or the chemical form of the nutrient. The mobility of nutrients within the soil is governed by a number of factors including mass flow of water, adsorption capacity of the soil and soil pH. The chemical form of nutrients within the soil also determines the extent of availability, for instance Fe(II) is available for plant uptake but Fe(III) is not. The mobility and availability of nutrients within soils has been comprehensively reviewed by Shuman (1991), Jungk (1991) and Marschner (1995).

Nutrient deficiencies in plants are widespread in many agricultural regions of the world and can result in reduced growth or yield of many crop species. Losses from environmental stresses, such as water and nutrient deficiency are enormous and have been estimated at 75% of the potential for US agriculture (Boyer, 1982). Some crops manage to maintain significant yields under nutrient stress but the grain is so low in certain nutrients that problems arise for farmers who resow grain the following year. Reduced nutrient content of seeds can result in low germination rates, reduced seedling vigor and susceptibility to soil-borne root pathogens (Huber, 1980; Bell *et al.*,1989; Graham and Webb, 1991). Fertilizers have commonly been used to reduce losses from nutrient stress but the increasing cost of fertilizers, government restrictions of fertilizer applications to minimize environmental damage and the increased use of marginal land (Sattelmacher *et al.*, 1994) has led to renewed interest in the physiology of nutrient stress.

Researches aiming to improve yield or quality on poor soils use the natural genetic diversity in plants to investigate the mechanisms responsible for tolerance to nutrient deficiencies. For example, cereal rye is tolerant of Cu-deficient soils, while wheat and oat are not. By isolating the gene to a specific section of chromosome, in this case it was found on the long arm of chromosome 5 R of rye (Graham *et al.*, 1987) and introducing it to wheat and oat, it was possible to double the yields of wheat and oat on Cu-deficient soils.

Marschner (1995) has suggested a number of possible mechanisms by which plants deal with nutrient deficiencies and these can be separated into two main categories. Plants may differ in the method used to acquire nutrients so that they increase the amount of nutrients taken up into the plant thus avoiding nutrient stress. Plants may change the morphology of their root system, and there may be inherent differences between species or responses to nutrient deficiency, so that the plant is better able to take up nutrients. Plants may also form symbiotic associations that aid the uptake of nutrients from the atmosphere or soil, such as with *Rhizobia* and mycorrhizas. There may also be differences in the physiology and biochemistry of the plant, such as increases in the affinity of specific nutrient uptake systems on the plasmamembrane. The plant may also modify the rhizosphere (secretion of chelating agents, reducing compounds, etc.) to increase the availability of nutrients. The different mechanisms in the root system used to alleviate nutrient deficiency stress by increasing the uptake of nutrients are summarized in Figure 10.1.

Plants also deal with nutrient stress by increasing the efficiency at which the nutrient is used within the plant. This may occur by increasing the efficiency at which the nutrient is transported (both long- and short-distance transport), or is compartmented within cells. Retranslocation of nutrients under nutrient stress within the shoot may also differ between species and cultivars. By identifying efficient traits for particular nutrients and locating the gene(s) responsible, it may be possible to introduce these to cultivars or species with other desirable attributes that are inefficient at acquiring or utilizing that particular nutrient.

This chapter will detail specific mechanisms that plants have evolved to alleviate nutrient deficiency stress. In particular, we will cover mechanisms of stress resistance to phosphorus, iron and zinc deficiencies, as examples. These deficiencies are common and widespread throughout the world and much research has been undertaken to understand the mechanisms employed to overcome or withstand these stresses. Readers are directed to several review articles dealing with other aspects of nutrient deficiency and plant responses to nutrient stress (nitrogen-Chapin, 1991; boron-Shelp *et al.*, 1995; see also Marschner, 1995). The essential role of nutrients in the biochemistry of plants has been comprehensively reviewed by Marschner (1995).

WHOLE PLANT RESPONSES TO NUTRIENT DEFICIENCY STRESS

Shoot/Root Ratios

Plants respond to nutrient deficiency stress in a number of ways and this differs depending upon the nutrient stress imposed and the function of the nutrient within the plant. On a whole plant basis, decreases in the shoot/root ratio are common in plants suffering form P, N, S and Zn deficiencies. Phosphorus deficiency retards the growth of the shoots much more than the roots by affecting the leaf expansion rates and leaf surface area (Fredeen *et al.*, 1989). The roots

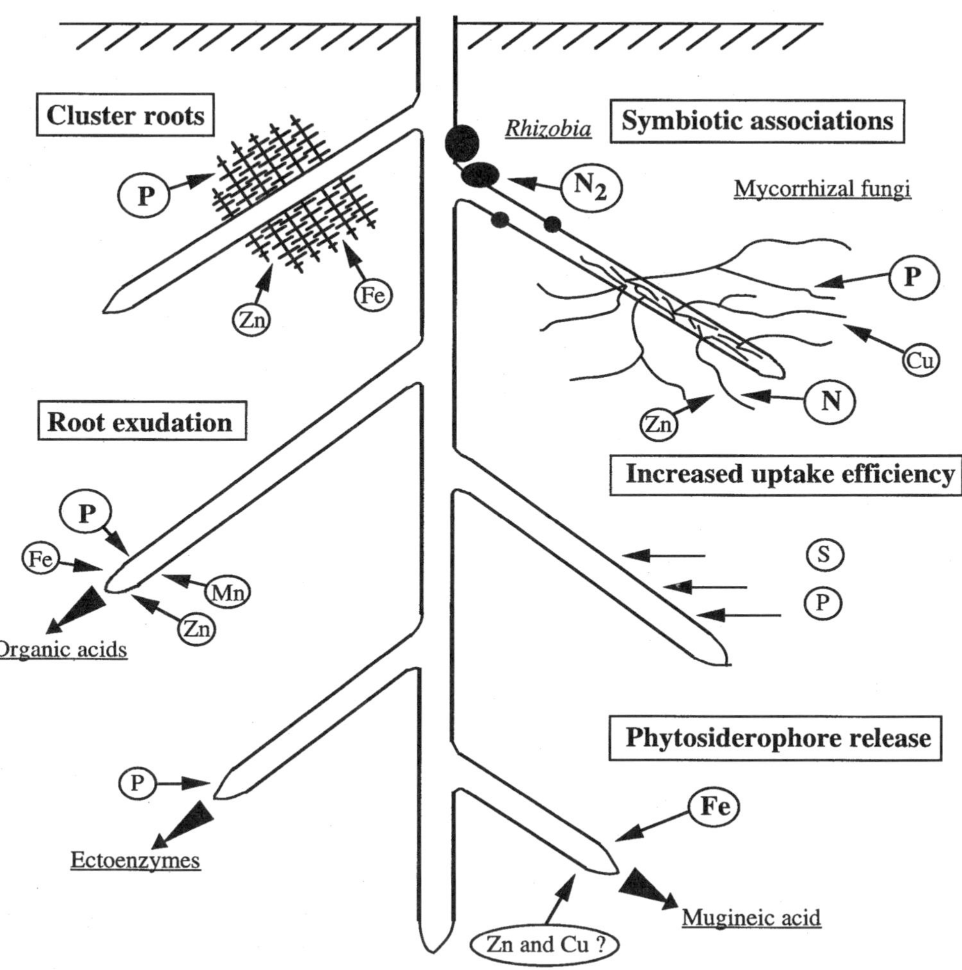

Figure 10.1. Summary of the mechanisms of resistance in the roots of plants to nutrient deficiency stress. The nutrient deficiency alleviated by each mechanism is circled, with size relating to importance for each mechanism in alleviating nutrient deficiency for that nutrient.

continue to grow normally as the partitioning of photosynthates to the roots is increased (Fredeen *et al.*, 1989). In bean plants, the partitioning of carbohydrates to the roots increased from 16% of the total photosynthates to 23% (Cakmak *et al.*, 1994b).

Nitrogen deficiency restricts leaf canopy development by decreasing the rate of leaf initiation and the development of existing sink leaves, causing the accumulation of sucrose and starch in source leaves. The accumulation of carbohydrates in the leaves is followed by increased translocation of carbohydrates to the root system and subsequent increased growth of the roots (Rufty *et al.*,1982).

Sulphur deficiency reduces the leaf area by reducing the size and number of cells therefore reducing the chlorophyll content per leaf and thus photosythesis. Starch may accumulate in the source leaves due to impaired carbohydrate metabolism or low demand at sink sites (Marschner, 1995). There is no evidence that sulphur deficiency has a specific effect on root growth.

Plants respond to Zn deficiency in a manner similar to P deficiency, that is root growth is less affected than shoot growth, however, root growth may actually be enhanced compared to Zn-sufficient roots (Cumbus, 1985; Webb and Loneragan, 1990; Rengel and Graham, 1995a;). Enhanced root growth may be due to increase partitioning of carbohydrates to the roots under Zn stress (Pearson and Rengel, unpublished).

With some nutrient deficiencies, root growth is inhibited more than shoot growth, such as with Mg, K, Fe and Mn deficiencies. In the case of Mg deficiency, the export of carbohydrates from the source to the sink is thought to be impaired (Cakmak *et al.*, 1994b), while K is important for the transport of sucrose within the phloem and the rate of cell extension and thus affects the source-sink relationship.

Root elongation is inhibited in Fe-deficient plants, while root hair formation is stimulated (Marschner, 1995). Furthermore, transfer cells are formed either in the rhizodermis or in the hypodermis and are part of the Fe-deficiency induced changes associated with the Strategy I response to Fe stress (discussed later). Mn deficiency is thought to reduce root growth through its role in the production of IAA (Abbott, 1967) and not due to altered source-sink ratios (Pearson and Rengel, unpublished).

Little has been reported on the effect of Ni, Mo and Cl deficiencies on plant growth and it has been suggested that there is no clear evidence that plants suffer these deficiencies (see Marschner, 1995). Boron deficiency is fairly widespread throughout the world but is thought to affect reproduction (and therefore yield) rather than vegetative growth, in particular in cereals (see Shelp *et al.*, 1995 for complete review).

Remobilization

During senescence and seed development, much of the nutrient reserves within the leaves, stem and roots are remobilized to actively growing tissue, such as new leaves or developing seeds. Remobilization from the leaves must occur via the phloem whereas nutrients remobilized from the stem and roots can occur via the xylem. Since the mobility of nutrients within the phloem differs so does the extent of remobilization. Phloem immobile nutrients such as Mn and B are not remobilized from leaves (Pearson and Rengel, 1994; Shelp *et al.*, 1995). Of the phloem mobile nutrients, Zn, P and N were found to be remobilized to the greatest extent (>78%) in hard red spring wheat with less than 20% remobilization for K, Ca, Cl and Fe (Miller *et al.*, 1993).

There have been only a few reports on the effect of nutrient deficiency stress on the remobilization of nutrients from the vegetative tissues. In wheat suffering Zn deficiency, Zn was remobilized to a greater extent from the leaves compared to plants with a sufficient supply of Zn (Pearson and Rengel, 1994). On the other hand, Fe remobilization from the primary leaf in bean was shown to increase from 20% to 34% during leaf senescence but remobilization during severe Fe deficiency was only small. This was suggested to be due to the requirement of Fe in organic structures (e.g. chloroplasts) that are required for leaf function (even under severe Fe deficiency) but are available for remobilization during leaf senescence (Zhang *et al.*, 1995). There has been little work presented that investigates whether cultivars of the same species differ in the extent of remobilization of certain nutrients.

PLANT MECHANISMS TO ALLEVIATE NUTRIENT DEFICIENCY STRESS

Nutrients in General

Plants have adapted to nutrient deficiency stress in a variety of ways, some of which are morphological adaptation to increase the ability of the plant to take up nutrients, such as cluster roots. Plants may also release chemical compounds into the soil environment to increase the efficiency at which nutrients are taken up or increase the number of soil nutrient pools available for uptake. Yet further still, plants have evolved a mechanism to alleviate nutritional stress by symbiotically associating with micro-organisms, such as legumes with *Rhizobium* and most terrestrial plants with mycorrhizal fungi. As an example, biological nitrogen fixation on a global basis is thought to account for 139–170 million tonnes of N per year while fertilization with nitrogen amounts to only 65 million tonnes per year (Peoples and Craswell, 1992). Most of the mechanisms described below are activated predominantly when the plant is under stress and are absent when the supply of nutrients is optimal.

In this review, we will mostly focus on mechanism of resistance to phosphorus, iron and zinc deficiencies, with readers directed to Chapin (1991) and Marschner (1995) for reviews of plant responses to nitrogen deficiency. As for other nutrients, plants generally only suffer from magnesium or calcium deficiencies in acid soils or when the plant is also suffering aluminium stress (see Rengel, 1992) and there appear to be no reports of plant responses to these deficiencies. Sulphur deficiency is not common, as plants can receive S from the atmosphere, but some plants do respond to sulphur deficiency by increasing the capacity for S uptake by increasing the V_{max} of the plasma membrane transporter (Hawkesford and Belcher, 1991). In young seedlings, 8 to 15-fold increases have been observed in sulphate uptake following S deprivation (Lee, 1982; Clarkson and Saker, 1989). It was further shown that the synthesis of a 36-kDa polypeptide increased in sulphate-starved roots; this polypeptide might be a component of the plasma membrane sulphate transport system (Hawkesford and Belcher, 1991).

Mechanisms of resistance to manganese and copper deficiency include exudation of organic acids (see P deficiency) and possible mobilization and uptake by phytosiderophores (see Fe deficiency). There have been no reports of plant mechanisms activated to alleviate Ni, B or Mo deficiencies and these deficiencies can be considered less common.

Phosphorus Deficiency

Phosphorus deficiency stress, as with other nutrient stresses, can be alleviated by either increasing the efficiency of utilization of P or efficiency of P uptake by the plant. Not much is understood about utilization efficiency and the prospects for increasing fertilizer efficiency through better utilization of P within the plant are thought to be poor (see Sattelmacher *et al.*, 1994). Three broad categories exist in some plants to increase the uptake of P under P deficiency: chemical, physical and through symbiotic associations. Plants may secrete or exude chemical compounds into the rhizosphere or soil environment to increase the uptake of P or increase the solubility of P in the soil. Secondly, the plant may alter the geometry or architecture of its root system to aid the uptake of P. Lastly, the plant may associate with microorganisms to facilitate the uptake of P. Plants may also increase the expression of high affinity P uptake systems on the plasma membranes when the plant is suffering P stress (see Clarkson and Lüttge, 1991), but this is thought to be of only minor importance for nutrients with low mobility in soils.

Root exudation

Some plants respond to P deficiency by releasing exudates from the root system, comprising of high or low molecular weight solutes. Among the high molecular weight exudates are mucilage and ectoenzymes while the low molecular weight exudates include organic acids, sugars and phenolics.

Mucilage, which is mainly composed of polysaccharides, is mostly found around the apical zones of roots and is thought to perform a number of functions, including protection of apical cells from desiccation, lubrication as the root moves through the soil, and most relevant here, facilitate the uptake of nutrients (Marschner, 1995). The contact between mucilage and soil particles is important for the uptake of P, as well as micronutrients. Phosphate is mobilized from the root-soil interface by phosphate desorption from clay surfaces by the polygalacturonic acid component of mucilage (Matar *et al.*, 1967). Mucilage only supplies the plant with a small fraction of the total P, whereas with nutrients such as Fe and Zn, much greater proportions are possible.

Low molecular weight (LMW) root exudates are primarily composed of sugars and organic acids, but also phenolics and amino acids (such as phytosiderophores-see later). The organic acids are suggested to be excreted from the apical root cells by a H^+- coupled cotransport system (see Marschner, 1995). Sugars, unloaded from the phloem, build up to high concentrations, resulting in leakage of sugars into the surrounding environment (Jones and Darrah, 1993). Organic acids are involved in the mobilization of Mn and Fe, as well as sparingly soluble phosphates. When malic acid is oxidized at the surface of MnO_2 (Mn^{4+}), Mn^{2+} is released. The Mn^{2+} is then chelated, preventing reoxidation, and is more mobile within the rhizosphere (Jauregui and Reisenauer, 1982). Sparingly soluble phosphates, such as iron or aluminimum phosphates, can be mobilized by the lowering of the pH of the rhizosphere, as well as desorption from sesquioxide surfaces by anion (ligand) exchange (Gerke, 1992). The organic acids form more stable complexes with the Fe and Al, allowing the phosphate to form more soluble complexes that can be taken up by the root.

Organic acids excreted by the root may differ in their ability to solubilize or mobilize insoluble phosphates. Some plants may have adapted to specific soil conditions to the extent that they excrete specific organic acids that mobilize P with the greatest efficiency. For example, Dye (1995) suggested that plants that release exudates with a high proportion of tartrate are more efficient at removing P from the soil, because tartrate occupies high affinity adsorption sites in the soil, thereby reducing the amount of added P fixed in the soil. Plants that excrete large proportions of citrate have access to fixed P because citrate dissolves adsorption sites at which P is fixed. Strom *et al.* (1994) compared the exudation of organic acids between calcifuge and calcicol plants. Calcicol species were found to excrete much more oxalic and citric acids when suffering P deficiency compared to calcifuge species which may allow access to phosphates and Fe in limestone soils (Strom *et al.*, 1994).

Most agricultural soils contain between 30–70% of the total soil P as organic P which is unavailable to the plant (Marschner, 1995). Organic P can be hydrolyzed by ectoenzymes, releasing the P to be taken up by the root. This was demonstrated experimentally by the mineralization of organic P compounds by enzymes, such as acid and alkaline phosphatases, phosphodiesterase and phytase, which were added to soil and extracts analyzed for orthophosphate (Bishop *et al.*, 1994). Ectoenzymes are derived from roots, fungi (in particular mycorrhizal fungi) and bacteria and consist of acid and alkaline phosphatases. Exudation of phosphatases by roots is from the apical region (Tadano and Sakai, 1991) and has been shown to increase under P stress.

Tadano *et al.*, (1993) demonstrated increased acid phosphatase activity in nine crop species grown under P-deficient conditions; the effect was particularly evident in lupin. Sakai and Tadano (1993) also demonstrated increased activity of acid phosphatase in lupin, rice, cabbage, tomato and sugar beet. They further showed that the activity of acid phosphatases were decreased by pH above 4.5

(in rice, cabbage and sugar beet) and high Al^{3+} levels in the soil. However, acid phosphatase activity was not sensitive to high concentrations of K^{+}, Mg^{2+}, Zn^{2+} and Cu^{2+} in the soil solution (Sakai and Tadano, 1993).

Of the many isozymes of acid phosphatase found in the leaves and roots of lupin only one was secreted into the rhizosphere. The molecular weight of this isozyme was estimated to be 72-kDa by SDS-PAGE and 140-kDa by gel filtration and was thus considered to be a homo-dimer (Ozawa *et al.*, 1995).

Mycorrhizal hyphae also play a major role in the release of acid and alkaline phosphatases into the soil environment to mineralize organic P, however, fungal species differ in the amount of acid phosphatase excreted (see Joner and Jakobsen, 1995).

The release of root cap cells, sugars, amino acids, flavonoids and mucilage by roots under deficiencies, in particular P, not only have a direct effect on the mobilization of nutrients but also an indirect effect on the rhizosphere microbial populations. The rhizosphere microorganisms feed on low molecular weight exudates and sloughed-off cells as their carbon source and produce chelators and organic acids which act similarly to the root exudates in mobilizing phosphates to the root surface (Marschner, 1995).

Cluster roots

Plants also deal with P deficiency with changes to the morphology and distribution of the roots within the soil, so as to increase the absorptive surface area of the root system. In general, there is increased partitioning of carbohydrates to the root system under P deficiency which is coincident with the initiation of lateral root formation and the root system becoming finer. It is thought that carbohydrates are partitioned within the root system so that sites within the root system with higher uptake of nutrients receive more carbohydrates (Marschner, 1995) together with IAA stimulating lateral root development (Thoms and Sattelmacher, 1990). Root hair development is also stimulated by P deficiency. In some plant species, such as tomato, root hairs are absent when the plant is supplied with high levels of P while root hairs are abundant under P deficiency (Föhse and Jungk, 1983).

Cluster roots are examples of physical alterations to the structure and architecture of the root system to overcome nutrient deficiency. For example, proteoid roots, which are characteristic of most members of the Proteacea, are dense clusters of rootlets that form along lateral roots. Another major type are the "non-proteoid-like" root clusters, which include dauciform, capillaroid and stalagmiform roots, which occur in the Cyperaceae, Restionaceae and the eucalypts. The various types of cluster roots vary in their length of axis and length and density of rootlets and root hairs and whether or not they are simple or compound (branched). Members of the Proteacea, as well as other cluster root forming species are adapted to soils with low levels of P and generally do not form associations with mycorrhizal fungi.

Root clusters can make up to 80% of the dry weight of the root system (Lamont, 1982) although generally less than 40% (see Dinkelaker *et al.*, 1995) and are most abundant when the supply of P to the plant is deficient (Dinkelaker *et al.*, 1995 and references therein). These root systems are also formed under deficient Fe supply (Marschner *et al.*, 1987; White and Robson, 1989) and deficient Zn supply (Dinkelaker *et al.*, 1995). The critical concentration of P in a nutrient solution for suppression of root clusters varies and has been shown to be 32 μM for *Myrica cerifera* (Louis *et al.*, 1990) and 256 μM for *Myrica gale* (Crocker and Schwintzer, 1993). Foliar applications of P also suppress the formation of cluster roots indicating that the internal P status of the plant determines cluster formation and not the P status of the soil (Marschner *et al.*, 1987; Louis *et al.*, 1990).

Cluster root development occurs at sites of localized supply of nutrients in deficient soil and can be found throughout the soil profile although cluster root density generally decreases with depth (Lamont, 1973; Dinkelaker *et al.*, 1995). The formation of root clusters is also controlled by plant demand because formation increases as plants age, which is suggested to be due to higher internal requirement of nutrients (see Dinkelaker *et al.*, 1995). It is not known what role hormones play in the formation of cluster roots but hormone balance generally alters when plants are suffering stress (Chapin, 1991).

Cluster roots have several advantages over other root systems, such as higher rates of P uptake (per unit dry weight, which has been attributed to increased root surface area, higher metabolic activity, and the ability to mobilize sparingly soluble P reserves in the rhizosphere (Dinkelaker *et al.*, 1995 and references therein). Uptake rates for P have been shown to be 2 to 13 times higher in proteoid roots compared with non-proteoid roots per unit dry weight (see Lamont, 1982). Comparing the efficiency of proteoid roots to finer, more evenly distributed roots has led to the implication that the movement of P to the roots is somehow enhanced in the vicinity of the proteoid root (Gardner *et al.*, 1983).

Proteoid roots have been associated with localized increased concentrations of organic acids, reducing agents, chelating agents and hydrogen ions in the rhizosphere (Gardner *et al.*, 1983). The increased efficiency of the cluster roots was thought to be linked to the exudation of large amounts of organic acids. Sparingly soluble calcium phosphate was shown to be dissolved in the rhizosphere of proteoid roots which was attributed to the excretion of citric acid (Gardner *et al.*, 1982). Although little is known about the excretion of organic acids, they may be metabolically-coupled with excretion of H^+, probably via a carrier—or channel-mediated anion export system coupled with a plasma-membrane bound H^+–ATPase (Dinkelaker *et al.*, 1995). Gardner *et al.* (1983) proposed that citrate excreted by the proteoid roots reacts with Fe(III) in the soil to form an Fe-citrate complex. Phosphate in the soil solution becomes involved in this reaction to form a polymer (probably ferric hydroxy phosphate citrate polymer). This reaction releases much of the citrate to react again with more Fe(III) in the soil while the equilibrium between the soil solid phase and the soil solution phase adjusts to release more phosphate into the solution. The polymer is thought to diffuse to the root surface where it is degraded by reducing agents and the phosphate and Fe(II) are taken up and hydrogen ions excreted (Gardner *et al.*, 1983).

The amount of organic acids excreted by the plant can represent as much as 23% of the total plant dry weight and can be found in the rhizosphere in concentrations around 50 μmol g^{-1} soil (Dinkelaker *et al.*, 1989). The composition of the organic acids excreted by the roots varies between species and environmental conditions. The roots of *Banksia integrifolia* excreted mostly citric acid, which amounted to 50% of the total, followed by malic acid and aconitic acid, which amounted to 18 and 17%, respectively (Grierson, 1992). The organic acids excreted also differ in the extent to which they can mobilize P from the soil. The mobilization of P was greatest with citric acid followed by oxalic acid, while malic and tartaric acids were moderately effective at mobilizing P. Acetic acid, succinic acid and lactic acid were the least effective at mobilizing P (Nagarajah *et al.*, 1970).

Root clusters, and in particular proteoid roots, are efficient at taking up P from P-deficient soil because they thoroughly exploit a small volume of soil with overlapping depletion zones and root exudates. Compared to other root responses to P-deficiency, including mycorrhizal associations, greater amounts of P are removed from the soil for a given volume. Much more work needs to be undertaken to investigate differences between species, in particular, in the types of organic acids excreted and their efficiency at mobilizing P. There appears to have been no investigations into the molecular basis of cluster roots and their function. Isolation of the gene(s) responsible for switching on the cluster formation may allow us to introduce it to commercially important plants to reduce the input of P fertilizers into the ecosystem.

Mycorrhizal fungi

Symbiotic associations with mycorrhizal fungi have received much attention in the last few decades, mainly due to the ability of the association to alleviate phosphorus deficiency. There are two main types of mycorrhizas: endomycorrhizas, those that live within root cortical cells and spread intercellularly within the root (for example—arbuscular mycorrhizas) and ectomycorrhizas, which mainly form on woody plants and form a mantle of hyphae around the surface of the root. The formation of these mycorrhizal associations occurs after recognition events take place when the host or the fungus, or both, produce signal molecules (Piche *et al.*, 1994). Colonization may also be dependent upon the carbohydrate status of the roots as more carbohydrates are partitioned to the roots under nutrient stress, such as P deficiency. Once the plants have a sufficient supply of nutrients, the mycorrhizal association is no longer required and in most cases, the extent of colonization within the root is reduced or eliminated and this has been shown to be associated with decreases in the level of carbohydrates (Thomson *et al.*, 1990; Pearson and Schweiger, 1993; Pearson *et al.*, 1994).

The most important function of mycorrhizas is the increased uptake of P from the soil, although mycorrhizas have been attributed with increased uptake of Zn (Faber *et al.*, 1990; Kothari *et al.*, 1991a; Wellings *et al.*, 1991; Bürkert and Robson, 1994) and Cu (Li *et al.*, 1991), and with decreased shoot Mn levels (Kothari *et al.*, 1991b; Lambert and Weidensaul, 1991). Mycorrhizal hyphae increase the absorptive surface area of the root system by spreading beyond the depletion zone created by the root hairs. This is most important when the hyphae spread beyond the depletion zone of a nutrient that is relatively immobile in soils, such as P. Mycorrhizas can increase the plant P content and shoot dry weight production several-fold compared to non-mycorrhizal plants when grown under P-deficient conditions (Harley and Smith, 1983). Daft and Nicolson (1972) observed a 5.5-fold increase in the growth of tomato when approximately half of its root system was colonized by a *Glomus* sp. on a soil deficient in P. However, when the soil P level was increased, the level of colonization by *Glomus* sp. decreased to a third of the root system and the difference in shoot dry weight was only 10% in favour of the mycorrhizal plants. At high soil P levels, there were no growth differences between the plants and the colonization by *Glomus* sp. had been reduced to less than 20% of the root length (Daft and Nicolson, 1972). In another example, as much as 80% of the total plant P in white clover was delivered by mycorrhizal fungi from as far as 10 cm away from the root when grown on a calcareous soil (Li *et al.*, 1991).

In arbuscular mycorrhizas, a thin space is created between the fungal plasma membrane and that of the root cortical cell when the hyphae invaginates a cortical cell to form an arbuscule (Smith and Smith, 1990). Through this space solutes obtained by the hyphae of fungus are exchanged for host carbohydrates which the fungus requires to grow. The amount of carbohydrates transferred to the host can be as much as 20% of the total plant photosynthate (Pang and Paul, 1980; Douds *et al.*, 1988; Jakobsen and Rosendahl, 1990; Pearson and Jakobsen, 1993a). The transfer of nutrients between the partners of the symbiosis must be bidirectional with carbohydrates transported from the host to the fungus and nutrients transported from the fungus to the host. Bidirectional transfer of carbohydrates for phosphates certainly occurs on a whole plant basis but has been assumed to occur across a single interface (Smith *et al.*, 1994). However, transfer may be spatially separated in arbuscular mycorrhizas, where phosphate is exchanged within the arbuscule while carbohydrates may be taken up by the intercellular hyphae (Smith and Smith, 1990; Pearson *et al.*, 1994). It is suggested that phosphate lost across the fungal plasma membrane is not reabsorbed and that high intracellular concentrations of phosphate maximize the efflux of phosphate into the fungus/root interface. The relatively low phosphate concentrations within the root cells and high phosphate translocating activity permit net movement of P into the plant (see Smith *et al.*, 1994). Similarly, carbohydrates may be lost from the root cells without reabsorption resulting in a net movement of carbohydrates to the fungus. However, it is suggested that efflux of P by the fungus may actually be promoted by opening channels or increases in the frequency or operation of membrane carriers (Patrick, 1989; Tester et al., 1992).

There are a number of ways mycorrhizal plants are better able to take up P from the soil. Mycorrhizal hyphae spread through the soil faster than roots, so are able to grow beyond the depletion zone of immobile nutrients faster than roots. The hyphae also have a higher surface area to dry weight ratio compared to roots, allowing more efficient acquisition of P from the soil per unit dry weight. Once the phosphate is taken up by the hyphae, the formation of polyphosphates minimizes the phosphate accumulation within the hyphae and allows further uptake, avoiding feedback regulation (Marschner and Dell, 1994). Furthermore, the rate of influx of P per unit length of hyphae is reported to be 2–6 times higher than that in roots (Jakobsen *et al.*, 1992a,b). Mycorrhizal hyphae also release acid phosphatases which catalyze the release of P from organic complexes (Straker and Mitchell, 1986; Ho, 1989; Hilger and Krause, 1989; Joner and Jakobsen, 1995). It has been suggested that mycorrhizal plants have access to sources of P not available to non-mycorrhizal plants (Bolan, 1991), such as sparingly soluble Ca-, Fe- and Al- phosphates, but there is a lack of convincing experimental evidence.

In general, most mycorrhizal fungi have the ability to increase the growth of the host when grown under P-deficient conditions. However, various species of fungi do differ in their ability to improve the growth of the host, with some species causing growth depressions. For instance, three mycorrhizal fungi were shown to vary greatly in carbon consumption and P transport when associated with cucumber. When efficiency was calculated as the amount of P(^{32}P) allocated to the plant per unit of carbon (^{14}C), it was found that one fungus, *Scutellospora calospora* (ratio 0.09) was far less efficient than the other two, *Glomus* sp. (0.15) and *Glomus caledonium*(2.5) and even more so when the amount of P transported to the plant per unit length of hyphae was calculated (*S. calospora*, 3.4; *Glomus* sp., 8.4; G. *caledonium*, 158.2) (Pearson and Jakobsen, 1993a). It was later shown that *G. caledonium* supplied almost the entire plant P to the host whilst *Glomus* sp. contributed 21% and *S. calospora* only 7% (Pearson and Jakobsen, 1993b). This work strongly supports the view that fungi from various regions and soil types should be isolated and screened for their ability to improve the growth of the host in an attempt to find the most compatible and efficient association.

Much more needs to be learnt about the recognition events that take place between the root and the fungus that allow formation to proceed and the functioning of the symbiosis on a cellular level. In particular, the exchange processes that occur at the fungal root interface and the carbon cost of the fungus. The ultimate objective may be to find a species of mycorrhiza that colonizes the host extensively and is able to transport large amounts of P to the host in exchange for a minimal amount of carbohydrates.

Fe Deficiency

Abundant literature exists on plant responses to Fe deficiency. Here, we will concentrate mostly on the new developments in the area of mechanisms of resistance to Fe deficiency. For more extensive reviews covering various other aspects of the Fe-deficiency stress, the reader is referred to Bienfait (1989), Römheld (1991), Guerinot and Yi (1994), Marschner and Römheld (1994), Moog and Brüggemann (1994), Mori (1994), and Welch (1995).

Generally, there are two types of resistance mechanisms: Strategy I, operating in dicots and non-gramineous monocots, and Strategy II, expressed in grasses.

Strategy I is characterized by increased reduction of Fe(III) to Fe(II) at the root-cell plasma membrane, acidification of the rhizosphere (enhanced net exudation of protons), enhanced exudation of reducing and/or chelating compounds into the rhizosphere, and changes of root histology and morphology (formation of rhizodermal transfer cells, induced root branching, more root hairs, etc.).

Strategy II involves increased exudation of metal chelators, phytosiderophores into the rhizosphere. The broader term, phytometallophores, has been suggested instead of the term phytosiderophores (Welch, 1995) because these metal chelators can bind not only Fe but other metals as well (Treeby *et al.*, 1989). However, we will continue to use the term phytosiderophores (PS) until the controversy of whether increased PS release under Zn deficiency is induced directly by Zn deficiency or indirectly by Zn-deficiency-induced Fe deficiency (see Walter *et al.*, 1994) is resolved.

The concept of two different strategies of Fe acquisition, especially from the standpoint of importance of Strategy II (enhanced PS release) for all grasses, has been questioned (e.g. Brown and Jolley, 1989; Crowley *et al.*, 1991). The critical examination of this controversy is beyond the scope of the review presented here; for a relevant, comprehensive discussion see Marschner and Römheld (1994). Suffice to say that recent results appear to confirm that even C_4 grasses do exude increased rates of PS under Fe deficiency, the effect which, however, may be evident only under axenic conditions (Von Wiren *et al.*, 1994, 1995a).

Strategy I

Iron (II) is the only form of Fe that Strategy I plants can absorb. Reduction of Fe(III) into Fe(II) is therefore of paramount importance. Strategy I species, as well as Strategy II graminaceous species, have the Fe-reducing capacity at the plasma membrane (Marschner and Römheld, 1994; Moor and Brüggemann, 1994; Welch, 1995). They all have the standard, plasma-membrane-bound Fe(III)-reductase capable of reducing ferricyanide (Bienfait, 1985); this enzyme is likely to be identical to the transmembrane redox pump (Marschner and Römheld, 1994).

Plants with the Strategy I response to Fe deficiency increase their capacity to reduce Fe at the root-cell plasma membrane. While it has been proposed that such an increased Fe-reducing capacity is due to inducible, plasma-membrane bound 'turbo' Fe(III)-reductase (Bienfait, 1985, 1989), present under Fe-sufficient conditions but greatly induced under Fe deficiency when it is capable of reducing a wide variety of Fe(III) chelates (Marschner and Römheld, 1994; Welch, 1995). Recent results question the existence of two different Fe(III)-reductase enzymes. It may only be an increased activity of one single type of membrane-bound Fe(III)-chelate reductase which brings about a greater reducing capacity under Fe-deficiency stress (Moog and Brüggemann, 1994; Bagnaresi and Pupillo, 1995). It is worth mentioning that an increased activity of Fe(III)-reductase may not only be a result of Fe deficiency but also of a specific regulation in mid-stages of the growth cycle under Fe-sufficient conditions (Grusak, 1995). In addition, an enlargement of the plasma membrane surface in the transfer cells formed under Fe deficiency may provide, albeit with some delay, an increased reducing capacity without necessarily invoking an existence and induction of the 'turbo' reductase activity (Bienfait, 1989). Further work is necessary to resolve an apparent controversy.

Acidification of rhizosphere, as an important part of the Fe- deficiency response of Strategy I plants, increases solubility of Fe-containing compounds. An increased net extrusion of protons is due to increased activity of the plasma-membrane-bound H-$^{+}$ATPase, resulting not only in the rhizosphere acidification but also in an increased transmembrane electrical potential and thus an increased driving force for Fe(II) uptake (Rabotti and Zocchi, 1994; Brancadoro *et al.*, 1995). The process of net proton extrusion is of special importance in the acquisition of Fe from calcareous soils which are also rich in bicarbonate which binds protons, prevents rhizosphere acidification and thus reduces effectiveness of the membrane-bound Fe(III)- reductase (Romera *et al.*, 1992).

Strategy I plants can modify biological and chemical conditions in the rhizosphere by extruding, reducing and/or chelating compounds. This process may not take place in nutrient solution-grown plants (Chaney and Bell, 1987) where relatively high rates of Fe(III) chelates are supplied because concentration of Fe(III) chelates in the vicinity of the plasma membrane-bound Fe(III)- reductase may be sufficiently high. Plants grown in Fe-deficient calcareous soil, however, may exude increased amounts of phenolics (Marschner and Römheld, 1994) and organic acids, like citric and malic (Landsberg, 1981; Brancadoro *et al.*,1995) and caffeic acids (Olsen *et al.*, 1981) which chelate/solubilize Fe(III) either directly or through acidifying action in the rhizosphere. However, Fe(III)-chelate would still need to be split and Fe(III) reduced at the root-cell plasma membrane because it is only ionic Fe(II) which can be taken up across the plasma membrane.

Strategy II

Gramineous species acquire Fe by releasing phytosiderophores (PS), non-proteinogenic amino acids with a high binding affinity for Fe, and taking up ferrated PS through a highly-specific transmembrane uptake system (Marschner and Römheld, 1994; Mori, 1994; Welch, 1995). An increased mobilization of Fe from a calcareous soil, even as far away from the root surface as 4 mm, demonstrated a high capacity of PS to mobilize Fe (Awad *et al.,* 1994).

The precursor for biosynthesis of all PS is methionine which is in a series of biosynthetic steps converted into nicotianamine and then 2'-deoxymugineic acid (DMA) (Mori and Nishizawa, 1987; Mori, 1994). From DMA, the biosynthetic pathway may diverge in different species (Ma and Nomoto, 1993a), resulting in different PS being exuded into the rhizosphere of different plant species. The PS are synthesized mainly in root tips, even though biosynthesis may occur in the meristematic tissue of the shoot as well (Walter *et al.,* 1995).

While all wheat genotypes tested to date (Mori and Nishizawa, 1987; Zhang *et al.,* 1989, 1991a, 1991b, Cakmak *et al.,* 1994a) as well as rice (Mori and Nishizawa, 1987) and maize (Mori, 1994) release only DMA, different oat and barley genotypes exude a wide variety of PS types (Mori, 1994). Rye, a species very tolerant to Fe deficiency in calcareous soils, exudes mugineic acid, hydroxymugineic acid (HMA) and DMA into the rhizosphere. It has been suggested that capacity of rye to exude HMA may be related to superior tolerance of rye to Fe deficiency when compared to wheat which exudes only DMA (I Cakmak, unpublished results).

It has been proposed that extrusion of PS into the rhizosphere is accompanied with K^+ efflux to balance the charges (Mori *et al.,* 1991; Mori, 1994); increased extrusion of PS and less Fe-deficiency-caused chlorosis was observed in oat genotypes supplied sufficient K compared to those deficient in K (Hughes *et al.,* 1992). In contrast, PS can be released into the apoplasm via a H^+ co-transport (Welch, 1995) or by exocytosis, together with various cell wall precursors (Marschner and Römheld, 1994).

An increased exudation of PS under Fe deficiency is confined to root tips and occurs in a distinct diurnal rhythm with a peak exudation after the onset of illumination, the light ensuring the continuous supply of assimilates from photosynthetically-active plant parts (Römheld, 1991). More detailed studies on the diurnal rhythm revealed that it is an increase in temperature during the light period, rather than the onset of light itself, which causes an increase in PS exudation (Takagi, 1991, as cited by Mori, 1994).

The rhizosphere microflora feed on PS (Darrah, 1991; Von Wiren *et al.,* 1993; Crowley and Gries, 1994); degradation of PS results in more severe Fe chlorosis (Mori, 1994; Von Wiren *et al.,*1993, 1994, 1995a). However, recent studies showed clearly that, while rhizosphere microflora can decrease effectiveness of PS in supplying plant with sufficient Fe, PS are important contributors of Fe to plants even under non-axenic conditions (Von Wiren *et al.,* 1994, 1995a). Such a result may be due to: (i) spatial separation of maximum microflora activity and the location of maximum PS release, the latter being confined mostly to the root tips which remained uncolonized by microorganisms for some time as roots grow and elongate (Römheld, 1991; Von Wiren *et al.,* 1993), (ii) diurnal rhythm in release of PS (Mori, 1994), and (iii) the fact that PS represent only a small proportion of organic material exuded into the rhizosphere, whereby non-PS organic material decreases and delays microbial degradation of PS. Results of the computer simulation indicated that the pulsed release of PS may be the most important mechanism allowing an active role of PS in supplying plants with Fe in non-axenic media (Crowley and Gries, 1994).

The Fe(III)-PS complex is taken up intact across the root-cell plasma membrane via a specific transport system in an energy- dependent process (Römheld and Marschner, 1986). The transporter recognizes the stereostructure of the Fe(III)-PS complex (Oida *et al.*, 1989) in preference to similar complexes with Cu(II), Zn(II), Co(II) and Co(III) (Ma *et al.*, 1993). However, these metal cations can decrease the rate of Fe(III)-PS uptake by competing with Fe(III) for binding to PS consistent with the stability constants of these metals with PS (Ma and Nomoto, 1993b).

It appears that more than one form of the Fe(III)-PS transporter may exist in the root-cell plasma membrane. Von Wiren *et al.* (1995b) found two kinetically distinct components of the Fe(III)-PS uptake (a saturable, high-affinity component at low concentrations and a linear component at higher concentrations), suggesting that either two transporters are present, or that one transporter may assume multiple structural and/or functional forms. Current work in several laboratories around the world is aimed at molecular characterization of the Fe(III)-PS transporter.

Differences between species and genotypes within species

Within both Strategy I and Strategy II plants, various species and genotypes within species differ in Fe efficiency (Jolley and Brown, 1994). As a response to Fe deficiency, Fe-efficient species and genotypes induce strong metabolic and structural changes that allow them to grow and yield better under deficiency conditions than Fe-inefficient species and genotypes. Fe-efficient soybean genotype accumulates more apoplasmic Fe in roots than Fe-inefficient one; the apoplasmic Fe pool is important in Fe nutrition of shoots (Longnecker and Welch, 1990). Among woody plants with Strategy I, *Vitis berlandieri* and *V. vinifera* show greater rhizosphere acidification, exude greater amounts of organic acids and have a greater capacity to reduce Fe(III) chelates in comparison with Fe-inefficient *V. riparia* (Brancadoro *et al.*, 1995).

Mutants of tomato (*chloronerva*, Stephan and Grün, 1989) and pea (*brz*, Grusak *et al.*, 1990) show typical Fe-deficiency responses (increased rates of Fe(III) reduction, increased rhizosphere acidification and increased exudation of reducing/chelating compounds into the rhizosphere) regardless of Fe availability in the medium. Further studies with these mutants will unveil important information about the Fe-deficiency responses and may allow molecular characterization of individual elements of the Strategy I adaptive systems.

Among Strategy II plants, the relative effectiveness in PS release, and thus Fe efficiency, decreases in order: barley > maize > sorghum (Römheld and Marschner, 1990) or oat > maize (Takagi, 1976; Mori, 1994). Fe-efficient genotypes of grass species release greater amounts of PS into the rhizosphere than Fe-inefficient ones (e.g. Qats, Brown and Jolley, 1989; Hansen and Jolley, 1995).

In contrast, there was no difference in the amount of PS released under Fe deficiency between two maize genotypes (Von Wiren *et al.*, 1994), but an Fe-inefficient one had almost 10-fold lower V_{max} of the saturable component of uptake, arising from either decreased activity or lower number of membrane transporters in its root-cell plasma membrane in comparison to the Fe-efficient genotype (Von Wiren *et al.*, 1995b).

Zinc Deficiency

Reduced supply of Zn is usually associated with an increased accumulation of P in shoot tissue to the extent that it may be difficult to differentiate between Zn deficiency and P toxicity (Loneragan and Webb, 1993, and references therein). Plants grown at deficient Zn supply continue to increase P concentration in shoots (Cakmak and Marschner, 1986), mimicking a behavior of P-deficient plants re-supplied with P (Drew and Saker, 1984). It is, therefore, important for studies on Zn deficiency to manipulate P supply to minimize P accumulation in shoot and leaf tissue of plants subjected to Zn deficiency. Parker (1993) and Gries *et al.* (1995) used hydroxyapatite as a slow-release source of P, while we employed the technique of adding small amounts of P frequently enough to match plant demands during the growth period (M S Wheal and Z Rengel, unpublished results) to prevent accumulation of P under Zn deficiency.

Research emphasis has recently been directed at understanding Zn deficiency and identifying plant characteristics responsible for adaption to Zn-deficient soils through investigating the genotypic differences in tolerance of crop species to low Zn soils. It has been shown that the genotypic variation in yield of wheat grown on low Zn soils can vary considerably. Grain yield of Excalibur, a Zn-efficient genotype, was more than three times greater than the yield of Durati and Kamilaroi, both inefficient genotypes, when grown on low Zn soils (Graham *et al.*, 1992). Other crop species, such as spinach, potato, navy bean, tomato, sorghum, maize and oats also show considerable variation in Zn efficiency (see Graham and Rengel, 1993).

The wheat genotype Excalibur, Zn-efficient under field conditions, was inefficient when grown in nutrient solution (Rengel and Graham, 1995a). When grown in Zn-deficient soil, Excalibur produced greater total length of small diameter ($<$ 0.3 mm) roots compared to other genotypes (Dong *et al.*, 1995). Smaller diameter roots have a higher surface area to dry weight ratio and are therefore able to explore soil more thoroughly and take up nutrients better than thicker roots, this trait, however, confers no apparent advantage in a nutrient solution culture system (Rengel and Graham, 1995a, b). It was suggested that changes to the geometry of the roots may not be the only mechanism of Zn efficiency and that different mechanisms may be expressed to a different extent under various growing conditions (Graham and Rengel, 1993).

More detailed experimentation has shown that wheat genotypes differ in a number of characteristics. Rengel (1995a) has recently demonstrated that the activity of carbonic anhydrase, principally responsible for the hydration of CO_2, was 2-fold higher in the Zn-efficient genotype, Warigal, compared to the Zn-inefficient Durati, when grown under Zn-deficient conditions. Even though the role of carbonic anhydrase in photosynthesis is not confirmed, it appears likely that such a role would be more important in wheat because of its inherently low activity of carbonic anhydrase in comparison with other plant species (Makino *et al.*, 1992).

Partitioning of carbohydrates is frequently altered by nutrient deficiency (e.g. Fredeen *et al.*, 1989; Cakmak *et al.*, 1994b). However, no information of that kind for Zn deficiency appears to have been published previously. In our recent study, Zn-inefficient Durati wheat allocated more carbohydrates to the root system early in the Zn-deficiency stress compared to the Zn-efficient Warigal (Pearson and Rengel, unpublished). Such a result might have been a consequence of growth of Durati leaf cells being affected more by Zn deficiency, thus reducing the sink in leaves and correspondingly increasing the amount of carbohydrates available for translocation to roots (similarly to P deficiency in soybean, Fredeen *et al.*, 1989). This increased allocation of carbohydrates to roots of Zn-deficient Durati resulted in a relatively larger root system in advanced stages of the Zn-deficiency stress compared to Warigal (Rengel and Graham, 1995a).

Zn uptake

Recent kinetic experiments have shown that Zn uptake in wheat genotypes is regulated by de-repression. Following a period of growth under Zn deficiency conditions, ability of wheat plants to take up Zn upon its re-supply increases (Rengel and Graham, 1996). Consequently, the rate of Zn uptake is inversely related to symplasmic Zn concentration (Z Rengel, unpublished results). It remains to be elucidated in the future how internal Zn concentration regulates activity of membrane transporters for Zn.

Differences in Zn uptake kinetics were noted among genotypes of rice, sorghum and maize differing in Zn efficiency (for the review see Graham and Rengel, 1993). However, these studies have been done using physiologically extreme concentrations of Zn in solution (e.g. from 10 to 250 μM, Bowen, 1986). When the chelate-buffered nutrient solution with appropriately low Zn activities was used, no difference in Zn uptake was noted between two tomato genotypes of unknown Zn efficiency (Parker *et al.*, 1992). In contrast, in the study with 12 wheat genotypes differing in Zn efficiency, there was a tendency for Zn-efficient genotypes to have a greater Zn uptake rate per unit of root, but the difference was not significant (Rengel and Graham, 1995b). In a more recent study, where two wheat genotypes exhibiting the greatest difference in Zn efficiency were grown at Zn^{2+} activities most likely to resemble those found in soil solution, Zn-efficient Warigal had a greater rate of Zn uptake per unit of root compared to Zn-inefficient Durati (Rengel and Graham, 1996).

Zn deficiency and the plasma membrane

Recent results have shown that there is *de novo* biosynthesis of a 34–kDa polypeptide in the root-cell plasma membrane fraction of Zn-efficient Warigal but not of Zn- inefficient Durati wheat grown without Zn for 18 d (Rengel and Hawkesford, 1995). *De novo* biosynthesis of the 34–kDa polypeptide was repressed by the addition of Zn to nutrient solution in which plants were grown. This is the first record of differential effects of Zn (and Zn deficiency) on polypeptide biosynthesis in the root-cell plasma membrane of different genotypes. Further work is in progress, but it appears tempting to postulate a role for the 34–kDa polypeptide in the increased capacity of Zn-efficient genotypes to take up Zn following a period of exposure to solutions containing no added Zn.

Zinc ions bind strongly to sulfhydryl groups (Chvapil, 1973), thus stabilizing the plasma membrane proteins by preventing oxidation of sulfhydryl groups to disulfides (cf. Welch and Norvell, 1993). Under Zn deficiency, Zn-efficient Warigal wheat maintained the amount of sulfhydryl groups in the root-cell plasma membrane while Zn-inefficient Durati suffered a significant loss of sulfhydryl groups under the same conditions (Rengel, 1995b).

Sulfhydryl groups, being involved in ion-channel gating and other processes regulating membrane permeability and ion transport (see Welch, 1995), can be important for structural and functional integrity of the root-cell plasma membrane (Welch and Norvell, 1993). Loss of sulfhydryl groups from the root-cell plasma membrane may result in increased leakage of organic and mineral components from Zn-deficient roots (Welch *et al.*, 1982; Cakmak and Marschner, 1988a; Welch and Norvell, 1993). In contrast, Cakmak and Marschner (1988b) suggested that enhanced production of superoxide radicals, due to reduced activity of superoxide dismutase (SOD) as well as to increased activity of NADPH oxidase, is a cause of structural and functional damage to the plasma membranes of root cells in Zn-deficient plants. In our experiments with wheat genotypes differing in Zn efficiency, however, no decrease in SOD activity in Zn-deficient roots and no difference between genotypes differing in Zn efficiency in SOD activity was observed (M.S Wheal and Z Rengel, unpublished results), putting more emphasis on confirmed decrease in the amount of sulfhydryl groups under Zn deficiency, especially in the Zn-inefficient wheat genotype (Rengel 1995b), as a causc of presumed decreased structural and functional integrity of the root-cell plasma membranes.

It appears that the concentration of reactive sulfhydryl groups bound to the external surface of the root-cell plasma membranes is many orders of magnitude greater than the amount of bound Zn (Rengel, 1995b). Such a result indicates that a small amount of Zn may be sufficient to protect from oxidation only those sulfhydryl groups that are strategically placed on relevant plasma membrane proteins (possibly at the apoplastic mouth of ion channels or other transport proteins; Kochian, 1993, Welch 1995). This suggestion is supported by an apparent requirement of Zn in the external, root-bathing medium for preventing membranes from becoming leaky (cf. Welch and Norvell 1993).

Zn circulation within the plants

Visible signs of Zn deficiency appear in fully expanded leaves of Zn-inefficient Durati wheat, even when the concentration of Zn in the leaves is similar to that in Zn- efficient genotypes showing no visible signs of Zn deficiency (Cakmak *et al.*, 1994b; Rengel and Graham, 1995a). These observations indicate that (i) the form (complexation) and/or compartmentation of Zn in root and leaf cells may be important for fulfilment of its physiological role, and (ii) that Zn may be utilized or compartmented differently within cells of genotypes differing in Zn efficiency.

Work is currently under way in our Laboratory to investigate whether the remobilization of Zn during grain development differs between genotypes that differ in Zn efficiency. It has been observed that the loading of foliarly-applied Zn into the leaf phloem is greater in Zn-efficient Warigal than in Zn-inefficient Durati, probably due to differences in the efficiency of the membrane transporters involved in phloem loading (J.N Pearson, unpublished results).

Zn and mycorrhizas

Mycorrhizas play an important role in Zn uptake (Faber *et al.*, 1990, Wellings *et al.*, 1991, Sharma *et al.*, 1992, Bürkert and Robson, 1994). An increased absorption area brought about by mycorrhizal hyphae is beneficial in uptake of Zn because it is an immobile nutrient, transported in soils principally by diffusion. Experiments with mycorrhizal and non-mycorrhizal maize in nutrient solution also showed beneficial effect of mycorrhizas in Zn uptake (Sharma *et al.*, 1992).

There have been no investigations assessing a possible role of mycorrhizas in bringing about genotypic differences in Zn efficiency. Working with mycorrhizas and Zn deficiency has proved difficult because the primary function of the symbiosis is P uptake and high soil P levels reduce mycorrhizal colonization. This is especially obvious in case of cereals where colonization is exceedingly low unless soil P is low (Baon *et al.*, 1992). Thus, to obtain sufficient levels of colonization of cereal roots and investigate Zn uptake by the symbiosis, the plants must also be grown at low soil P, which makes interpreting an interaction between Zn responses and P responses difficult (J.N Pearson, unpublished results).

Phytosiderophores and Zn acquisition

Phytosiderophores (PS) may be important in the acquisition of Zn from the soil (Treeby *et al.*, 1989). It was recently demonstrated that PS release from roots of a number of plant species was enhanced by Zn-deficiency stress (Zhang *et al.*, 1989, 1991a, b), the effect, however, could not be shown for barley grown in the Zn-deficient chelate-buffered solution (Gries *et al.*, 1995).

In the case of wheat, Zn-deficiency-related increase in PS release appears to be greater in the Zn-efficient genotype Aroona than in the Zn-inefficient Durati (Cakmak *et al.*, 1994a; Walter *et al.*, 1994). It, therefore, appears that the increased release of PS may not be caused by Zn deficiency directly, but rather as a response to impaired Fe translocation from the root to the shoot under Zn deficiency, or in other words Zn-deficiency-induced Fe deficiency. An application of Fe-citrate to the leaves to relieve hidden, physiological Fe deficiency arising from the imbalance in Fe circulation within the plant depressed PS release under Zn deficiency (Walter *et al.*, 1994). It should, however, be stressed that the the total Fe pool in leaf or root cells may not necessarily be quantitatively the same as the physiologically-active Fe pool, thus raising a possibility that plants exhibiting total Fe concentrations above the critical level actually suffer from physiological Fe deficiency that may result in initiation of the Fe deficiency response.

A different Zn-efficient wheat genotype, Warigal, showed a greater rate of net Fe uptake under Zn-deficiency conditions (Rengel and Graham, 1995b) and also released greater amounts of PS (Z Rengel, V Römheld and H Marschner, unpublished results) compared to Zn-inefficient Durati. Under Zn deficiency, Fe concentrations in Warigal roots increased due to depressed translocation of Fe from roots to shoots (Rengel and Graham, 1995b). Such a Zn-deficiency-induced change in Fe circulation within the plant could not be observed in Zn-inefficient Durati, indicating that increased Zn efficiency may be related to increased disturbance in Fe uptake and translocation, resulting in increased PS release.

There appears to be a positive relationship between the amount of PS released under Zn deficiency and Zn efficiency of various cereal genotypes (I Cakmak, personal communication). However, release of PS by soil-grown plants under Zn deficiency and an unequivocal proof that PS play a role in mobilization and uptake of Zn from Zn-deficient soils has yet to be shown. This is especially important because PS have a greater affinity for Fe than for Zn (*e.g.* 2'–deoxymugineic acid, DMA, has a 2-fold higher stability constant for Fe than for Zn; see Murakami *et al.*, 1989).

CONCLUSION

The ability of a plant species or cultivar to withstand or avoid nutrient deficiency stress has important implications for agriculture in general. It is evident that many and varied mechanisms are employed by plants to alleviate nutrient deficiency stress, such as root exudation and symbiotic associations. Little progress has been made in understanding the molecular mechanisms of resistance to nutrient stress. However, once understood there remains the possibility that these characteristics may be introduced to important agricultural crops. Cultivars may be produced that grow more effectively in the marginal agricultural regions where nutrient deficiencies are common, and may also lead to decreased fertilizer use and reduced farming costs. The role of phytosiderophores may prove extremely important in the uptake of micronutrients by grasses. Since the seeds of grasses are the staple food for most people (rice, wheat and barley), there appears to be much potential in increasing the yield of these crops under deficient conditions. Firstly, it must be determined whether phytosiderophores aid the uptake of micronutrient cations other than iron. Additionally, the potential use of cluster roots to resist nutrient deficiency stress may also have great benefits if the mechanism that triggers cluster root formation can be elucidated and transferred to commercially important plants.

ACKNOWLEDGEMENTS

We would like to thank Dr David Saunders (Department of Plant Science, University of Adelaide) for help in preparing this manuscript. Our research has been funded by the Australian Research Council and the Grains Research and Development Corporation.

REFERENCES

Abbott A.J. (1967) Physiological effects of micronutrient deficiencies in isolated roots of *Lycopersicum esculentum*. *New Phytologist*, **66**, 419–437.

Awad F., Römheld V. and Marschner H. (1994) Effect of root exudates on mobilization in the rhizosphere and uptake of iron by wheat plants. *Plant and Soil*, **165**, 213–218.

Bagnaresi P. and Pupillo P. (1995) Characterization of NADH- dependent Fe^{3+}-chelate reductases of maize roots. *Journal of Experimental Botany*, **46**, 1497–1503.

Baon J.B., Smith S.E., Alston A.M. and Wheeler R.D. (1992) Phosphorus efficiency of three cereals as related to indigenous mycorrhizal infection. *Australian Journal of Agricultural Research*, **43**, 479–491.

Bell R.W., McClay L., Plaskett D., Dell B. and Loneragan J.F. (1989). Germination and vigour of black gram (*Vigna mungo* (L.) Hepper) seed from plants grown with and without boron. *Australian Journal of Agricultural Research*, **40**, 273–279.

Bienfait H.F. (1985) Regulated redox processes at the plasmalemma of plant root cells and their function in iron uptake. *Journal of Bioenergetics and Biomembranes*, **17**, 73–83.

Bienfait H.F. (1989) Prevention of stress in iron metabolism of plants. *Acta Botanica Neerlandica*, **38**, 105–129.

Bishop M.L., Chang A.C. and Lee R.W.K. (1994) Enzymatic mineralization of organic phosphorus in a volcanic soil in Chile. *Soil Science*, **157**, 238–243.

Bolan N.S. (1991) A critical review of the role of mycorrhizal fungi in the uptake of phosphorus by plants. *Plant and Soil*, **134**, 189–207.

Bowen J.E. (1986) Kinetics of zinc uptake by two rice cultivars. *Plant and Soil*, **94**, 99–107.

Boyer J.S. (1982) Plant productivity and environment. *Science*, **218**, 443–448.

Brancadoro L., Rabotti G., Scienza A. and Zocchi G. (1995) Mechanisms of Fe-efficiency in roots of *Vitis* spp. in response to iron deficiency stress. *Plant and Soil*, **171**, 229–234.

Brown J.C. and Jolley V.D. (1989) Plant metabolic responses to iron deficiency stress. *BioScience*, **39**, 546–551.

Bürkert B. and Robson A.D. (1994) ^{65}Zn uptake in subterranean clover (*Trifolium subterraneum* L.) by three vesicular- arbuscular mycorrhizal fungi in a root-free sandy soil. *Soil Biology and Biochemistry*, **26**, 1117–1124.

Cakmak I., Gulut K.Y., Marschner H. and Graham R.D. (1994a) Effect of iron and zinc deficiency on phytosiderophore release in wheat genotypes differing in zinc efficiency. *Journal of Plant Nutrition*, **17**, 1–17.

Cakmak I., Hengeler C. and Marschner H. (1994b) Partitioning of shoot and root dry matter and carbohydrates in bean plants suffering from phosphorus, potassium and magnessium deficiency. *Journal of Experimental Botany*, **45**, 1245–1250.

Cakmak I. and Marschner H. (1986) Mechanism of phosphorus induced zinc deficiency in cotton. I. Zinc deficiency enhanced uptake rate of phosphorus. *Physiologia Plantarum*, **68**, 483–490.

Cakmak I. and Marschner H. (1988a) Increase in membrane permeability and exudation in roots of zinc deficient plants. *Journal of Plant Physiology*, **132**, 356–361.

Cakmak I. and Marschner H. (1988b) Enhanced superoxide radical production in roots of zinc-deficient plants. *Journal of Experimental Botany*, **39**, 1449–1460.

Cakmak I. and Marschner H. (1990) Decrease in nitrate uptake and increase in proton release in zinc deficient cotton, sunflower and buckwheat plants. *Plant and Soil*, **129**, 261–268.

Chaney R.L. and Bell P.F. (1987) Complexity of iron nutrition: Lessons from plant-soil interaction research. *Journal of Plant Nutrition*, **10**, 963–994.

Chapin F.S. III (1991) Integrated responses of plants to stress. A centralized system of physiological responses. *Bio Science*, **41**, 29–36.

Chvapil M. (1973) New aspects on the biological role of zinc: a stabilizer of macromolecules and biological membranes. *Life Science*, **13**, 1041–1049.

Clarkson D.T. and Lüttge U. (1991) Mineral nutrition: Inducible and repressible nutrient transport system. *Progress in Botany*, **52**, 61–83.

Clarkson D.T. and Saker L.R. (1989) Sulphate influx in wheat and barley roots becomes more sensitive to specific protein-binding reagents when plants are sulphate-deficient. *Planta*, **178**, 249–257.

Crocker L.J. and Schwintzer C.R. (1993) Factors affecting formation of cluster roots in *Myrica gale* seedlings in water culture. *Plant and Soil*, **152**, 287–298.

Crowley D.E. and Gries D. (1994) Modeling of iron availability in the plant rhizosphere. In *Biochemistry of Metal Micronutrients in the Rhizosphere*, edited by J.A. Manthey, D.E. Crowley and D.G. Luster, pp. 199–223. Boca Raton: Lewis Publishers.

Crowley D.E., Wang Y.C., Reid C.P. and Szaniszlo P.J. (1991) Mechanisms of iron acquisition from siderophores by microorganisms and plants. *Plant and Soil*, **130**, 179–198.

Cumbus I.P. (1985) Development of wheat roots under zinc deficiency. *Plant and Soil*, **83**, 313–316.

Daft M.J. and Nicolson T.H. (1972) Effect of *Endogone mycorrhiza* on plant growth. IV. Quantitative relationship between growth of the host and the development of the endophyte in tomato and maize. *New Phytologist*, **71**, 287–295

Darrah P.R. (1991) Models of the rhizosphere. I. Microbial population dynamics around a root releasing soluble and insoluble carbon. *Plant and Soil*, **133**, 187–199.

Dinkelaker B., Hengeler C. and Marschner H. (1995). Distribution and function of proteoid roots and other root clusters. *Botanica Acta*, **108**, 183–200.

Dinkelaker B., Römheld V. and Marschner H. (1989) Citric acid excretion and precipitation of calcium citrate in the rhizosphere of white lupin *(Lupinus albus L.). Plant, Cell and Environment*, **12**, 285–292.

Dong B., Rengel Z. and Graham R.D. (1995) Effects of herbicide chlorsulfuron on growth and nutrient uptake parameters of wheat genotypes differing in Zn efficiency. *Plant and Soil*, **173**, 275–282.

Douds D.D., Johnson C.R. and Koch K.E. (1988) Carbon cost of the fungal symbiont relative to net leaf P accumulation in a split- root VA mycorrhizal symbiosis. *Plant Physiology*, **86**, 491–496.

Drew M.C. and Saker L.R. (1984) Uptake and long-distance transport of phosphate, potassium and chloride in relation to internal ion concentrations in barley: evidence of non-allosteric regulation. *Planta*, **160**, 500–507.

Dye C. (1995) Effect of citrate and tartarate on phosphate adsorption by amorphous ferric hydroxide. *Fertilizer Research*, **40**, 129–134.

Faber B.A., Zasoski R.J., Burau R.G. and Urio K. (1990) Zinc uptake by corn as affected by vesicular-arbuscular mycorrhizae. *Plant and Soil*, **129**, 121–130.

Fredeen A.L., Rao I.M. and Terry N. (1989) Influence of phosphorus nutrition on growth and carbon partitioning in *Glycine max* (L). Merr. *Plant Physiology*, **89**, 225–230.

Föhse D. and Jungk A. (1983) Influence of phosphate and nitrate supply on root hair formation of rape, spinach and tomato plants. *Plant and Soil*, **74**, 359–368.

Gardner W.K., Barber D.A. and Parbery D.G. (1983) The acquisition of phosphorus by *Lupinus albus* L. III. The probable mechanism by which phosphorus movement in the soil/root interface is enhanced. *Plant and Soil*, **70**, 107–124.

Gardner W.K., Parbery D.G. and Barber D.A. (1982) The acquisition of phosphorus by *Lupinus albus* L. II. The effect of varying phosphorus supply and soil type on some characteristics of the soil/root interface. *Plant and Soil*, **68**, 33–41.

Gerke J. (1992) Orthophosphate and organic phosphate in the soil solution of four sandy soils in relation to pH - evidence for a humic-Fe(Al)phosphate complexes. *Communications in Soil Science and Plant Analysis*, **23**, 601–612.

Graham R.D., Ascher J.S., Ellis P.A.E., and Shepherd K.W. (1987) Transfer to wheat of the copper efficiency factor carried on the rye chromosome arm 5RL. *Plant and Soil*. **99**, 107–114.

Graham R.D., Ascher J.S. and Hynes S.C. (1992) Selecting zinc- efficient cereal genotypes for soils of low zinc status. *Plant and Soil*, **146**, 241–250.

Graham R.D. and Rengel Z.(1993) Genotypic variation in zinc uptake and utilization by plants. In *Zinc in Soils and Plants*, edited by A.D. Robson, pp. 107–118. Dordrecht: Kluwer Academic Publishers.

Graham R.D. and Webb M.J. (1991) Micronutrients and plant disease resistance and tolerance in plants. In *Micronutrients in Agriculture*, edited by J.J. Mortvedt, F.R. Cox, L.M. Shuman and R.M. Welch, pp. 329–370. Madison, WI: Soil Science Society of America Book Series No.4.

Grierson P.F. (1992) Organic acids in the rhizosphere of *Banksia integrifolia* L. f. *Plant and Soil*, **144**, 259–265.

Gries D., Brunn S., Crowley D.E. and Parker D.R. (1995) Phytosiderophore production in relation to micronutrient metal deficiencies in barley. *Plant and Soil*, **172**, 299–308.

Grusak M.A. (1995) Whole-root iron(III)-reductase activity throughout the life cycle of iron-grown *Pisum sativum* L. (Fabaceae): relevance to the iron nutrition of developing seeds. *Planta*, **197**, 111–117.

Grusak M.A., Welch R.M. and Kochian L.V. (1990) Physiological characterization of a single-gene mutant of *Pisum sativum* exhibiting excess iron accumulation. I. Root iron reduction and iron uptake. *Plant Physiology*, **93**, 976–981.

Guerinot M.L. and Yi Y. (1994) Iron: nutritious, noxious, and not readily available. *Plant Physiology*, **104**, 815–820.

Hansen N.C. and Jolley V.D. (1995) Phytosiderophore release as a criterion for genotypic evaluation of iron efficiency in oat. *Journal of Plant Nutrition*, **18**, 455–465.

Harley J.L. and Smith S.E. (1983) *Mycorrhizal Symbiosis*. London: Academic Press.

Hawkesford M.J. and Belcher A.R. (1991) Differential protein synthesis in response to sulphate and phosphate deprivation: Identification of possible components of plasma membrane transport system in cultured tomato roots. *Planta*, **185**, 323–329.

Hilger A.B. and Krause H.H. (1989) Growth characteristics of *Laccaria laccata* and *Paxillus involutus* in liquid culture media with inorganic and organic phosphorus sources. *Canadian Journal of Botany*, **67**, 1782–1789.

Ho I. (1989) Acid phosphatase, alkaline phosphatase, and nitrate reductase activity of selected ectomycorrhizal fungi. *Canadian Journal of Botany*, **67**, 750–753.

Huber D.M. (1980) The role of mineral nutrition in defense. In *Plant Disease, Vol. V*, edited by J.G. Harsfall and E.B. Cowling, pp. 381–406. New York: Academic Press.

Hughes D.F., Jolley V.D. and Brown J.C. (1992) Role of potassium in the iron-stress response mechanism of iron-efficient oat. *Soil Science Society of America Journal*, **56**, 830–835.

Jakobsen I., Abbott L.K. and Robson A.D. (1992a) External hyphae of vesicular-arbuscular mycorrhizal fungi associated with *Trifolium subterraneum*. I. Spread of hyphae and phosphorus inflow into the roots. *New Phytologist*, **120**, 371–380.

Jakobsen I., Abbott L.K. and Robson A.D. (1992b) External hyphae of vesicular-arbuscular mycorrhizal fungi associated with *Trifolium subterraneum*. II. Hyphal transport of ^{32}P over defined distances. *New Phytologist*, **120**, 509–516.

Jakobsen I. and Rosendahl L. (1990) Carbon flow into soil and external hyphae from roots of mycorrhizal cucumber plants. *New Phytologist*, **115**, 77–83.

Jauergui M.A. and Reisenauer H.M. (1982) Dissolution of oxides of manganese and iron by root exudate components. *Soil Science Society of America Journal*, **46**, 314–317.

Jolley V.D. and Brown J.C. (1994) Genetically controlled uptake and use of iron by plants. In *Biochemistry of Metal Micronutrients in the Rhizosphere*, edited by J.A. Manthey, D.E. Crowley and D.G. Luster, pp. 251–266. Boca Raton: Lewis Publishers.

Joner E.J. and Jakobsen I. (1995) Growth and extracellular phosphatase activity of arbuscular mycorrhizal hyphae as influenced by soil organic matter. *Soil Biology and Biochemistry*, **27**, 1153–1159.

Jones D.L. and Darrah P.R. (1993) Re-absorption of organic compounds by roots of *Zea mays* L. and its consequences in the rhizosphere. II. Experimental and model evidence for simultaneous exudation and re-adsorption of soluble C compounds. *Plant and Soil*, **153**, 47–59.

Jungk A. (1991) Dynamics of nutrient movement at the soil-root interface. In *Plant Roots: The Hidden Half*, edited by J. Waisel, A. Eshel and U. Kafkafi, pp. 455–481. New York: Marcel Dekker.

Kochian L.V. (1993) Zinc absorption from hydroponic solutions by plant roots. In *Zinc in Soils and Plants*, edited by A D Robson, pp. 45–57. Dordrecht: Kluwer Academic Publishers.

Kothari S.K., Marschner H. and Römheld V. (1991a) Contribution of VA mycorrhizal hyphae in acquisition of phosphorus and zinc by maize grown in a calcareous soil. *Plant and Soil*, **131**, 177–185.

Kothari S.K., Marschner H. and Römheld V. (1991b) Effect of a vesicular-arbuscular mycorrhizal fungus and rhizosphere micro-organisms on manganese reduction in the rhizosphere and manganese concentrations in maize. *New Phytologist*, **117**, 649–655.

Lambert D.H. and Weidensaul T.C. (1991) Element uptake by mycorrhizal soybean from sewage-sludge-treated soil. *Soil Science Society of America Journal*, **55**, 393–397.

Lamont B. (1973) Factors affecting the distribution of proteoid roots within the root system of two *Hakea* species. *Australian Journal of Botany*, **21**, 165–187.

Lamont B. (1982) Mechanisms for enhancing nutrient uptake in plants with particular reference to mediterranean South Africa and Western Australia. *Botanical Review*, **48**, 597–689.

Landsberg E.C. (1981) Organic acid synthesis and release of hydrogen ions in response to Fe-deficiency stress of mono- and dicotyledonous plant species. *Journal of Plant Nutrition*, **3**, 579–591.

Lee R.B. (1982) Selectivity and kinetics of ion uptake of barley plants following nutrient deficiency. *Annals of Botany*, **50**, 429–449.

Li X.L, George E. and Marschner H. (1991) Extension of the phosphorus depletion zone in VA-mycorrhizal white clover in a calcareous soil. *Plant and Soil*, **136**, 41–48.

Loneragan J.F. and Webb M.J. (1993) Interactions between zinc and other nutrients affecting the growth of plants. In *Zinc in Soils and Plants*, edited by A D Robson, pp. 119–134. Dordrecht: Kluwer Academic Publishers.

Longnecker N. and Welch R.M. (1990) Accumulation of apoplastic iron in plant roots: a factor in the resistance of soybeans to iron deficiency induced chlorosis? *Plant Physiology*, **92**, 17–22.

Louis I., Racette S. and Torrey J.G. (1990) Occurrence of cluster roots on *Myrica cerifera* L. (Myricaceae) in water culture in relation to phosphorus nutrition. *New Phytologist*, **115**, 311–317.

Ma J.F., Kusano G., Kimura S. and Nomoto K. (1993) Specific recognition of mugineic acid-ferric complex by barley roots. *Phytochemistry*, **34**, 599–603.

Ma J.F. and Nomoto K. (1993a) Two related biosynthetic pathways of mugineic acids in gramineous plants. *Plant Physiology*, **102**, 373–378.

Ma J.F. and Nomoto K. (1993b) Inhibition of mugineic acid-ferric complex uptake in barley by copper, zinc and cobalt. *Physiologia Plantarum*, **89**, 331–334.

Makino A., Sahashita H., Hidema J., Mae, T., Ojima K. and Osmond B. (1992) Distinctive responses of ribulose–1,5–bisphosphate carboxylase and carbonic anhydrase in wheat leaves to nitrogen nutrition and their possible relationships to the CO_2-transfer resistance. *Plant Physiology*, **100**, 1737–1743.

Marschner H. (1995) *Mineral Nutrition of Higher Plants*. 2nd Edition. London: Academic Press.

Marschner H. and Dell B. (1994) Nutrient uptake in mycorrhizal symbiosis. In *Management of Mycorrhizas in Agriculture Horticulture and Forestry*, edited by A.D. Robson, L.K. Abbott and N. Malajczuk, pp. 89–102. Dordrecht: Kluwer Academic Press.

Marschner H. and Römheld V. (1994) Strategies of plants for acquisition of iron. *Plant and Soil*, **165**, 261–274.

Marschner H., Römheld V. and Cakmak I. (1987) Root-induced changes of nutrient availability in the rhizosphere. *Journal of Plant Nutrition,* **10**, 1175–1184.

Matar A.E., Paul J.L. and Jenny H. (1967) Two phase experiments with plants growing in phosphate-treated soil. *Proceedings of the Soil Science Society of America,* **31,** 235–237.

Miller R.O., Jacobsen J.S. and Skogley E.O. (1993) Aerial accumulation and partitioning of nutrients by hard red spring wheat. *Communications in Soil Science and Plant Analysis*, **24**, 2389–2407.

Moog P.R. and Brüggemann W. (1994) Iron reductase systems on the plant plasma membrane - A review. *Plant and Soil*, **165**, 241–260.

Mori S. (1994) Mechanisms of iron acquisition by graminaceous (strategy II) plants. In *Biochemistry of Metal Micronutrients in the Rhizosphere*, edited by J.A. Manthey, D.E. Crowley and D.G. Luster, pp.225–249. Boca Raton: Lewis Publishers.

Mori S. and Nishizawa N. (1987) Methionine as a dominant precursor of phytosiderophores in graminaceous plants. *Plant, and Cell Physiology*, **28**, 1081–1092.

Mori S., Nishizawa N., Hayashi H., Chino M., Yoshimura E. and Ishihara J. (1991) Why are young rice plants highly susceptible to iron deficiency? *Plant and Soil*, **130**, 143–156.

Murakami T., Ise K., Hayakawa M., Kamei S. and Takagi S. (1989) Stabilities of metal complexes of mugineic acids and their specific affinities for iron (III). *Chemistry Letters*, pp. 2137–2140.

Nagarajah S., Posner A.M. and Quirk J.P. (1970) Competitive adsorption of phosphate with polygalacturonate and other organic anions on kaolinite and oxide surfaces. *Nature*, **228**, 83–85.

Oida F., Ota N., Mino Y., Nomoto K. and Sugiura Y. (1989) Stereospecific iron uptake mediated by phytosiderophore in graminaceous plants. *Journal of the American Chemical Society*, **111**, 3436–3437.

Olsen R.A., Bennett J.H., Blume D. and Brown J.C. (1981) Chemical aspects of the Fe stress response mechanisms in tomatoes. *Journal of Plant Nutrition*, **3**, 905–921.

Ozawa K., Osaki M., Matsui H., Honma M. and Tadano T. (1995) Purification and properties of acid phosphatase secreted from lupin roots under phosphorus-deficiency conditions. *Soil Science and Plant Nutrition*, **41**, 461–469.

Pang P.C. and Paul E.A. (1980) Effects of vesicular-arbuscular mycorrhiza on ^{14}C and ^{15}N distribution in nodulated faba beans. *Canadian Journal of Soil Science*, **60**, 241–250.

Parker D.R. (1993) Novel nutrient solutions for zinc nutrition research: buffering free $zinc^{2+}$ with synthetic chelators and P with hydroxyapatite. In *Plant Nutrition - from Genetic Engineering to Field Practice*, edited by N J Barrow, pp. 677–680. Dordrecht: Kluwer Academic Publishers.

Parker D.R., Aguilera J.J. and Thomason D.N. (1992) Zinc- phosphorus interactions in two cultivars of tomato (*Lycopersicon esculentum* L.) grown in chelator-buffered nutrient solutions. *Plant and Soil*, **143**, 163–177.

Patrick J.W. (1989) Solute efflux from the host at plant- microorganism interfaces. *Australian Journal of Plant Physiology*, **16**, 53–67.

Pearson J.N., Abbott L.K. and Jasper D.A. (1994) Phosphorus, soluble carbohydrates and the competition between two arbuscular mycorrhizal fungi colonizing subterranean clover. *New Phytologist*, **127**, 101–106.

Pearson J.N. and Jakobsen I. (1993a) Symbiotic exchange of carbon and phosphorus between cucumber and three arbuscular mycorrhizal fungi. *New Phytologist*, **124**, 481–488.

Pearson J.N. and Jakobsen I. (1993b). The relative contribution of hyphae and roots to phosphorus uptake by arbuscular mycorrhizal plants, measured by dual labelling with ^{32}P and ^{33}P. *New Phytologist*, **124**, 489–494.

Pearson J.N. and Rengel Z. (1994) Distribution and remobilization of Zn and Mn during grain development in wheat. *Journal of Experimental Botany*, **45**, 1829–1835.

Pearson J.N. and Schweiger P. (1993) *Scutellospora calospora* (Nicol. & Gerd.) Walker & Sanders associated with subterranean clover: dynamics of colonization, sporulation and soluble carbohydrates. *New Phytologist*, **124**, 215–219.

Peoples M.B. and Craswell E.T. (1992) Biological nitrogen fixation: investments, expectations and actual contributions to agriculture. *Plant and Soil*, **141**, 13–39.

Piche Y., Simon L. and Seguin A. (1994) Genetic manipulation in vesicular-arbuscular mycorrhizal fungi. In *Management of Mycorrhizas in Agriculture, Horticulture and Forestry*, edited by AD Robson, LK Abbott and N Malajczuk, pp. 171–178. Dordrecht: Kluwer Academic Press.

Rabotti G. and Zocchi G. (1994) Plasma membrane-bound H^+-ATPase and reductase activities in Fe-deficient cucumber roots. *Physiologia Plantarum*, **90**, 779–785.
Rengel Z. (1992) Role of calcium in aluminium toxicity. *New Phytologist*, **121**, 499–513.
Rengel Z. (1995a) Carbonic anhydrase activity in leaves of wheat genotypes differing in Zn efficiency. *Journal of Plant Physiology*, **147**, 216–226.
Rengel Z. (1995b) Sulfhydryl groups in root-cell plasma membranes of wheat genotypes differing in Zn efficiency. *Physiologia Plantarum*, **95**, 604–612.
Rengel Z. and Graham R.D. (1995a) Wheat genotypes differ in Zn efficiency when grown in chelate-buffered nutrient solution. I. Growth. *Plant and Soil*, **173**, 307–316
Rengel Z. and Graham R.D. (1995b) Wheat genotypes differ in Zn efficiency when grown in chelate-buffered nutrient solution. II. Nutrient uptake. *Plant and Soil*, **176**, 317–324.
Rengel Z. and Graham R.D. (1996) Uptake of zinc from chelate- buffered nutrient solutions by wheat genotypes differing in Zn efficiency. *Journal of Experimental Botany*, **47**, 217–226.
Rengel Z. and Hawkesford M.J. (1995) Biosynthesis of a 34–kDa polypeptide in root-cell plasma membrane is differentially regulated in two wheat genotypes differing in Zn efficiency. In: *10th Int. Workshop on Plant Membrane Biology*, Regensburg, Aug 1995, R44.
Romera F.J., Alcantara E. and De la Guardia M.D. (1992) Effects of bicarbonate, phosphate and high pH on the reducing capacity of Fe-deficient sunflower and cucumber plants. *Journal of Plant Nutrition*, **15**, 1519–1530.
Rufty Jr T.W., Raper C.D. and Jackoson W.A. (1982) Nitrate uptake, root and shoot growth, and ion balance during acclimation to root-zone acidity. *Botanical Gazzette*, **143**, 5–14.
Römheld V. (1991) The role of phytosiderophores in acquisition of iron and other micronutrients in graminaceous species: an ecological approach. *Plant and Soil*, **130**, 127–134.
Römheld V. and Marschner H. (1986) Mobilization of iron in the rhizosphere of different plant species. In *Advances in Plant Nutrition, Vol. 2*, edited by B. Tinker and A. Läuchli, pp. 155–204. New York: Praeger Scientific.
Römheld V. and Marschner H. (1990) Genotypic differences among gramineaceous species in release of phytosiderophores and uptake of iron phytosiderophores. *Plant and Soil*, **123**, 147–153.
Sakai H. and Tadano T. (1993) Characteristics of response of acid phosphatase secreted by the roots of several crops to various conditions in the growth media. *Soil Science and Plant Nutrition*, **39**, 437–444.
Sattelmacher B., Horst W.J. and Becker H.C. (1994) Factors that contribute to genetic variation for nutrient efficiency of crop plants. *Zeitschrift für Pflanzenerährung und Bodenk*, **157**, 215–224.
Sharma A.K., Srivastava P.C., Johri B.N. and Rathore V.S. (1992) Kinetics of zinc uptake by mycorrhizal (VAM) and non-mycorrhizal corn (*Zea mays* L.) roots. *Biology and Fertility of Soils*, **13**, 206–210.
Shelp B.J., Marentes E., Kitheka A.M. and Vivekanandan P. (1995) Boron mobility in plants. *Physiologia Plantarum*, **94**, 356–361.
Shuman L.M. (1991) Chemical forms of micronutrients in soils. In *Micronutrients in Agriculture*, edited by J.J. Mortvedt, F.R. Cox, L.M. Shuman and R.M. Welch, pp. 113–144. Madison, WI: Soil Science Society of America.
Smith S.E., Gianinazzi-Pearson V., Koide R. and Cairney J.W.G. (1994) Nutrient transport in mycorrhizas: structure, physiology and consequences for efficiency of the symbiosis. In *Management of Mycorrhizas in Agriculture, Horticulture and Forestry*, edited by A.D. Robson, L.K. Abbott and N. Malajczuk, pp. 103–114. Dordrecht: Kluwer Academic Press.
Smith S.E. and Smith F.A. (1990) Structure and function of the interfaces in biotrophic symbioses as they relate to nutrient transport. *New Phytologist*, **114**, 1–38.
Stephan U.W. and Grün M. (1989) Physiological disorders of the nicotianamine-auxotroph tomato mutant chloronerva at different levels of iron nutrition. II. Iron deficiency response and heavy metal metabolism. *Biochemie und Physiologie der Pflanzen*, **185**, 189–200.
Straker C.J. and Mitchell D.T. (1986) The activity and characterization of acid phosphatases in endomycorrhizal fungi of Ericaceae. *New Phytologist*, **104**, 243–256.
Strom L., Olsson T. and Tyler G. (1994) Differences between calcifuge and acidifuge plants in root exudation of low-molecular organic acids. *Plant and Soil*, **167**, 239–245
Tadano T., Ozawa K., Sakai H., Osaki M. and Matsui H. (1993) Secretion of acid phosphatase by the roots of crop plants under phosphorus-deficient conditions and some properties of the enzyme secreted by lupin roots. *Plant and Soil*, **156**, 95–98.

Tadano T. and Sakai H. (1991) Secretion of acid phosphatase by roots of several crops species under phosphorus-deficient conditions. *Soil Science and Plant Nutrition,* **37**,129–140.

Takagi S. (1976) Naturally-occurring iron-chelating compounds in oat and rice root washings. I. Activity measurements and preliminary characterization. *Soil Science and Plant Nutrition*, **22**, 423–433.

Tester M.A., Smith F.A. and Smith S.E. (1992) The role of ion channels in controlling solute exchange in mycorrhizal associations. In *Mycorrhizas in Ecosystems*, edited by D.J. Read, D.H. Lewis, A.H. Fitter and I.J. Alexander, pp. 348–351. Oxon UK: CAB International.

Thoms K. and Sattelmacher B. (1990) Influence of nitrate placement on morphology and physiology of maize (Zea mays) root systems. In *Plant Nutrition - Physiology and Applications*, edited by M L van Beusichem, pp. 29–32. Dordrecht: Kluwer Academic Press.

Thomson B.D., Robson A.D. and Abbott L.K. (1990) Mycorrhizas formed by *Gigaspora calospora* and *Glomus fasciculatum* on subterranean clover in relation to soluble carbohydrate concentration in roots. *New Phytologist*, **114**, 217–225.

Treeby M., Marschner H. and Römheld V. (1989) Mobilization of iron and other micronutrient cations from a calcareous soils by plant-borne, microbial, and synthetic metal chelators. *Plant and Soil*, **114**, 217–226.

Von Wiren N., Marschner H. and Römheld V. (1995b) Uptake kinetics of iron-phytosiderophores in two maize genotypes differing in iron efficiency. *Physiologia Plantarum*, **93**, 611–616.

Von Wiren N., Mori S., Marschner H. and Römheld V. (1994) Iron inefficiency in maize mutant ys1 (*Zea mays* L. cv yellow- stripe) is caused by a defect in uptake of iron phytosiderophores. *Plant Physiology*, **106**, 71–77.

Von Wiren N., Römheld V., Morel J. L., Guckert A. and Marschner H. (1993) Influence of microorganisms on iron acquisition in maize. *Soil Biology and Biochemistry*, **25**, 371–376.

Von Wiren N., Römheld V., Shioiri T. and Marschner H. (1995a) Competition between micro-organisms and roots of barley and sorghum for iron accumulated in the root apoplasm. *New Phytologist*, **130**, 511–521.

Walter A., Pich A., Scholz G., Marschner H. and Römheld V. (1995) Effects of iron nutritional status and time of day on concentrations of phytosiderophores and nicotianamine in different root and shoot zones of barley. *Journal of Plant Nutrition*, **18**, 1577–1593.

Walter A., Römheld V., Marschner H. and Mori S. (1994) Is the release of phytosiderophores in zinc-deficient wheat plants a response to impaired iron utilization? *Physiologia Plantarum*, **92**, 493–500.

Webb M.J. and Loneragan J. (1990) Zinc translocation to wheat roots and its implications for a phosphorus/zinc interaction in wheat plants. *Journal of Plant Nutrition*, **13**, 1499–1512.

Welch R. M. (1995) Micronutrient nutrition of plants. *CRC Critical Reviews in Plant Sciences,* **14**, 49–82.

Welch R.M. and Norvell W.A. (1993) Growth and nutrient uptake by barley (*Hordeum vulgare* L. cv Herta): Studies using an N-(2- hydroxyethyl) ethylenedinitrilotriacetic acid-buffered nutrient solution technique. *Plant Physiology*, **101**, 627–631.

Welch R. M., Webb M.J. and Loneragan J.F. (1982) Zinc in membrane function and its role in phosphorus toxicity. In *Plant Nutrition 1982, Proceedings of the Ninth International Plant Nutrition Colloquium*, edited by A. Scaife, pp. 710–715. Farnham House, Farnham Royal, UK: Commonwealth Agricultural Bureau.

Wellings N.P., Wearing A.H. and Thompson J.P. (1991) Vesicular- arbuscular mycorrhizae (VAM) improve phosphorus and zinc nutrition and growth of pigeonpea in a vertisol. *Australian Journal of Agricultural Research*, **42**, 835–845.

White P.F. and Robson A.D. (1989) Rhizosphere acidification and Fe^{3+} reduction in lupins and peas: iron deficiency in lupins is not due to a poor ability to reduce Fe^{3+}. *Plant and Soil*, **119**, 163–175.

Zhang C., Römheld V. and Marschner H. (1995) Retranslocation of iron from primary leaves of bean plants grown under iron deficiency. *Journal of Plant Physiology*, **146**, 268–272.

Zhang F.S., Römheld V. and Marschner H. (1989) Effect of zinc deficiency in wheat on the release of zinc and iron mobilizing root exudates. *Zeitschrift für Pflanzenernährung und Bodenkunde*, **152**, 205–210.

Zhang F.S., Römheld V. and Marschner H. (1991a) Diurnal rhythm of release of phytosiderophores and uptake rate of zinc in iron- deficient wheat. *Soil Science and Plant Nutrition*, **37**, 671–678.

Zhang F.S., Römheld V. and Marschner H. (1991b) Release of zinc mobilizing root exudates in different plant species as affected by zinc nutritional status. *Journal of Plant Nutrition*, **14**, 675–686.

11. MECHANISMS OF PLANT RESISTANCE TO TOXICITY OF ALUMINIUM AND HEAVY METALS

ZDENKO RENGEL

Soil Science and Plant Nutrition, Faculty of Agriculture, The University of Western Australia, Nedlands, Perth WA 6907, Australia

INTRODUCTION

While Al is the third most abundant element in the Earth's crust, heavy metals occur in comparatively small amounts under natural conditions. However, heavy metals can be introduced into the environment in larger amounts as pollutants and retained in soils in a largely irreversible manner (Tiller, 1989). In many acid soils throughout the world, Al is the strongest growth-limiting factor (Foy, 1988), possibly affecting up to 70% of the world's arable land that is potentially usable for food and biomass production (Haug and Caldwell, 1985, and references therein). As a consequence, literature on Al toxicity and resistance in plants has almost exclusively dealt with crop plants, while research on most heavy metal pollutants is mainly conducted on non-cultivated plants (e.g. re-vegetation of mine spoils and other polluted sites). The notable exception is Cd, a potential toxin in the human diet, toxicity of which has usually been investigated on crop plants, since at least 70% of Cd intake by humans comes from plant-derived foods (Wagner, 1993).

Solubilization of Al-containing minerals as well as heavy-metal-containing compounds is enhanced in acidic environments (Reddy *et al.*, 1995). Soil acidification results from some common agricultural practices (Bolan *et al.*, 1991) as well as from acid precipitation (Nouri and Reddy, 1995), making the problem of Al or heavy metal toxicity likely to increase in time and space.

Many heavy metals are essential for plants and animals (micronutrients: Cu, Zn, Fe, Mn, Mo, Ni, and Co) and become toxic only when a concentration limit is exceeded. Heavy metals that are important environmental polllutants include Cd, Pb, Cu, Zn, and Ni (Tiller, 1989), all belonging to the group of borderline (intermediate) metals according to Nieboer and Richardson (1980). In addition to those pollutants, this review will consider Al and Mn, which are frequently toxic under certain environmental conditions.

This chapter will deal with mechanisms of resistance to Al and heavy metals. Metal ion speciation, toxicity symptoms, and uptake of Al and heavy metals as a basis for understanding resistance mechanisms will also be covered. For other aspects of the toxicity of Al and heavy metals to crops and wild plants, the reader is referred to other recent reviews (Baker and Proctor, 1990; Reddy and Prasad, 1990; Steffens 1990; Cumming and Tomsett, 1992; Ernst *et al.*, 1992; Jackson and Alloway, 1992; Rengel, 1992a; Chaney, 1993; Macnair, 1993; Meharg, 1993;

Wagner, 1993; Godbold, 1994; Delhaize and Ryan, 1995; Horst, 1995; Kochian, 1995; Rengel *et al.*, 1995; Taylor, 1995).

TERMINOLOGY

The term 'heavy metals' is commonly used in literature to describe all metals with the atomic number >20, generally excluding the alkali metals and alkali earths (Tiller, 1989). Even though 'heavy' does not have any precise meaning with respect to either gravity or atomic weight, attempts to make classification of metals based on related atomic properties and solution chemistry (oxygen-seeking, N or S-seeking, and borderline, Nieboer and Richardson, 1980) did not seem to have taken hold.

When present in concentrations exceeding a certain threshold limit, metal ions become stress factors that can, at least in some genotypes, reduce vigor, inhibit growth and/or cause death. *Resistance* refers to a plant reaction which allows survival and reproduction in the stressful environment. Resistance can be achieved by *avoidance*, when protective mechanisms prevent uptake of a metal ion, and *tolerance*, which is conferred by mechanisms allowing survival of the internal stress arising from metal ions that were taken up (Baker, 1987). While metal ions are almost inevitably taken up given sufficiently long exposure, plants posses a range of exclusion mechanisms that represent either true avoidance, such as selective root growth into less contaminated soil horizons (Baker and Proctor, 1990; Marschner, 1991; Hairiah *et al.*, 1992, 1994; Turner and Dickinson, 1993), or detoxification of metal ions in the rhizosphere by complexation (Merckx *et al.*, 1986; Miyasaka *et al.*, 1991; Christiansen-Weniger *et al.*, 1992; Delhaize *et al.*, 1993b; Basu *et al.*, 1994b; Jones and Darrah, 1995; Pellet *et al.*, 1995; Ryan *et al.*, 1995a, b). The term 'resistance' should therefore be used in preference to 'tolerance' because it has broader meaning and includes avoidance as well as tolerance. The distinction between resistance and tolerance is especially important if a case of Al where it has clearly been shown that exclusion mechanisms, like avoidance (Hairiah *et al.*, 1992, 1993, 1994) and complexation (detoxification) of Al in the rhizosphere (e.g. Delhaize *et al.*, 1993b), may represent important components of plant survival under conditions of the Al stress.

IONIC SPECIATION

The most comprehensive work on the importance of ionic speciation of metal ions to their potential toxicity was done with Al, even though many questions remained unanswered (Kinraide, 1991; Rengel, 1992a, and reference therein). Among other toxic metals, speciation of Sc is poorly characterized, similarly to Ga, which however hydrolyzes more readily than Sc or Al (Reid *et al.*, 1996). Increasing pH causes a shift from predominantly Sc^{3+} at pH below 4.2 to almost exclusively $Sc(OH)^{2+}$ at pH 5–5.5. Lanthanum does not hydrolyze in acidic media (Kinraide *et al.*, 1992). Speciation of La and Ce is not affected by an increase in pH from 4.5 to 5.5, while a decrease in Cr^{3+} activity and precipitation of $Cr(OH)_3$ occurred with a rise in pH for one unit to 5.5 (Rengel, 1994).

Experimental approaches aimed at quantifying various ionic species of Al are plagued with numerous problems (for the review see Bertsch, 1990) because of the complex chemistry of Al whose chemical speciation in dilute solutions undergoes dramatical changes upon small changes in pH and is strongly dependent on ionic strength as well as the activity of other ions (Kinraide, 1991). The exact chemical speciation of dissolved Al remains elusive for the present experimental techniques, even though limited success was achieved by ^{27}Al NMR spectroscopy and ion chromatography (for reviews see Rengel, 1992a, 1996) as well as by size-exclusion chromatography (Kerven *et al.*, 1995). Researchers therefore used various computer programs to calculate the activities of individual ionic species in aqueous systems based on an assumed equilibrium condition and on the set of thermodynamic constants for the reactions under study (Rengel, 1992a; Parker *et al.*, 1995, and references therein). Computational speciation of complex Al-containing aqueous solutions yields, however, only rough estimates of activities of various Al complexes.

Difficulties inherent in ionic speciation of Al in solutions are augmented when speciation of relatively complex nutrient solutions in which plants are grown is attempted. There may be a significant difference between ionic composition of the bulk solution and the solution present in the root-cell apoplasm where formation of highly toxic polynuclear Al complexes (Parker *et al.*, 1989) may occur if high local concentration of Al and/or localized high pH create favorable conditions for Al polymerization (Rengel, 1992a). Therefore, not only the identity of Al complexes in contact with the root-cell plasma membrane and the time-course of their transfer into the cytosol, but the unequivocal confirmation of relative toxicities of various Al species, are all beyond the limits of the current experimental techniques.

TOXICITY SYMPTOMS

Sequence of Appearance

Al- and heavy metal-stressed plants in nature grow and develop in the environment with a constant presence of toxic ions, experiencing the cumulative toxicity damage. Different toxicity symptoms are expressed during different developmental stages, pointing to various levels and degrees of adaptation to the stressful environment (see Godbold, 1994). Generally, some symptoms of the metal toxicity appear after a short-term exposure (measurable within minutes or even seconds of exposure to a particular metal) followed by the long-term responses that are measured hours or days after imposition of stress. The long-term responses are not necessarily caused by the toxic metal directly but may rather be a consequence of metal-related impairment of numerous other biochemical and physiological processes.

Aluminium as well as heavy metal ions are toxic to plants at sufficiently high concentrations. Toxicity symptoms induced by Ga, In and La (Clarkson, 1965) and Sc (Yang *et al.*, 1989) were similar to those induced by Al. All toxic metal ions primarily restrict root growth (Macnair, 1983; Burke *et al.*, 1990; Godbold and Kettner, 1991; Taylor *et al.*, 1991, 1992; Ernst *et al.*, 1992; Kinraide *et al.*, 1992; Rengel, 1992a; Wheeler *et al.*, 1993; Brune *et al.*, 1994a; Macfie *et al.*, 1994; Golan-Goldhirsh *et al.*, 1995; Rengel and Graham, 1995), even though there are reports showing almost no effects of heavy metals on root growth for a range of plant species (Guo *et al.*, 1995) and even a stimulation of root growth by Al ions in a case of Al-accumulator species (Tsuji *et al.*, 1994, and references therein).

Growth, as inherently complex parameter, may not be the most suitable one to measure when searching for the primary lesion of metal toxicity because it is reasonable to assume that a number of physiological and biochemical processes in the plant cell have been affected prior to the resulting growth impedance. Earliest symptoms of Al and heavy metal toxicity in different plant systems occur between a few seconds and the first 30 min of exposure. These symptoms are reduction of net Ca^{2+} uptake, blockage of the plasma membrane-embedded Ca^{2+}-channels, a decrease in net uptake of Mg^{2+} and NO_3^-, reduction in K^+ efflux, accumulation of callose, exudation of malate, increased biosynthesis of phytochelatins, transcription of the heat-shock genes, and biosynthesis of the heat-shock proteins (for references see Rengel, 1992a; Neumann *et al.*, 1994; Delhaize and Ryan, 1995; Rengel *et al.*, 1995; Wollgiehn and Neumann, 1995).

Given sufficiently long exposure to Al or heavy metals, a myriad of different symptoms of toxicity appear on both roots and shoots (Foy *et al.*, 1978; Foy, 1988; Godbold and Hüttermann, 1988; Becerril *et al.*, 1989; Rengel, 1992a; Brune *et al.*, 1994b, 1995; Godbold, 1994; Macfie *et al.*, 1994; Fontes and Cox, 1995; Moustakas *et al.*, 1995; Sharma *et al.*, 1995). Toxicity symptoms appearing on shoots are likely to be a consequence of root growth inhibition rather than a direct effect of toxic metals on shoots (e.g. Godbold and Hütterman, 1988).

Interactions between Al and Heavy Metals

When present in the same growing medium (as might occur in the soil solution, especially with an increase in acidity), various toxic metal ions interact to produce a range of additive, multiplicative, synergistic and antagonistic responses (Taylor, 1989). Aluminium ions acted synergistically with ferrous iron to halt growth of the *Nicotiana tabacum* suspension-cultured cells insensitive to either ion separately (Yamamoto *et al.*, 1994). The same cells, however, were sensitive to Cu toxicity, the effect that was alleviated by an addition of Al ions. No interaction between Al and Cd was noted in that system (Yamamoto *et al.*, 1994). At the whole plant level, Al also alleviated Cu toxicity in *Triticum aestivum* (Hiatt *et al.*, 1963) and Cd toxicity in Cd-sensitive *Holcus lanatus* (McGrath *et al.*, 1980) but did not

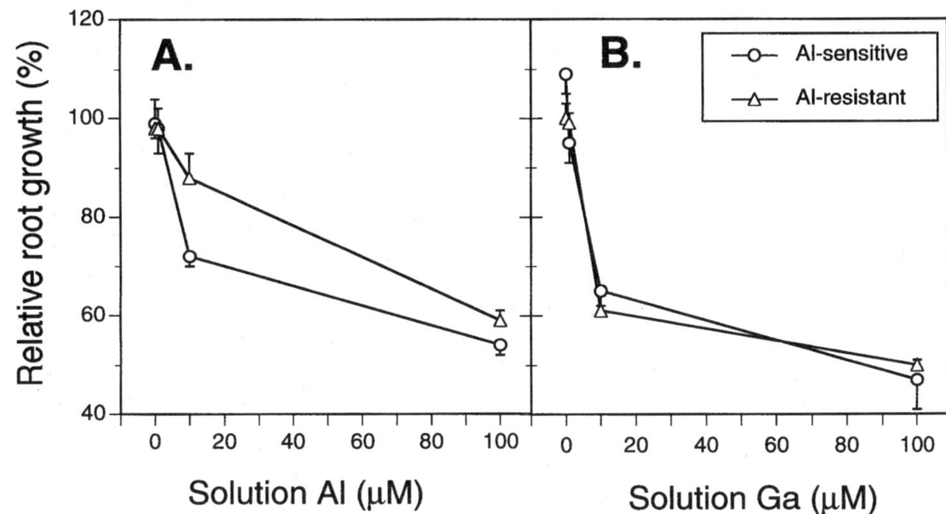

Figure 11.1. Effects of Al (A) and Ga (B) on root elongation of *Triticum aestivum* genotypes differing in resistance to Al. Al-resistant and Al-sensitive lines were near isogenic and were derived from the cross between Warigal (sensitive) and Waalt (resistant) genotypes. Plants were grown in 0.2 mM $CaCl_2$, pH 4.5, for 5 d, when Al or Ga treatment was imposed. Root elongation was measured at day 8.

influence response of *Triticum aestivum* to Ni toxicity (Taylor *et al.*, 1992). In contrast, Cd, Mn and Zn partly alleviated the Ni toxicity stress in the same system (Taylor *et al.*, 1992). Antagonistic effects between toxic metals may indicate similar site and/or mode of action, suggesting a possibility of the same resistance mechanisms providing protection from multiple metal toxicity stress. In contrast, various rhizotoxic metal ions affected *Triticum aestivum* genotypes differing in Al resistance to the same extent (Wheeler *et al.*, 1993; see also Figure 11.1), indicating that different resistance mechanisms may be operating for toxicity of different metals.

Net uptake of $^{45}Ca^{2+}$ by *Amaranthus tricolor* protoplasts was blocked by Al (Rengel and Elliott, 1992a, b; Rengel *et al.*, 1995) as well as by well-known Ca^{2+} channel blockers La and Gd (Rengel, 1994). However, the presence of Al ions alleviated La-caused inhibition and had no effect on Cd- or Ce-related inhibition of net $^{45}Ca^{2+}$ uptake (Rengel, 1994). These ion interactions were independent of ionic speciation effects.

Location of the Primary Toxic Lesion

A number of authors speculated on the internal (symplasmic) (Haug, 1984; Haug and Shi, 1991; Lazof *et al.*, 1994a) or external (apoplasmic) character of the primary lesion in the Al and heavy metal toxicity syndrome (e.g. Blamey *et al.*, 1990a; Rengel, 1992a, b; Meharg, 1993; Horst, 1995; Keltjens, 1995; Rengel *et al.*, 1995), but an unequivocal experimental proof for either one is still lacking.

The internal primary toxic lesion would either require uptake of metal ions across the root-cell plasma membrane or the perception of the signal of metal toxicity in the apoplasm and transduction of that signal to the internal cellular responsive elements (possibly via Ca^{2+} as a secondary messenger system, Bennett and Breen, 1991; Cumming and Tomsett, 1992; Rengel, 1992b; Rengel *et al.*, 1995; or via phosphoinositide-associated signal transduction, Haug *et al.*, 1994). Answering this question, and indeed the wider dilemma of the internal vs. external primary lesion of Al and heavy metal toxicity, hinges on answering the question whether uptake of these metals across the root-cell plasma membrane is a necessary prerequisite for the first symptoms of toxicity to occur.

With an increse in the solution pH, Sc toxicity to *Chara corallina* cells decreased together with cell wall binding of Sc, while its uptake across the plasma membrane increased (Reid *et al.*, 1996; see also Figure 11.2), indicating (i) $Sc(OH)^{2+}$, as a form that is readily transported into the cytoplasm, is not a toxic species because its activity increases with an increase in pH, (ii) Sc^{3+} is likely to be the toxic species because its activity in solution as well as binding to the cell wall increase with a decrease in pH, and (iii) binding of Sc^{3+} to the cell wall is likely to be the primary toxic lesion.

If the primary lession is cytoplasmic, the resistance should be conferred by the ability to exclude toxic metal from the symplasm, detoxify it upon eventual entry, or tolerate the presence of potentially toxic amounts of metal in the symplasm. If the primary lesion is apoplasmic, resistance mechanisms would need to exclude toxic ions from the cell wall and membrane structures and protect their physiological functions. An additional hypothesis has been put forward recently for Al: the primary lesion takes place at the membrane structures involved in net Ca^{2+} uptake at the root tip but consequences of this primary lesion are effected in the cytoplasm as disturbed cell Ca^{2+} homeostasis and reduced total Ca^{2+} concentration; differential tolerance to Al would rely on variation in the ability to maintain net Ca^{2+} uptake and cell Ca^{2+} homeostasis unaltered by Al, possibly due to exclusion of Al from the critical membrane sites (Rengel *et al.*, 1995).

Toxicity of metal ions is a complex phenomenon. It would be difficult to imagine that the myriad of various toxicity-related effects described to date (see Ernst *et al.*, 1992; Rengel, 1992a; Meharg, 1993; Taylor, 1995; Rengel, 1996) would stem from only one lesion. It appears likely that Al and heavy metals would cause a number of lesions in a time-dependent manner and that importance of a particular lesion may be related to growing conditions (ionic composition of the growing medium, temperature, plant species, genotype, age, etc.).

UPTAKE OF METAL IONS

Reported rates of uptake of highly charged rhizotoxic metal ions are fairly high (*cf.* Zhang *et al.*, 1989; Ernst *et al.*, 1992; Meharg, 1993; Lazof *et al.*, 1994a; Vazquez *et al.*, 1995) and generally similar to those expected for monovalent cations. However, it is to be expected that the negatively charged cell wall will accumulate large amounts of these cations because of negative potential difference (PD, approximately – 100 mV) between the Donnan phase and the solution bathing the root (Walker and Pitman, 1976). For a trivalent cation, even a conservative estimate of the PD of –60 mV would cause a 1000-fold accumulation in the cell wall in comparison with the solution (Reid *et al.*, 1996). Given that highly charged cations are notoriously difficult to desorb from the cell wall, there is a possibility that high rates of uptake of trivalent cations into plant

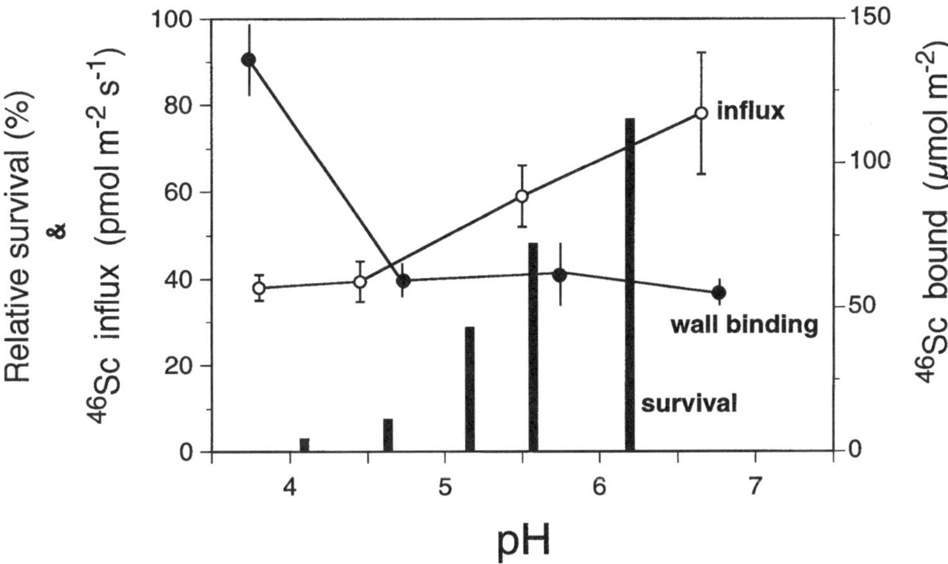

Figure 11.2. Influence of pH on influx of ^{46}Sc into internodal cells (○), binding of ^{46}Sc to isolated cell walls (•) of *Chara corallina*, and the survival of internodal cells (vertical bars) after 5 days in the Artificial Pond Water supplemented with 5 μM Sc (n=18 cells). Modified after Reid *et al.* (1996).

cells actually represent accumulation in the cell wall and not necessarily in the symplasm.

Using giant algal cells of *Chara corallina*, where physical separation of the cell wall and the cytoplasm can be achieved surgically, Reid *et al.* (1996) showed that the amounts of cell wall-bound Al, Sc and Ga, even after stringent and lengthy desorption procedures, exceed the amounts in the cytoplasm by 10 to 60 times. These results cast serious doubts on previously reported high rates of uptake of these metal ions by plant roots and higher plant cells due to inability to separate cell wall from the cytoplasm when one deals with comparatively small cells of higher plants or such complex tissues as plant roots.

Comparison of uptake rates of metal ions of different charge showed that divalent and monovalent cations, respectively, are taken up at rates two and three orders of magnitude greater than those for Al or Ga (Reid *et al.*, 1996; Rengel, 1996), when physical separation of the cell wall and the cytoplasm was performed prior to measurements. Scandium was taken up into the cytoplasm at rates similar to those of divalent cations, indicating that $Sc(OH)^{2+}$ might be the form that crosses the plasma membrane (Reid *et al.*, 1996, see also Figure 11.2).

Uptake of Aluminium

Aluminium, as a polyvalent cation under acidic conditions, binds strongly to fixed negative charges in the Donnan Free Space of the root-cell apoplasm (Rengel and Robinson, 1989a; Jentschke *et al.*, 1991; Eeckhaoudt *et al.*, 1992; Rengel, 1992a; Ownby, 1993; Horst, 1995; Keltjens, 1995). These negative charges are mostly free carboxyl groups of pectic material (Blamey *et al.*, 1993). As a consequence, no intracellular Al could be detected by the energy-dispersive X-ray microanalysis in a number of plant species after several hours of exposure to Al (Huett and Menary, 1980), and after 24 h of exposure in *Avena sativa* (Marienfeld and Stelzer, 1993) or *Triticum aestivum* roots (Ownby, 1993). Even after 10-d exposure, the intracellular amount of Al in *Avena sativa* roots was relatively small (Marienfeld *et al.*, 1995).

Kinetic studies showed biphasic uptake of Al by Triticum aestivum; an initial short, rapid, saturable phase was followed by a slower, linear phase (Pettersson and Strid, 1989; Zhang and Taylor, 1989, 1990; Archambault et al., 1996). The initial rapid phase may be accounted for by accumulation of easily exchangeable Al in the apoplasm because most of that Al pool was removed after a 30-min desorption of roots in citric acid (Zhang and Taylor, 1989). A linear phase of Al uptake may represent the transport of Al into the symplasm, based on the classical interpretation of the kinetic data (Pettersson and Strid, 1989; Zhang and Taylor, 1989), or metabolism-dependent binding of Al in the apoplasm (Zhang and Taylor, 1990). However, given that previous studies underestimated the contribution of apoplasmically-bound Al, which may account for as much as 99.99% of Al taken up (Taylor et al., 1996), and that complete desorption of Al from the apoplasm cannot be achieved by any of the desorption methods published thus far (Reid et al., 1996; Rengel, 1996), it is likely that the symplasmic component of Al uptake was overestimated. In other words, while Al inevitably enters the symplasm, the rate of uptake is slow and amounts crossing the plasma membrane are exceedingly small, putting more emphasis on the notion that the primary Al toxic lesion is extracellular.

Lazof *et al.* (1994a, b) have recently reported detection of symplasmic Al in *Glycine max* root apices exposed to Al for as little as 30 min. Such a claim would indicate that intracellular (symplasmic) primary lesion of Al is possible, if not likely (Kochian, 1995). The analysis of techniques used by Lazof and coworkers to desorb Al from the cell wall prior to employing the Secondary Ion Mass Spectrometry for determining Al in thin cryosections of root apices led to suggestions (Reid *et al.*, 1996; Rengel, 1996) that the claim of detection of the symplasmic Al is equivocal because it appears unlikely that complete desorption of the cell wall Al was achieved by 10 mM citrate during 30-min rinse, raising a possibility of contamination of the symplasm by apoplasmic Al and/or incorrect interpretation of the location of the Al signal. Further work is required to shed more light on this possible controversy.

Genotypic Differences in Metal Accumulation

The Al-sensitive genotypes may accumulate more (Horst *et al.*, 1983; Rincón and Gonzales, 1992; Delhaize *et al.*, 1993a; Lazof *et al.*, 1994b) or less Al (Aniol, 1983) in their root apices and also less Al in shoots (Massot *et al.*, 1992) when compared to the Al-resistant genotypes. Cell suspension of Al-sensitive *Phaseolus vulgaris* genotype accumulated more Al when compared to the Al-resistant one (McDonald-Stephens and Taylor, 1995). However, in a number of studies, no difference was found between the Al-sensitive and Al-resistant genotypes (Wallace *et al.*, 1982; Zhang and Taylor, 1989; Pettersson and Strid, 1989; Tice *et al.*, 1992; Archambault *et al.*, 1996). This apparent inconsistency in results might have been due to a number of reasons (see Rengel, 1996), the most important probably being (i) usage of longer root sections (i.e. Wallace *et al.*, 1982; Zhang and Taylor, 1989; Archambault *et al.*, 1996) or even whole roots (Aniol, 1983; Pettersson and Strid, 1989) compared to using a terminal 2-mm portion only (Rincón and Gonzales, 1992), and (ii) testing non-isogenic genotype lines which differ in number of traits, most of which may not be specifically associated with the differential resistance to Al. In that respect, the most respectable results should be those of Delhaize *et al.* (1993a) who used near-isogenic lines differing in Al resistance at a single locus. They concluded that Al accumulates in greater amounts in root tips of the Al-sensitive genotype of *Triticum aestivum*, indicating that at least one of the mechanisms of Al resistance (e.g. malate exudation into the rhizosphere) is connected with exclusion of Al from the root tips (Delhaize *et al.*, 1993a, b). These results have subsequently been confirmed in a range of *Triticum aestivum* genotypes (Basu *et al.*, 1994b; Ryan *et al.*, 1995b).

There may be a difference in apoplasmic binding of Al between genotypes differing in Al resistance (Rengel and Robinson, 1989a; Blamey *et al.*, 1990a; Keltjens, 1995). The Al-sensitive genotype of *Triticum aestivum* apparently had Al more tightly bound as only 30% of the total amount was released from the 0–2 mm root tips during 6 h in the Al-free nutrient solution compared with the Al-resistant genotype which released twice as much Al under the same conditions (Rincón and Gonzales, 1992). No difference between the two genotypes were found for the more distal root portions, which is in accordance with the findings that Al toxicity is brought about (Ryan *et al.*, 1993) and Al resistance mechanisms initiated (Ryan *et al.*, 1995a) when the root tips, but not more distal regions, were exposed to Al.

Differential capacity of genotypes differing in the level of Al resistance to bind Al in the root apoplasm may be a cellular property. McDonald-Stephens and Taylor (1995) found that cell suspensions of Al-resistant and Al-sensitive *Phaseolus vulgaris* genotypes bound initially the same amount of Al, but after desorption of cells in citric acid, cell suspensions of the Al-sensitive genotype retained a greater amount of Al than those of the Al-resistant genotype.

Since avoidance/exclusion resistance mechanisms appear to be more important for resistance to Al than to resistance to heavy metals where detoxification of metal ions in the symplasm confers resistance, the difference between heavy metal-sensitive and -resistant genotypes in uptake of metals is not so clear-cut. Differences in the rate of metal uptake are unlikely to be a basis of differential resistance to heavy metals. Zn-resistant genotypes of *Silene vulgaris* may take up Zn at rates similar to those of the sensitive genotypes (Harmens *et al.*, 1993a). Resistant genotypes sometimes contained larger amounts of metal in a detoxified form than sensitive genotypes (e.g. globular deposits of Zn-phytate are more abundant in Zn-resistant than Zn-sensitive ecotypes of *Deschampsia caespitosa*, Van Steveninck *et al.*, 1987) while sometimes no difference could be found (e.g. for *Betula* ssp., Denny and Wilkins, 1987).

RESISTANCE MECHANISMS

There are basically two different types of mechanisms that may confer resistance to toxicity of metal ions: (i) avoidance, involving various ways to prevent toxic ions to reach their target binding sites, and (ii) tolerance to metal ions which entered the symplasm. While under certain conditions one mechanism may be expressed to a high degree (e.g. malate extrusion in Al-resistant *Triticum aestivum* genotype, Delhaize *et al.*, 1993b; Basu *et al.*, 1994b; Ryan *et al.*, 1995a, b), conflicting reports about some mechanisms can frequently be found (at least when different species were considered, Blamey *et al.*, 1990b). Nevertheless, several generalizations are proposed here:

1) physiological mechanisms by which a certain level of metal resistance is achieved differ among genotypes of the same species as well as among different species,
2) more than one physiological and/or biochemical mechanism may be responsible for the level of metal resistance in a particular genotype,
3) expression of a certain mechanism of resistance as well as relative effectiveness of various mechanisms are highly dependent on environmental factors,
4) increased resistance to metals observed in one genotype compared with another involves additional mechanisms not present in the less resistant genotype, and
5) the effects of differential metal resistance mechanisms present in a particular genotype may at least be partly additive.

The possible interaction between the resistance mechanism that becomes effective first (such a mechanism may not be situated at the level of cytohistological organization that comes into contact with metal ions first) and other resistance mechanisms may render the mechanistic explanation of the observed phenotypic resistance difficult. It would be too simplistic to expect that the additive effect of two or more resistance mechanisms needs to conform to the simple mathematics of the additive interaction; all possible outcomes between around zero and 100% additivity may be expected depending on resistance mechanisms concerned and environmental conditions tested.

The possible mechanisms conferring resistance to toxic metal ions that will be discussed in the present review include (Figure 11.3):

1) Physical avoidance of contaminated areas,
2) Exudation of complexing agents into the rhizosphere,
3) Binding in the cell wall,
4) Efflux of metal ions from the symplasm,
5) Prevention of upward transport of metal ions into above-ground parts,
6) Complexation with various ligands in the symplasm,
7) Transport of metal-ligand complexes into vacuole,
8) Storage of metal ions in the vacuole by complexation with vacuolar ligands, and
9) Formation of metal-resistant enzymes to minimize the internal injury caused by toxicity.

The first two resistance mechanisms mentioned are true avoidance/exclusion mechanisms. The mechanisms numbered three to five can either be avoidance or tolerance types depending on the location of the primary toxic injury: if the primary lesion is symplasmic, all three mechanisms in this group would be avoidance-type mechanisms, but if the primary lesion is at the plasma membrane, transport of metals into the symplasm is expected to cause a damage already and efflux of metal ions back into the apoplasm will not represent a true avoidance/exclusion type of the resistance mechanism.

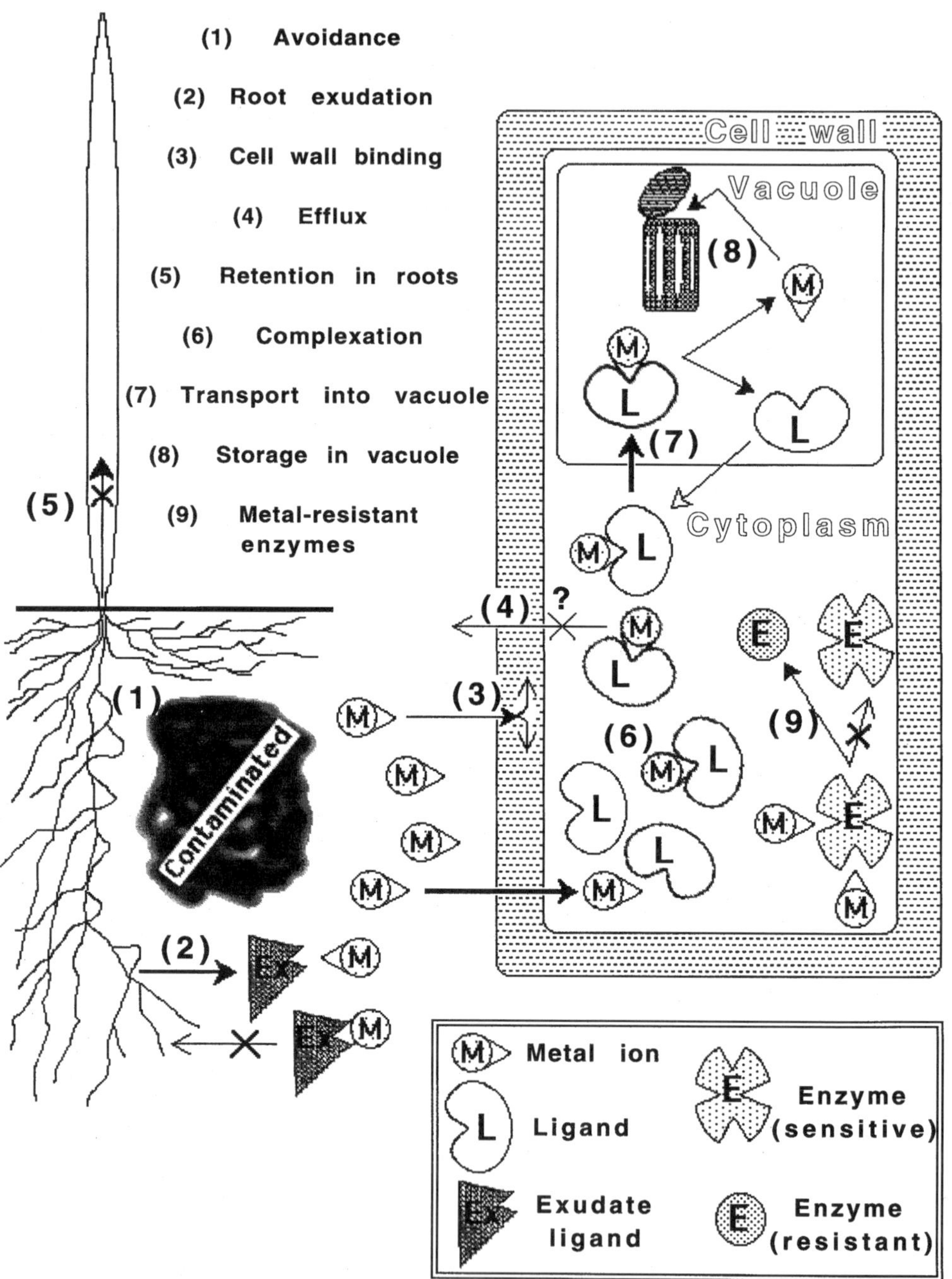

Figure 11.3. Various mechanisms of resistance to metal toxicity.

The last four resistance mechanisms listed above are true tolerance mechanisms allowing plants to minimize deleterious effects of metal ions which entered the symplasm. If metal ions are left uncomplexed and hence un-detoxified in the symplasm, the resulting toxicity may be due to (i) blocking of the essential biological functional groups of biomolecules, (ii) displacing the essential metal ion in biomolecules, and (iii) modifying the active conformation of biomolecules.

Physical Avoidance

It is considered common knowledge that plants growing in soils with spatial differences in concentrations of heavy metals and/or Al would avoid to grow roots into areas where these concentrations reached toxic levels (e.g. plants do not grow roots into acidic, Al-toxic subsoil horizons, Foy, 1988). An interesting, research in that area was done by Hairiah and co-workers who studied Al resistance of *Mucuna pruriens*, leguminous fodder and the cover crop in humid tropics. Split-root experiments in the nutrient solution showed the preferential root growth into the side not supplemented with toxic concentration of Al ions (Hairiah *et al.*, 1992), despite the fact that *Mucuna* roots were moderately resistant to Al in a homogeneous solution system (Hairiah *et al.*, 1990). For acid soils, it is likely that Al-induced P deficiency causes roots to grow preferentially toward localized P supply (Hairiah *et al.*, 1993), even though the avoidance reaction did not occur when N was supplied as the mixture of NO_3^- and NH_{4+} forms that would be expected to occur in acid soils (Hairiah *et al.*, 1994).

Root Exudation

pH changes in the rhizosphere

Since ionic speciation of most toxic metals is dependent on pH (Rengel, 1994; Parker *et al.*, 1995, Reid *et al.*, 1996), it is expected that an increase in the rhizosphere pH would cause a decrease in activity of highly charged monomeric cations (prevalently 3+) that are the most likely toxic species. Al-resistant genotypes of various crops increase pH of the solution they are grown in (Taylor and Foy, 1985; Foy, 1988; Rengel and Robinson, 1989b) or pH of the rhizosphere soil (Kirlew and Bouldin, 1987), thus decreasing solubility and changing ionic speciation of Al toward less toxic hydroxy-Al species, resulting in an exclusion of Al from the plant symplasm (Taylor, 1991).

An increase in the rhizosphere pH could arise from greater uptake of anions than cations (Taylor and Foy, 1985) or from decreased proton extrusion. However, most of the proton extrusion occurs by the more mature root parts causing the bulk solution pH changes (Kochian, 1995) as measured in all studies mentioned above. In contrast, the root tips (which are the site of Al toxicity and the only root parts sensitive to Al, Ryan *et al.*, 1993) generally show net H^+ influx rather than efflux (Ryan *et al.*, 1992).

Comparison of the two *Triticum aestivum* genotypes differing in resistance to Al yielded no difference in the rhizosphere pH as measured around the more mature root parts by the pH microelectrode (Miyasaka *et al.*, 1989). A number of other reports also failed to provide convincing evidence for changes in the rhizosphere pH being in a core of resistance to Al toxicity (Taylor, 1991, and references therein), prompting Kochian (1995) to conclude that this exclusion mechanism is not likely to be important mechanism of Al resistance. Differential changes of solution pH, if any, by different genotypes are probably a consequence, rather than a cause, of differential resistance to Al (Rengel and Robinson, 1989b).

Exudation of organic acids

Active efflux of Al-chelating substances into the rhizosphere may keep Al outside the symplasm. An increased organic acid exudation as an Al-resistance mechanism was shown in *Phaseolus vulgaris* (Miyasaka *et al.*, 1991), *Triticum aestivum* (Delhaize *et al.*, 1993b; Basu *et al.*, 1994b; Ryan *et al.*, 1995a, b), and *Zea mays* (Pellet *et al.*, 1995), but not in *Lotus* species (Blamey *et al.*, 1990b).

The importance of organic acid exudation as a mechanism of resistance to Al is obvious from the experiments where supplementation of the growing medium by organic acids alleviated Al toxicity in a range of plant species (Hue *et al.*, 1986; Ownby and Popham, 1989; Shuman *et al.*, 1991; Delhaize *et al.*, 1993b; Basu *et al.*, 1994b). Similar effects were observed *in vitro* where an addition of citrate restored growth of the cell lines of *Daucus carota* in Al-supplemented suspension (Ojima *et al.*, 1984). Adding malate to the medium containing plasma membrane vesicles isolated from *Pisum sativum* alleviated Al-caused inhibition of Mg^{2+}-dependent, K^+-stimulated ATPase activity (Matsumoto and Yamaya, 1986).

Ojima *et al.* (1984) were the first to show that exudation of citrate by plant cells might be a mechanism of resistance to Al toxicity. They showed that the putative Al-resistant cell line of *Daucus carota* releases more citrate into the medium than non-selected cell lines. However, more careful studies have showed that the Al-resistant line they used was not sensitive to ionic Al because it was selected in the presence of precipitated Al phosphate at the relatively high pH rather than in the presence of ionic Al at acidic pH (Koyama *et al.*, 1988; see also Taylor, 1995). Therefore, the putative Al-resistant cell line (Ojima *et al.*, 1984) is tolerant to Al-related low P availability rather than to toxic Al ions (Koyama *et al.*, 1988). It should be mentioned that citrate, known to increase P availability from Fe and Al phosphates, is also released by intact roots exposed to P deficiency (Lipton *et al.*, 1987; see also Pearson and Rengel, 1996).

The first report showing exudation of citrate into the rhizosphere of intact plants under Al toxicity stress was by Miyasaka *et al.* (1991). They reported that under Al stress the Al-resistant genotype of *Phaseolus vulgaris* exuded 70-fold more citrate than under control conditions and also 10 times more citrate than the Al-sensitive genotype grown with Al. The Al stress did not cause an increase in citrate release in the Al-sensitive genotype. Miyasaka *et al.* (1991) could not separate effects of toxic Al ions from low P availability as causes for increased citrate release under the Al stress because plants were grown in solutions containing both Al and P for 8 d. In a more recent study in which the Al-resistant genotype of *Zea mays*, grown in the presence of toxic Al ions, exuded more citrate into the medium than the Al-sensitive genotype (Pellet *et al.*, 1995), no P deficiency could have occurred because plants were pre-grown for 5 d in the solution containing no Al and sufficient P before being transferred into the simple-salt treatment solution with Al and without P for 20 h. Similarly, the absence of inorganic P in the growing medium could not elicit malate exudation to simulate the effect of the presence of toxic Al ions (Delhaize *et al.*, 1993b). So, it can be concluded that exudation of citrate, at least in *Zea mays* and *Triticum aestivum*, represents an important and specific component of resistance to Al.

The best documented example yet of exudation of organic acids into the rhizosphere of plants grown under the Al stress is the exudation of malate by *Triticum aestivum* (Delhaize *et al.*, 1993b; Basu *et al.*, 1994b; Ryan *et al.*, 1995a, b). Such exudation of malate under Al stress was accompanied by efflux of K^+ to balance charges (Ryan *et al.*, 1995a). Malate efflux was stimulated by Al exposure only in the Al-resistant genotypes (Delhaize *et al.*, 1993b; Rayan *et al.*, 1995a, b); the efflux was accompanied by a continuous *de novo* biosynthesis (Basu *et al.*, 1994a) allowing for the malate content in root apices to be the same in both Al-resistant and Al-sensitive genotypes grown with or without Al (Delhaize *et al.*, 1993b). So, it appears it is efflux of malate across the plasma membrane of root tip cells, rather than biosynthesis of malate, which is differentially influenced by Al in genotypes differing in Al resistance (Ryan *et al.*, 1995a). A hypothesis which remains to be proven is that Al causes greater efflux of malate by stimulating anion channels in the plasma membrane of root tips of the Al-resistant genotypes (Ryan *et al.*, 1995a). A possibility cannot be excluded that such an effect is not caused by Al ions directly, but rather indirectly via effects on some other cellular signalling pathway (*cf.* Rengel, 1992b; Haug *et al.*, 1994; Rengel *et al.*, 1995). Similarly, it is possible that the primary effect of Al is to stimulate malate biosynthesis in the Al-resistant more than in the Al-sensitive genotype, with an increased malate exudation by the Al-resistant genotype being only a consequence of increased intracellular malate concentration.

Increased exudation of malate by root apices of the Al-resistant *Triticum aestivum* genotype is a relatively quick response. It could be detected after 30 min (Delhaize *et al.*, 1993b) or even as early as after 15 min of Al exposure (Ryan *et al.*, 1995a). In contrast, the earliest reported differential inhibition of root elongation caused by Al in different genotypes is after 1 h, even though other responses may be faster (see Rengel, 1992a for references). It appears that an increase in malate exudation occurring in the Al-resistant genotype upon exposure to Al is sufficiently fast to account for differential resistance to Al between different genotypes.

Using a near-isogenic pair of *Triticum aestivum* genotypes differing in Al resistance at a single locus, an increased exudation of malate was correlated well with the Al tolerance trait in a population of seedlings segregating for Al resistance (Delhaize *et al.*, 1993b). In addition, Ryan *et al.* (1995b) compared 36 *Triticum aestivum* genotypes and found a significant positive relationship between the amount of malate extruded under Al stress and the level of resistance to Al toxicity. Exudation of organic acids by the root tips of Al-exposed seedlings, as an important aspect of resistance to Al toxicity, is also confirmed by the fact that the same Al activity that stimulated maximum citrate exudation by *Zea mays* root apices elicited the maximum difference in sensitivity between the Al-resistant and the Al-sensitive genotypes (Pellet *et al.*, 1995).

Aluminium ions are toxic only to the root tip cells; exposure of more mature root parts to Al does not result in Al toxicity (Ryan *et al.*, 1993). It is therefore reasonable to expect that Al resistance mechanisms either operate in the root tip, or are at least aimed at protecting the root tip from the Al injury (Kochian, 1995). Exudation of malate by *Triticum aestivum* (Delhaize *et al.*, 1993b; Ryan *et al.*, 1995a, b) and citrate by *Zea mays* root tips (Pellet *et al.*, 1995) is confined to the root tip. Calculations showed that the concentration of exuded organic acids at the root tip might be sufficiently high to complex and therefore detoxify Al ions around the root tips, especially if mucilage that surrounds the root tip and increases the thickness of the unstirred layer at the root tip surface is considered (Pellet *et al.*, 1995; Ryan *et al.*, 1995b). Such a protection would be required for a relatively short period of time during which root tip cells would mature and become less sensitive to Al (Basu *et al.*, 1994b). Therefore, exudation of a relatively small amount of organic acids, concentrated in a small soil volume at the root tip, is energetically effective means of overcoming inhibition of growth due to Al toxicity.

It is indicative that Al-related increase in exudation of organic acids by root tips cannot be mimicked by exposure to other toxic metals. Rhizotoxic concentrations of La and ferric iron (Delhaize *et al.*, 1993b) as well as Ga, Sc, In, La, Mn, Zn, and highly toxic Al_{13} polymeric cation with 7+ charge (Ryan *et al.*, 1995a) failed to mimick increased malate exudation by *Triticum aestivum* root tips as achieved by Al. In accordance with such a result, various pairs of *Triticum aestivum* genotypes showing differential resistance to Al were affected by La to the same extent (Parker *et al.*, 1988; Kinraide *et al.*, 1992; Delhaize *et al.*, 1993b), indicating that resistance mechanisms other than increased malate exudation might operate in a case of La toxicity. Moreover, it is apparent that different resistance mechanisms might operate in plants to protect them from toxicity of monomeric Al as compared to polymeric Al_{13} cations (see Ryan *et al.*, 1995a).

Exudation of phosphate

In addition to organic acids, exudation of other Al-chelating compounds into the rhizosphere (e.g. phosphate) might potentially have a role in Al resistance. The putative phosphate efflux was induced by exposure of *Beta vulgaris* (Lindberg, 1990) or *Zea mays* roots (Pellet *et al.*, 1995) to toxic levels of Al ions. It was suggested that such phosphate efflux may be energy-dependent, especially in the

Al-resistant genotype (Lindberg, 1990), which indicated an involvement in resistance mechanisms.

Phosphate ions effluxed into the apoplasm and/or rhizosphere could bind Al into various non-toxic Al-phosphate complexes. Using photoelectron X-ray spectroscopy, Millard *et al.* (1990) showed that more Al-phosphate accumulates on the root surface of the Al-resistant compared to the Al-sensitive *Hordeum vulgare* genotype, apparently making a protective barrier to minimize Al penetration into the root. Further work in that area is warranted.

Apoplasm as a Site of Resistance Mechanisms

Binding in the cell wall

As most toxic metals are polyvalent cations under acidic conditions, they effectively screen negative charges present in the Donnan Free Space of the root apoplasm (Clarkson and Sanderson, 1971; Rengel and Robinson, 1989a; Ernst *et al.*, 1992; Rengel, 1992a). These negative charges are mainly free carboxyl groups of pectic substances, which are mostly polymers of galacturonic acid (Blamey *et al.*, 1993). Affinity of polygalacturonic acids for metal cations decreases in the order Al>Pb>Cr>Cu>Zn≈Mn (Ernst *et al.*, 1992; Rengel, 1992a; Wang *et al.*, 1992a).

Differences in the methylation of the polygalacturonic acid component of pectin account for differences in the total number of negative exchange sites (root cation exchange capacity, CEC) available for binding metal cations (Knight *et al.*, 1961). It was suggested that lower CEC (i.e. relatively more strongly methylated pectines) would serve a protective role in the Al stress because higher activities of Al ions would be needed to precipitate highly methylated pectines (Blamey *et al.*, 1990a, 1993). However, consideration should be given to the argument (Kinzel, 1982) that the ionic interactions between pectic substances and, at least, Al ions are unlikely to happen *in vivo* because, generally, Al is present in an ionic form under acidic conditions while pectines are ions under neutral conditions.

The cell wall exchange properties may influence availability of metal ions for uptake across the plasma membrane, their diffusion rates in the apoplasm, the chemical and electrical environment of the plasma membrane and its constituents, and function of the cell wall enzymes (Allan and Jarrel, 1989). Differences in the electrical and chemical properties of the cell wall constituents may therefore influence selective uptake and differential resistance to toxicity of metal cations. A relatively lower ratio of COO^- to COOH groups in the root cell walls of the Mn-resistant *Nicotiana tabacum* genotype compared to the Mn-sensitive one was used to explain lower accumulation of Mn in roots and greater resistance to Mn toxicity in that genotype (Wang *et al.*, 1992a).

It has repeatedly been suggested in earlier literature that one of the mechanisms of differential Al resistance among genotypes of various plant species is dependent on root CEC (lower root CEC was associated with more resistant genotypes) (Vose and Randall, 1962; Mugwira and Elgawhary, 1979; Rengel and

Robinson, 1989b; Blamey *et al.*, 1990a). However, some studies found the opposite relationship (Allan *et al.*, 1990) or, at best, no relationship between root CEC and Al resistance in several species (Wagatsuma and Ezoe, 1985).

The negative sites in the cell wall would be relatively quickly saturated by metal ions, and continuous biosynthesis of the cell wall (if root growth is not inhibited) would be needed to provide new binding sites for metal immobilization. Dynamics of these processes have yet to be thoroughly studied in *vivo*. Other than that, there are two main obstacles for accepting lower CEC in the apoplasm as an important mechanism of resistance to Al and heavy metals (apart from concerns about methodological merit of techniques used to determine root CEC as outlined by Kochian, 1995): the first one is that an increased binding capacity of the cell wall may not decrease activity of free metal ions if the soil solution, the cell wall solution, and the cell wall matrix are at equilibrium (Ernst *et al.*, 1992), and the second is more specific for Al toxicity where the important site of Al toxicity is the root apex, while root CEC was always measured on the whole roots.

A specific role of the cell wall in the Al and heavy metal resistance cannot be excluded. Cell wall-bound phosphatases in metal-resistant populations may form less stable complexes with toxic metals (and therefore retain a greater activity) compared with sensitive populations (Cox and Thurman, 1978). It should also be borne in mind that the inner layers of the cell wall are tightly apressed against the plasma membrane, and that not only the chemical but also mechanical environment (see discussion on load-bearing bonds, Rengel, 1992a) in which the plasma membrane as a whole, and transport proteins and enzymes embedded in it in particular, operate may be significantly influenced by the amount of metal ions bound in the apoplasm.

Role of the plasma membrane in metal toxicity

Rhizotoxicity of Al and other minerals can be interpreted based on the membrane-surface electrical potential (Kinraide *et al.*, 1992, 1994; Kinraide, 1994), pointing to an extracellular (apoplasmic) primary toxic lesion. Negatively charged membrane surfaces attract and accumulate cations proportionately to their cation charge, inducing toxicity. A decrease in the membrane-surface electrical potential by charge screening or binding of non-toxic cations decreases toxicity of Al and other rhizotoxic mineral ions (Kinraide *et al.*, 1992).

The Gouy-Chapman-Stern model was successfuly used to explain various features of metal toxicity (Kinraide, 1994). The model relates the charge density and membrane-surface electrical potential to activities of toxic ions in a bulk solution, taking into account changes in the charge density brought about by ion binding at the membrane surface (Kinraide, 1994). Differential cell surface charge may therefore be the basis of differential genotypic sensitivity to metal ions (Kinraide *et al.*, 1992). However, pairs of *Triticum aestivum* genotypes differing in Al resistance did not differ in resistance to La (Kinraide *et al.*, 1992), Ga (Figure 11.1) and to a range of other heavy metal ions (Wheeler *et al.*, 1993), indicating that mechanisms of resistance to toxic metal ions may not be related to the effects on membrane surface charges.

Role of ion channels in resistance to Al

Since it was found that Al inhibits net Ca^{2+} flux more in the Al-sensitive than in the Al-resistant *Triticum aestivum* genotype (Huang *et al.*, 1992, 1993, 1995) and that Al ions block the plasma membrane Ca^{2+} channels by binding to the verapamil-specific channel-receptor site as well as by interfering with the action of GTP-proteins that are involved in regulation of transmembrane Ca^{2+} fluxes (Rengel and Elliott, 1992a, b; Rengel, 1994; Piñeros and Tester, 1995; Rengel *et al.*, 1995), it was hypothesized that differential resistance to Al might be due to differential ability of Ca^{2+}-channel proteins from genotypes differing in Al resistance to withstand presence of Al ions without suffering blockage (Rengel, 1992a, b). Recent reports, however, indicate that this might not be the case because Al blocks equally strongly Ca^{2+} channels in the plasma membrane vesicles isolated from roots of Al-resistant and Al-sensitive *Triticum aestivum* genotypes (Sasaki *et al.*, 1994b; Huang *et al.*, 1995). However, in these two studies, the plasma membrane vesicles were isolated from the whole roots (the size of the root segments used was not reported in Huang *et al.*, 1995, but according to a similar study /see Huang *et al.*, 1994/ it was likely to be whole roots). Since a strong case was established that the likely mechanisms of Al resistance reside in the root part which seems to perceive the Al toxicity stress (the top few millimeteres of the root, Kochian, 1995; Ryan *et al.*, 1995a), it appears necessary to test the effect of Al on Ca^{2+} channels isolated from the plasma membrane of the root apical cells of genotypes differing in resistance to Al before definite conclusion is made about the role of Ca^{2+} channels in differential Al resistance.

As for responses of other plasma membrane-embedded ion channels, Al decreased K^+ efflux from the root cap cells of Al-resistant *Phaseolus vulgaris* genotype causing membrane depolarization to a greater extent than in Al-sensitive genotype (Olivetti *et al.*, 1995). Similar differential effects of Al on K^+ efflux were observed in *Triticum aestivum* genotypes differing in Al resistance, even though whole roots rather than just root cap cells were used (Sasaki *et al.*, 1994a, 1995). In contrast, small K^+ efflux detected at the *Triticum aestivum* root tips by ion-sensitive microelectrodes was not influenced by Al (Ryan *et al.*, 1992). Further work is necessary to elucidate the role of transmembrane K^+ fluxes in initiation of Al toxicity and in differential genotypic resistance to Al.

Efflux of Metal Ions from the Symplasm

An active mechanism (exclusion or efflux of Al) to keep Al outside the symplasm has been suggested (Zhang and Taylor, 1989; Lindberg, 1990). Given a relatively high pH in the cytoplasm (ranging from about 6.0 to 7.5, Guern *et al.*, 1991, even though peripheral layers might be more acidic, Ross, 1992) and the abundance of cytoplasmic ligands with a high affinity for Al (Rengel, 1992a), it should be considered that (i) majority of Al would be complexed into relatively stable and cumbersomely large complexes which would be unlikely candidates for efflux across the plasma membrane, and (ii) that free Al^{3+} ion, presumably the most toxic Al species (Kinraide, 1991), which would potentially use some common, non-specific cation pathway for efflux, would be present in low, sub-nanomolar concentrations (Martin, 1988). No pathway for extrusion of intracellular Al has been described yet, casting doubts on Al efflux as a proposed mechanism of resistance and raising a possibility of Al accumulation in cells over prolonged time.

Significant efflux of Sc was found from the *Chara corallina* cells exposed to toxic concentrations of that metal. Efflux was correlated with the activity of the divalent $Sc(OH)^{2+}$ form; the trivalent Sc would be present at extremely low concentrations in the cytoplasm due to relatively high pH (Reid *et al.*, 1996).

Compartmentation of Metals at the Whole Plant Level

Plants growing in environments containing toxic levels of metal ions inevitably accumulate these ions over time because no avoidance/exclusion mechanism is perfect. While there is an obvious advantage in sequestering these toxic metals into plant parts where the least damage due to their accumulation is likely to occur (frequently that means retaining metals in roots and/or older leaves soon to be shed to minimize exposure of sensitive, photosynthetically-active above-ground tissues, Wolterbeek *et al.*, 1988; Ernst *et al.*, 1992; Rengel, 1992a; Harmens *et al.*, 1993a; Wheeler and Power, 1995; Rengel, 1996), there is no evidence that compartmentation mechanisms at the whole plant level are more than just a consequence of resistance mechanisms operating at the cellular level (like storage in root-cell vacuoles, which would reduce transport of metals to shoots).

Metallophytes and metal-resistance genotypes frequently hyperaccumulate metals in leaves rather than in roots (Rengel, 1992a; Harmens *et al.*, 1993b; Vazques *et al.*, 1995). Metals are complexed into non-toxic forms and stored in vacuoles of leaf cells.

Recent experiments have shown that there is a genetic variability in accumulation of Cd in kernels of various *Helianthus annuus* genotypes (Li *et al.*, 1995a, b), possibly involving differences in uptake, sequestration and/or transport of Cd toward the kernels. Breeding for low-Cd kernels was judged feasible (Li *et al.*, 1995b). It remains to be seen whether a specific genetic basis of differences in heavy metal compartmentation at the whole plant level can be found in other plant species under conditions of heavy metal toxicity.

Internal Complexation

Phytochelatins

Higher plants, algae and fungi exposed to the heavy-metal toxicity stress synthesize non-protein thiols with the general structure (γ-glutamyl-cysteinyl)$_n$-glycine, where n=2–11. These compounds have been named phytochelatins, cadystins, γ-glutamyl peptides, or $(\gamma\text{-EC})_nG$'s (Reddy and Prasad, 1990; Steffens, 1990; Ernst *et al.*, 1992). Here, they will be called phytochelatins (PCs) for simplification, even though recent research has revealed not only new forms of metal-binding compounds in some plant species (like hydroxymethyl-phytochelatins or $(\gamma\text{-Glu-Cys})_n$-Ser in some *Poaceae*, Klapheck *et al.*, 1994), but also different forms $[(\gamma\text{-EC})_n$, $(\gamma\text{-EC})_nE$ and $(\gamma\text{-EC})_nG]$ in the same root system (Meuwly *et al.*, 1995; Rauser and Meuwly, 1995).

PCs can bind heavy metals, thereby decreasing the activity of free metal ions in cytosol. The affinity of PCs for heavy metals differs: for Cu it is independent of *n*, while the affinity for Cd increases with an increase in *n* (Matsumoto *et al.*, 1990). The heavy metal-PC complex may be transported across the tonoplast into the vacuole (Vögeli-Lange and Wagner, 1990) and may dissociate there, indicating that the final storage form of heavy metals is not in the complex with PCs (De Knecht *et al.*, 1995) but probably as an organic acid complex (Wang *et al.*, 1991; 1992b).

Biosynthesis of PCs is catalyzed by phytochelatin synthase, an enzyme that is (i) constitutively expressed in higher plant cells, and (ii) regulated by cytosolic activity of free heavy metal ions (Löffler *et al.*, 1989). Roots are at least partly autonomous in synthesizing glutathione required for PCs (Rüegsegger and Brunold, 1992, De Knecht *et al.*, 1995).

Despite the early claims that increased resistance to Cd relied on an increased synthesis of PCs (Steffens *et al.*, 1986), it appears that differential production of PCs is a consequence, rather than a cause, of differential heavy metal resistance (Ernst *et al.*, 1992; Harmens *et al.*, 1993b; Brune *et al.*, 1994a). In addition, synthesis of PCs can be induced by any metal, both in resistant and non-resistant genotypes, while examples of multiple resistances and co-resistance to increased levels of metals to which studied plants were not exposed are rare (Watmough and Dickinson, 1995, and references therein). Moreover, Cd-resistant *Silene vulgaris* plants produce three times less PCs than Cd-sensitive ones when challenged with Cd toxicity (De Knecht *et al.*, 1995), confirming that an increased accumulation of PCs is unlikely to be a cause of increased resistance to heavy metals.

Organic acids

Organic acids have frequently been implied in a true tolerance type of resistance to heavy metals and Al in the symplasm. Organic acids were present in greater amounts in the Al-resistant than the Al-sensitive genotypes of *Pisum sativum* (Klimashevskii and Chernysheva, 1980), *Sorghum bicolor* (Cambraia *et al.*, 1983), *Zea mays* (Suhayda and Haug, 1986), and *Hordeum vulgare* (Foy *et al.*, 1987). In *Phaseolus vulgaris*, the Al stress caused an overall decline in the organic acid content but the Al-resistant genotype still maintained higher concentrations than the Al-sensitive genotype (Lee and Foy, 1986). Different results were obtained for *Triticum aestivum* (Foy *et al.*, 1990; Delhaize *et al.*, 1993b) and *Zea mays* (Pellet *et al.*, 1995) where similar concentrations of organic acids were found in roots of genotypes differing in Al resistance upon exposure to the Al stress; however, release of organic acids into the medium was higher in the Al-resistant compared to the Al-sensitive genotypes (see the section on *Exudation of organic acids*).

Supplementing organic acids to the cell-free medium *in vitro* prevented Al-caused conformational changes to calmodulin, restoring its regulatory function and thus activity of specific calmodulin-dependent enzymes (Suhayda and Haug, 1984, 1985, 1986). Alleviating action decreased in the order: citric > oxalic > malic > tartaric (Suhayda and Haug, 1984; 1985). While the importance of Al-calmodulin interaction as the primary lesion in Al toxicity as suggested by Haug and coworkers (Siegel and Haug, 1983; Haug, 1984; Suhayda and Haug, 1984, 1985) is questionable (Rengel, 1992a; Kochian, 1995, and references therein), these results indicate that organic acids present in plant cells have a potential to counteract or prevent deleterious effects caused by Al ions in the symplasm.

Increased concentration of citrate in roots correlated well with increased resistance to Zn in *Deschampsia caespitosa* (Godbold *et al.*, 1984). Increased leaf concentration of citrate was related to greater accumulation of Ni in 17 Ni-accumulator species (Lee *et al.*, 1978), and elevated leaf concentration of malate was implied in increased resistance to Zn toxicity in *Silene cucubalis, Rumex acetosa, Agrostis tenuis and Thlaspi alpestre* (Mathys, 1980). Increased concentrations of vacuolar citrate may be important in binding and immobilizing excess Zn in the vacuole during the Zn toxicity stress (Wang *et al.*, 1991, 1992b). In contrast, in roots of *Triticum aestivum* exposed to Mn toxicity, concentrations of citrate and malate were either higher in Mn-sensitive than Mn-resistant genotypes (Burke *et al.*, 1990) or were not different between the two classes of genotypes (Macfie *et al.*, 1994). Moreover, Mn toxicity caused an increase in concentrations of aconitate, α-ketoglutarate and succinate in leaves of Mn-sensitive *Triticum aestivum* genotype only (Macfie *et al.*, 1994). Similar results were reported for *Triticum aestivum* grown under the Al toxicity stress (Scott *et al.*, 1991).

Since Al and heavy metals influence a number of metabolic pathways, both increases and decreases in synthesis of particular organic acids can be observed, as unused substrates of a disabled pathway are diverted to other, non-disabled pathways (Macfie *et al.*, 1994). Given that biosynthetic pathways for all organic acids share common precursors, it is more important to study dynamics of the distribution of these precursors in various organic acid pools than the final size of the pool; moreover, measuring only one or two organic acids may provide information of doubtful validity. In that respect, study by Macfie *et al.* (1994) should be taken as an example of a conclusive study that showed increased biosynthesis of various organic acids as a response to Mn toxicity, but no involvement of organic acids in differential resistance to Mn toxicity.

Polypeptides

Biosynthesis of the heat-shock proteins (HSPs) is increased under heavy-metal stress (Czarnecka *et al.*, 1984; Edelman *et al.*, 1988; Delhaize *et al.*, 1989); HSPs protect proteins and the cellular membranes against irreversible damage caused by the stress and are therefore important components of the heavy metal resistance (Neumann *et al.*, 1994). Increased biosynthesis of HSPs and their accumulation in the cytoplasm and the nuclei under heavy metal stress occurred in *Lycopersicon peruvianum* cells without measurable effects on biosynthesis of the house-keeping proteins, a finding that is in contrast with the situation under the heat shock (Wollgiehn and Neumann, 1995).

While there appears to be some similarity in the mode of action of the heat stress and heavy metal toxicity (Czarnecka *et al.*, 1984; Edelman *et al.*, 1988; Delhaize *et al.*, 1989; Neumann *et al.*, 1994; Wolgiehn and Neumann, 1995), Al stress has a different physiological basis because no induction of hsp genes coding for HSPs (Richards and Gardner, 1994) or biosynthesis of HSPs was found under conditions of Al toxicity in *Triticum aestivum* (Picton *et al.*, 1991). Early reports to the contrary (Czarnecka *et al.*, 1984) should be viewed with caution because Al was applied to *Glycine max* seedlings at pH 6.0, indicating that factors other than Al toxicity might have influenced results. More plant species, however, would have to be tested before a definite conclusion about the relationship between the heat shock response and Al toxicity is made.

The induction of the hsp gene transcription in *Lycopersicon peruvianum* cells was rapid, occurring after only 1 min of the either heat shock or heavy metal stress (Wolgiehn and Neumann, 1995). Maximum rates were achieved after only 2 to 5 min. Given that the transcription of ribosomal RNA (rRNA) in *Lycopersicon peruvianum* cells was not deleteriously affected by heavy metal stress (Wollgiehn and Neumann, 1995), the ribosomes and the protein-synthesizing cell system remained largely unaffected and capable of translating the hsp gene transcripts into the HSPs.

In some experimental systems, exposure to Al (Liu and Jiang, 1993) or heavy metals (Liu *et al.*, 1994; Gabara *et al.*, 1995) decreased the size or changed shape of nucleoli, the organelles in which ribosomal DNA is transcribed into rRNA, indicating reduced production of rRNA (e.g. in root tip cells of *Allium sativum*, Liu and Jiang, 1993; or *Allium cepa*, Liu *et al.*, 1994; and in cortex cells of *Pisum sativum*, Gabara *et al.*, 1995). Furter work is required to clarify effects of Al and heavy metals on functioning of nucleoli and transcription of rDNA into rRNA under metal toxicity because these effects may impinge directly on protein biosynthesis.

Exposure to Al or heavy metals can change the pattern of polypeptide biosynthesis. While excess Zn in the rooting medium did not result in changes in polypeptide profile in mesophyll and epidermis protoplasts, mesophyll vacuoles and chloroplasts of *Hordeum vulgare*, there was an increased accumulation of several apoplasmic polypeptides with the molecular mass between 15 and 35 kDa (Brune *et al.*, 1994b). Apoplasmic polypeptides are likely to represent those loosely bound to the cell wall and the plasma membrane. Further work is required to ascertain whether increased accumulation of apoplasmic polypeptides is due to their increased biosynthesis or decreased degradation, but preliminary experiments indicate that it is likely to be increased biosynthesis (Brune *et al.*, 1994b). No information exists as yet on the function of these apoplasmic poplypeptides whose accumulation is influenced by excess Zn nor on their possible involvement in resistance to Zn toxicity.

Exposure of plants to Al may induce or depress biosynthesis of specific polypeptides (Aniol, 1984; Clouse *et al.*, 1991; Delhaize *et al.*, 1991; Ownby and Hruschka, 1991; Picton *et al.*, 1991; Rincón and Gonzales, 1991; Basu *et al.*, 1994a). It is indicative that all these reports were on *Triticum aestivum*; reports describing Al effects on polypeptide biosynthesis in other species are rare (e.g. Campbell *et al.*, 1994).

Some of Al-caused changes in polypeptide biosynthesis appear in the Al-resistant genotypes of *Triticum aestivum* (Ownby and Hruschka, 1991; Basu *et al.*, 1994a) or Al-resistant clones of *Medicago sativa* only (Campbell *et al.*, 1994) but not in the Al-sensitive counterparts. These changes in the polypeptide biosynthesis may therefore be connected with the resistance mechanisms. However, the confirmation of such a role hinges on sequencing and further functional characterization of these polypeptides, the work that is yet to appear in the literature.

Biosynthesis of two microsomal polypeptides with the molecular mass of 51 kDa (or possibly one polypeptide with the two structural forms) was induced by the Al stress in the Al-resistant but not in the Al-sensitive genotype of *Triticum aestivum* (Basu *et al.*, 1994a). No changes in the patterns of the cytoplasmic polypeptides were induced by Al in either genotype in the same study, while, in contrast, Ownby and Hruschka (1991) found differences between the Al-resistant and the Al-sensitive genotype being expressed more in cytoplasmic polypetides than in the microsomal ones. In both these studies the Al treatment lasted for at least 24 h; therefore, changes in the polypeptide patterns are unlikely to have been a cause for differential Al resistance (because genotypic differences in the response to Al can be observed in not more than 1–2 h, Rengel, 1992a; Kochian, 1995), but rather a consequence of the differential Al-related stress response.

Seven genes induced by an exposure of *Triticum aestivum* genotypes to Al have been cloned (Snowden *et al.*, 1993; Richards *et al.*, 1994). These genes encode putative metallothionein-like proteins, phenylalanine-ammonia lyase, proteinase inhibitors, and asparagine synthetase. Six of these genes were induced by Al in both Al-sensitive and Al-resistant genotype; the remaining gene, *wali2*, was induced by Al in the Al-sensitive genotype only and may therefore indicate a greater level of stress that genotype suffered. These seven genes are therefore unlikely to be involved in Al resistance. As with determination of polypeptide patterns, attempts should be made to determine effects of Al on gene induction much earlier in the Al-stress treatment to precede the onset of measurable Al-toxicity symptoms and differential Al resistance.

Transport and Storage of Metals in the Vacuole

Vacuoles of plants exposed to toxicity of heavy metals accumulate especially Zn and Ni, and to a lesser degree Cu, Pb and Cd (Wang *et al.*, 1991, 1992b, Ernst *et al.*, 1992). Organic acids (like malate in a case of Zn, Mathys, 1980) and PCs (e.g. for Cd, Vögeli-Lange and Wagner, 1990) are complexes formed in the cytosol and transported across the tonoplast into the vacuole. Once in the vacuole, metal ions are complexed with other compounds for storage. These storage compounds are frequently organic acids (Wang *et al.*, 1991, 1992b), but may be other compounds that specifically bind some metal ions (e.g. phytic acid binds Zn but not Cd in the vacuole of a number of crop species, Van Steveninck *et al.*, 1994, while complexes other than phytate and rich in S may bind Zn in vacuoles of hyperaccumulator *Thlaspi caerulescens*, Vazquez *et al.*, 1995).

Increased resistance to heavy metals may be due to an increased ability to sequester the metals from the cytosol into vacuole and store them there in non-toxic complexes. Such a resistance mechanism would be more important for Zn and Ni than for the other metals (Ernst *et al.*, 1992). However, even for Al which is primarily toxic to meristematic cells of the root tip, vacuolar sequestration of Al ions that eventually entered the symplasm may be an important resistance mechanism because of (i) the presence of substantial vacuoles even in meristematic cells (Patel *et al.*, 1990; Davies *et al.*, 1992), and (ii) activities of tonoplast transport systems, as one of likely mechanisms required for increased capacity to sequester metals into the vacuole (Ernst *et al.*, 1992), change substantially upon exposure to Al (Matsumoto, 1991; Kasai *et al.*, 1992, 1993).

While saturation of the vacuolar storage capacity is likely to lead to plant death from metal toxicity, some crop species (like *Triticum aestivum* and *Oryza sativa*, but not *Secale cereale*) possess significant capacity to increase vacuolar volume, even in the meristematic cells, when challenged with Zn toxicity (Davies *et al.*, 1992). Increased vacuolarization enables a decrease in cytotoxic concentrations in the cytosolic compartment. The Zn-resistant *Festuca rubra* genotype had a greater vacuolar volume in meristematic cells without exposure to toxic Zn concentrations, and showed less increase in vacuolarization under Zn toxicity, when compared to the Zn-sensitive genotype (Davies *et al.*, 1991), indicating that Zn-induced vacuolarization may be an adaptive response of Zn-sensitive genotypes rather than an important Zn-resistance mechanism that differentiates genotypes of *Festuca rubra*.

Metal-resistant Enzymes

Root apical cells of Al-resistant genotypes of *Triticum aestivum* have a greater activity of NAD kinase than Al-sensitive genotypes (Slaski, 1989, 1995). These differences became even more pronounced upon exposure to Al toxicity (Slaski, 1989). Similar results were obtained for other cereals (*Hordeum vulgare*, *Avena sativa*, and *Secale cereale* but not for legumes *Lupinus albus* and *Pisum sativum* in which Al stress caused a decrease in NAD activity) (Slaski, 1990). An increase in NAD activity in cereals exposed to the Al stress was abolished by cycloheximide, indicating that *de novo* biosynthesis of proteins is required (Slaski, 1989). Given that NAD kinase is involved in synthesis of $NADP^+$, a compound that is rate-limiting in numerous physiological processes, genotypes capable of maintaining the higher level of that enzyme, especially under stress, may be more resistant to effects of Al toxicity.

An existence of metal-tolerant forms of cytosolic enzymes could not be shown in a number of studies involving toxicity of heavy metals (Ernst *et al.*, 1992). However, cell wall enzymes may have differential sensitivity to heavy metal cations (Woolhouse and Walker, 1981). More work is required in this area.

CONCLUSIONS

The plant kingdom, as we know it today, has evolved in the continuous contact with metal ions. Under appropriate environmental conditions, Al and heavy metal toxicity stress may be a powerful selection factor. However, it is reasonable to assume that random genetic alterations that led to the phenotypic expression of resistance to metals have given birth to different mechanisms of resistance, more than one of which may operate in a given genotype. The level of resistance of a particular genotype is therefore a net result of additive effects of different mechanisms of resistance that may be operating at different levels of plant organization (metabolic, physiological, structural, or developmental). In addition, a certain level of phenotypic resistance to Al and heavy metals may be quite non-specific as a result of general adaptability to stress environments.

Distinguishing between relative contribution of a particular resistance mechanism to the overall phenotypic expression of metal resistance did not appear to have been attempted as yet. For doing so, it would be necessary to test genotypes of differential resistance at various levels of organization (organelles, protoplasts, cells, tissues, organs) before trying to partition the whole-plant phenotypic resistance into its components. When dealing with resistance of crop plants to Al and Mn, such an approach may be useful for plant breeders because identification of various single components of the resistance mechanism may be followed by the breeding program that will be aimed at compounding all single mechanisms (or a practically possible number of them) into a single genotype (Yeo and Flowers, 1986; Rengel and Jurkic, 1992).

Within a large number of proposed resistance mechanisms and frequently contradicting reports about their significance, a unifying picture is starting to emerge, at least for resistance to Al, where exclusion of Al from the symplasm by complexation with organic acid exuded into the rhizosphere appears to be an important resistance mechanism for a number of plant species. In contrast, resistance to heavy metals appears to be conferred mostly by true tolerance types of resistance mechanisms resulting in complexation and detoxification of symplastically located metal ions and their final storage in the vacuole.

ACKNOWLEDGEMENTS

I thank colleagues who sent me their papers submitted for publication. Dr. R. Gardner, University of Auckland, New Zealand, kindly provided wheat seeds. The financial support from the Australian Research Council is acknowledged.

REFERENCES

Allan D.L. and Jarrel W.M. (1989) Proton and copper adsorption to maize and soybean root cell walls. *Plant Physiology,* **89**, 823–832.

Allan D.L., Shann J.R. and Bertsch P.M. (1990) Role of root cell walls in iron deficiency of soybean (*Glycine max*) and aluminium toxicity of wheat (*Triticum aestivum*). In *Plant Nutrition Physiology and Applications*, edited by M.L. Van Beusichem, pp. 345–349. Dordrecht: Kluwer Academic Publishers.

Aniol A. (1983) Aluminium uptake by roots of two winter wheat varieties of different tolerance to aluminium. *Biochemie und Physiologie der Pflanzen*, **178**, 11–20.

Aniol A. (1984) Induction of aluminum tolerance by low doses of aluminum in the nutrient solution. *Plant Physiology*, **76**, 551–555.

Archambault D.J., Zhang G. and Taylor G.J. (1996) A comparison of the kinetics of Al uptake in roots of an Al-resistant cultivar and an Al-sensitive cultivar of *Triticum aestivum* L. using different Al sources. A revision of the operational definition of symplastic Al. physiologia plantarum, in press.

Baker, A.J.M. (1987) Metal tolerance. *New Phytologist*, **106 (Suppl.)**, 93–111.

Baker A.J.M. and Proctor J. (1990) The influence of cadmium, copper, lead, and zinc on the distribution and evolution of metallophytes in the British Isles. *Plant Systematics & Evolution*, **173**, 91–108.

Basu A., Basu U. and Taylor G.J. (1994a) Induction of microsomal membrane proteins in roots of an aluminum-resistant cultivar of *Triticum aestivum* L. under conditions of aluminum stress. *Plant Physiology*, **104**, 1007–1013.

Basu U., Godbold D. and Taylor G.J. (1994b) Aluminum resistance in *Triticum aestivum* associated with enhanced exudation of malate. *Journal of Plant Physiology*, **144**, 747–753.

Becerril J.M., Gonzales-Murua C., Munoz-Rueda A. and De Felipe M.R. (1989) Changes induced by cadmium and lead in gas exchange and water relations of clover and lucerne. *Plant Physiology and Biochemistry*, **27**, 913–918.

Bennet R.J and Breen C.M. (1991) The aluminium signal: new dimension to mechanisms of aluminium tolerance. *Plant and Soil*, **134**, 153–166.

Bertsch P.M. (1990) Aluminum speciation: methodology and applications. In *Advances in Environmental Science*, Vol. 4. *Acidic Precipitation: Soil, Aquatic Processes, and Lake Acidification*, edited by S.A. Norton, S.E. Lindberg and A.L. Page, pp. 63–105. Berlin: Springer-Verlag.

Blamey F.P.C., Asher C.J., Kerven G.L. and Edwards D.G. (1993) Factors affecting aluminium sorption by calcium pectate. *Plant and Soil*, **149**, 87–94.

Blamey F.P.C., Wheeler D.M., Christie R.A. and Edmeades D.C. (1990a) Variation in aluminum tolerance among and within *Lotus* lines. *Journal of Plant Nutrition*, **13**, 745–755.

Blamey F.P.C., Wheeler D.M., Edmeades D.C. and Christie R.A. (1990b) Independence of differential aluminum tolerance in *Lotus* on changes in rhizosphere pH or excretion of organic ligands. *Journal of Plant Nutrition*, **13**, 713–728.

Bolan N.S., Hedley M.J. and White R.E. (1991) Processes of soil acidification during nitrogen cycling with emphasis on legume based pasture. *Plant and Soil*, **134**, 53–63.

Brune A. and Dietz K.J. (1995) A comparative analysis of element composition of roots and leaves of barley seedlings grown in the presence of toxic cadmium, molybdenum, nickel, and zinc concentrations. *Journal of Plant Nutrition*, **18**, 853–868.

Brune A., Urbach W. and Dietz K.J. (1994a) Compartmentation and transport of zinc in barley primary leaves as basic mechanisms involved in zinc tolerance. *Plant, Cell and Environment*, **17**, 153–162.

Brune A., Urbach W. and Dietz K.J. (1994b) Zinc stress induces changes in apoplasmic protein content and polypeptide composition of barley primary leaves. *Journal of Experimental Botany*, **45**, 1189–1196.

Burke D.G., Watkins K. and Scott B. (1990) Manganese toxicity effects on visible symtoms, yield, manganese levels and organic acid levels in tolerant and sensitive wheat cultivars. *Crop Science*, **30**, 275–280.

Cambraia J., Galvani F.R. and Estevao M.M. (1983) Effects of aluminum and organic acid, sugar and amino acid composition of the root system of sorghum (*Sorghum bicolor* L. Moench). *Journal of Plant Nutrition*, **6**, 313–322.

Campbell T.A., Jackson P.R. and Xia Z.L. (1994) Effects of aluminum stress on alfalfa root proteins. *Journal of Plant Nutrition*, **17**, 461–471.

Chaney R.L. (1993) Zinc phytotoxicity. In *Zinc in Soils and Plants*, edited by A.D. Robson, pp. 135–150. Dordrecht: Kluwer Academic Publishers.

Christiansen-Weniger C., Groneman A.F. and Van Veen J.A. (1992) Associative N_2 fixation and root exudation of organic acids from wheat cultivars of different aluminium tolerance. *Plant and Soil*, **139**, 167–174.

Clarkson D.T. (1965) The effect of aluminium and some other trivalent metal cations on cell division in the root apices of *Allium cepa*. *Annals of Botany*, **29**, 309–315.

Clarkson D.T. and Sanderson J. (1971) Inhibition of the uptake and long-distance transport of calcium by aluminium and other polyvalent cations. *Journal of Experimental Botany*, **22**, 837–851.

Clouse J., Rincon M. and Gonzales R.A. (1991) Aluminum associates with wheat root proteins. *Plant Physiology*, **96**, S–111.

Cox R.M. and Thurman D.A. (1978) Inhibition by zinc of soluble and cell wall acid phosphatases of zinc-tolerant and non-tolerant clones of *Anthoxanthum odoratum*. *New Phytologist*, **80**, 17–32.

Cumming J.R. and Tomsett A.B. (1992) Metal tolerance in plants: signal transduction and acceleration mechanisms. In *Biogeochemistry of Trace Metals*, edited by D.C. Adriano, pp. 329–364. Boca Raton: Lewis Publishers.

Czarnecka E., Edelman L., Schöffl F. and Key J.L. (1984) Comparative analysis of physical stress responses in soybean seedlings using cloned heat shock cDNAs. *Plant Molecular Biology*, **3**, 45–58.

Davies K.L., Davies M.S. and Francis D. (1992) Zinc-induced vacuolation in root meristematic cells of cereals. *Annals of Botany*, **69**, 21–24.

Davies M.S., Francis D. and Thomas J.D. (1991) Rapidity of cellular changes induced by zinc in a zinc tolerant and non- tolerant cultivar of *Festuca rubra* L. *New Phytologist*, **117**, 103–108.

De Knecht J.A., Van Baren N., Ten Bookum W.M., Sang H.W.W F, Koevoets P.L.M., Schat H. *et al.* (1995) Synthesis and degradation of phytochelatins in cadmium-sensitive and cadmium-tolerant *Silene vulgaris*. *Plant Science*, **106**, 9–18.

Delhaize E., Craig S., Beaton C.D., Bennet R.J., Jagadish V.C. and Randal P.J. (1993a) Aluminum tolerance in wheat (*Triticum aestivum* L.). I. Uptake and distribution of aluminum in root apices. *Plant Physiology*, **103**, 685–693.

Delhaize E., Higgins T.J.V. and Randall P.J. (1991) Aluminium tolerance in wheat: Analysis of polypeptides in the root apices of tolerant and sensitive genotypes. In *Plant-Soil Interactions at Low pH*, edited by R.J. Wright, V.C. Baligar and R.P. Murrmann, pp. 1071–1079. Dordrecht: Kluwer Academic Publishers.

Delhaize E., Robinson N.J. and Jackson P.J. (1989) Effect of cadmium on gene expression in cadmium-tolerant and cadmium-sensitive *Datura inoxia* cells. *Plant Molecular Biology*, **12**, 487–497.

Delhaize E. and Ryan P.R. (1995) Aluminum toxicity and tolerance in plants. *Plant Physiology*, **107**, 315–321.

Delhaize E., Ryan P.R. and Randall P.J. (1993b) Aluminum tolerance in wheat (*Triticum aestivum* L.) II. Aluminum-stimulated excretion of malic acid from root apices. *Plant Physiology*, **103**, 695–702.

Denny H.J. and Wilkins D.A. (1987) Zinc tolerance in *Betula* ssp. II. Microanalytical studies of zinc uptake into root tissues. *New Phytologist*, **106**, 525–534.

Edelman L., Czarneska E. and Key J.L. (1988) Induction and accumulation of heat shock-specific poly (A^+) RNAs and proteins in soybean seedlings during arsenite and cadmium treatments. *Plant Physiology*, **86**, 1048–1056.

Eeckhaoudt S., Vandeputte D., Van Praag H., Van Grieken R. and Jacob W. (1992) Laser microprobe mass analysis (LAMMA) of aluminum and lead in fine roots and their ectomycorrhizal mantles of Norway spruce (*Picea abies* (L.) Karst.). *Tree Physiology*, **10**, 209–215.

Ernst W.H.O., Verkleij J.A.C. and Schat H. (1992) Metal tolerance in plants. *Acta Botanica Neerlandica*, **41**, 229–248.

Fontes R.L.F. and Cox F.R. (1995) Effects of sulfur supply on soybean plants exposed to zinc toxicity. *Journal of Plant Nutrition*, **18**, 1893–1906.

Foy C.D. (1988) Plant adaptation to acid, aluminum-toxic soils. *Communications in Soil Science and Plant Analysis*, **19**, 959–987.

Foy C.D., Chaney R.L. and White M.C. (1978) The physiology of metal toxicity in plants. *Annual Review of Plant Physiology*, **29**, 511–566.

Foy C.D., Lee E.H., Coradetti C.A. and Taylor G.J. (1990) Organic acids related to differential aluminum tolerance in wheat (*Triticum aestivum*) cultivars. In *Plant Nutrition - Physiology and Applications*, edited by M.L. Van Beusichem, pp. 381–389. Dordrecht: Kluwer Academic Publishers.

Foy C.D., Lee E.H. and Wilding S.B. (1987) Differential aluminum tolerances of two barley cultivars related to organic acids in their roots. *Journal of Plant Nutrition*, **10**, 1089–1101.

Gabara B., Krajewska K. and Stecka E. (1995) Calcium effect on number, dimension and activity of nucleoli in cortex cells of pea (*Pisum sativum* L.) roots after treatment with heavy metals. *Plant Science*, **111**, 153–161.

Godbold D.L. (1994) Aluminium and heavy metal stress: from the rhizosphere to the whole plant. In *Effects of Acid Rain on Forest Processes*, edited by D.L. Godbold and A. Hüttermann, pp. 231–264. New York: Wiley-Liss.

Godbold D.L. and Hüttermann A. (1988) Inhibition of photosynthesis and transpiration in relation to mercury-induced root damage in spruce seedlings. *Physiologia Plantarum*, **74**, 270–275.

Godbold D.L., Horst W.J., Collins J.C., Thurman D.A. and Marschner H. (1984) Accumulation of zinc and organic acids in roots of Zn-tolerant and non-tolerant ecotypes of *Deschampsia caespitosa*. *Journal of Plant Physiology*, **116**, 59–69.

Godbold D.L. and Kettner C. (1991) Lead influences root growth and mineral nutrition of *Picea abies* seedlings. *Journal of Plant Physiology*, **139**, 95–99.

Golan-Goldhirsh A., Mozafar A. and Oertli J.J. (1995) Effect of ascorbic acid on soybean seedlings grown on medium containing a high concentration of copper. *Journal of Plant Nutrition*, **18**, 1735–1741.

Guern J., Felle H., Mathieu Y. and Karkdjian A. (1991) Regulation of intracellular pH in plant cells. *International Review of Cytology*, **127**, 111–173.

Guo Y., Schulz R. and Marschner H. (1995) Genotypic differences in uptake and distribution of cadmium and nickel in plants. *Angewandte Botanik*, **69**, 42–48.

Hairiah K., Stulen I. and Kuiper P.J.C. (1990) Aluminium tolerance of the velvet beans *Mucuna pruriens* var. utilis and *Mucuna deeringiana*. I. Effects of aluminium on growth and mineral composition. In *Plant Nutrition - Physiology and Applications*, edited by M.L. Van Beusichem, pp. 365–374. Dordrecht: Kluwer Academic Publishers.

Hairiah K., Stulen I., Van Nordwijk M. and Kuiper P.J.C. (1994) Al avoidance and Al tolerance of *Mucuna pruriens* var. utilis: Effects of a heterogeneous root environment and the nitrogen form in the root environment. *Plant and Soil*, **167**, 67–72.

Hairiah K., Van Noordwijk M., Stulen I. and Kuiper P.J.C. (1992) Aluminium avoidance by *Mucuna pruriens*. *Physiologia Plantarum*, **86**, 17–24.

Hairiah K., Van Noordwijk M., Stulen I., Meijboom F.W. and Kuiper P.J.C. (1993) Phosphate nutrition effects on aluminium avoidance of *Mucuna pruriens* var utilis. *Environmental and Experimental Botany*, **33**, 75–83.

Harmens H., Den Hartog P.R., Ten Bookem W.M. and Verkleij J.A.C. (1993a) Increased zinc tolerance in *Silene vulgaris* (Moench) Garcke is not due to increased production of phytochelatins. *Plant Physiology*, **103**, 1305–1309.

Harmens H., Gusmao N.G.C.P.B., Den Hartog P.R., Verkleij J.A.C. and Ernst W.H.O. (1993b) Uptake and transport of zinc in zinc-sensitive and zinc-resistant *Silene vulgaris*. *Journal of Plant Physiology*, **141**, 309–315.

Haug A. (1984) Molecular aspects of aluminum toxicity. *CRC Critical Reviews in Plant Sciences*, **1**, 345–373.

Haug A.R. and Caldwell C.R. (1985) Aluminum toxicity in plants: the role of the root plasma membrane and calmodulin. In *Frontiers of Membrane Research in Agriculture*, Beltsville Symp. 9, edited by J.B. John St., E. Berlin and P.C. Jackson, pp. 359–381. Totowa: Rowman & Allanheld.

Haug A. and Shi B. (1991) Biochemical basis of aluminium tolerance in plant cells. In *Plant-Soil Interactions at Low pH*, edited by R.J. Wright, V.C. Baligar and R.P. Murrmann, pp. 839–850. Dordrecht: Kluwer Academic Publishers.

Haug A. Shi B. and Vitorello V. (1994) Aluminum interaction with phosphoinositide-associated signal transduction. *Archives of Toxicology*, **68**, 1–7.

Hiatt A.J., Amos D.F. and Massey H.F. (1963) Effect of aluminum on copper sorption by wheat. *Agronomy Journal*, **55**, 284–287.

Horst W.J. (1995) The role of the apoplast in aluminium toxicity and resistance of higher plants: a review: *Zeitschrift für Pflanzenernährung und Bodenkunde*, **158**, 419–428.

Horst W.J, Wagner A. and Marschner H. (1983) Effect of aluminium on root growth, cell-division rate and mineral element contents in roots of *Vigna unguiculata* genotypes. *Zeitschrift für Pflanzenphysiologie*, **109**, 95–103.

Huang J.W., Grunes D.L. and Kochian L.V. (1993) Aluminum effects on calcium ($^{45}Ca^{2+}$) translocation in aluminum-tolerant and aluminum-sensitive wheat (*Triticum aestivum* L.) cultivars. *Plant Physiology*, **102**, 85–93.

Huang J.W., Grunes D.L. and Kochian L.V. (1994) Voltage-dependent Ca^{2+} influx into right-side-out plasma membrane vesicles isolated from wheat roots: characterization of a putative Ca^{2+} channel. *Proceeding of the National Academy of Sciences USA*, **91**, 3473–3477.

Huang J.W., Grunes D.L. and Kochian L.V. (1995) Aluminium and calcium transport interactions in intact roots and root plasmalemma vesicles from aluminium-sensitive and tolerant wheat cultivars. *Plant and Soil*, **171**, 131–135.

Huang J.W., Shaff J.E., Grunes D.L. and Kochian L.V. (1992) Aluminum effects on calcium fluxes at the root apex of aluminum- tolerant and aluminum-sensitive wheat cultivars. *Plant Physiology*, **98**, 230–237.

Hue N.V., Craddock G.R. and Adams F. (1986) Effect of organic acids on aluminum toxicity in subsoils. *Soil Science Society of America Journal*, **50**, 28–34.

Huett D.O. and Menary R.C. (1980) Aluminium distribution in freeze-dried roots of cabbage, lettuce and kikuyu grass by energy dispersive X-ray analysis. *Australian Journal of Plant Physiology*, **7**, 101–111.

Jackson A.P. and Alloway B.J. (1992) The transfer of cadmium from agricultural soils to the human food chain. In *Biochemistry of Trace Metals*, edited by D.C. Adriano, pp. 109–152. Chicago: Lewis Publishers.

Jentschke G., Schlegel H. and Godbold D.L. (1991) The effect of aluminium on uptake and distribution of magnesium and calcium in roots of mycorrhizal Norway spruce seedlings. *Physiologia Plantarum*, **82**, 266–270.

Jones D.L. and Darrah P.R. (1995) Influx and efflux of organic acids across the soil-root interface of *Zea mays* L. and its implications in rhizosphere C flow. *Plant and Soil*, **173**, 103–109.

Kasai M., Sasaki M., Tanakamaru S., Yamamoto Y. and Matsumoto H. (1993) Possible involvement of abscisic acid in increases in activities of two vacuolar H^+ pumps in barley roots under aluminum stress. *Plant and Cell Physiology*, **34**, 1335–1338.

Kasai M., Sasaki M., Yamamoto Y. and Matsumoto H. (1992) Aluminum stress increases K^+ efflux and activities of ATP- and PP_i-dependent H^+ pumps of tonoplast-enriched membrane vesicles from barley roots. *Plant and Cell Physiology*, **33**, 1035–1039.

Keltjens W.G. (1995) Magnesium uptake by Al-stressed maize plants with special emphasis on cation interactions at root exchange sites. *Plant and Soil*, **171**, 141–146.

Kerven G.L., Ostatek-Boczynski Z., Edwards D.G., Asher C.J. and Oweczkin J. (1995) Chromatographic techniques for the separation of Al and associated organic ligands present in soil solution. In *Plant-Soil Interactions at Low pH*, edited by R.A. Date, N.J. Grundon, G.E. Rayment and M.E. Probert, pp. 47–52. Dordrecht: Kluwer Academic Publishers.

Kinraide T.B. (1991) Identity of the rhizotoxic aluminum species. *Plant and Soil*, **134**, 167–178.

Kinraide T.B. (1994) Use of a Gouy-Chapman-Stern model for membrane-surface electrical potential to interpret some features of mineral rhizotoxicity. *Plant Physiology*, **106**, 1583–1592.

Kinraide T.B., Ryan P.R. and Kochian L.V. (1992) Interactive effects of Al^{3+}, H^+, and other cations on root elongation considered in terms of cell-surface electrical potential. *Plant Physiology*, **99**, 1461–1468.

Kinraide T.B., Ryan P.R. and Kochian L.V. (1994) Al^{3+}-Ca^{2+} interactions in aluminum rhizotoxicity. II. Evaluating the Ca^{2+}-displacement hypothesis. *Planta*, **192**, 104–109.

Kinzel H. (1982) *Pflanzenökologie und Mineralstoffwechsel*. Stuttgart: Ulmer.

Kirlew P.W. and Bouldin D.R. (1987) Chemical properties of the rhizosphere in an acid subsoil. *Soil Science Society of America Journal*, **51**, 128–132.

Klapheck S., Fliegner W. and Zimmer I. (1994) Hydroxymethyl-phytochelatins [(γ-glutamylcysteine)$_n$-serine] are metal-induced peptides of the *Poaceae*. *Plant Physiology*, **104**, 1325–1332.

Klimashevskii E.L. and Chernysheva N.F. (1980) Content of organic acids and physiologically active compounds in plants differing in their susceptibility to the toxicity of Al^{3+}. *Soviet Agricultural Science*, **2**, 5–7.

Knight A.H., Crooke W.M. and Inkson R.H.E. (1961) Cation-exchange capacities of tissues of higher and lower plants and their related uronic acid contents. *Nature*, **192**, 142–143.

Kochian L.V. (1995) Cellular mechanisms of aluminum toxicity and resistance in plants. *Annual Review of Plant Physiology and Plant Molecular Biology*, **46**, 237–260.

Koyama H., Okawara R., Ojima K. and Yamaya T. (1988) Re-evaluation of characterisitics of a carrot cell line previously selected as aluminum-tolerant cells. *Physiologia Plantarum*, **74**, 683–687.

Lazof D.B., Goldsmith J.G., Rufty T.W. and Linton R.W. (1994a) Rapid uptake of aluminum into cells of intact soybean root tips. A microanalytical study using Secondary Ion Mass Spectrometry. *Plant Physiology*, **106**, 1107–1114.

Lazof D.B., Rincón M., Rufty T.W., Mactown C.T. and Carter T.E. (1994b) Aluminum accumulation and associated effects on $^{15}NO_3^-$ influx in roots of two soybean genotypes differing in Al tolerance. *Plant and Soil*, **164**, 291–297.

Lee E.H. and Foy C.D. (1986) Aluminum tolerance of two snapbean cultivars related to organic acid content evaluated by high performance liquid chromatography. *Journal of Plant Nutrition*, **9**, 1481–1498.

Lee J., Reeves R.D., Brooks R.R. and Jaffre T. (1978) The relation between nickel and citric acid in some nickel-accumulating plants. *Phytochemistry*, **17**, 1033–1035.

Li Y.M., Chaney R.L., Schneiter A.A. and Miller J.F. (1995a) Genotypic variation in kernel cadmium concentration in sunflower germplasm under varying soil conditions. *Crop Science*, **35**, 137–141.

Li Y.M., Chaney R.L., Schneiter A.A. and Miller J.F. (1995b) Combining ability and heterosis estimates for kernel cadmium level in sunflower. *Crop Science*, **35**, 1015–1019.

Lindberg S. (1990) Aluminium interactions with K^+ ($^{86}Rb^+$) and $^{45}Ca^{2+}$ fluxes in three cultivars of sugar beet (*Beta vulgaris*). *Physiologia Plantarum*, **79**, 275–282.

Lipton D.S., Blanchar R.W. and Blevins D.G. (1987) Citrate, malate, and succinate concentration in exudates from P-sufficient and P-stressed *Medicago sativa* L. seedlings. *Plant Physiology*, **85**, 315–317.

Liu D. and Jiang W. (1993) Effects of aluminium ion on root growth, cell division, and nucleoli of garlic (*Allium sativum* L.), *Environmental Pollution*, **82**, 295–299.

Liu D., Jiang W., Guo L., Hao Y., Lu C. and Zhao F. (1994) Effects of nickel sulfate on root growth and nucleoli in root tip cells of *Allium cepa*. *Israel Journal of Plant Science*, **42**, 143–148.

Löffler S., Hochberger A., Grill E., Winnacker E.L. and Zenk M.H. (1989) Termination of the phytochelatin synthase reaction through sequestration of heavy metals by the reaction product. *FEBS Letters*, **258**, 42–46.

Macfie S.M., Cossins E.A. and Taylor G.J. (1994) Effects of excess manganese on production of organic acids in Mn-tolerant and Mn-sensitive cultivars of *Triticum aestivum* L. (wheat). *Journal of Plant Physiology*, **143**, 135–144.

Macnair M.R. (1983) The genetic control of copper tolerance in the yellow monkey flower, *Mimulus guttatus*. *Heredity*, **50**, 283–293.

Macnair M.R. (1993) The genetics of metal tolerance in vascular plants. *New Phytologist*, **124**, 541–559.

Marienfeld S., Lehmann H. and Stelzer R. (1995) Ultrastructural investigations and EDX-analyses of Al-treated oat (*Avena sativa*) roots. *Plant and Soil*, **171**, 167–173.

Marienfeld S. and Stelzer R. (1993) X-ray microanalyses in roots of Al-treated *Avena sativa* plants. *Journal of Plant Physiology*, **141**, 569–573.

Marschner H. (1991) Mechanisms of adaptation of plants to acid soils. *Plant and Soil*, **134**, 1–20.

Martin R.B. (1988) Bioinorganic chemistry of aluminum. In *Metal Ions in Biological Systems, Vol. 24, Aluminum and its Role in Biology*, edited by H. Sigel, pp. 1–57. New York: Marcel Dekker.

Massot N., Poschenrieder C. and Barcelo J. (1992) Differential response of three bean (*Phaseolus vulgaris*) cultivars to aluminium. *Acta Botanica Neerlandica*, **41**, 293–298.

Mathys W. (1980) Zinc tolerance in plants. In *Zinc in the Environment, Part II: Health Effects*, edited by J.O. Nriagu, pp. 415–437. New York: John Wiley and Sons.

Matsumoto H. (1991) Biochemical mechanism of the toxicity of aluminium and the sequestration of aluminium in plant cells. In *Plant-Soil Interactions at Low pH*, edited by R.J. Wright, V.C. Baligar and R.P. Murrmann, pp. 825–838. Dordrecht: Kluwer Academic Publishers.

Matsumoto H. and Yamaya T. (1986) Inhibition of potassium uptake and regulation of membrane-associated Mg^{2+}-ATPase activity of pea roots by aluminum. *Soil Science and Plant Nutrition*, **32**, 179–188.

Matsumoto Y., Okado Y., Min K.S., Onosaka S. and Tanaka K. (1990) Amino acids and peptides. XXVII. Synthesis of phytochelatin-related peptides and examination of their metal-binding properties. *Chemical and Pharmaceutical Bulletin*, **38**, 2364–2368.

McDonald-Stephens J.L. and Taylor G.J. (1995) Kinetics of aluminum uptake by cell suspensions of *Phaseolus vulgaris* L. *Journal of Plant Physiology*, **145**, 327–334.

McGrath S.P., Baker A.J.M., Morgan A.N., Salmon W.J. and Williams M. (1980) The effects of interactions between cadmium and aluminium on the growth of two metal-tolerant races of *Holcus lanatus* L. *Environmental Pollution, Series A*, **23**, 267–277.

Meharg A.A. (1993) The role of plasmalemma in metal tolerance in angiosperms. *Physiologia Plantarum*, **88**, 191–198.

Merckx R., Van Ginkel J.H., Sinnaeve J. and Cremers A. (1986) Plant-induced changes in the rhizosphere of maize and wheat. II. Complexation of cobalt, zinc and manganese in the rhizosphere of maize and wheat. *Plant and Soil*, **96**, 95–107.

Meuwly P., Thibault P., Schwan A.L. and Rauser W.E. (1995) Three families of thiol peptides are induced by cadmium in maize. *The Plant Journal*, **7**, 391–400.

Millard M.M., Foy C.D., Coradetti C.A. and Reinsel M.D. (1990) X-ray photoelectron spectroscopy surface analysis of aluminum ion stress in barley roots. *Plant Physiology*, **93**, 578–583.

Miyasaka S.C., Buta J.G., Howell R.K. and Foy C.D. (1991) Mechanism of aluminum tolerance in snapbeans. Root exudation of citric acid. *Plant Physiology*, **96**, 737–743.

Miyasaka S.C., Kochian L.V., Shaff J.E. and Foy C.D. (1989) Mechanisms of aluminum tolerance in wheat. An investigation of genotypic differences in rhizosphere pH, K^+, and H^+ transport, and root-cell membrane potentials. *Plant Physiology*, **91**, 1188–1196.

Moustakas M., Ouzounidou G. and Lannoye R. (1995) Aluminum effects on photosynthesis and elemental uptake in an aluminum-tolerant and non-tolerant wheat cultivar. *Journal of Plant Nutrition*, **18**, 669–683.

Mugwira L.M. and Elhawhary S.M. (1979) Aluminum accumulation and tolerance of triticale and wheat in relation to root cation exchange capacity. *Soil Science Society of America Journal*, **43**, 736–740.

Neumann D., Lichtenberger O., Günther D., Tschiersch K. and Nover L. (1994) Heat shock proteins induce heavy-metal tolerance in higher plants. *Planta*, **194**, 360–367.

Nieboer E. and Richardson D.H.S. (1980) The replacement of the non-descript term 'heavy metals' by a biologically and chemically significant classification of metal ions. *Environmental Pollution, Series B*, **1**, 3–26.

Nouri P.A. and Reddy G.B. (1995) Influence of acid rain and ozone on heavy metals under loblolly pine trees: a field study. *Plant and Soil*, **171**, 59–62.

Ojima K., Abe H. and Ohira K. (1984) Release of citric acid into the medium by aluminum-tolerant carrot cells. *Plant and Cell Physiology*, **25**, 855–858.

Olivetti G.P., Cumming J.R. and Etherton B. (1995) Membrane potential depolarization of root cap cells precedes aluminum tolerance in snapbean. *Plant Physiology*, **109**, 123–129.

Ownby J.D. (1993) Mechanisms of reaction with aluminium-treated roots. *Physiologia Plantarum*, **87**, 371–380.

Ownby J.D. and Hruschka W.R. (1991) Quantitative changes in cytoplasmic and microsomal proteins associated with aluminium toxicity in two cultivars of winter wheat. *Plant, Cell & Environment*, **14**, 303–309.

Ownby J.D. and Popham H.R. (1989) Citrate reverses the inhibition of wheat root growth caused by aluminum. *Journal of Plant Physiology*, **135**, 588–591.

Parker D.R. (1995) Root growth analysis: an underutilised approach to understanding aluminium rhizotoxicity. *Plant and Soil*, **171**, 151–157.

Parker D.R., Chaney R.L. and Norvell W.A.(1995) Chemical equilibrium models: applications to plant nutrition research. In *Chemical Equilibrium and Reaction Models*, edited by R.H. Loeppert, A.P. Schwab and S. Goldberg, pp. 163–200. Madison, WI: Soil Science Society of America.

Parker D.R., Kinraide T.B. and Zelazny L.W. (1988) Aluminum speciation and phytotoxicity in dilute hydroxy-aluminium solutions. *Soil Science Society of America Journal*, **52**, 438–444.

Parker D.R., Kinraide T.B. and Zelazny L.W. (1989) On the phytotoxicity of polynuclear hydroxy aluminum complexes. *Soil Science Society of America Journal*, **53**, 789–796.

Patel D.D., Barlow P.W. and Lee R.B. (1990) Development of vacuolar volume in the root tip of pea. *Annals of Botany*, **65**, 159–169.

Pearson J.N. and Rengel Z. (1996) Mechanisms of plant resistance to nutrient deficiency stress. In *Mechanisms of Environmental Stress Resistance in Plants*, edited by A.S. Basra, Amsterdam: Harwood and R.K. Basra, pp. 213–240. Academic Publishers.

Pellet D.M., Grunes D.L. and Kochian L.V. (1995) Organic acid exudation as an aluminum tolerance mechanism in maize (*Zea mays* L.). *Planta*, **196**, 788–795.

Pettersson S. and Strid H. (1989) Effects of aluminium on growth and kinetics of K^+ (^{86}Rb) uptake in two cultivars of wheat (*Triticum aestivum*) with different sensitivity to aluminium. *Physiologia Plantarum*, **76**, 255–261.

Picton S.J., Richards K.D. and Gardner R.C. (1991) Protein profiles in root-tips of two wheat (*Triticum aestivum* L.) cultivars with different tolerance to aluminum. In *Plant-Soil Interactions at Low pH*, edited by R.J. Wright, V.C. Baligar and R.P. Murrmann, pp. 1063–1070. Dordrecht: Kluwer Academic Publishers.

Rauser W.E. and Meuwly P. (1995) Retention of cadmium in roots of maize seedlings. Role of complexation by phytochelatins and related thiol peptides. *Plant Physiology*, **109**, 195–202.

Reddy G.N. and Prasad M.N.V. (1990) Heavy metal-binding proteins/peptides: occurrence, structure, synthesis and functions. A review. *Environmental and Experimental Botany*, **30**, 251–264.

Reddy K.J., Wang L. and Gloss S.P. (1995) Solubility and mobility of copper, zinc and lead in acidic environments. *Plant and Soil*, **171**, 53–58.

Reid R.J., Rengel Z. and Smith F.A. (1996) Membrane fluxes and comparative toxicities of the trivalent cations aluminium, scandium and gallium. *Journal of Experimental Botany*, in press.

Rengel Z. (1992a) Role of calcium in aluminium toxicity. *New Phytologist*, **121**, 499–513.

Rengel Z. (1992b) Disturbance of Ca^{2+} homeostasis as a primary trigger in the Al toxicity syndrome. *Plant, Cell & Environment*, **15**, 931–938.

Rengel Z. (1994) Effects of Al, rare earth elements and other metals on net $^{45}Ca^{2+}$ uptake by *Amaranthus* protoplasts. *Journal of Plant Physiology*, **143**, 47–51.

Rengel. Z. (1996) Uptake of aluminium by plant cells. *New Phytologist*, in press.

Rengel Z. and Elliott D.C. (1992a) Aluminium inhibits net $^{45}Ca^{2+}$ uptake by *Amaranthus* protoplasts. *Biochemie und Physiologie der Pflanzen*, **188**, 177–186.

Rengel Z. and Elliott D.C. (1992b) Mechanism of aluminum inhibition of net $^{45}Ca^{2+}$ uptake by *Amaranthus* protoplasts. *Plant Physiology*, **98**, 632–638.

Rengel Z. and Graham R.D. (1995) Importance of seed Zn content for wheat growth on Zn-deficient soil. I. Vegetative growth. *Plant and Soil*, **173**, 259–266.

Rengel Z. and Jurkic V. (1992) Genotypic differences in wheat Al tolerance. *Euphytica*, **62**, 111–117.

Rengel Z. and Robinson D.L. (1989a) Aluminum and plant age effects on adsorption of cations in the Donnan free space of ryegrass roots. *Plant and Soil*, **116**, 223–227.

Rengel Z. and Robinson D.L. (1989b) Aluminum effects on growth and macronutrient uptake by annual ryegrass. *Agronomy Journal*, **81**, 208–215.

Richards K.D. and Gardner R.C. (1994) The effect of aluminium treatment on wheat roots: expression of heat shock, histone and SHH genes. *Plant Science*, **98**, 37–45.

Richards K.D., Snowden K.C. and Gardner R.C. (1994) wali6 and wali7. Genes induced by aluminum in wheat (*Triticum aestivum* L.) roots. *Plant Physiology*, **105**, 1455–1456.

Rincón M. and Gonzales R.A. (1991) Induction of protein biosynthesis by aluminum in wheat (*Triticum aestivum* L.) root tips. In *Plant-Soil Interactions at Low pH*, edited by R.J. Wright, V.C. Baligar and R.P. Murrmann, pp. 851–858. Dordrecht: Kluwer Academic Publishers.

Rincón M. and Gonzales R.A. (1992) Aluminum partitioning in intact roots of aluminum-tolerant and aluminum-sensitive wheat (*Triticum aestivum* L.) cultivars. *Plant Physiology*, **99**, 1021–1028.

Ross W. (1992) Confocal pH topography in plant cells - acidic layers in the peripheral cytoplasm and the apoplast. *Botanica Acta*, **105**, 253–259.

Rüegsegger A. and Brunold C. (1992) Effect of cadmium on γ-glutamylcysteine synthesis in maize seedlings. *Plant Physiology*, **99**, 428–433.

Ryan P.R., Delhaize E. and Randall P.J. (1995a) Characterization of Al-stimulated efflux of malate from the apices of Al-tolerant wheat roots. *Planta*, **196**, 103–110.

Ryan P.R., Delhaize E. and Randall P.J. (1995b) Malate efflux from root apices and tolerance to aluminium are highly correlated in wheat. *Australian Journal of Plant Physiology*, **22**, 531–536.

Ryan P.R., DiTomaso J.M. and Kochian L.V. (1993) Aluminium toxicity in roots: an investigation of spatial sensitivity and the role of the root cap. *Journal of Experimental Botany*, **44**, 437–446.

Ryan P.R., Shaff J.E. and Kochian L.V. (1992) Aluminum toxicity in roots. Correlation among ionic currents, ion fluxes, and root elongation in aluminum-sensitive and aluminum-tolerant wheat cultivars. *Plant Physiology*, **99**, 1193–1200.

Sasaki M., Kasai M., Yamamoto Y. and Matsumoto H. (1994a) Comparison of the early response to aluminum stress between tolerant and sensitive wheat cultivars: root growth, aluminum content and efflux of K^+. *Journal of Plant Nutrition*, **17**, 1275–1288.

Sasaki M., Kasai M., Yamamoto Y. and Matsumoto H. (1995) Involvement of plasma membrane potential in the tolerance mechanism of plant roots to aluminium toxicity. *Plant and Soil*, **171**, 119–124.

Sasaki M., Yamamoto Y. and Matsumoto H. (1994b) Putative Ca^{2+} channels of plasma membrane vesicles are not involved in the tolerance mechanism of aluminum in aluminum tolerant wheat (*Triticum aestivum* L.) cultivar. *Soil Science and Plant Nutrition*, **40**, 709–714.

Scott R., Hoddinott J., Taylor G.J. and Briggs K. (1991) The influence of aluminum on growth, carbohydrate, and organic acid content of an aluminum-tolerant and an aluminum-sensitive cultivar of wheat. *Canadian Journal of Botany*, **69**, 711–716.

Sharma D.C., Chatterjee C. and Sharma C.P. (1995) Chromium accumulation and its effects on wheat (*Triticum aestivum* L. cv. HD 2204) metabolism. *Plant Science*, **111**, 145–151.

Shuman L.M., Wilson D.O. and Ramseur E.L. (1991) Amelioration of aluminum toxicity to sorghum seedlings by chelating agents. *Journal of Plant Nutrition*, **14**, 119–128.

Siegel N. and Haug A. (1983) Calmodulin-dependent formation of membrane potential in barley root plasma membrane vesicles: A biochemical model of aluminum toxicity in plants. *Physiologia Plantarum*, **59**, 285–291.

Slaski J.J. (1989) Effect of aluminium on calmodulin-dependent and calmodulin-independent NAD kinase activity in wheat (*Triticum aestivum* L.) root tips. *Journal of Plant Physiology*, **133**, 696–701.

Slaski J.J. (1990) Response of calmodulin-dependent and calmodulin-independent NAD kinase to aluminium in root tips from various cultivated plants. *Journal of Plant Physiology*, **136**, 40–44.

Slaski J.J. (1995) NAD^+ kinase activity in root tips of nearly isogenic lines of wheat (*Triticum aestivum* L.) that differ in their tolerance to aluminium. *Journal of Plant Physiology*, **145**, 143–147.

Snowden K.C. and Gardner R.C. (1993) Five genes induced by aluminum in wheat (*Triticum aestivum* L.) roots. *Plant Physiology*, **103**, 855–861.

Steffens J.C. (1990) The heavy metal-binding peptides of plants. *Annual Review of Plant Physiology and Plant Molecular Biology*, **41**, 553–575.

Steffens J.C., Hunt D.F. and Williams B.G. (1986) Accumulation of non-protein metal-binding polypeptides (γ-glutamyl-cysteinyl)$_n$-glycine in selected cadmium-resistant tomato cells. *Journal of Biological Chemistry*, **261**, 13879–13882.

Suhayda C.G. and Haug A. (1984) Organic acids prevent aluminum-induced conformational changes in calmodulin. *Biochemical and Biophysical Research Communications*, **119**, 376–381.

Suhayda C.G. and Haug A. (1985) Citrate chelation as a potential mechanism against aluminum toxicity in cells: the role of calmodulin. *Canadian Journal of Biochemistry and Cell Biology*, **63**, 1167–1175.

Suhayda C.G. and Haug A. (1986) Organic acids reduce aluminum toxicity in maize root membranes. *Physiologia Plantarum*, **68**, 189–195.

Taylor G.J. (1989) Multiple metal stress in *Triticum aestivum*. Differentiation between additive, multiplicative, antagonistic, and synergistic effects. *Canadian Journal of Botany*, **67**, 2272–2276.

Taylor G.J. (1991) Current views of the aluminum stress response: the physiological basis of tolerance. *Current Topics in Plant Biochemistry and Physiology*, **10**, 57–93.

Taylor G.J. (1995) Overcoming barriers to understanding the cellular basis of aluminium resistance. *Plant and Soil*, **171**, 89–103.

Taylor G.J. and Foy C.D. (1985) Mechanisms of aluminum tolerance in *Triticum aestivum* (wheat). I. Differential pH induced by winter cultivrs in nutrient solutions. *American Journal of Botany*, **72**, 695–701.

Taylor G.J., Hunter D.B., Stephens J., Bertsch P.M., Elmore D., Rengel Z. *et al.* (1996) Direct measurement of Al transport across the plasma membrane of Chara corallina. In *Proceedings of the 4th International Symposium on Plant-Soil Interactions at Low pH*, Belo Horizonte, Brazil, March 1996.

Taylor G.J., Stadt K.J. and Dale M.R.T. (1991) Modelling the phytotoxicity of aluminum, cadmium, copper, manganese, nickel, and zinc using the Weibull frequency distribution. *Canadian Journal of Botany*, **69**, 359–367.

Taylor G.J., Stadt K.J. and Dale M.R.T. (1992) Modelling the interactive effects of aluminum, cadmium, manganese, nickel and zinc stress using the Weibull frequency distribution. *Environmental and Experimental Botany*, **12**, 281–293.

Tice K.R., Parker D.R. and DeMason A. (1992) Operationally defined apoplastic and symplastic aluminum fractions in root tips of aluminum-intoxicated wheat. *Plant Physiology*, **100**, 309–318.

Tiller K.G. (1989) Heavy metals in soils and their environmental significance. *Advances in Soil Science*, **9**, 113–142.

Tsuji M., Kuboi T. and Konishi S. (1994) Stimulatory effects of aluminum on the growth of cultured roots of tea. *Soil Science and Plant Nutrition*, **40**, 471–476.

Turner A.P. and Dickinson N.M. (1993) Survival of *Acer pseudoplatanus* L. (sycamore) seedlings on metalliferous soils. *New Phytologist*, **123**, 509–522.

Van Steveninck R.F.M., Babare A., Fernando D.R. and Van Steveninck M.E. (1994) The binding of zinc, but not cadmium, by phytic acid in roots of crop plants. *Plant and Soil*, **167**, 157–164.

Van Steveninck R.F.M., Van Steveninck M.E., Fernando D.R., Horst W.J. and Marschner H. (1987) Deposition of zinc phytate in globular bodies in roots of *Deschampsia caespitosa* ecotypes: a detoxification mechanism? *Journal of Plant Physiology*, **131**, 247–257.

Vazquez M.D., Poschenrieder C., Barcelo J., Baker A.J.M., Hatton P. and Cope G.H. (1994) Compartmentation of zinc in roots and leaves of the zinc hyperacccumulator *Thlaspi caerulescens* J & C Presl. *Botanica Acta*, **107**, 243–250.

Vögeli-Lange R. and Wagner G.J. (1990) Subcellular localization of cadmium and cadmium-binding peptides in tobacco leaves. *Plant Physiology*, **92**, 1086–1093.

Vose P.B. and Randall P.J. (1962) Resistance to aluminium and manganese toxicity in plants related to variety and cation exchange capacity. *Nature*, **196**, 85–86.

Wagatsuma T. and Ezoe Y. (1985) Effect of pH on ionic species of aluminum in medium and on aluminum toxicity under solution culture. *Soil Science and Plant Nutrition*, **31**, 547–561.

Wagner G.J. (1993) Accumulation of cadmium in crop plants and its consequences to human health. *Advances in Agronomy*, **51**, 173–212.

Walker N.A. and Pitman M.G. (1976) Measurement of fluxes across membranes. In *Encyclopedia of Plant Physiology*, edited by U. Lüttge and M.G. Pitman, pp. 93–98. Berlin: Springer- Verlag.

Wallace S.U., Henning S.J. and Anderson I.C. (1982) Elongation, Al concentration, and hematoxylin staining of aluminum-treated wheat roots. *Iowa State Journal of Research*, **57**, 97–106.

Wang J., Evangelou B.P., Nielsen M.T. and Wagner G.J. (1991) Computer-simulated evaluation of possible mechanisms for quenching heavy metal ion activity in plant vacuoles. I. Cadmium. *Plant Physiology*, **97**, 1154–1160.

Wang J., Evangelou B.P. and Nielsen M.T. (1992a) Surface chemical properties of purified root cell walls from two tobacco genotypes exhibiting different tolerance to manganese toxicity. *Plant Physiology*, **100**, 496–501.

Wang J., Evangelou B.P., Nielsen M.T. and Wagner G.J. (1992b) Computer-simulated evaluation of possible mechanisms for sequestering metal ion activity in plant vacuoles. II. Zinc. *Plant Physiology*, **99**, 621–626.

Watmough S.A. and Dickinson N.M. (1995) Multiple metal resistance and co-resistance in *Acer pseudoplatanus* L. (Sycamore) callus cultures. *Annals of Botany*, **76**, 465–472.

Wheeler D.M. and Power I.L. (1995) Comparison of plant uptake and plant toxicity of various ions in wheat. *Plant and Soil*, **172**, 167–173.

Wheeler D.M., Power I.L. and Edmeades D.C. (1993) Effect of various metal ions on growth of two wheat lines known to differ in aluminium tolerance. In *Plant Nutrition - from Genetic Engineering to Field Practice*, edited by N.J. Barrow, pp. 723–726. Dordrecht: Kluwer Academic Publishers.

Wollgiehn R. and Neumann D. (1995) Stress response of tomato cell cultures to toxic metals and heat shock: differences and similarities. *Journal of Plant Physiology*, **146**, 736–742.

Wolterbeek H.T., Van der Meer A. and De Bruin M. (1988) The uptake and distribution of cadmium in tomato plants as affected by ethylenediaminetetraacetic acid and 2,4-dinitrophenol. *Environmental Pollution*, **55**, 301–315.

Yamamoto Y., Chang Y.C., Ono K. and Matsumoto H. (1994) Effects of aluminum on the toxicity of iron (II), copper and cadmium in suspension-cultured tobacco cells. *Bulletin of the Research Institute for Bioresources Okayama University*, **2**, 181–190.

Yang C.S., Schaedle M. and Tepper H.B. (1989) Phytotoxicity of scandium in solution culture of loblolly pine (*Pinus taeda* L.) and honey locust (*Gleditchia triacanthos* L.). *Environmental and Experimental Botany*, **29**, 155–164.

Yeo A.R. and Flowers T.J. (1986) Salinity resistance in rice (*Oryza sativa* L.) and a pyramiding approach to breeding varieties for saline soils. *Australian Journal of Plant Physiology*, **13**, 161–173.

Zhang G. and Taylor G.J. (1989) Kinetics of aluminum uptake by excised roots of aluminum-tolerant and aluminum-sensitive cultivars of *Triticum aestivum* L. *Plant Physiology*, **91**, 1094–1099.

Zhang G. and Taylor G.J. (1990) Kinetics of aluminum uptake in *Triticum aestivum* L. Identity of the linear phase of aluminum uptake by excised roots of aluminum-tolerant and aluminum-sensitive cultivars. *Plant Physiology*, **94**, 577–584.

12. MECHANISMS OF MECHANICAL STRESS RESISTANCE IN PLANTS

RUSSELL SCOTT JONES

Mountain State Environmental Services, 120 Cloverdale Heights Charles Town, West Virginia 25414, USA.

INTRODUCTION

All plant life, from the loftiest redwood to the tiniest unicellular alga, is continually buffeted by dynamic, physical forces at every stage of growth. Land dwelling plants experience dynamic, physical forces in the form of wind and precipitation, from the passage of animals and machinery, and by the abrasion of elongating plant roots by soil particles. Aquatic plants are additionally subjected to agitation by waves, tides, and currents, or by water movement in river and stream beds. The continual, yet discontinuous, shaking, rubbing, bending, twisting, and vibrating induces in plants a condition known as mechanical stress. Plant responses to mechanical stress include a variety of morphological and physiological changes that may differ in degree and complexity depending upon the manner in which the stress was received by the plant. The most common and visible plant responses are reductions in stem height and leaf area, and increases in stem diameter (Figure 12.1). Physiological, biochemical, and molecular responses are less apparent, but all contribute to the observed changes in plant size, mass, and architecture. Morphological changes that occur in plants exposed to mechanical stress may play a significant role in adaptation, ecology, and evolution, particularly in windy environments.

Any shaking, rubbing, twisting, bending, or vibrational stimuli applied to a plant may be received singly or simultaneously. For example, a light breeze may only cause slight leaf fluttering, but as wind velocity increases, the entire shoot may begin swaying, twisting, and bending and adjacent leaves and branches start rubbing together. Similarly, the impact of raindrops upon exposed shoots simultaneously will induce both rubbing and shaking stresses. Plants growing along roadsides and highway medians may receive vibrational stimuli from vehicular traffic propagated through the soil to the root zone. When grown in controlled-environment facilities, plants will also be exposed to mechanical vibrations from fans, pumps, etc. Except for vibration, the morphological and physiological responses to mechanical stress are similar. Vibrational effects are less clearly understood and have not been investigated in detail.

Figure 12.1. Seismic stress effects on growth of greenhouse-grown, vegetative soybean. The plant on the left was undisturbed. The plant on the right received seismic stress for 20 consecutive days. Seismic stress was applied by placing the plant on a rotatory shaker twice-daily, for 10-minute intervals, at 190-240 rpm (Jones and Mitchell 1989, unpublished).

TERMINOLOGY

Mechanical stress is a condition induced in plants by exposure to non-injurious, dynamic forces that cause the physical displacement of individual plant organs and/or the entire plant body (i.e., the plant is mechanically stressed). The term *mechanically-induced stress*, or *MIS*, also has been used to describe this physiological condition (Biddington, 1986). Certain, unique growth and developmental changes occur in response to mechanical stress. The term *thigmo*morphogenesis (Jaffe, 1973) is generally used to describe the growth and developmental responses of plants to the rubbing, twisting, and/or bending of plant organs caused by *physical contact*. *Seismo*morphogenesis (Mitchell *et al.*, 1975) describes plant responses specifically induced by *shaking*. The term *vibric* (Coe and Mitchell, 1989) has been used to describe plant stress induced by vibrational stimuli. Growth and developmental changes occurring in response to vibration could be described as *vibro*morphogenesis. These definitions are not all encompassing. Thigmic rubbing, twisting, or bending of a plant organ usually causes other parts of the plant body to experience some degree of shaking. Conversely, seismic shaking of a whole plant may result in the twisting and bending of individual plant parts, and the rubbing together of adjacent ones. A precise definition for vibric forces has not been formulated, although plant studies using vibrational stimuli have been conducted on a limited basis (Quirk and Freese, 1976; Coe and Mitchell, 1989). Furthermore, the intensity at which increasingly severe vibric stress becomes seismic stress has not been clearly defined. Since some clarification on terminology is needed, the following definitions are suggested here to delineate the *qualitative* nature of vibric and seismic stresses: (i) vibrational stress occurs only when mechanical stimuli cause a plant organ to be physically displaced no more than a distance equal to the thickness and/or diameter of the plant organ receiving the stimuli (microdisplacement); (ii) seismic stress occurs only when mechanical stimuli cause a physical displacement greater than the thickness and/or diameter of the plant organ receiving the mechanical stimuli (macrodisplacement); and (iii) vibric and seismic stimuli may be applied/received simultaneously in the same or different plant organs. For example, greenhouse-grown plants may receive seismic stimuli via air movement by circulating fans, and also receive vibric stimuli propagated through the floor or greenhouse bench via fan motor vibration. The preceding are only a working definitions and may be refined as more information pertaining to the unique responses of seismic, thigmic, and vibric stresses are reported in the literature. For purposes of this discussion, the terms thigmic, seismic, and vibric will refer to the *manner in which mechanical stress is applied* to plants rather than to any specific stress-induced phenomena. All plant mechanical stress will probably cause some deformation of the affected plant cells, regardless of how the stress is applied.

METHODS USED TO AVOID AND/OR SIMULATE ENVIRONMENTAL MECHANICAL STRESS FOR EXPERIMENTAL PURPOSES

The objective of all plant mechanical stress research is to elucidate the manner by which plants respond and adapt to dynamic, physical forces in the environment, with particular emphasis on the wind. Two approaches may be used to study wind effects: avoidance and mimicry. The avoidance approach seeks to reduce or eliminate the dynamic physical forces imposed by the wind, such as swaying or twisting. Wind-sway of trees may be reduced by the attachment of guy wires (Jacobs, 1954; Larson, 1965; Holbrook and Putz, 1989) or by crowding trees in dense stands (Holbrook and Putz, 1989). Caution must be exercised when interpreting the data from experiments using the "attachment" or "crowding" protocols. Trees supported by guy wires or other supporting devices will receive thigmic stress at the points where guy wires are attached and localized increases in stem diameters will occur. Jacobs (1954) observed that trunk diameters of Monterey pine (*Pinus radiata*) were ca. 24% larger just above guy wire supports than the diameters of unsupported trees at equivalent heights, even though the average diameter of the free swaying trees was greater overall. In addition, the effects of thigmic stress may not be localized to the region of thigmic contact. Erner *et al.* (1980) have reported evidence for an unidentified, translocatable substance produced in beans in response to thigmic stress. If such a translocatable "thigmic stress factor" can be confirmed, experiments using trees supported by guy wires or other means of support may require reinterpretation.

When crowding trees into dense stands to reduce wind-induced stress, the effects of mutual shading must be taken into account. Shading effects on stem diameter and height growth of trees are much greater than those resulting from reduced wind sway (Holbrook and Putz, 1989) and may mask growth responses resulting soley from wind avoidance.

Non-woody crop species have been protected from wind-induced stress by the use of natural or artificial windbreaks. Natural windbreaks may be created by planting strips of a relatively short-statured crop (beets, peas, soybeans, etc.) within strips of tall-statured crops, such as corn or sunflowers (Rosenberg, 1966; Senthinathan *et al.*, 1984), and larger crop areas may be protected by mature belts of trees (Ogbuehi and Brandle, 1982). Artificial windbreaks have been constructed using snow fences, slat cribbing, and solid board fences (Bagley, 1964; Rosenberg, 1966; Frank *et al.*, 1974; Radke and Hagstrom, 1974). Caution must be exercised when reporting and comparing the data from experiments using natural and artificial windbreaks. Different barrier types will have different levels of permeability to the wind (Bagley, 1964; Radke and Hagstrom, 1974) and may permit varying degrees of wind stress upon the "sheltered" plants. Bagley (1964) observed that field-grown tomato and bean plants growing immediately adjacent to a slat fence were smaller than plants growing in rows a few feet away. Complete avoidance of wind-induced stress by using wind-proof enclosures in the field may be impossible to accomplish without compromising other environmental growth factors, such as light intensity, temperature, humidity, and CO_2 concentration. Field experiments also may be hampered by the inability to adequately separate the effects of wind-induced stress on plant growth from the growth responses induced by other environmental variables. Variations in light intensity (Jones and Mitchell, 1990), temperature (Hunt and Jaffe 1980), and water availability (Frank *et al.*, 1974) have been shown to modify plant responses to artificial and wind-induced mechanical stress.

Several approaches have been used to mimic the effects of wind upon plant growth in the laboratory. The most direct approach is to increase the air stream velocity passing around the experimental plants. Early investigators of wind stress on plant growth used fans or rudimentary wind tunnels consisting of an open-ended glass jar with one end attached to a commercial ventilator fan (Hill, 1920; Finnel, 1928). Data reported from different laboratories using these methods were difficult to compare because of a lack of uniformity in experimental material, growth environments, and the degree of stress (i.e. wind velocity) applied (Wadsworth, 1959). Although plant growth was generally reduced by artificial wind stress, these experiments also were hampered by the inability to rigorously control the environmental conditions. Wind velocities also tended to be relatively high (up to 15 mph) and young seedlings usually suffered from some degree of desiccation. Therefore, some of the wind-induced growth effects may have been masked by water stress. Improvements in instrumentation in more recent years permitted the application of a range of controlled, unidirectional airstream velocities under controlled environment conditions. Hunt and Jaffe (1980) mounted an oscillating, variable speed fan within a controlled-environment chamber to simulate wind stress on kidney beans. Controlled-environment wind tunnels also have been used to test the effects of wind on growth and water relations of Fraser fir (*Abies fraseri*) (Rees and Grace, 1981) and fescue (*Festuca arundinacea*) grass (Grace and Russell, 1981). One problem that arises from the use of fans and/or wind tunnels involves the inability to separate plant growth responses resulting solely from dynamic wind action from those growth responses induced by the altered microclimate surrounding the wind-stressed plants. As air stream velocities around plants change, changes in evapotranspiration, water potential, humidity, and photosynthesis also will occur (Nobel, 1974). These environmental parameters may increase or decrease depending upon the plant species used, plant moisture status, the wind velocity applied, and the duration of exposure to wind stress.

Whole plants or individual plant organs may be subjected to dynamic physical forces, with little effect on microclimate, by direct application of thigmic, seismic, or vibric stimuli. Thigmic stress may be applied manually by gently rubbing, bending, and/or twisting plant organs between thumb and forefinger (Turgeon and Webb, 1971; Mitchell *et al.*, 1975; Biro *et al.*, 1980; Jones and Mitchell, 1988), brushing with stiff paper (Biddington and Dearman, 1987; Wurr and Fellows, 1987), or by gentle rubbing with a hard object such as a pipette or a syringe needle (Huberman and Jaffe, 1986). Jaffe (1980) has devised a "thigmostimulator" that uses a small roller to apply known amounts of force (measured in newtons) to plant stems. Thigmic stimuli may be applied to specific regions of the plant body, with little disturbance to the remainder of the plant body. Thigmic stress episodes are brief (usually no more than ca. 20 seconds), and only one or two episodes daily are required to produce visible growth responses. In contrast to thigmic stress, seismic stress generally must be applied to the whole plant. Seismic stimuli are applied by manually shaking or rotating individual pots containing one or more plants (Steuchek and Gordon, 1975) or by placing potted plants on motorized shaking platforms (Mitchell *et al.*, 1975; Jones *et al.*, 1990). Twice-daily episodes of seismic stress for periods of 10–15 minutes generally produce visible growth effects in about a week. Vibrational stimuli have been applied to pine trees by fitting the boles with cushioned ring-collars attached to small induction motors (Quirk and Freese, 1976), and by attachment of commercial vibrators to the bases of potted peas (Coe and Mitchell, 1989).

Beyl and Mitchell (1977) constructed automatic mechanical oscillatory shaking (AMOS) devices for the purpose of applying automatic, uniform seismic and thigmic stimuli to many plants simultaneously. Each AMOS consisted of an adjustable sliding rack mounted within a stationary frame that could be attached to a standard greenhouse bench. The heights of greenhouse-grown chrysanthemum cultivars were reduced 16–30% by four minutes of daily application of AMOS treatment for 28 days. Another form of mechanical stress application involves the use of water sprays to plant foliage. Height of greenhouse-grown tomatoes were reduced up to 40% when subjected daily to 10 seconds of gentle water sprays for a period of 14 weeks (Wheeler and Salisbury, 1979). For short-term laboratory experiments, Knight *et al.* (1992) mimicked wind-induced stress by forcing variable amounts of air through syringes.

Automated and/or mechanical methods of applying thigmic, seismic, and vibric stresses are preferred over manual methods because the degree and duration of applied stress is quantifiable, repeatable in the degree and duration of stress, and easily duplicated by different investigators in different laboratories. From a commercial standpoint, automated methods of mechanical stress application may reduce dependence on costly and increasingly regulated plant growth retardants to inhibit the "leginess" that commonly occurs in greenhouse-grown ornamental and food species.

ECOLOGICAL AND ADAPTIVE SIGNIFICANCE OF MECHANICAL STRESS RESPONSES

One of the most pervasive dynamic physical forces in the environment is the wind. Terrestrial vegetation is continually buffeted by winds of constantly changing direction and velocity. Wind velocity tends to be greater at higher altitudes and windward slopes, and decreases in valleys and leeward slopes. Exposure to wind-induced mechanical stress is not necessarily detrimental, and may provide some measure of hardiness against environmental extremes. Trees will show distinct morphological changes with altitude and exposure to variable wind stress. Lawton (1982) observed that for trees growing in a tropical mountain rainforest, trunk and twig diameters increased, and tree height decreased with proximity to the crests of ridges. Internodes of the trees also tended to be shorter on the windward side of the tree, giving the tree the appearance of leaning into the wind. This compact growth habit resulting from wind-induced stresses, enhanced the ability of the trees to withstand greater wind stress at higher altitudes. Flag-formation in tree crowns also is a beneficial adaptation for trees growing in a windy environment. Telewski and Jaffe (1986) reported that reduced crowns in flag-formed trees significantly decreased the wind-speed-specific drag and, therefore, reduced potentially injurious loads upon the trunk.

Weaker stems may result when vegetation is protected from wind stress. When larch *(Larix spp.)*, sweet gum *(Liquidambar styraciflua)* and Monterey pine *(Pinus radiata)* saplings were supported by guy wires, the trunks tended to be taller and thinner when compared to unsupported trees (Jacobs, 1954; Larson, 1965; Holbrook and Putz, 1989). When the trees were released from support, they became less stable than their unsupported neighbors and more susceptible to blowdown by strong wind gusts. Artificial shading further enhanced the taller and thinner growth habit of wire-supported sweet gum. These tall, thin trees exceeded what was described as the "critical buckling height", above which the trees were prone to lodging (Holbrook and Putz, 1989), and had growth habits similar to sweet gum trees growing within dense stands of pine. Tateno (1991) reported that daily rubbing, brushing, or shaking of two-year-old mulberry trees decreased stem height and increased the "lodging safety factor". The lodging safety factor was defined by Tateno and Bae (1990) as the ratio of "critical lodging load" to fresh leaf weight. The critical lodging load is defined as the minimum leaf fresh weight required to cause the lodging of a plant, and is analogous to the critical buckling height described by Holbrook and Putz (1989).

The ecological significance of these observations is that dense stands of trees, mutually shaded and mutually protected from the wind, may be extremely susceptible to breakage from excessive load-inducing phenomena (strong wind gusts, heavy snow accumulation, etc.) as the trees approach their critical buckling height and/or critical lodging load. When an area of dense, lodging susceptible trees are subjected to blowdown, species diversity may change. Increased light penetration in the area of the blowdown will favor new growth of high light intensity-requiring plant species that had previously been suppressed by low light levels in the understory. Stands with more widely separated trees that permit greater light penetration and a moderate amount of wind stress will be less likely to exceed the critical buckling height and/or critical lodging load and have an increased lodging safety factor. Tree stands having greater lodging safety factors, therefore, will be less susceptible to blowdown and the species mix will tend to remain constant provided all other environental variables remain within acceptable limits.

The changes in growth habit and architecture of herbaceous plants in response to wind-induced physical agitation are similar to those observed for woody species. Outdoor-grown soybeans (Pappas and Mitchell, 1985), eggplant and soybeans (Latimer *et al.*, 1986), and kidney beans (Hunt and Jaffe, 1980), were reduced in stem height and leaf area, and had a tendency toward increased stem diameter when compared to the same species grown in greenhouses, and thereby protected from the mechanical effects of wind and precipitation. Shorter (and sometimes thicker) stems, reduced leaf area, changes in leaf angle, and flag formation (in trees) all serve to reduce the surface area of the plant exposed to the wind.

From an agroecological stand point, it seems desirable to expose crop species to some degree of physical agitation to "harden" the stems and improve resistance to lodging. In practice, the benefits derived from stress-induced improvements in lodging resistance may be offset by reductions in harvestable yield. Several investigators have reported that protection of crop plants from the wind by use of natural or artificial windbreaks increases growth and yield. Greater growth and productivity have been reported for field-grown fescue (Grace and Russell, 1982), soybeans (Sturrock, 1969, 1974; Radke and Burrows, 1970; Radke and Hagstrom, 1973; Frank *et al.*, 1974; Ogbuehi and Brandle, 1981), sugar beets (Rosenberg, 1966), snap beans and tomatoes (Bagley, 1964), and wheat (Pelton, 1967; Frank and Willis, 1972; Skidmore *et al.*, 1974) when protected from wind-induced stress. When water was not limiting, the increased growth and yield of windbreak-protected crops were generally attributed to increased water use efficiency, rather than to any other direct effects of physical agitation. Increased water use efficiency was related to decreased evaporative demand, higher relative humidity, higher daytime temperatures and lower night-time temperatures in the sheltered areas. However, one study noted that the leaf area profile of wind-protected soybeans permitted more light penetration into the lower canopy (Ogbuehi and Brandle, 1982). Sheltered soybeans had less leaf area density in the upper two-thirds of the canopy than did the sheltered plants. Ogbuehi and Brandle (1982) inferred that the improved light environment coupled with a non-limiting water supply increased yields of sheltered soybeans. These data agree with previous reports that show a direct relationship between supplemental light intensity in the lower canopy and increased yield of field-grown soybeans (Johnson *et al.*, 1969). However, when water is a limiting factor, windbreaks may enhance drought effects. Windbreak-sheltered dryland soybeans showed an increase in early vegetative growth when compared to exposed plants, but later growth was inhibited because of more rapid soil water depletion (Frank *et al.,* 1974).

Increased light penetration to lower leaves within the canopy does not fully explain the effects of windbreak protection upon soybean yield. Jones and Mitchell (1992) subjected greenhouse-grown soybeans to brief (10 min), twice-daily episodes of seismic stress using a modified rotatory shaker to simulate wind stress. Between stress applications, potted soybeans plants were placed on a greenhouse bench such that mutual shading was kept to a minimum, and an automatic watering system assured that water and nutrients were not limiting. They observed that when stress was applied after anthesis, two components of yield (pods/plant and seeds/pod) were reduced significantly, whereas individual seed weights were unaffected (Figure 12.2). Yield components of plants exposed to seismic stress prior to anthesis also were relatively unaffected when compared to non-stressed plants. These observations indicate that other unidentified, yield-related, physiological processes are likely perturbed when plants are subjected to wind-induced and artificial shaking. One explanation may be that mechanical stress induces a repartitioning of photosynthates away from reproductive tissues toward vegetative tissues as a means of increasing the structural strength of stems and petioles.

The compact growth habit exhibited by mechanically stressed plants generally results from reduced internode and petiole elongation, increased stem and petiole diameters, reduced leaf area, and increased leaf thickness, although there

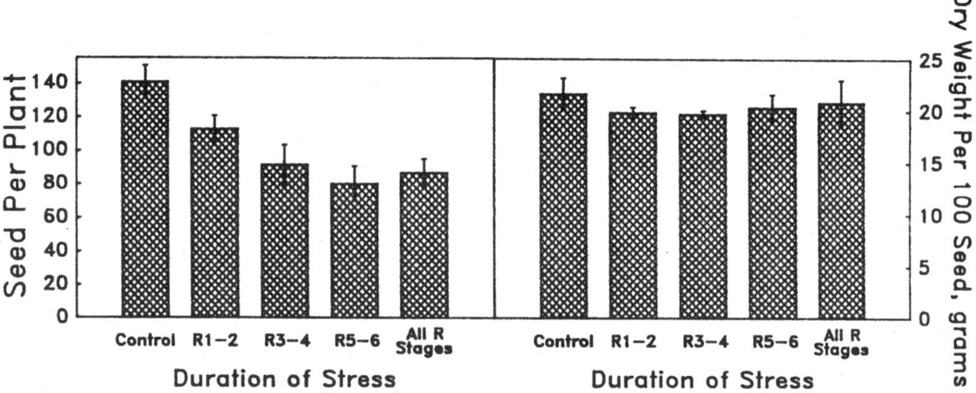

Figure 12.2. Soybean yield components after twice-daily seismic stress treatments applied during reproductive growth stages R1-2 (first flower to full flower), R3-4 (beginning pod to full pod), R5-6 (beginning seed to full seed), and R1-6 (first flower to full seed). Seed weight per plant was highly correlated (data not shown) with seed number per plant and pod number per plant (data used for bar graph from Jones and Mitchell, 1992).

are a few exceptions. For example, petiole diameters of tomato seedlings (Heuchert *et al.,* 1983) and stem internode diameters of soybeans (Jones *et al.,* 1990) were relatively unaffected, or reduced (Pappas and Mitchell, 1985a; Latimer *et al.,* 1986) when the plants were exposed to mechanical stress. The reasons for these contradictory growth responses for stems and petioles are not clear, but may result from species specific responses to mechanical stress.

Even within a species, plant response may be variable. Steuchek and Gordon (1975) reported that dark-grown coleoptiles of a lodging-susceptible oat cultivar *(Avena sativa* cv. Arthur) were insensitive to mechanical stress, whereas coleoptiles from a lodging-resistant cultivar (Blue Boy) were growth-inhibited after exposure to brief periods (1–2 min) of seismic stress. Telewski and Jaffe (1986) showed that 23 genetically different strains of loblolly pine each varied in the degree to which stems shortened and thickened, needle area was reduced, and stem flexibility was decreased by brief (20 sec), daily, thigmic bending over a six month period. The genotypic responses to mechanical stress described above demonstrate that genetic variation in the ability of a plant to perceive and respond to mechanical stimuli may be a factor in species adaptation to windy environments.

MECHANICAL STRESS EFFECTS ON GROWTH AND DEVELOPMENT

Plant growth is visibly inhibited by mechanical stress within 24 hours to several days depending upon the species that has been stressed, its stage of growth, the type of stress applied (thigmic or seismic), and the environmental conditions under which the stress was applied (Jaffe, 1973, 1976; Jones and Mitchell, 1988, 1990). Using an instrument as simple as a metric ruler, measurable stem growth inhibition can be observed in dark-grown soybean ca. 1 hour after a single thigmic episode (Jones and Mitchell, 1989b). With angular position-sensing transducers (Mitchell and Coe, 1991) or auxanometric techniques (Jaffe, 1973), it can be shown that plant growth is inhibited within a few seconds to a few minutes after brief exposure to mechanical stress. The kinetics of the growth response are complex and not well understood. Jaffe (1973) reported that growth of thigmic stressed green beans *(Phaseolus vulgaris* L.) transiently increased three minutes after stress application, stopped completely by six minutes after stress application, and growth did not resume for up to 18 minutes. In contrast, thigmic stress immediately inhibited growth of dark-grown peas (*Pisum sativum)* and soybeans (Mitchell and Coe, 1991), with no initial transient growth stimulation. Peas resumed growth within 10 minutes after the thigmic episode, and had recovered to 70% of the pre-stress growth rate after 30 minutes. However, Mitchell and Coe (1991) also reported that not all soybeans responded similarly to thigmic stress. In a preliminary experiment, ca. 60% of 'Century 84' soybean seedlings exposed to thigmic stress showed growth rates at 36% of the non-stressed controls, whereas growth of the remainder of the stressed seedlings was unaffected. In those soybeans that were growth inhibited, the response was rapid and growth rates were depressed for up to seven hours after stress application (Mitchell and Coe, 1991). Clearly, further investigation is needed to resolve these apparently contradictory data.

In the preceding discussion, it can be deduced that both short- and long-term responses are involved in the morphological and physiological changes that occur in plants after exposure to mechanical stress. Short-term response mechanisms may even interact in synergistic fashion with previously induced long-term responses to reinforce or enhance the morphological and physiological effects of mechanical stress. Mechanical stress-induced stem growth inhibition results primarily from an inhibition of internode elongation, although in some species such as wild cucumber (*Bryonia*) and tomato (*Lycopersicon esculentum* Mill.), mechanical stress also reduces the number of new nodes produced (Jaffe, 1973; Mitchell, 1975). Stress effects on stem growth are most dramatic in those stem regions having the greatest potential for further elongation (i.e. the zone of maximum elongation), although all extension growth is ultimately reduced by mechanical stress. In light-grown plants, the most rapidly elongating regions are the uppermost internodes of stems, whereas in dark-grown seedlings (Figure 12.3A-C), a region immediately below the apical hook of a hypocotyl (e.g soybean, sunflower) or the uppermost internode of an epicotyl (e.g. pea) show the most extension growth (Jones and Mitchell, 1988). Mechanical stress effects on leaf expansion have not been well characterized except to show a general pattern of leaf area inhibition and increased specific leaf weight.

The short- and long-term response mechanisms responsible for stress-induced growth inhibition include biophysical mechanisms, involving water relations and

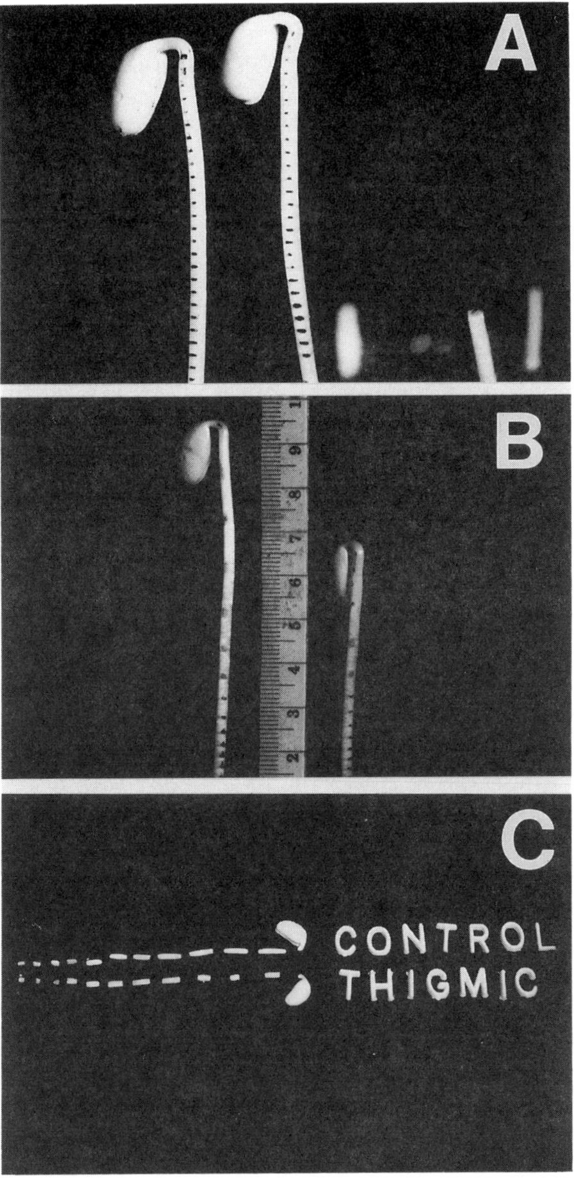

Figure 12.3. (A) One-day-old, dark-grown soybean seedlings, marked at 2-mm increments with water-based India ink (from Jones and Mitchell, 1988a.) (B) The same seedlings 24 hours after the seedling on the right received thigmic stress as 20, gentle strokes between thumb and forefinger at the apical hook; compare the extension growth between the ink marks on the hypocotyl of each seedling (from Jones and Mitchell, 1988a.) (C) Comparison of sections excised from each hypocotyl; excisions were made at the base of each ink mark. Note that the zone of maximum growth extension was in a region that originally comprised the uppermost 6-mm of the hypocotyls (Jones and Mitchell, 1988, unpublished).

physical restraints to cell expansion, and biosynthetic mechanisms, involving the stimulation or inhibition of new tissue formation, enzyme synthesis, hormone production, etc.

Biophysical Mechanisms

Plant growth occurs mostly by absorption of water by plant cells. When the resulting turgor pressure exceeds a certain threshold level, cell walls deform, cells expand, and plant size increases (Cosgrove, 1987). If turgor is reduced below this threshhold level, cell expansion and growth may be inhibited. Rapid turgor loss within enlarging cells may be a mechanism responsible for the observed rapid growth inhibitions induced by mechanical stimuli. Mechanical stress effects upon turgor are readily apparent. Figure 12.4 demonstrates the effects of 10 minutes of seismic stress upon turgor in a greenhouse-grown soybean. The drooping leaves and petioles of the stressed plants indicate that water absorption and/or transport has been perturbed. The author has observed similar responses in leaves of sunflower *(Helianthus annuus* L.) and rapeseed *(Brassica napus)* following 5–10 minutes exposure to seismic stress, and in the petals of petunia flowers exposed to ca. 10 seconds of a gentle water spray from a garden hose (Jones, unpublished). Unfortunately, these observations do not answer the central question of whether rapid stress-induced growth inhibition results from rapid stress-induced turgor loss in the cells of enlarging tissues. Turgor in enlarging tissues of intact plants can be measured using thermocouple psychometry (Boyer *et al.*, 1985), or in individual cells of intact plants using a micropressure probe (Steudle *et al.*, 1982), but no one has yet used these techniques to assess the effects of mechanical stress upon turgor and plant growth. Less direct methods have been used to infer the involvement of mechanical stress-induced turgor changes and growth inhibition. Myers and Mitchell (1992) used an experimental system consisting of dark-grown soybean seedlings grown in a solution of 5 mM Ca^{2+} containing variable concentrations of K^+ (1 to 10 mM). They (Myers and Mitchell, 1992) reported that a single episode of thigmic stress applied to apical hook of a soybean seedling resulted in an immediate decrease in elongation rate (up to 85%) followed by a slower recovery phase that lasted several hours. The recovery rate was enhanced 60% in seedlings growing with 10 mM K^+ in the nutrient solution. Furthermore, seedlings grown with K^+ had a more negative solute potential in the elongation zone of the hypocotyl when compared to control seedlings 24 hours after stress application. Myers and Mitchell (1992) suggested that an osmotic adjustment in the cells of the elongation zone occurred in response to the thigmic stress.

One confounding factor that must be addressed in any experiment involving mechanical stress effects on turgor, is that the growth of expanding cells may be inhibited without corresponding changes in the turgor of these cells. Cosgrove (1990) demonstrated that there were no significant turgor pressure differences between cells on the upper and lower sides of a gravistimulated cucumber stem, even though cells on the upper side of the stem expanded more slowly than cells on the lower side. Boyer (1988) observed that water potential (and therefore, turgor) in expanding tissues changes slowly in response to environmental stimuli because they contain many small cells with poor vascular development. However, these expanding cells may be very sensitive to water potential changes in the xylem. For example, when a corn (*Zea mays* L.) seedling was excised at its base, growth and xylem water potential decreased immediately, whereas water potential in the growth zone remained unchanged (Boyer *et al.,* 1985). These rapid changes in xylem water potential also occur when transpiration changes, or water transport from the roots is reduced (Boyer, 1988), such as by excision or water stress. Transpiration and water conduction from the roots also are reduced by mechanical stress (Beyl and Mitchell, 1983; Pappas and Mitchell, 1985b). Since the xylem is a continuous hydraulic system within the plant (Boyer, 1988), the

Figure 12.4. Seismic stress effects on whole plant turgor. Soybean plant on the right received 10 minutes of seismic stress applied using a modified rotatory shaker at 190 rpm. Compare drooping leaves and petioles of the stressed plant to the upright leaves and petioles of the non-stressed plant on the left (Jones and Mitchell 1989, unpublished).

water potential change would be rapidly transmitted throughout the plant, and growth would be inhibited. The mechanism by which changes in xylem water potential inhibit growth without altering the turgor of expanding cells is not known.

The mechanism by which mechanical stimuli may influence cell turgor and/or xylem water potential to produce inhibitory growth responses also is unknown. Beyl and Mitchell (1983) reported that xylem exudation rates of detopped sunflowers that had been seismic- and thigmic-stressed were 64% and 40%, respectively, of detopped control plants. Since there was no difference in root mass between control and treatment plants, it was suggested that reduced exudation was due to reduced root pressure. Although xylem water potential in the stressed plants was apparently altered, the mechanism of control was not evident. Edwards and Pickard (1987) suggest that any bending, flexing, or pressure exerted upon plant tissues by mechanical stimuli may cause the opening of "stretch-activated" ion channels in the plasma membrane. When suction pressure was applied to a portion of an isolated plasma membrane, discrete electrical voltage pulses were observed, and each pulse represented fluxes of ions through individual channels in response to pressure changes (Edwards and Pickard, 1987). Rapid, outward fluxes of osmotically active ions (e.g. K^+), mediated by "stretch-activated" ion channels, could provide a basis for rapid turgor changes in expanding plant cells and subsequent growth inhibition. This hypothesis is attractive because it does not depend on xylem water potential changes to induce the mechanical stress response. The experiments described by Edwards and Pickard (1987) used isolated protoplasts derived from cultured cells of tobacco stem pith; therefore, caution should be taken in using these studies as a model for intact plants. Rapid changes in electrical resistance also have been observed in bean internodes (Jaffe, 1976a) and pea epicotyls (Pickard, 1971) in response to thigmic stress, but whether these electrical phenomena represent stress-induced opening of "stretch-activated" ion channels is unknown.

Another level of control in the mechanical stress response may be exerted in the interaction between the inner and outer tissues of plant organs. In rapidly enlarging organs, the cell walls of epidermal cells tend to be thicker and less extensible than the cell walls of the parenchyma tissues that they enclose. Forces produced by the expansion of inner tissues are balanced by the contractile forces of the outer epidermis (Kutschera *et al.*, 1987; Kutschera, 1989). Therefore, the extension of an internode or petiole, or the expansion of a leaf will be limited by the rate at which epidermal cells expand, and the expansion rate will be limited by the extensibility of the epidermal cell walls. Does mechanical stress-induced growth inhibition result from decreased cell wall extensibility? As a short-term response, probably not. Jones and Mitchell (1989b) observed that cell wall extensibility, as measured by an Instron-type device (Cleland, 1984), apparently increased up to 54% in dark-grown soybean hypocotyls 24 hours after a single thigmic episode. The elastic (non-growth) component of extensibility was most affected by mechanical stress and elongation growth was reduced 40–50% in the same time period (Jones and Mitchell, 1989b). Microscopic studies revealed that the epidermal cell lengths of stressed hypocotyls were approximately half to two-thirds the lengths of epidermal cells of non-stressed hypocotyls (Jones and Mitchell, 1989b). Since the original extensibility measurements were conducted on equivalent lengths of excised hypocotyl sections (Jones and Mitchell, 1989b), it became clear that the apparent 54% extensibility increase occurred because more cells were being extended in stressed hypocotyls by the Instron-extension. Therefore, wall extensibility of individual cells in the growth zones of thigmic-and non-stressed soybean hypocotyls probably did not differ, but the stressed hypocotyls had greater elastic extensibility per unit length because there were more cells in the vertical profile. Thus, the growth-inhibited cells retained their potential for expansion, as measured by Instron-extension. Jones and Mitchell (1989b) have termed this response as the "unextended 'Slinky' model", after the child's spring toy (Figure 12.5). In this model, the cells in the stem of an unstressed plant would be equivalent to a spring pulled to its maximum extension, with little potential for further elastic extension. In this fully extended state, the vertical profile stem cells would be susceptible to plastic deformations when subjected to severe bending stress. In the partially extended state, the stem cells would retain the potential for further elastic extension and would be less susceptible to plastic deformations when subjected to bending or swaying stress. The following two studies using light-grown plants tend to support the "Slinky" model.

Heuchert and Mitchell (1983) observed that hypocotyls, first internodes, and petioles of seismic-stressed tomatoes were shorter and had significantly greater elasticity, shear strength, and/or rupture strength than non-stressed controls. When Jones and Mitchell (1992) subjected greenhouse-grown soybeans to brief, daily episodes of seismic stress, they observed that soybeans stressed only during vegetative growth, or throughout the both vegetative and reproductive growth stages had shorter stems and remained upright until harvest at physiological maturity, without any means of artificial support (Figure 12.6A & B). Non-stressed plants, and plants that received seismic stress only after the onset of reproductive growth, had taller, thinner stems that had to be staked for vertical support by the time the pods had filled. When support was removed, the soybeans bent completely over under the weight of their pods and leaves (Figure 12.6A & B). Unfortunately, no detailed anatomical studies were conducted on either the tomato or soybean stems used for the experiments described above. Biro *et al.* (1980) reported that inhibited stem growth of thigmic-stressed wax beans was associated with reduced elongation of epidermal and cortical cells as well as fewer numbers of cells in the vascular tissue and pith, but conducted no measurements of stem strength or elasticity. Only a few mechanical stress studies have combined limited anatomical examinations with measurements of stem physical properties. Stems of Fraser fir were shorter and showed increased elasticity as a result of wind exposure or manual shaking (Telewski and Jaffe, 1986), when compared to non-stressed trees. The decrease in stem elongation was associated with decreased tracheid length and increased grain angle of the wood (Telewski and Jaffe, 1986).

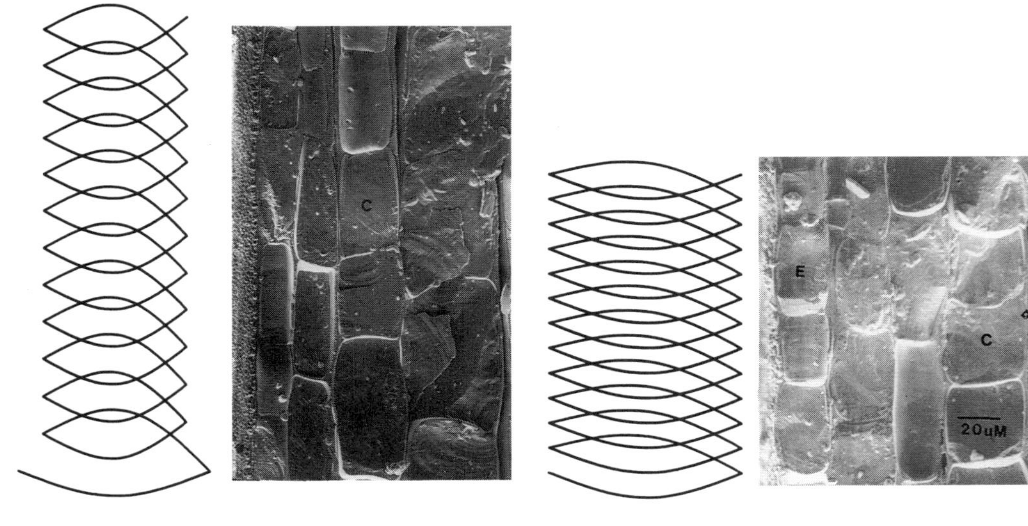

Figure 12.5. The Unextended "Slinky' Model". Electron micrographs of freeze-fractured, longitudinal sections excised from corresponding growth zones of dark-grown soybean hypocotyls. Samples were obtained 24 hours after a brief application of thigmic stress to the stressed seedling. Epidermal (E) cells are on the far left of each photograph. Epidermal and coritcal (C) cells are ca. 30% to 50% longer in the non-stressed hypocotyl and are equivalent to a spring pulled to its maximum extension (i.e. the cells are fully expanded) and are more susceptible to plastic deformations under bending or twisting stress. Hypocotyl cells of the stressed seedling are equivalent to a partially extended spring (i.e. the cells are partially expanded) and retain the potential for further elastic expansion (Jones and Mitchell, unpublished).

Mechanical stress does not always result in reduced cell lengths. Grace and Russell (1976) observed that when fescue seedlings were subjected to continuous artificial wind stress using fans, leaves were smaller, thinner, and "stiffer" than those on protected plants but epidermal cell lengths were the same. In contrast to the "Slinky" model, leaves of wind-stressed fescue showed almost complete elasticity. When stressed leaves were bent 40–50 degrees from the horizontal, they returned to almost the original unbent position, whereas non-stressed leaves retained ca. 12 degrees of residual plastic distortion (Grace and Russell, 1976). Contradictory data also are available for trees. When sweet gum saplings were guy-wired to prevent wind-induced sway, no differences were observed in wood elasticity when compared to unprotected trees even though the unprotected trees were shorter and thicker (Holbrook and Putz, 1989). According to the "Slinky" model, wind-stressed trees should have demonstrated increased elasticity.

It may be concluded from the discussion above, that adaptive advantages may be conferred by having plant organs with increased extensibility per unit length. The increased whole organ extensibility would permit a greater range of bending or swaying when plants are abruptly exposed to high mechanical loads, such as during wind gusts. Clearly additional data are needed from a wider variety of plant species collected under more uniform experimental conditions. Future investigations must combine in-depth anatomical studies with complete measurements of tissue physical strength as well as examining the hydrostatic forces responsible for driving cell expansion.

Biosynthetic Mechanisms

Both seismic and thigmic stimuli reduce stem elongation and increase the ratio of stem diameter to length, but stem thickening generally occurs only in response to thigmic stress (Biro *et al.*, 1980; Jaffe 1980). Stem diameters are usually unaffected or reduced by seismic stimuli when compared to non-stresssed controls (Heuchert *et al.*, 1983; Pappas *et al.*, 1986; Latimer and Mitchell, 1985a; Jones *et al.*, 1990). The basis for these different responses is not known. Where stem thickening does occur, it presumably results from a stress-induced radial increase in cell expansion and/or cell number. Unfortunately, only a few anatomical studies have been conducted. Biro *et al.* (1980) showed that increased radial growth of thigmic-stressed bean internodes resulted from increased cortical cell expansion and greater cell division activity in the vascular cambium resulting in greater secondary xylem production. Thigmic-stressed Fraser fir saplings had 50% and 129% more tracheids in the axes perpendicular and parallel, respectively, to the direction of thigmic bending than non-stressed controls (Telewski and Jaffe, 1986). More recently, it was reported that sunflower seedlings subjected to controlled bending for 60 seconds per day over a six week interval had thicker and shorter stems with a higher proportion of collenchyma to cortex than unstressed controls (Patterson, 1992).

In trees, particularly in windy environments, similar eccentric trunk development has been observed to occur parallel to the prevailing wind direction

Figure 12.6. Seismic stress effects on lodging resistance of 'McCall' soybeans. (A) plants attached to upright supports. (B) Plants released from upright supports. Non-stress = undisturbed controls; Continuous = plants stressed twice-daily throughout all vegetative and reproductive growth stages until the end of R6 (full seed); V only = plants stressed only vegetative growth; R only = plants stressed only during reproductive growth stages R1 (first flower) through R6 (Jones and Mitchell, 1992).

(Jacobs, 1953; Lawton, 1982; Telewski and Jaffe, 1986) providing a streamlining effect, as well as increasing structural support parallel to the direction of the wind load. The increased radial growth tends to reduce whole tree flexibility, partially offsetting stress-induced increases in wood elasticity (Telewski and Jaffe, 1986), therefore, the stems are more rigid. However, increased stem rigidity should not be confused with brittleness. Stems of tomato seedlings that had been seismic-stressed for 20 minutes, twice-daily for a period of 10 days could be repeatedly bent into a U-shape without breaking, whereas the stems of non-stressed controls snapped at less than 90% bending (Heuchert *et al.*, 1983).

Changes in cell wall chemical composition have been reported to occur in response to mechanical stimuli. Unfortunately, these studies have involved only a limited number of species, at different ages, and in different environments. In addition, different aspects of cell wall chemistry were usually measured in these studies. Responses also may vary according to the type of mechanical stimulus applied. In the tomato experiment described above by Heuchert *et al.* (1983), stems of seismic-stressed seedlings had 15% more cellulose than non-stressed controls, but concentrations of hemicellulose, lignin, and silica did not change appreciably. Loblolly pines subjected to thigmic bending showed no change in cellulose or lignin content, but did increase the amount of organic solvent-extractable substances in the stems (Telewski and Jaffe, 1981). Leaves of wind-stressed fescue seedlings demonstrated increases in silica content and marginal schlerenchyma. Elemental contents also vary in response to mechanical stimuli. Tomatoes subjected to brief thigmic episodes daily for 14 days showed decreased stem K, increased stem P and Mg, and unchanged amounts of plant N and Ca, and leaf K compared to non-stressed controls (Adler and Wilcox, 1987). Growth zones of dark-grown soybean hypocotyls have been reported to accumulate significant amounts of Ca^{45} within 24 hours of a single thigmic episode (Jones and Mitchell, 1988b). It is difficult to make any general conclusions from the observations discussed above, except that in a species dependent manner the content of certain cell wall components increases in response to mechanical stimuli. The increased content of cell wall components is related to the measured increases in stem structural strength.

In trees, the formation of reaction wood is a common response to when stems are bent or tilted away from a vertical orientation for long periods of time as a result of strong prevailing winds, heavy snow burdens, or an unstable root environment. In dicotyledonous trees, reaction wood is known as "tension wood" and consists of non-lignified cells of reduced width and number on the upper sides of stems or branches that have been bent downward (pulling them back up) and is associated with cell contraction. In monocotyledonous species. reaction wood is known as "compression wood" and consists of dense lignified cells on the lower sides of stems or branches that have been bent downward (pushing them back up) and is associated with cell expansion (Scurfield, 1973; Little and Jones, 1980). Its main function appears to be that of restoring the tree to a vertical orientation. It has been suggested that reaction wood formation also occurs in response to mechanical stimulation (Jaffe, 1980). However, Scurfield (1973) demonstrated experimentally that reaction wood does not form in response to mechanical stimuli *per se* but to a change in the vertical orientation of the cambial initials. Formation of reaction wood is, therefore, probably more of a response to gravity than it is to physical agitation.

Role of Plant Hormones

Neel and Harris (1971) first proposed that plant hormones might mediate plant responses to physical agitation. It is now generally accepted that plant hormones are important components in the integrated responses of plants to mechanical stimuli. Only Parkhurst and Pearman (1972) have attempted to explain proposed hormonal mechanisms for mechanical stress as purely biophysical phenomena. While biophysical phenomena are undoubtedly involved in the mechanical stress response (see discussion above), they also are influenced by, and interact with plant hormones. The role of hormones in the collective adaptive responses of plants to environmental mechanical stress is not well understood. Most investigations have emphasized the ability to mimic or inhibit mechanical stress responses by exogenous applications of hormones, hormone analogs, and/or their inhibitors using dark-grown, etiolated seedlings or excised sections, or light-grown plants receiving localized, exogenous applications that exceed physiological concentrations. Few studies have measured endogenous hormone-like activity or actual hormone levels in response to mechanical stress episodes, and when these studies have been conducted, usually only single hormone effects are investigated. Most studies have emphasized the effects of ethylene and auxin, and to a lesser extent, abscisic acid and gibberellins. No one has yet proposed or investigated a potential role for cytokinins in mechanical stress responses. Stress-induced developmental lags (e.g. formation of new stem nodes) observed for some species seem to imply that cytokinin activity may have been affected (i.e. cell division in apical meristems possibly was reduced).

Based on the limited information available, the activity of growth promoters (auxins, cytokinins, gibberellins) is generally inhibited by mechanical stimuli, whereas the activity of growth inhibitors (ethylene, abscisic acid) is generally stimulated. An exception to these general observations is in the potential for stress-induced accumulation of the growth promoter, auxin, to stimulate activity of the growth inhibitor, ethylene. In the following discussion, the activity of each hormone and its role in the integrated responses to mechanical stress will be discussed separately. However, the reader is cautioned that mechanical stimuli probably affect the activities of all plant hormones, and that plant response to mechanical stress, in part, reflects the complex interactions of these endogenous growth substances amongst themselves and with secondary messengers upon growth and development.

Auxin

Rapid inhibition of internode elongation is a consistent response to mechanical stress, and anatomical studies have demonstrated that this inhibition results primarily from an inhibition of cell expansion (Biro *et al.*, 1980; Jones and Mitchell, 1989b). Auxin stimulates cell enlargement, and in many species, there is a good correlation between the relative growth of plant organs and relative auxin content (Moore, 1979). Auxin promotes cell wall loosening, which subsequently decreases wall resistance to stretching, and ultimately permits the wall to yield to pressure generated by turgor. The mechanism of wall loosening is uncertain and apparently is different in monocots and dicots. Expansion of *Avena* (a monocot) coleoptiles is promoted by acidic pH and may involve enzymatic cleavage of glycosidic bonds in the wall, whereas expansion of soybean (a dicot) hypocotyls is promoted by removal of wall calcium and/or low pH (Cleland, 1987). The different mechanisms may be explained on the basis of the different chemical composition of monocot and dicot cell walls. In monocots, cell wall cross-links between two ferulic acid moieties can be created via the action of peroxidase (Fry, 1986) and the resulting diferulate bridges may reduce the elasticity of the cell wall. Similarly, peroxidase-mediated linking of tyrosine moieties into isodityrosine bridges may reduce the extensibility of dicot cell walls (Fry, 1986). Mechanical stress-induced, peroxidase-mediated, cell wall cross-linking as a factor in stress-induced growth inhibition has yet to be confirmed experimentally. Mechanical stress has been shown to increase peroxidase activity in wild cucumber and cucumber fruit (Boyer *et al.*, 1979; De Jaegher *et al.*, 1985; Miller and Kelley, 1989) and was associated with decreased auxin levels.

The common link is that low pH appears to stimulate cell expansion in both monocots and dicots, and that auxin has been demonstrated to promote H^+ excretion from plant cells of many species (Cleland and Rayle, 1978), lending support to the "acid-growth" theory of auxin action. Regardless of the mechanisms involved, it would seem a reasonable hypothesis that mechanical stress-induced growth inhibition may result, in part, from an inhibition of auxin activity. Is this hypothesis valid, and if so, how is auxin activity reduced? These questions are explored below.

It is well established that tissue auxin levels are reduced by mechanical stress. Hofinger *et al.* (1979) showed that auxin was not detectable in the lower internodes of wild cucumber stems following thigmic stress, although it was present in undisturbed plants. Neel and Harris (1972), citing Boyer (1967), reported that thigmic stroking of wild cucumber leaves resulted in shorter internodes and lowered auxin concentrations, when compared to undisturbed plants, and that normal growth rate returned when exogenous auxin was applied to the stressed plants. Similarly, when epicotyl sections excised from dark-grown, mechanically-stressed pea seedlings were immersed in auxin-containing solutions, elongation growth was partially restored to levels shown by sections from non-stressed seedlings (Mitchell, 1977). Conversely, Erner and Jaffe (1982) reported an increase in auxin-like substances in the stems of green bean plants following brief thigmic stress episodes daily for 10 consecutive days. Two mechanisms are likely to explain the observed responses of auxin to mechanical stress: (i) inhibition of auxin transport; and (ii) increased metabolism of auxin. The two mechanisms are not mutually exclusive.

In his experiments with excised pea epicotyls, Mitchell (1977) observed that both seismic and thigmic stress inhibited elongation and that the same response was obtained by removal of the apical hook. This inhibition of auxin-dependent growth was time dependent, requiring six hours from the time of stress treatment to the time excision to produce a significant inhibitory response. The observations are consistent with the hypothesis that decreased movement of growth promoters from the apical hook occurred in response to mechanical stress. Additional experiments showed that when a single episode of thigmic stress was applied, transport of radiolabeled auxin through excised pea epicotyls was inhibited 40–50% within 24 hours, the same amount that elongation growth was inhibited in the same time period (Mitchell, 1977, 1990). Similar inhibition of basipetal auxin transport (ca. 60%) was observed in excised stems of light-grown, thigmic-stressed bean plants (Erner and Jaffe, 1982). These observations are consistent with those of Hofinger *et al.* (1979) who reported that auxin was not detectable in wild cucumber internodes below the region of thigmic stress application. If auxin transport is inhibited, it must either accumulate above the region of the transport blockage, be rapidly metabolized, or become "bound" and inactivated. Increased auxin accumulation would stimulate synthesis of the growth inhibitor ethylene (Yang and Hoffman, 1984) which would contribute to reduced internode elongation and increased radial expansion. Application of other ethylene-stimulating substances (e.g. ethephon) also mimics the morphological effects of mechanical stress in some plant species (see Ethylene discussion below).

Experiments with wild cucumber suggest that mechanical stress stimulates the activity of soluble and cell wall-bound peroxidases (Boyer *et al.*, 1979; De Jaegher *et al.*, 1985) and that this activity is associated with reduced internode elongation. Within 48 hours of a brief thigmic episode, pre-existing basic peroxidases increased in activity, and a new peroxidase were detected in stressed internodes (Boyer *et al.*, 1979). Since peroxidases are known to metabolize endogenous auxin (Ray, 1960), these reports support the hypothesis that an auxin-peroxidase system is involved plant response to mechanical stress. Boyer *et al.*, (1983) also observed that the stress-induced stimulation of peroxidase activity was accompanied by reduced ethylene evolution from the rubbed internodes, implying that for those species that show reduced auxin activity, any auxin-induced ethylene effects would be reduced.

It appears that at least three mechanical stress-auxin systems are potentially involved in mechanical stress-induced plant growth inhibition. The first is demonstrated in plants such as wild cucumber (Boyer *et al.*, 1979, 1983) where mechanical stimuli increase the activity of soluble and/or cell wall bound peroxidases. Increased peroxidase activity results in decreased auxin content, which subsequently inhibits cell expansion. The second system is demonstrated in plants such as beans (Erner and Jaffe, 1982) where endogenous auxin accumulates as a result of mechanical stress. The localized increase in auxin content stimulates ethylene synthesis which subsequently inhibits internode elongation and increases radial expansion. Mechanical stress-induced radial stem expansion is associated with increased radial cell division in the vascular cambium (Biro *et al.*, 1980; Telewski and Jaffe, 1981). Increased ethylene levels also inhibit basipetal auxin transport to the elongating zones of the stem. Formation of reaction wood in trees also may reflect an accumulation of auxin in bent or tilted stems (Scurfield, 1973). The third system may involve a little known, and incompletely characterized interaction with gibberellins (see discussion of Gibberrellins below).

The first system inhibits growth by reducing auxin activity. The basis for inhibited basipetal auxin transport is not known but may involve inhibition of phloem transport as a result of stress-induced callose synthesis (Jaffe *et al.*, 1985). Evidence supporting the hypothesis that mechanical stress causes transient blockages to phloem transport, comes from experiments by Jaeger *et al.* (1988) who reported that 1–2 minutes of gentle seismic, thigmic, or vibric stress inhibited ^{11}C-transport in cotton plants exposed to $^{11}CO_2$-labeled air. Endogenous auxin can move readily in the phloem (Moore, 1979). The second system inhibits growth by stimulating biosynthesis of the growth-inhibiting hormone, ethylene (see discussion of Ethylene below). Evidence for a third mechanical stress-auxin system is less certain, but from the available literature it appears that stress-induced inhibition of auxin activity may result from an inhibition of gibberellin (GA) activity. When pea stems are treated with exogenous GA, elongation growth and tissue auxin content increased; when pea stems were treated with S-3307, a GA biosynthesis inhibitor, both stem growth and auxin content were reduced (Hamilton and Law, 1987; Law and Hamilton, 1984; Lantican and Muir, 1969).

Gibberellins

Gibberellins (GAs) stimulate both cell division and cell expansion. The most dramatic GA effect is the promotion of stem elongation in genetic dwarfs, the rosette stages of certain biennials, and in species requiring vernalization and/or long day photoperiodic induction prior to the onset of reproductive growth (Moore, 1979). Similar to auxins, GAs may promote cell expansion by increasing cell wall extensibility. Adams *et al.* (1975) showed that GA-induced increases in wall extensibility of *Avena* internodes were closely associated with GA-stimulated elongation growth, as determined by Instron extensometry. It has been suggested that GA-stimulated polysaccharide hydrolases may participate in cell wall loosening (Carpita, 1987), but this hypothesis requires further investigation. The mechanism(s) whereby GAs may increase wall extensibility are unknown, but it is interesting to note that recent experiments by Hamilton and Law (1987) suggest that GA stimulates auxin activity.

It also has been suggested that GA-enhanced cell elongation may result from decreased osmotic potential in cells of the enlargement zones, thereby promoting water uptake and increasing turgor, without any effect on wall extensibility. When cucumber hypocotyls were treated with GA, wall extensibility was unaffected even though elongation growth occured (Cleland *et al.*, 1968), and it was inferred that growth could only have resulted from decreased osmotic potential. Support for this level of control comes from experiments with

'Alaska' pea. Hormone-induced lowering of cell osmotic potential in the subhook region (the zone of maximum elongation) of 'Alaska' peas was correlated with growth and the amount of osmotic solutes (soluble sugars, amino acids, K^+ ions) in the cell sap (Miyamoto and Kamisaki, 1988). The cell sap of unstressed controls had a substantially lower concentration of osmotic solutes. However, unless the GA-induced turgor force exceeds a certain wall yield threshold, the cell will not expand (refer to Biophysical Mechanisms section in this chapter). Therefore, regardless of the effects on turgor, the primary level of control by GA on elongation is still probable in the cell wall.

Despite the demonstrated involvement of GAs in elongation growth, few reports have investigated a role for this hormone in mechanical stress-induced growth inhibition. Reduced GA activity has been implicated in mechanical stress-induced senescence (Salisbury, 1963), dormancy (Neel and Harris 1971), and in reduced stem hollowing and increased pithiness of beans (Takano *et al.*, 1995). Thigmic and seismic stimuli also have been shown to rapidly reduce the content of GA-like substances in xylem exudates of sunflower and extracts of bean seedlings after the plants had been stressed for several days (Suge, 1978; Beyl and Mitchell, 1983). Unfortunately, rigorous time-course experiments, comparing the onset of mechanical stress-induced growth inhibition with endogenous GA levels have not been conducted. It is also not known whether stress-induced reductions in GA-like activity result from masking by inhibitors, conjugation of GA to biologically inactive metabolites, or an inhibition of GA biosynthesis (Beyl and Mitchell, 1983). Further research is needed in this area.

Cytokinins

Prior to the publication of this manuscipt chapter, there have been six reviews specifically addressing plant response and adaptation to environmental mechanical stress (Jaffe and Biro, 1979; Jaffe, 1980, 1985; Biddington, 1986; Latimer, 1991), as well as an update of previous work by Jaffe (Jaffe and Forbes, 1993). None of these reviews have surveyed any literature, nor even addressed any potential involvement for cytokinins in the mechanical stress response. Apart from one report that seismic and thigmic stress increased and decreased, respectively, the amount of cytokinin-like activity in exudates of detopped sunflowers (Beyl and Mitchell, 1983), no direct information regarding the effects of mechanical stress on endogenous cytokinin activity is available. This is a susprising revelation, since a few of the observed morphological responses to mechanical stress could conceivably be attributed to changes in cytokinin activity. Mechanical stress is known to stimulate radial cell division in stem internodes (Biro *et al.*, 1980), inhibit leaf area increase (Jones and Mitchell, 1988), and promote foliar senescence (Giridhar and Jaffe, 1988). Cytokinins are known to promote cell division, stimulate leaf area increase, and retard foliar senescence (Moore, 1979). Cytokinins also interact with auxin in a complex manner to control cell division and differentiation (Skoog and Miller, 1957). The observed morphological changes usually have been attributed to auxins and ethylene, but the potential role of cytokinins in mechanical stress-induced growth responses has largely been ignored. The lag in the development of new stem nodes by some species in response to thigmic and seismic stimuli implies inhibited cell division in apical meristems, and as discussed above, stress-induced stem thickening results from increased radial cell division as well as cell expansion. These observations infer a potential decrease and increase in cytokinin activity, respectively, in response to mechanical stimuli. Clearly, additional research is needed to determine whether cytokinins have a role in the morphological and physiological responses of plants to mechanical stress.

Abscisic Acid

Abscisic acid (ABA) participates in many developmental events during the life cycle of the plant and is regarded as a general growth inhibitor. Elevated ABA levels in plant tissues promote organ abscission, seed and bud dormancy, senescence, turgor loss in guard cells, and inhibit transpiration and respiration (Owen and Napier, 1988). Exogenous ABA application has been shown to reduce elongation growth and cell wall extensibility of excised maize coleoptiles, counteract auxin-stimulated increases in wall extensibility and elongation growth (Kutschera and Schöpfer, 1986a,b), and reduce both stem elongation and diameter of intact green bean plants (Erner and Jaffe, 1982). Only a few studies conducted on a limited number of species have examined the effects of mechanical stress upon tissue ABA content, despite its obvious relevance to mechanical stress-induced growth inhibition. Two-and three-fold increases in ABA content have been reported in tissues of thigmic-stressed rice and green beans, respectively (Jeong and Ota 1980; Erner and Jaffe, 1982), but it was unchanged in thigmic-stressed cauliflower and eggplant (Biddington, 1986; Latimer and Mitchell, 1988).

Mechanical stress has been observed to promote dormancy in sweetgum seedlings (Neel and Harris, 1971), senescence in excised oat leaves and intact wild cucumber (Boyer *et al.*, 1983; Giridhar and Jaffe, 1988), increase stomatal resistance and decrease transpiration in soybeans (Pappas and Mitchell, 1985b), cause slight decreases in respiration of tomatoes (Mitchell *et al.*, 1975), reduce auxin-stimulated elongation of excised pea epicotyls (Mitchell, 1977), and reduce stem elongation and diameter of soybean plants (Latimer *et al.*, 1986). Based on the limited data available, it is not certain whether there is a role for ABA in the plant response to mechanical stress. Although growth and certain physiological processes are affected by both mechanical stress and exogenous ABA treatment, the potential for correlative physiological changes that might result from stress-induced increases in ABA, have not been investigated. If ABA does participate in mechanical stress responses, it may be manifested as subtle effects upon membranes and/or secondary messengers such as Ca^{2+} (Owen and Napier, 1988).

Ethylene

Ethylene has a broad regulatory role in the life cycle of plants and influences all phases of growth and development. Although it is generally considered a growth inhibitor, ethylene also promotes RNA and protein synthesis, root and root hair initiation, anthocyanin synthesis, flower initiation, fruit ripening, leaf epinasty, release of apical dominance, and the release of seed and bud dormancy (Beyer *et al.,* 1984). This gaseous hormone has long been associated with mechanical stress responses, particularly in its effects on the vegetative morphology of plants. Ethylene was first implicated as a factor in mechanical stress in experiments using dark-grown peas. When elongation of dark-grown peas was physically impeded by glass beads or foam rubber, stem diameter increased, elongation decreased, and stem growth was altered in a horizontal direction (Goeschl *et al.,* 1966), an example of the "triple response" of plants to ethylene exposure. These growth and morphology changes were accompanied by large amounts of ethylene evolution. Biro *et al.* (1980) reported that thigmic stress-induced inhibition of bean stem elongation resulted from reduced elongation of epidermal and cortical cells and reduced cell numbers in the vascular tissue and pith. Increased stem diameter was caused by increased radial expansion of the cortical cells and increased secondary xylem. Application of the ethylene-releasing agent, 2-chloroethyl phosphonic acid (ethephon) mimicked the morphological effects of thigmic stress on bean stems (Biro *et al.,* 1980). Ethylene-induced lateral cell expansion has been reported to result from a reorientation of cellulose microfibrils in a longitudinal direction (Eisinger, 1983), which restricts longitudinal elongation. In nature, increased shoot radial growth, mediated by stress-induced ethylene evolution, enables a seedling to emerge through compacted or crusted soils by increasing its lifting capacity, and the horizontal growth habit permits the seedling shoot to find a less restrictive pathway from which to emerge from the soil. More recent studies have shown that exogenous application of ethylene will mimic some of the morphological effects of mechanical stress in many plant species, including leaf epinasty, stem thickening, and reduced stem elongation (Jaffe 1973, 1976b; Barker, 1979; Biro *et al.,* 1980). Inhibitors of ethylene action (Ag^{++}, CO_2) or ethylene synthesis (aminoethoxyvinylglycine (AVG), aminooxyacetic acid, Co^{++}, hypobaric treatment) generally prevent or nullify mechanical stress-induced morphological changes (Biro and Jaffe, 1984; Prasad and Cline, 1985; Biddington and Dearman, 1986; Boyer *et al.,* 1986), although there are exceptions. For example, application of applied as $AgNO_3$, AVG did not reverse mechanical stress-induced stem growth inhibition in cucumber and bean plants (Huberman and Jaffe, 1981; Takahashi and Suge, 1980).

The time lag between the application of mechanical stress and the evolution of measurable ethylene levels indicates that stress-induced morphological responses mediated by ethylene are probably long-term phenomena, and do not contribute to rapid stress-induced growth effects. Reported time lags for stress-induced ethylene evolution are 10 minutes in wild cucumber, 45–60 minutes in beans, 2 hours in cucumber hypocotyls, and ca. 4 hours in pine trees and Fraser firs (Biro and Jaffe, 1984; Telewski and Jaffe, 1986b; Takahashi and Jaffe, 1990), whereas growth inhibition occurs within seconds after a mechanical stress episode (Coe and Mitchell, 1985; Myers and Mitchell, 1992).

Only the shoots of seedlings physically impeded prior to emergence appear to demonstrate all three components of the ethylene "triple response". Schwarzbach *et al.* (1992) provide data indicating that even the apparent ethylene- mediated induction of diagravitropism in physically impeded emergent pea seedlings results from the seedlings following the path of least resistance through the soil, rather than from stress-induced ethylene production. There are no reports indicating that seismic, thigmic, or vibric stresses will induce an emerged, upright plant to grow horizontally as a result of stress-induced ethylene evolution. Stem diameters of some species may either be unaffected or reduced following mechanical stress. Although an increase in ABA activity or a decrease in GA activity inhibit stem elongation, only ethylene mimics the effect of mechanical stress upon certain species (such as beans and peas) by simultaneously inhibiting stem elongation and increasing radial expansion. These responses imply a stronger role for ABA and GA (see preceding sections) in mediating the stem diameter response of "insensitive" species than for ethylene, although other substances may be involved. Alternatively, ethylene-mediated, mechanical stress- induced morphological changes may be localized to the plant regions where the stress was applied. Biro and Jaffe (1984) observed that when one internode of a bean plant is thigmic stressed, no ethylene is evolved from other internodes even though elongation is inhibited. Extracellular solutions obtained from thigmic stressed bean stems mimicked the effects of thigmic stress (inhibited stem elongation and increased diameter) in non-stressed beans and the effects were decreased by application of AVG (Takahashi and Jaffe, 1984). The extracellular solution from stressed plants also induced the synthesis of phytoalexin-like substances and ethylene evolution. Takahashi and Jaffe (1984) suggested that unidentified elicitors were formed in response to mechanical stress and contribute to the observed morphological changes. The identity of these stress-induced elicitors has not been confirmed, but they may be cell wall fragments produced in response to mechanical stress that serve as signals inducing subsequent ethylene synthesis (Yang and Hoffman,1984).

Moss *et al.* (1988) presented evidence that neither ethylene nor ABA were involved with the morphological response of maize roots to physical impedance. When the roots of maize seedlings were grown in a container of glass spheres, physical impedance decreased root length, increased diameter, and stimulated a greater than two-fold increase in ethylene evolution when compared to control roots. Exposure of control roots to ethylene simulated the effects of impedance. Application of norbornadiene (an ethylene activity inhibitor) and AVG nullified the effects of exogenous ethylene but did not change the growth habit of impeded roots, even though ethylene evolution was reduced below that of unimpeded roots. Root ABA titer was unaffected by mechanical impedance (Moss *et al.*, 1986). Since specific inhibitors of ethylene action and synthesis failed to overcome the effects of mechanical impedance, the involvement of ethylene in the response of maize roots to mechanical stress, at least for the cultivar used in this study, is unlikely; ABA also was ruled out. However, no other causal agent was proposed. Caution should be taken when interpreting these data since the observations may or may not hold true for root systems of other maize cultivars or for other species.

Stress-induced ethylene evolution apparently results from increased synthesis of the ethylene precursor, 1-aminocyclopropane-1-carboxylic acid (ACC) (Yang and Hoffman, 1984) and the synthesis of ACC is regulated by the enzyme, ACC synthase (Yu *et al.*, 1979a). Therefore, mechanical stress-induced ethylene evolution must involve a stimulation of ACC synthase activity. Since elevated auxin levels are known to stimulate ACC synthase activity in plants (Yu *et al.*, 1979b), it has been assumed that supraoptimal endogenous auxin levels may be responsible for mechanical stress-induced ethylene evolution. However, it is difficult to reconcile these data with the observations that mechanical stress usually (but not always) reduces endogenous auxin content in plant growth zones (see Auxin discussion above). In addition, adequate auxin must be present in plant tissues for stress-induced ethylene to induce radial cell expansion (Osborne, 1982).

The mechanisms whereby ethylene mediates mechanical stress growth responses are not well defined and are apparently different for different species. Stress-induced ethylene activity may involve a complex interaction with auxin, whose endogenous activity also is mediated by stress, possibly by stress-activated peroxidase activity (see Auxin discussion above).

Role of Calcium and Calmodulin

Calcium (Ca^{2+}) is an important regulator of cell metabolism (Hepler and Wayne 1985; Kauss 1987) and has a role as a secondary messenger in plants (Trewavas and Knight, 1994). The largest Ca^{2+} concentrations are found in the wall space with high concentrations found at the exterior surface of the plasmalemma (Demarty *et al.*, 1984). There is a steep gradient of Ca^{2+} across the plasmalemma because of the relatively low (ca. 10^{-8}-10^{-6} M) cytoplasmic Ca^{2+} concentration of the plant cell (Poovaiah, 1985). Environmental stimuli such as light and gravity have been shown to induce an asymmetric Ca^{2+} distribution in the growth zones of roots and shoots (Slocum and Roux, 1983; Lee and Evans, 1985). Subsequent cell expansion is reduced in the region of elevated Ca^{2+} concentration, and photo- and gravitropic curvatures occur toward the side of the plant organ with elevated Ca^{2+}. When plant tissues are exposed to high levels of exogenous Ca^{2+}, stem elongation is inhibited and stem diameter is increased (Burstrom, 1968) in a manner similar to that observed in response to mechanical stimuli. Although the mode of action is unclear, mechanical stress-induced growth inhibition and tropic curvature induced by light or gravity, apparently share similar transduction and response mechanisms with respect to Ca^{2+}. Jones and Mitchell (1989a) suggested that growth inhibition caused by mechanical stress may be accompanied by a differential Ca^{2+} redistribution across the cell enlargement zone, similar to the Ca^{2+} redistribution that occurs during phototropism and gravitropism. It has also been suggested that gravitropism may be a "specific adaptation" to mechanical stress (Trewavas and Knight, 1994).

When Ca^{2+} was applied exogenously to the growth zones of dark-grown soybean hypocotyls, elongation growth is inhibited up to 28% within 24 hours (Jones and Mitchell, 1989a). Stem growth inhibition was accompanied by a visible increase in stem diameter and was similar to growth responses caused by thigmic stress. Jones and Mitchell (1989a) observed that exogenous Ca^{2+} was effective in reducing growth only when Ca^{2+} ionophore A23187 was present in the treatment solution. Since Compound A23187 is known to enhance the permeability of membranes to Ca^{2+}, it was suggested that Ca^{2+} ions must enter the cytoplasm to cause the observed growth effects, and that direct effects of Ca^{2+} upon the cell wall were less likely to affect growth. Additional experiments showed that exogenous application of La^{3+} produced growth inhibitory effects similar to those observed for Ca^{2+} and thigmic stress (Jones and Mitchell, 1989a). Unlike Ca^{2+}, La^{3+} does not enter the cytoplasm and is restricted to the cell wall space (Thompson *et al.*, 1973). The capacity of Ca^{2+} and La^{3+} to cause similar growth effects while having differing permeabilities in membranes, suggests that some Ca^{2+}-sensitive sites of growth inhibitory activity are located on the surface of the plasmalemma.

Direct effects of free Ca^{2+} on cell walls should not be discounted. In isolated bean stem cell walls, Ca^{2+}-induced growth inhibition was associated with the binding affinity of the cation to cell walls (Tepfer and Taylor, 1980). They suggested that putative cation binding sites in bean cell walls inhibited stem growth either by preventing wall-loosening enzymes from acting upon their substrates or by reducing wall extensibility despite the activity of wall-loosening enzymes. Cell walls also may be rendered less extensible by Ca^{2+}-mediated stimulation of wall-synthesizing enzymes, such as β-1,3-glucan synthase. This plasmalemma-bound enzyme is usually activated by wounding, but also may be activated by non-damaging physical stress (Kauss, 1987). The synthesis of callose, and possibly cellulose, are dependent on β-1,3-glucan synthase activity (Delmer *et al.*, 1985; Waldmann *et al.*, 1988). Deposition of callose and cellulose on the inner surfaces of cell walls, via mechanical stress-induced, Ca^{2+}-stimulated enzyme activity, may be a mechanism whereby Ca^{2+} contributes to reduced wall extensibilty and subsequent long-term growth inhibition. Both callose and cellulose have been reported to increase in stems of beans and tomatoes, respectively, when the plants were subjected to mechanical stress (Heuchert *et al.*, 1983; Jaffe *et al.*, 1985).

Mimicry of mechanical stress effects on stems with relatively high concentrations of exogenous Ca^{2+} does not necessarily indicate that endogenous Ca^{2+} participates in mechanical stress-induced growth inhibition. To test this hypothesis, Jones and Mitchell (1989b) incubated dark-grown soybean seedlings in hydroponic solutions containing Ca^{45} immediately after a single, brief thigmic stress episode. After 24 hours, radioactivity in the growth zones of thigmic stressed seedlings was substantially greater than that found in non-stressed seedlings. Since measurable growth inhibition occured more rapidly (< 1 hour) than radiolabel accumulation in the growth zones (ca. 24 hours), it was assumed that Ca^{2+} accumulation was a consequence of growth inhibition, not the cause. However, Ca^{2+} recently absorbed by the roots is not necessary for Ca^{2+}-mediated, mechanical stress-induced growth inhibition. Other pools of Ca^{2+} may be found sequestered in the wall space and cytoplasmic organelles of the cells of the growth zone (Hepler and Wayne, 1985).

If endogenous Ca^{2+} does participate in stress-induced growth inhibition, then removal of free Ca^{2+} from the growth zones of stems should partially nullify the effects of mechanical stress. Application of the Ca^{2+}-specific chelator ethyleneglycol-bis-(β-aminoethylether)-N, N, N', N'-tetraacetic acid (EGTA) nullified the effects of thigmic stress on dark-grown soybean stems up to 36% if applied immediately after stress application (Jones and Mitchell, 1989a). Similar to La^{3+}, EGTA does not penetrate the plasmalemma (Hepler and Wayne, 1985) and therefore Ca^{2+} was presumably removed only from the wall space. Chelation of free Ca^{2+} in the wall space presumably would prevent its binding to putative inhibitory sites located on the plasmalemma.

Physical perturbation of the plasmalemma is known to cause increased Ca^{2+} uptake into cells (Kauss, 1987; Knight *et al.*, 1992), possibly by opening "stretch activated" ion channels (Edwards and Pickard, 1987), permitting an influx of Ca^{2+} into the cytoplasm. However, correlating short-term changes in cytosolic Ca^{2+} content with rapid, mechanical stress -induced growth inhibition has been problematic. A powerful technique for detecting changes in the levels of cytosolic Ca^{2+} in response to mechanical stress has recently been developed. Knight *et al.* (1991, 1993) genetically transformed a species of tobacco (*Nicotiana plumbaginifolia*) to luminously visualize cytosolic Ca^{2+} levels by emitting blue light. The transgenic tobacco plants were engineered to express apoaequorin, a protein produced by a marine jellyfish; the blue light is emitted in a dose-dependent manner when Ca^{2+} ions bind to apoaequorin. Using this system, Knight *et al.* (1992) and Haley *et al.* (1995) were able to measure short-term changes in cytosolic Ca^{2+} levels in response to mechanical stress. They first demonstrated that wind-induced stem bending caused immediate, but transient increases in cytosolic Ca^{2+} and only occurred when the stem was actually in motion (Knight *et al.* 1992). Repeated bending reduced plant sensitivity to physical displacement and the degree of cytosolic Ca^{2+} response was diminished (Knight *et al.,* 1992). Protoplasts isolated from these genetically transformed seedlings also demonstrated rapid, transient increases in cytosolic Ca^{2+} of up to 10 μM following mechanical stimulation by swirling in solution (Haley *et al.,* 1995). The magnitude of the cytosolic Ca^{2+} increase in the protoplasts was proportional to the strength of the swirling.

The subsequent elevation of cytoplasmic Ca^{2+} levels would trigger a variety of Ca^{2+}-sensitive mechanisms within the plant cell, particularly those mediated by calmodulin (Poovaiah, 1985; Trewavas and Knight, 1994 and references therein). Calmodulin is a Ca^{2+}-binding, regulatory protein found in plants and animals, that is activated by elevated cytosolic levels of Ca^{2+}. The resulting Ca^{2+}-calmodulin complex elicits the activity of many Ca^{2+}-sensitive enzymes. Calmodulin activity also has been shown to affect the asymmetric growth inhibition of gravitationally stimulated plant tissues (Biro *et al.,* 1982; Björkmann and Leopold, 1987). Jones and Mitchell (1989a) presented evidence that calmodulin also may participate similarly in Ca^{2+}-mediated, mechanical stress-induced growth inhibition. When they treated thigmic-stressed, dark-grown soybean seedlings with the calmodulin antagonists calmidizolium (1μM), chlorpromazine

(25–75 μM), and 48/80 (400 ppm) immediately after a thigmic stress episode, growth inhibition was nullified 23%, 50%, and 35%, respectively (Jones and Mitchell, 1989a). The ability of three different calmodulin antagonists to partially nullify stress-induced growth inhibition is consistent with the hypothesis that endogenous Ca^{2+} participates in the mechanical stress response. Further evidence for the involvement of calmodulin in mechanical stress-induced growth effects was presented by Braam and Davis (1990) who were able to induce the expression of calmodulin and four calmodulin-related genes in *Arabidopsis* with water sprays, thigmic stress, and wounding. These data suggest that the Ca^{2+}-calmodulin system mediates the mechanical stress growth phenomena at the signal transduction level as well as at the growth response level. Calmodulin-related genes were identified as TCH (touch-induced). The calmodulin-related, TCH gene family has not yet been identified in other plant species, so it is not known if this is a generalized mechanical stress signal transduction/response mechanism in plants or a pathway unique to this strain of *Arabidopsis*. The possible existence of genes that respond specifically to mechanical stimuli may pave exciting new roads into the landscape of plant mechanical stress research.

Calcium and/or the Ca^{2+}-calmodulin system may interact with plant hormones to mediate their effects on growth and development. Calmodulin inhibitors have been reported to inhibit auxin-stimulated elongation of corn and oat coleoptiles (Raghothama *et al.*, 1983). Elevated Ca^{2+} levels inhibit GA-stimulated stem elongation of lettuce seedlings and enhances GA-stimulated α-amylase activity (Hepler and Wayne, 1985). In mung bean hypocotyls, Ca^{2+} stimulates ethylene evolution in the presence of cytokinins (Lau and Yang, 1976). In a variety of plants, membrane permeability to Ca^{2+} may be enhanced by ABA, stimulating calmodulin activity by the influx of Ca^{2+} into the cytoplasm, which subsequently mediates physiological responses of plants to ABA (Owen and Napier, 1988). Auxin, GA, ethylene, ABA, Ca^{2+}, and calmodulin have been directly implicated in the growth and morphological changes that occur in response to mechanical stimuli, but no coordinated framework integrating the activities of the hormones with each other and with the Ca^{2+}-calmodulin system has been postulated.

The available data suggest a role for Ca^{2+} and calmodulin in mechanical stress-induced growth inhibition. The endogenous source of Ca^{2+} used by plant cells to transduce the mechanical stress response is not known but may reside in the wall space and/or is sequestered within cellular organelles. Radioactive Ca^{45} absorbed by the roots readily accumulates in the growth zones of thigmic stressed, dark-grown soybean seedlings, but since Ca^{2+} accumulates as a consequence of the growth inhibition, it is not an immediate cause of the growth inhibitory response. However, excess Ca^{2+} accumulation in the growth zones may contribute to long-term growth inhibition. In response to mechanical stimuli, sequestered Ca^{2+} released from cells in seedling growth zones may be redistributed to sites of growth-inhibitory activity. Release and transport of Ca^{2+} likely takes place as a result of the opening of ion channels in the plasmalemma and/or in the membranes of organelles and in the subsequent activation of calmodulin-mediated Ca^{2+}-sensitive enzymes. Calmodulin and calmodulin-related TCH genes may be involved in the signal transduction and response pathway. Regardless of the mechanism(s), it appears that some deformation of the plant cell, either by compression and/or tension, coupled with an increase in cytosolic Ca^{2+} contributes in some manner to the induction of mechanical stress related growth effects.

ENVIRONMENTAL INFLUENCE ON MECHANICAL STRESS RESPONSES

The interaction of environmental factors with mechanical stress has not been well characteized. On a limited basis, only the effects of variable photosynthetic photon flux (PPF) on soybeans, of photomorphogenic radiation on green beans, and of temperature on green beans have been investigated for their capacity to modify the effects of mechanical stress on plant growth (Jaffe, 1976a,b; Hunt and Jaffe, 1980; Pappas and Mitchell, 1985a; Jones *et al.*, 1990).

Light Intensity

During early experiments characterizing the nature of seismic stress effects on greenhouuse-grown tomatoes, it was observed that plants were less responsive to the stress when it was applied during late spring or summer than during late autumn or winter months (Mitchell and Heuchert, 1983). Later experiments showed that greenhouse-grown soybeans had large reductions in dry weight and leaf area when exposed to seismic stress under shading conditions of 31% to 41% of full sunlight (Pappas and Mitchell, 1985a) and it was hypothesized that variations in solar irradiance could modify the mechanical stress respose. Unfortunately, because greenhouse environmental conditions tended to be variable, it was uncertain what artifacts may have been created by large hour-to-hour and day-to-day fluctuations in PPF and temperature.

Jones *et al.* (1990) were the first to design experiments to quantify the light levels at which plants demonstrate maximum responsiveness to mechanical stress. They reported that PPF level could influence the type and extent of soybean response to seismic stress under controlled-environment conditions. Different plant parts responded differently to twice-daily applications of seismic stress as PPF varied between 135 μmoles and 592 μmoles m^{-2} s^{-1} (ca. 7 and 30% of full sunlight, respectively) over a 20 day experimental period (Figure 12.7A-D). Soybean stem length was significantly reduced by stress at 135 μmoles m^{-2} s^{-1} but the effect diminished at higher PPFs. In contrast, leaf area was insensitive to stress treatment at the lowest PPF level, but was progressively inhibited as PPF increased. Statistically significant reductions in shoot fresh and dry weights occured only at 295 μmoles m^{-2} s^{-1} (ca. 15% of full sunlight). Root length and mass were unaffected by seismic stress at all PPF levels tested. The effects of seismic stress on leaf area probably had the most important consequences for long-term plant growth. Since plant productivity is dependent upon total leaf area (Potter

Figure 12.7. Effect of 15 days of twice-daily seismic stress episodes (produced on a rotatory shaker beginning on 5 days after emergence (DAE), for 10 minutes per episode, at 190-240 rpm) on the appearance of 20-day-old soybean plants. Plants were grown at mean photosynthetic photon fluxes (PPF) of: (A) 135, (B) 295, (C) 471, and (D) 592 μ-moles m^{-2} s^{-1}, respectively (Jones *et al.*, 1990).

and Jones, 1977), stress-induced inhibition of leaf expansion would reduce the development of photosynthetic capacity and ultimately reduce biomass accumulation over the life cycle of the plant (Jones *et al.,* 1990). Transient reductions in net photosynthesis probably would have no long-term growth effects (Pappas and Mitchell, 1985b). Plants grown at PPF below 295 µmoles m^{-2} s^{-1}, were so limited by the effects of reduced photosynthesis and leaf area, that seismic stress had little additional influence on biomass accumulation. For plants grown at PPF greater that 295 µmoles m^{-2} s^{-1}, seismic stress effects may have been masked by the growth effects of high PPF. It was postulated that if the stress was more severe, or applied for longer time periods, it may have overcome the masking effects of elevated PPF (Jones *et al.,* 1990).

The experiments described above were limited both in duration and in the range of light levels investigated. The data also gave no indication whether the changes in stem and leaf growth were brought about by light-mediated, stress effects on cell division, cell expansion, or both.

Light Quality

The effects of light quality on the mechanical stress response has received even less attention than the effects of light intensity. When dark-grown bean seedlings were thigmic-stressed or briefly exposed to red light, stem elongation was inhibited 27% and 19%, respectively (Jaffe, 1976a,b). Exposure of the seedlings to far-red light completely reversed the inhibitory effects of red light, but had no effect on thigmic-stressed seedlings. These data implied that phytochrome was probably not directly involved in mechanical stress responses, but it was suggested that phytochrome and thigmic stimuli may affect the same growth retarding mechanisms (Jaffe, 1976a,b).

The mode by which mechanical stress interacts with light is probably very complex and may involve PPF-dependent photosynthetic and non-photosynthetic light dependent processes. Further research is needed to identify the nature of the light-mediated processes influenced by mechanical stress and the manner whereby they interact to modify plant growth. An investigation of the comparative stress responses of sun and shade plants may serve to elucidate the underlying basis for the interaction of light level and/or quality with mechanical stress.

Temperature

Variable temperature influences the degree of the mechanical stress response. Jaffe (1976b) grew green beans in controlled-environment chambers at variable temperatures between 21° and 30° C and observed that thigmic-stressed plants had optimal growth at ca. 28° C whereas unstressed plants grew optimally at 25° C. When the growth effects resulting from thigmic stress were computed (producing a third response curve representing the difference between the two experimental curves), it was shown that the effects of thigmic stress on green beans were optimal at ca. 24° C. Further calculations indicated that the thigmic response had a thermal coefficient of 1.7 between 21° and 24° C and that it may have one or more energy-requiring, rate limiting steps (Jaffe, 1976b). In a later study, it was observed that thigmic response in green beans was more highly correlated with low night temperatures than with relatively higher day temperatures. The effects of thigmic stress were substantially reduced below 16° C (Hunt and Jaffe, 1980).

CONCLUSIONS

Although a vast body of work has been generated since the early 1970s, the underlying biochemical and biophysical bases whereby plants respond and adapt to environmental mechanical stimuli have not been clearly defined. Stress-induced changes in membrane permeability, cell water status, cell wall extensibility, phytohormone activity, ion fluxes, transport, and photosynthesis are certainly involved, but the complex manner in which these systems interact to produce the observed growth responses is far from certain.

The suite of plant responses to environmental mechanical stimuli contains both short-term and long-term components. Short-term physiological changes that accompany rapid, stress-induced growth inhibition include, but are not limited to rapid plant turgor loss, loss of GA-like activity, callose synthesis, bursts of ethylene evolution (in some species), Ca^{2+} flux into the cytoplasm, activation of TCH genes, and rapid (but transient) decline in net photosynthesis and transpiration. Other phenomena such as changes in membrane permeability, opening of stretch-activated ion channels, and loss of cell turgor in growth zones may be inferred from the data obtained from certain experiments but have not yet been confirmed as contributing to, or participating in mechanical stress responses. Activation of short-term response mechanisms also may serve as triggers activating other long-term growth-inhibitory mechanisms. Rapid turgor loss may induce ABA synthesis, and membrane deformation may stimulate ion fluxes, water movement, and ethylene evolution.

Long-term physiological changes that accompany reduced plant growth and biomass accumulation include inhibited cell expansion, loss of photosynthetic capacity due to reduced leaf area, stimulation of cell division in the cambium (in some species), blockages to phloem transport, inhibited transport and/or activity of auxin, supraoptimal accumulation of auxin in growth zones, Ca^{2+} accumulation in growth zones, increased peroxidase and β-1,3-glucan synthase activity, increased ethylene evolution and ABA activity, changes in cell wall chemistry, and reduced wall extensibility. The capacity for mechanical stimuli to induce inhibitory elicitor-like activity, inhibit wall loosening processes, and cause the binding of ions to growth inhibitory sites on cell walls and membranes remains speculative, although some experimental evidence indicates that these processes may occur.

There also are several other unanswered questions. What are the initial events triggering rapid growth inhibition? Why do some species increase stem diameter when exposed to mechanical stress when stems of other species contract, or are unresponsive? How do light and temperature interact to modify mechanical stress responses? How do short- and long-term response mechanisms interact to modify plant growth and development? What other gene families are activated by mechanical stimuli? Do gravitropism, phototropism, and mechanical stress phenomena share similar receptors, transduction, and response pathways?

The effects of mechanical stimuli upon plants have been noted anecdotally since the late 19th century when Darwin (1881) first observed that pea roots changed the amount and direction of their growth when exposed to thigmic stress. Mechanical stress-induced growth effects on trees, herbaceous plants, and agricultural crop species were usually attributed to wind-induced phenomena that were unrelated to physical agitation. Specific attention was not devoted to understanding the nature of the phenomena in its own right until the early 1970s. The importance of plant responses to mechanical stimuli has evolved from being an annoying curiosity, whose effects investigators should avoid when conducting other physiological experiments, to a legitimate area of plant physiology research. Mechanical stress-induced phenomena appear to influence virtually every other physiological process occurring in plants, from the biophysical, to the biochemical and molecular.

An understanding of the nature of plant responses to environmental mechanical stimuli may provide the means to improve the quantity and quality of agricultural production as well as providing a clearer understanding of the manner in which all plant life responds and adapts to environmental stress.

REFERENCES

Adams P. A., Montague M. J., Tepfer, M., Rayle D. J., Ikuma, H. and Kaufman P. B. (1975) Effect of gibberellic acid on the plasticity and elasticity of *Avena* stem segments. *Plant Physiology*, **56**, 757–760.

Adler P. R. and Wilcox G. E. (1987) Influence of thigmic stress or chlormequat chloride on tomato morphology and elemental uptake. *Journal of Plant Nutrition*, **10**, 831–840.

Bagley W. T. (1964) Response of tomatoes and beans to windbreak shelter. *Journal of Soil and Water Conservation*, **19**, 71–73

Barker J. E. (1979) Growth and wood properties of *Pinus radiata* in relation to applied ethylene. *New Zealand Journal of Forest Science*, **9**, 15–19.

Beyer E., Morgan P. W. and Yang S. F. (1984) Ethylene. In *Advanced Plant Physiology*, edited by M.B. Wilkins, pp. 111–126. Marshfield, MA: Pitman Publishing Inc.

Beyl C. A. and Mitchell C. A. (1977) Automated mechanical stress application for height control of greenhouse chrysanthemum. *HortScience*, **12**, 575–577.

Beyl C. A. and Mitchell C. A. (1983) Alteration of growth, exudation rate, and endogenous hormone profiles in mechanically dwarfed sunflower. *Journal of the American Society for Horticultural Science*, **108**, 257–262.

Biddington N. L. (1986) The importance of mechanically-induced stress in plants - a review. *Plant Growth Regulation*, **4**, 103–123.

Biddington N. L. and Dearman A. S. (1986) A comparison of the effects of mechanically-induced stress, ethephon and silver thiosulphate on the growth of cauliflower seedlings. *Plant Growth Regulation*, **4**. 33–41.

Biro R. L., Hale C. C., Weigand O. S. and Roux. S. J. (1982) Effects of chlorpromazine on gravitropism in *Avena* coleoptiles. *Annals of Botany*, **50**, 735–747.

Biro R. L., Hunt E. R., Erner Y., and Jaffe M. J. (1980) Thigmomorphogenesis: Changes in cell division and elongation in the internodes of mechanically-perturbed or ethrel-treated bean plants. *Annals of Botany*, **45**, 655–664.

Biro R. L. and Jaffe M. J. (1984) Thigmorphogenesis: ethylene evolution and its role in the changes observed in mechanically perturbed bean plants. *Physiologia Plantarum*, **62**, 289–296.

Björkann T. and Leopold A. C. (1987) Effect of inhibitors of auxin transport and of calmodulin on a gravisensing-dependent current in maize roots. *Plant Physiology*, **84**, 847–850.

Boyer J. S. (1987) Hydraulics, wall extensibility, and wall proteins. In *Physiology of Cell Expansion during Growth, Proceedings of the Second Annual Penn State Symposium in Plant Physiology*, edited by D.J. Cosgrove and D.P. Kneivel, pp.109–121. Rockville, MD: American Society of Plant Physiologists, 109–121.

Boyer J. S., Cavalieri A. J. and Schulze E.D. (1985) Control of the rate of cell enlargement: excision, wall relaxation, and growth-induced water potentials. *Planta*, **163**, 527–543.

Boyer N. (1967) Modifications de la croissance d la tige de Bryone *(Bryonia dioica)* a la suite d' irritations tactiles. *CR Academy of Sciences Paris*, **264**, 2114–2117.

Boyer N., Chappelle B., and Gaspar T. (1979) Lithium inhibition of the thigmomorphogenetic response in *Bryonia dioica*. *Plant Physiology*, **63**, 1215–121.

Boyer N., De Jaegher G., Bon M-C, and Gaspar T. (1986) Cobalt inhibition of thigmomorphogenesis in *Bryonia dioica*: a possible role and mechanism of ethylene production. *Physiologia Plantarum*, **67**, 552–556.

Boyer N., Desbeiz M., Hofinger M. and Gaspar T. (1983) Effect of lithium on thigmorphogenesis in *Bryonia dioica*. Ethylene production and sensitivity. *Plant Physiology*, **72**, 522–525.

Braam J. and Davis R. W. (1990) Rain-, wind-, and touch-induced expression of calmodulin and calmodulin-related genes in *Arabidopsis*. *Cell*, **60**, 357–364.

Burstrom H. G. (1968) Calcium and plant growth. *Biological Reviews*, **43**, 287–316

Carpita N. C. (1987) The biochemistry of "growing" cell walls. In *Physiology of Cell Expansion during Growth, Proceedings of the Second Annual Penn State Symposium in Plant Physiology*, edited by D.J. Cosgrove and D.P. Kneivel, pp.28–45. Rockville MD: American Society of Plant Physiologists.

Cleland R. E. (1984) The Instron technique as a measure of intermediate past wall extensibility. *Planta*, **160**, 514–520.

Cleland R. E. (1987) The mechanism of wall loosening and wall extension. In *Physiology of Cell Expansion during Growth, Proceedings of the Second Annual Penn State Symposium in Plant Physiology*, edited by D.J. Cosgrove and D.P. Kneivel, pp.18–27, Rockville, MD: American Society of Plant Physiologists.

Cleland R. E. and Rayle D. L. (1978) Auxin, H^+-excretion, and cell elongation. *Botanical Magazine Tokyo Special Issue*, **1**, 125–139.

Cleland R.E., Thompson M. L., Rayle D. L., and Purves W. K. (1968) Differences in effects of gibberellins and auxins on wall extensibility of cucumber hypocotyls. *Nature*, **219**, 510–511.

Coe L. L. and Mitchell C. A. (1989) Kinetic responses of growth, auxin transport, and ethylene evolution by 'Alaska' pea to mechanical stress. *Plant Physiology Supplement*, **89**, 619 (Abstr.)

Cosgrove D. J. (1987) Linkage of wall extension with water and solute uptake. In *Physiology of Cell Expansion during Growth, Proceedings of the Second Annual Penn State Symposium in Plant Physiology*, edited by D.J. Cosgrove and D.P. Kneivel, pp.88–100 Rockville, MD: American Society of Plant Physiologists.

Cosgrove D. J. (1990) Gravitropism of cucumber hypocotyls: biophysical mechanism of altered plant growth. *Plant, Cell, and Environment*, **13**, 235–241.

Darwin, C. (1881) *The Power of Movement in Plants*. New York: D. Appleton and Company.

Delmer D. P., Cooper G., Alexander D., Cooper J., and Hayashi, T. (1985) New approaches to the study of cellulose biosynthesis. *Journal of Cell Science (Suppl.)*, **2**, 33–50.

Demarty M., Morvan C., and Thellier M. (1984) Calcium and the cell wall. *Plant, Cell, and Environment*, **7**, 441–444.

Edwards K. L. and Pickard B. G. (1987) Detection and transduction of physical stimuli in plants. In *The Cell Surface in Signal Transduction*, edited by H. Greppin, B. Millet, and E. Wagner, pp.41–66. Berlin: Springer-Verlag.

Eisinger W. (1983) Regulation of pea internode expansion by ethylene. *Annual Review of Plant Physiology*, **34**, 225–240.

Erner Y., Biro R., and Jaffe M. J. (1980) Thigmomorphogenesis: Evidence for a translocatable thigmomorphogentic factor induced by mechanical perturbation of beans (*Phaseolus vulgaris*). *Physiologia Plantarum*, **50**, 21–25.

Erner Y. and Jaffe M. J. (1982) Thigmorphogenesis: The involvement of auxin and abscisic acid in growth retardation due to mechanical stress. *Plant and Cell Physiology*, **23**, 935–941.

Finnel H. H. (1928) The effect of wind on plant growth. *Journal of the American Society for Agronomy*, **20**, 1206–1210.

Frank A. B., Harris D. G. and Willis W. O. (1974) Windbreak influence on water relations, growth, and yield of soybeans. *Crop Science*, **14**, 761–765.

Giridhar G. and Jaffe M. J. (1988) Thigmomorphogenesis: XXIII. Promotion of foliar senescence by mechanical perturbation of *Avena sativa* and four other species. *Physiologia Plantarum*, **74**, 473–480.

Goeschl J. D., Rappaport L., and Pratt H. K. (1966) Ethylene as a factor regulating the growth of pea epicotyls subjected to physical stress. *Plant Physiology*, **41**, 877–884.

Grace J. and Russell G. (1976) The effect of wind on grasses: III. Influence of continous drought or wind on anatomy and water relations in *Festuca arundinacea* Schreb. *Journal of Experimental Botany*, **2**, 268–278.

Grace J. and Russell G. (1982) The effect of wind and a reduced supply of water on the growth and water relations of *Festuca arundinacea* Schreb. *Annals of Botany*, **49**, 217–225.

Haley A., Russell A. J., Wood, N. Allan A. C., Knight M., Campbell A. K. *et al.* (1995) The effects of mechanical signaling on plant cell cytosolic calcium. *Proceedings of the National Academy of Sciences, USA*, **92**, 4124–4128.

Hamilton R, H. and Law D.M. (1987) The effect of S-3307 on the IAA levels in elongating stem of Alaska pea. In *Physiology of Cell Expansion during Growth, Proceedings of the Second Annual Penn State Symposium in Plant Physiology*, edited by D.J. Cosgrove and D.P. Kneivel, pp. 313–314 Rockville, MD: American Society of Plant Physiologists.

Hepler P. K. and Wayne R. O. (1985) Calcium and plant development. *Annual Review of Plant Physiology*, **36**, 397–439.

Heuchert J. C., Marks J. S., and Mitchell C. A. (1983) Strengthening of tomato shoots by gyratory shaking. *Journal of the American Society of Horticultural Science*, **10**, 801–805.

Hill L. (1921) The growth of seedlings in wind. *Proceedings of the Royal Society, Part B*, **92**, 28–31.

Hofinger M., Chappelle B., Boyer N., and Gaspar T. (1979) GC-MS identification and titration of IAA in mechanically perturbed *Bryonia dioica*. *Plant Physiology Supplement*, **63**, 52 (Abstr.).

Holbrook N. M. and Putz F. E. (1989) Influence of neighbors on tree form: Effects of lateral shade and prevention of sway on the allometry of *Liquidambar styraciflua* (sweet gum). *American Journal of Botany*, **76**, 1740–1749.

Huberman M. and Jaffe M. J. (1981) Morphological changes of mechanically-perturbed or ethylene treated bean plants. *Plant Physiology*, **67**, S-17 (Abstr.).

Huberman M. and Jaffe M. J. (1986) Thigmotropism in organs of the bean plant (*Phaseolus vulgaris* L.). *Annals of Botany* **57**, 133–137.

Hunt E. R. and Jaffe M. J. (1980) Thigmorphogenesis: The interaction of wind and temperature in the field on the growth of *Phaseolus vulgaris* L. *Annals of Botany*, **45**, 665–672.

Jacobs M.R. (1954) The effect of wind sway on the form and development of *Pinus radiata*. *Australian Journal of Botany*, **2**, 35–51.

Jaeger C. H., Goeschl J. D., Magnuson C. F., Fares Y., and Strain B. R. (1988) Short-term responses of phloem transport to mechanical perturbation. *Physiologia Plantarum*, **72**, 588–594.

Jaffe M. J. (1973) Thigmomorphogenesis: the response of plant growth and development to mechanical stimulation with special reference to *Bryonia dioica*. *Planta*, **114**, 143–157.

Jaffe M. J. (1976a) Thigmomorphogenesis: Electrical resistance and mechanical correlates of the early events of growth retardation due to mechanical stimulation in beans. *Zeitschrift für Pflanzenphysiologie*, **78**, 34–42.

Jaffe M. J. (1976b) Thigmorphogenesis: A detailed characterization of the response of beans (*Phaseolus vulgaris* L.) to mechanical stimulation. *Zeitschrift für Pflanzenphysiologie*, **77**,437–453.

Jaffe M. J. (1980) Morphogenetic responses of plants to mechanical stimuli or stress. *BioScience*, **30**, 239–243.

Jaffe M. J. (1985) Wind and other mechanical effects in the development and behavior of plants with special emphasis on the role of hormones. *Encyclopedia of Plant Physiology, New Series*, **11**, 444–484.

Jaffe M. J. and Biro R. (1979) Thigmorphogenesis: The effect of mechanical perturbation on the growth of plants with special reference to anatomical changes, the role of ethylene, and the interaction with other environmental stresses. In *Stress Physiology in Crop Plants*, edited by H. Mussel and R. Staples, pp.26–59. New York: John Wiley & Sons.

Jaffe M. J. and Forbes S. (1993) Thigmomorphogenesis: The effect of mechanical perturbation on plants. *Plant Growth Regulation*, **12**, 313–324.

Jaffe M. J., Huberman M., Johnson J., and Telewski F. W. (1985) Thigmomorphogenesis: The induction of callose formation and ethylene evolution by mechanical perturbation in bean stems. *Physiologia Plantarum*, **64**, 271–279.

Jeong Y. and Ota Y. (1980) A relationship between growth inhibition and abscisic acid content by mechanical stimulation in rice plant. *Japanese Journal of Crop Science*, **49**, 615–616 (Abstr.).

Jones R. S., Coe L. L., Montgomery L. and Mitchell C. A. (1990) Seismic stress response of soybean to different photosynthetic photon flux. *Annals of Botany*, **66**, 617–622.

Jones R. S. and Mitchell C. A. (1988a) A marking device for the identification and measurement of growth zones in seedlings with minimum mechanical stress. *Annals of Applied Biology*, **113**, 437–441.

Jones R. S. and Mitchell C. A. (1988b) Calcium-45 accumulation in the growth zone of mechanically-stressed soybean. *Plant Physiology Supplement*, **88**, (Abstr.).

Jones R. S. and Mitchell C. A. (1989a) Calcium ion involvement in growth inhibition of mechanically stressed soybean (*Glycine max*) seedlings. *Physiologia Plantarum*, **76**, 598–602.

Jones R. S. and Mitchell C. A. (1989b) Mechanical stress-induced changes in cell wall extensibility compliances. *Plant Physiology Supplement*, **89**, 619 (Abstr.).

Jones R. S. and Mitchell C. A. (1992) Mechanical stress effects on growth and yield of greenhouse-grown soybean (*Glycine max* L. Merr.). *Crop Science*, **32**, 404–408.

Kauss H. (1987) Some aspects of calcium dependent regulation in plant metabolism. *Annual Review of Plant Physiology*, **38**, 7–72.

Knight M. R., Campbell A. K., Smith S. M. and Trewavas A. J. (1991) Transgenic plant aequorin reports the effects of touch and cold shock and fungal elicitors on cytosolic calcium. *Nature*, **352**, 524–526.

Knight M. R., Read N. D., Campbell A. K., and Trewavas A. J. (1993) Imaging calcium dynamics in living plants using semi-synthetic recombinant aequorins. *The Journal of Cell Biology*, **121**, 83–90.

Knight M. R., Smith S. M., and Trewavas A. J. (1992) Wind-induced motion immediately increases cytosolic calcium. *Proceedings of the National Academy of Sciences, USA*, **89**, 4967–4971.

Kutschera U. (1989) Tissue stresses in growing plant organs. *Physiologia Plantarum*, **77**, 157–163.

Kutschera U., Berfeld R. and Schöpfer P. (1987) Cooperation of epidermis and inner tissues in auxin-mediated growth of maize coleoptiles. *Planta*, **170**, 16–18.

Kutschera U. and Schöpfer P. (1986a) *In-vivo* measurements of cell-extensibility in maize coleoptiles: Effects of auxin and abscisic acid. *Planta*, **169**, 437–442.

Kutschera U. and Schöpfer P. (1986b) Effect of auxin and abscisic acid on cell wall extensibility in maize coleoptiles. *Planta*, **167**, 527–535.

Lantican B. P. and Muir R. M. (1969) Auxin physiology of dwarfism in *Pisum sativum*. *Physiologia Plantarum*, **22**, 412–423.

Larson P. R. (1965) Stem formation in young *Larix* as influenced by wind and pruning. *Forest Science*, **11**, 212–242.

Latimer J. G. (1991) Mechanical conditioning for control of growth and quality of vegetable transplants. *HortScience*, **26**, 1456–1461.

Latimer J. G. and Mitchell C. A. (1988) Effects of mechanical stress or abscisic acid on growth, water status, and leaf abscisic acid content of eggplant seedlings. *Scientia Horticulturae*, **36**, 37–46.

Latimer J. G., Pappas T. and Mitchell C. A. (1986) Growth responses of eggplant to mechanical stress in greenhouse and outdoor environments. *Journal of the American Society for Horticultural Science*, **111**, 694–698.

Lau O. L. and Yang S. F. (1974) Synergistic effect of calcium and kinetin on ethylene production by the mung bean hypocotyl. *Planta*, **118**, 1–6.

Law D. M. and Hamilton R. H. (1984) Effects of gibberellic acid on endogenous indole-3-acetic acid and indoleacetylaspartic acid levels in a dwarf pea. *Plant Physiology*, **75**, 255-256.

Lawton R. O. (1982) Wind stress and elfin stature in a montane rain forest tree: An adaptive explanation. *American Journal of Botany*, **69**, 1224–1230.

Mitchell C. A. (1977) Influence of mechanical stress on auxin-stimulated growth of excised pea stem sections. *Physiologia Plantarum*, **41**, 129–134.

Mitchell C. A. (1990) Mechanical stress regulation of plant growth and development. In *1988–89 NASA Space Biology Accomplishments, NASA Technical Memorandum 4160*, edited by T. W. Halstead. Washington, DC: NASA Office of Space Science and Applications, Scientific and Technical Information Division.

Mitchell C. A. and Coe L. L. (1991) Mechanical stress regulation of plant growth and development. In *1989–90 NASA Space Biology Accomplishments, NASA Technical Memorandum 4258*, edited by T. W. Halstead. Washington, DC: NASA Office of Space Science and Applications, Scientific and Technical Information Division.

Mitchell C. A., Severson C. J., Wott J. A., and Hammer P. A. (1975) Seismomorphogenic regulation of plant growth. *Journal of the American Society for Horticultural Science*, **100**, 161–165.

Moore T. C. (1979) *Biochemistry and Physiology of Plant Hormones*. New York: Springer-Verlag.

Moss G. I., Hall K. C. and Jackson, M. B. (1988) Ethylene and the responses of roots of maize (*Zea mays* L.) to physical impedance. *New Phytologist*, **109**, 303–311.

Myers P. N. and Mitchell C. A. (1991) Investigations of the biophysical basis for mechanical stress inhibition of hypocotyl elongation by dark-grown soybean seedlings. *ASGSB Bulletin*, **6**, 98 (Abstr.).

Neel P. L. and Harris R. L. (1971) Motion-induced inhibition of elongation of dormancy in *Liquidambar*. *Science*, **173**, 58–59

Ogbuehi S. N. and Brandle J. R.. 1981. Influence of windbreak shelter on soybean production under rainfed conditions. *Agronomy Journal*, **73**, 625-628.

Ogbuehi S. N. and Brandle J. R. (1982) Influence of windbreak-shelter on soybean growth, canopy structure, and light relations. *Crop Science*, **22**, 269–273.

Osborne D. J. (1982) The ethylene regulation of cell growth in specific target tissues of plants. In *Plant Growth Substances 1982*, edited by P. F. Wareing, pp.279–290. London: Academic Press.

Owen J. H. and Napier J. A. (1988) Abscisic acid: new ideas on its role and mode of action. *Plants Today*, **1**, 55–58.

Pappas T. and Mitchell C. A. (1985a) Effects of seismic stress on the vegetative growth of *Glycine max* (L.) Merr. cv. Wells II. *Plant, Cell, and Environment*, **8**, 143–148.

Pappas T. and Mitchell C. A. (1985b) Influence of seismic stress on photosynthetic productivity, gas exchange, and leaf diffusive resistance of *Glycine max* (L.) Merrill cv Wells II. *Plant Physiology*, **79**, 285–289.

Patterson M. R. (1992) Role of mechanical loading in growth of sunflower (*Helianthus annuus*) seedlings. *Journal of Experimental Botany*, **43**, 933–939.

Pelton W. L. (1967) The effect of a windbreak on wind travel, evaporation, and wheat yield. *Canadian Journal of Plant Science*, **47**, 209–214.

Pickard B. G. (1971) Action potentials resulting from mechanical stimulation of pea epicotyls. *Planta*, **97**, 106–115.

Poovaiah B. W. (1985) Role of calcium and calmodulin in plant growth and development. *HortScience*, **20**, 347–354.

Potter J. R. and Jones J. W. (1977) Leaf partitioning as an important factor in growth. *Plant Physiology*, **59**, 10–14.

Prasad T. K. and Cline M. G. (1985) Mechanical perturbation-induced ethylene releases apical dominance in *Pharbitis nil* by restricting shoot growth. *Plant Science*, **41**, 217–222.

Quirk J. T. and Freese F. (1976) Effect of mechanical stress on growth and anatomical structure of red pine: stem vibration. *Canadian Journal of Forest Research*, **6**, 375–381.

Radke J. K. and Burrows W. C. (1970) Soybean plant response to temporary field windbreaks. *Agronomy Journal*, **62**, 424–429.

Radke J. K. and Hagstrom R. T. (1973) Plant-water measurements on soybeans sheltered by temporary corn windbreaks. *Crop Science*, **13**, 543–548.

Raghothama K. G., Mizrahi Y. and Poovaiah B. W. (1983) Effects of calmodulin inhibitors on auxin-induced elongation. *Plant Physiology Supplement*, **72**, 144.

Ray P. M. (1960) The destruction of indoleacetic acid. III. Relationships between peroxidase action and indoleacetic acid oxidation. *Archives of Biochemistry and Biophysics*, **87**, 19–30.

Rees D. J. and Grace J. (1980a) The effect of wind on extension growth of *Pinus contorta* Douglas. *Forestry*, **53**, 145–153

Rees, D. J. and Grace, J. (1980b). The effect of shaking on extension growth of *Pinus contorta* Douglas. *Forestry*, **53**, 154–166.

Rees D. J. and Grace J. (1981) The effect of wind and shaking on the water relations of *Pinus contorta* Douglas. *Physiologia Plantarum*, **51**, 222–228.

Rosenberg N. J. (1966) Influence of snow fence and corn windbreaks on microclimate and growth of irrigated sugar beets. *Agronomy Journal*, **58**, 469–475.

Schwarzbach D. A., Woltering E. J. and Saltveit M. E. (1992) Behavior of etiolated peas (*Pisum sativum* cv Alaska) when obstructed by a mechanical barrier. *Plant Physiology*, **98**, 769–773.

Scurfield G. (1973) Reaction wood: Its structure and function. *Science*, **179**, 647–655.

Senthinathan A., McVetty P. B. E. and Strobe E. H. (1984) Soybean plant response to annual crop windbreaks under rain-fed conditions. *Canadian Journal of Plant Science*, **64**, 879–883.

Skidmore E. L., Hagen L. J., Naylor D. G. and Teare I. D. (1974) Winter wheat response to barrier-induced microclimate. *Agronomy Journal*, **66**, 501–506.

Slocum R. D. and Roux S. J. (1983) Cellular and subcellular localization of calcium ions in gravi stimulated oat coleoptiles and its possible significance in the establishment of tropic curvature. *Planta*, **157**, 484–492.

Steudle E., Zimmermann U. and Zillikens J. (1982) Effect of cell turgor on hydraulic conductivity and elastic modulus of *Elodea* leaf cells. *Planta*, **154**, 371–380.

Sturrock J. W. (1970) Studies on the effect of wind reduction on soyabeans: I. A preliminary assessment. *New Zealand Journal of Agricultural Research*, **13**, 33–44.

Takahashi H. and Jaffe M. J. (1984) Thigmomorphogenesis: the relationship of mechanical perturbation to elicitor-like activity and ethylene production. *Physiologia Plantarum*, **61**, 405–411.

Takahashi H. and Suge H. (1980) Sex expression in cucumber plants as affected by mechanical stress. *Plant and Cell Physiology*, **21**, 303–310.

Takano M., Takahashi H. and Suge H. (1995) Mechanical stress and gibberellin: Regulation of hollowing induction in the stem of a bean plant, *Phaseolus vulgaris* L. *Plant and Cell Physiology*, **36**, 101–108.

Tateno M. (1991) Increase in lodging safety factor of thigmomorphogenically dwarfed shoots of mulberry tree. *Physiologia Plantarum*, **81**, 239–243.

Tateno M. and Bae K. (1990) Comparison of lodging safety factor on untreated and succinic acid 2,2-dimethylhydrazide-treated shoots of mulberry tree. *Plant Physiology*, **92**, 12–16.

Telewski F. and Jaffe M. J. (1981). Thigmomorphogenesis: Changes in the morphology and chemical composition induced by mechanical perturbation in 6-month-old *Pinus taeda* seedlings. *Canadian Journal of Forest Research*, **11**, 380–387.

Telewski F. W. and Jaffe M. J. (1986a) Thigmorphogenesis: Field and laboratory studies of *Abies fraseri* in response to wind or mechanical perturbation. *Physiologia Plantarum*, **66**, 211–218.

Telewski F. and Jaffe M. J. (1986b) Thigmomorphogenesis: the role of ethylene in the response of *Pinus taeda* and *Abies fraseri* to mechanical perturbation. *Physiologia Plantarum*, **66**, 227–233.

Tepfer M. and Taylor A. A. (1980) The interaction of divalent cations with pectic substances and their influence on acid-induced cell wall loosening. *Canadian Journal of Botany*, **59**, 1522–1525.

Thompson W. W., Platt K. A. and Campbell N. (1973) The use of lanthanum to delineate the apoplastic continuum of plants. *Cytobios*, **8**, 57–62.

Trewavas A. and Knight M. (1994) Mechanical signaling, calcium and plant form. *Plant Molecular Biology*, **26**, 1329–1341.

Turgeon R. and Webb J. A. (1971) Growth inhibition by mechanical stress. *Science*, **174**, 961–962.

Wadsworth R. M. (1959) An optimum wind speed for plant growth. *Annals of Botany*, **89**, 195–199.

Waldmann T., Jeblick W., and Kauss H. (1988) Induced net Ca^{2+} uptake in suspension-cultured plant cells. *Planta*, **173**, 88–95.

Wheeler R. M. and Salisbury F. B. (1979) Water spray as a convenient means of imparting mechanical stimulation to plants. *HortScience*, **14**, 270–271.

Victor T. S. and Vanderhoef L. N. (1975) Mechanical inhibition of hypocotyl elongation induces radial enlargement. *Plant Physiology*, **56**, 845–846.

Yang S. F. and Hoffman N. E. (1984) Ethylene biosynthesis and its regulation in higher plants. *Annual Review of Plant Physiology*, **35**, 155–189.

Yu Y., Adams B., and Yang S. F. (1979a) 1-aminocyclopropane-1-carboxylic acid synthase, a key enzyme in ethylene biosynthesis. *Archives of Biochemistry and Biophysics*, **198**, 280–286.

Yu Y., Adams B. and Yang S. F. (1979b) Regulation of auxin-induced ethylene production in mung bean hypocotyls. Role of 1-aminocyclopropane-1-carboxylic acid. *Plant Physiology*, **63**, 589–590.

13. MECHANISMS OF VIRUS RESISTANCE IN PLANTS

R.S.S. FRASER

Society For General Microbiology, Marlborough House, Basingstoke Road, Spencers Wood, READING, RG7 1AE, UK

INTRODUCTION

Plant viruses can cause severe losses of yield and product quality in many agricultural and horticultural crops. Some crops, such as potato, suffer from a large number of different viruses. Some viruses, such as cucumber mosaic virus (CMV), can affect a large number of economically important plant species. Virus infections can be symptomless as in the case of beet cryptic virus, but can cause economically significant loss of yield. Alternatively, the virus may cause visually prominent symptoms which can lead to the crop being disfigured and unmarketable. An example is the infection of courgettes by zucchini yellow mosaic virus (ZYMV).

There are many routes to control of diseases caused by plant viruses. However, unlike control of fungal pathogens, one method which is not available in practice for viral pathogens is the use of chemical pesticides. Numerous chemicals have been reported to control virus multiplication and the formation of disease symptoms (Cassells, 1983), but none has been adopted for practical use in crop production systems. Many of these chemicals have phytotoxic effects. They are also potentially damaging to animals, humans and the environment. Given the increasing stringency of regulations for the registration of chemical pesticides in many countries, it is unlikely that the future will see any significant introduction of chemical pesticides for viruses either.

The single case where chemical treatment of virus infections can have a practical value in disease management, is in the preparation of healthy propagation material from totally infected lines of crops such as garlic. The chemicals are applied to meristem tip or callus cultures in combination with plant hormone treatments, and often heat treatments, leading to virus free lines (Stace-Smith, 1990).

Virus diseases may also be avoided or controlled by planting certified virus-free material of vegetatively propagated crops such as potatoes or perennial trees, by testing seed lots for the absence of seed transmitted viruses, by using cultural practices designed to avoid transmission of mechanically transmitted viruses, and by seeking to prevent transmission of vector transmitted viruses by controlling the vector or its access to host plants. All of these methods can make a valuable contribution to practical disease control in specific favorable situations, but they do not offer the whole answer for all situations and are sometimes of little practical benefit. Overall, the greatest opportunities for meaningful control of virus diseases are through host plant resistance and that is the topic of this chapter.

TYPES OF RESISTANCE

Plant viruses show a very wide diversity of particle size, shape and structure, form and genetic structure of their genomes, and in mechanisms by which the genome is expressed and replicated (Goldbach *et al.*, 1990). They cause diverse forms of pathogenesis on their equally diverse host plants (Fraser, 1992). It would, therefore, be expected that plants would have evolved a variety of mechanisms of resistance to counter different types of viral attack. Before considering examples and mechanisms, it will be useful to consider overall types of resistance mechanism from a theoretical stand point. We can envisage three main types of mechanisms operating at different levels of host population complexity.

At the level of the plant species, if the whole species is immune to a particular virus it is said to display *non-host resistance* to that virus. Thus, for example, bean common mosaic virus (BCMV) is restricted to *Phaseolus vulgaris* and a few other leguminous hosts. All other species tested appear incapable of supporting multiplication or symptom development by BCMV (Bos, 1971).

In species which are normally susceptible to a particular virus, some cultivars, or breeding lines, may display heritable resistance to that virus. Sometimes this resistance has been introduced by crossing with resistant individuals of related wild species. This *cultivar resistance* is the form used by the plant breeder in practical crop protection and in many cases the genetic basis has been determined. However, it should be noted that resistance breaking (*virulent*) strains of the virus, with the ability to overcome particular resistance genes, have evolved in many cases.

Acquired resistance (sometimes called *induced resistance*) occurs in susceptible individuals of the species and is induced by a prior infection, or by a chemical or cultural treatment. Generally this type of resistance is not heritable. However, it has recently become possible to create a special form of this type of resistance by genetic transformation of susceptible plants with DNA copies of portions of the viral genome or associated nucleic acids. Under control of a suitable promoter to ensure expression of active RNA or protein moieties, these transgenes can confer resistance to infection by the whole virus. If they are stably integrated into the host chromosome they will be heritable and, in effect, a lasting form equivalent to cultivar resistance will have been created.

TARGETS OF THE RESISTANCE MECHANISMS

The pathogenic cycle of any plant virus is complex. It involves replication of virus particles causing the primary infection in each initially-infected cell, secondary cycles of replication resulting from spread of the virus from cell to cell and over long distances through the plant, and spread of the progenyparticles by passive means or through vector transmissions to initiate fresh cycles of infection in new plants. These "macro-economic"cycles of replication and spread each involve "micro-economic"molecular processes such as uncoating of the virus genome, expression of virus genes, replication of the genome, synthesis of virus proteins, assembly of progeny particles, molecular interaction with host components involved in replication and protein synthesis, molecular interaction with host components involved in cell-to-cell and long distance spread and molecular interaction with components of the possible vectors. Any of these macro- or micro-economic components might form the target for a host resistance mechanism. The potential targets are shown diagramatically in Figure 13.1 (Fraser, 1990b). This scheme also covers the three broad classes of resistance introduced above.

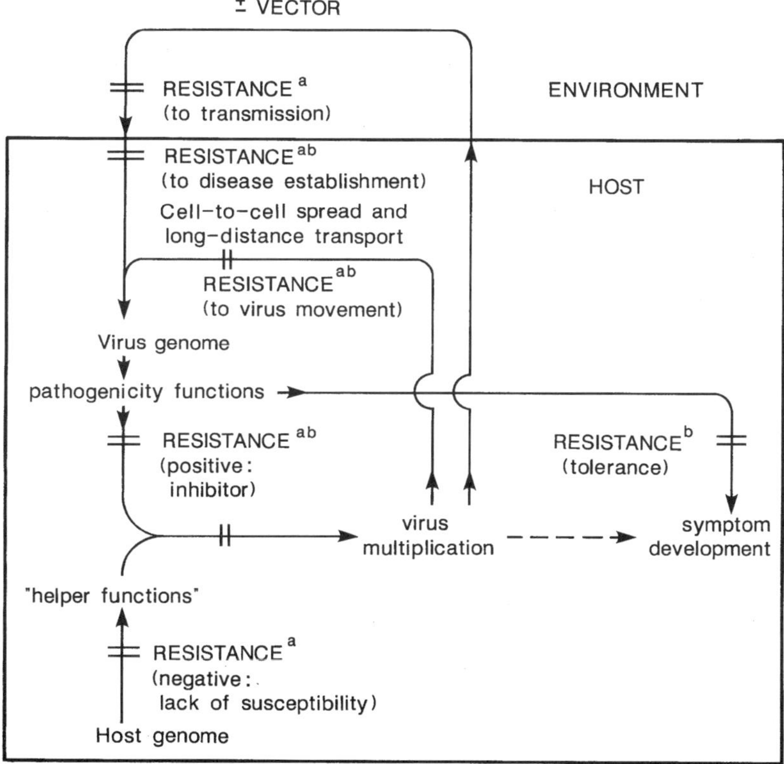

Figure 13.1. The replicative cycle of a plant virus showing various possible targets of cultivar resistance mechanisms. a) Targets which might also be involved in the determination of host range and non-host immunity. b) Targets that might also be involved in various types of acquired resistance mechanism. Modified from Fraser, 1990b.

The resistance mechanisms in Figure 13.1 operating at different target sites in the viral replicative cycle, may be classed into two types. *Positive* mechanisms are where the resistant plant contains some property which actively or directly inhibits some phase of the viral replicative cycle. Examples could include anything which reduces the plant's attractiveness to a vector of the virus: numerous well established types include hairyness, secretion by the plant of insect repelling pheromones, and antifeedants (Gibson *et al.*, 1984; Martin *et al.*, 1984). Inhibitors of viral replicase activity, inhibitors of symptom formation, or inhibitors of cell-to-cell spread of the virus, would also fall into this category. It should be noted that the inhibitory mechanism in the positive model could be constitutive (i.e. present in the plant prior to infection). However, the resistance mechanism could also be induced as a result of an initially successful infection. An example is the commonly occurring necrotic local-lesion response to viruses in which a small number of cells become infected around each primary inoculation point, but further spread of the virus is halted, generally within two to three days of the initial infection event. The implication is that early in infection there is a recognition event between plant and virus components which induces a resistance response via a signal transduction pathway. These mechanisms are considered in more detail in later sections.

The alternative form of resistance mechanism is the *negative* one, where resistant plants lack some component required by the virus to complete its replicative cycle. Examples could include a suitable host coded subunit of the functional virus replicase, or a host component to recognize the virus movement protein and allow modification of plasmodesmata to permit cell to cell spread. Such negative mechanisms should be constitutive and, if subjected to genetic analysis, should turn out to be controlled by recessive alleles.

NON-HOST RESISTANCE

What precisely is non-host resistance and how does it work? These are among the most intriguing questions in plant virology today. The answers are probably complex and it is likely that several mechanisms may be involved. Recent studies using protoplast systems and sequencing of virus genomes are beginning to illuminate the mechanisms in more detail.

The question of the mechanism of non-host immunity is the mirror image of an equally intriguing question — what controls the host range of a particular plant virus? Why do some viruses have very wide host ranges, while others appear able to infect only a very small number of closely related hosts? How is it that a given pair of viruses may be able to infect some hosts in common, but each virus may also be able to infect a separate range of other hosts inaccessible to the other virus (a phenomenon known as host range congruence). In a series of classical papers, Bald and Tinsley (1967a,b,c) postulated that viruses may produce various "pathogenicity factors" which are required for successful multiplication and spread. Plants produce various "susceptibility factors" which can be used by viruses to assist multiplication and spread. Only when there is functional compatibility between the appropriate virus and plant coded factors would successful plant pathogenesis be established. Given a suitable number of pathogenicity and susceptibility factors in each plant-virus interaction, and given a suitable level of variation in pathogenicity and susceptibility factors in all plants and viruses, it is clear that this model should be able to explain the great diversity of response in plant-virus interactions.

With the molecular genetic analysis of genome expression and function now well advanced for many groups of plant viruses (Goldbach *et al.*, 1990), a good deal is known about the viral pathogenicity factors and their functions, and it would certainly appear that a sufficient number of factors with sufficient variation in each, exists to satisfy the Bald and Tinsley model. The determinants of pathogenicity of barley stripe mosaic virus (BSMV) to oat have been mapped to the α RNA of the viral genome (Weiland and Edwards, 1994). There were six amino acid differences between BSMV isolates pathogenic and non-pathogenic on oats. Interestingly, the pathogenicity determinants mapped in an open reading frame with a sequence suggesting a methyltransferase/RNA helicase activity, but the non-pathogenic isolate appeared to be defective in cell-to-cell movement in oat, as it could replicate in oat protoplasts (Zheng and Edwards, 1990).

Where knowledge is more deficient is in understanding the diversity and variability in host susceptibility factors. Indeed, there is disappointingly little knowledge of how host-coded factors are specifically involved in virus replication and spread. The existing examples involve host-coded components of viral replicases, but more information is required (Hayes and Buck, 1990, Quadt *et al.*, 1993).

Many host range studies have merely involved mechanical inoculation of target species with the virus or attempts at vector infection. Failure of the species to develop visible symptoms or gross evidence of virus multiplication, was taken as an indicator of non-host status. However, closer examination in some plant-virus combinations revealed the phenomenon of subliminal infection where the virus appears able to multiply in a very small number of cells, presumably the results of the initial inoculation, but fails to spread to further cells in the leaf (Cheo and Garard, 1971; Sulzinski and Zaitlin, 1982). This mechanism was supported by reports that virus multiplication could be established by inoculation of protoplasts of some plants previously thought to be non-hosts for the viruses involved (Furusawa and Okuno, 1978). Experiments involving detection of subliminal infection and successful infection of protoplasts from ostensibly non-host species, are much more time-consuming than "inoculate and watch" tests of non-host immunity. Inevitably, then, very few host-virus combinations have been examined in the greater level of analytical detail. It is possible that the cells of more plant species are potentially hosts for more plant viruses than previously thought. It is also possible that failure of cell-to-cell spread after the initiation of a point infection is an important mechanism in many cases of apparent non-host resistance (van Loon, 1987). The difference between a true and complete non-host immunity, and one which operates by preventing cell-to-cell spread after allowing a minute amount of multiplication, can be quantified in semantic terms as precisely one cell.

An alternative mechanism proposed by Holmes (1955) is that non-host plants contain a large number of individual components inhibitory of virus formation, and that these operate in an additive manner to give complete supression of infection. It may be that plants have evolved gradual incremental resistance mechanisms which have led to viruses hosted during their evolutionary history being totally excluded from the species. Unfortunately there is no direct experimental evidence for this "pyramiding" approach to the exclusion of viruses. It is, however, intriguing that particular plant groups, such as the Pteridophytes and Gymnosperms appear to be remarkably free of virus infections. Alternatively, perhaps plant virologists have not looked hard enough.

CULTIVAR RESISTANCE

Genetics of Plant Virus Interactions

The first studies of mechanisms of resistance to plant viruses involved determination of the genetics of inheritance of the resistances as adjuncts to their use in breeding programmes. It is now more than half a century since the demonstration by Holmes (1938) that the hypersensitive resistance to tobacco mosaic virus (TMV) in certain tobacco lines was caused by a single dominant gene *N*. Since then a large number of crop-virus combinations have been examined. Reviews by Fraser (1986, 1990a, 1992b) summarize findings for 87 randomly chosen examples of crop species resistant to particular viruses. These reviews also consider various conceptual models of plant virus interactions at the genetic level.

Most of the examples in the survey involved resistance determined at a single genetic locus. In some crop species examined there were several sources of resistance, each capable of operating independently. Thus, in tomato, resistance to tomato mosaic virus (ToMV, a close relative to TMV) is controlled by genes at two loci, *Tm–1* and *Tm–2*. Two alleles exist at the *Tm–2* locus — *Tm–2* itself and $Tm\text{–}2^2$ (Hall 1980). In contrast, there are a few well documented cases where genes at different loci show epistatic effects. Thus, in resistance to BCMV in *Phaseolus vulgaris* the gene *bc–u* has little effect on its own, but is required for full expression of resistance conditioned by genes at each of three other loci designated *bc–1, bc–2, and bc–3* (Drijfhout, 1978).

In the survey a slight majority of the resistance alleles examined appeared to be dominant over susceptibility. About one quarter showed evidence of gene-dosage dependent effects, i.e. resistance was more effective in homozygous resistant plants than in heterozygotes. Somewhat less than one quarter of the sample showed resistance as recessive to susceptibility.

The range of dominance/recessiveness behavior of alleles for resistance to plant viruses is interesting and might suggest a variety of mechanisms. In previous reviews (Fraser, 1990a, 1992b), I have suggested that the dominant and gene-dosage dependent resistances may be associated with positive (inhibitor-producing) mechanisms, while the recessive forms might be associated with negative (lack of susceptibility) mechanisms. It was suggested that such mechanisms might be very difficult for the virus to overcome by mutation and should, therefore, prove very durable and effective against all isolates of the virus. It is interesting to compare this with resistance to fungal pathogens in plants, where recessiveness is very rare. However, one well characterized example, the *mlo* genes for resistance to powdery mildew in barley, have also been extremely durable and are effective against all isolates of the fungus (Jørgensen, 1992).

In the plant/virus survey referred to earlier, more than half of the examples of resistance genes had been overcome by virulent (resistance breaking) isolates of the virus involved. Many of the other genes did not appear to have been adequately tested by exposure to a large number of isolates of the virus. There are very few examples of resistance genes which have proved completely robust over a long period of exposure to their virus in practical crop protection situations in different areas, as well as in the virologist's laboratory. Notable examples include *bc–3* for BCMV resistance in *Phaseolus vulgaris*, and *Ry* for PVY resistance in potato (Jellis, 1992).

The *N* gene in tobacco remained robust for 50 years despite being used intensively in experimental virology and notably being challenged by numerous experiments involving viral mutagenesis. However, it has now been reported that a pepper isolate of TMV, Ob, will overcome this resistance (Csillery *et al.*, 1983). The $Tm\text{–}2^2$ resistance in tomato to ToMV has been outstandingly durable in commercial tomato cultivars. Rare resistance breaking isolates have been found (Fletcher, 1992) but these are unable to establish themselves fully on resistant crops and so far have posed no threat to the practical value of the resistance (Fraser, 1992).

MECHANISMS OF RESISTANCE

Broadly speaking, we still do not have a deep understanding of how most mechanisms of resistance to viruses in plants operate, although there are a few notable exceptions. Paradoxically, some of the most valuable information on resistance mechanisms is coming from studies of the virus component of the interaction. The ability to clone and sequence DNA copies of viral genomes and to construct infectious artificial recombinants between strains with different patterns of interaction with resistance genes has allowed identification of the viral function which interacts with the resistance gene product. Molecular genetic mapping of the host genome has also provided RFLP and RAPD markers which are sufficiently closely linked to known resistance genes (Haley *et al.*, 1994; Yu *et al.*, 1994) to make their isolation and sequencing likely in the very near future. So far, this approach has been used to isolate the first plant gene for fungus resistance (Martin *et al.*, 1994) and virus resistance genes should not be far behind. The recent identification of resistance and susceptible responses to a virus in the model plant species *Arabidopsis thaliana* (Dempsey *et al.*, 1993) with its small and intensively mapped genome, may offer a fast route to this objective.

Broadly speaking, the different types of genetic control of plant resistance to viruses seem to be associated with different types of mechanism (Fraser, 1990a). Dominant resistance is strongly associated with mechanisms which localize virus around the site of initial infection. The strongest cases of this are exemplified by *Tm–2* and *Tm–2*^{2} in tomato where inoculating virus is confined to a single infected cell in resistant plants (Nishiguchi and Motoyoshi, 1987). The implication is that the resistance operates by preventing the virus from modifying plasmodesmata to allow cell-to-cell spread, or by otherwise blocking the transport process. It is interesting that resistance-breaking isolates overcoming both alleles have been shown to be altered in the 30kDa viral movement protein which controls cell-to-cell spread (Meshi *et al.,* 1989; Calder and Palukaitis, 1992). Artificial recombinants between the resistance breaking and avirulent strains have conclusively shown that the resistance breaking property is located in the 30kDa protein.

A very common form of virus-localizing resistance is the hypersensitive response where the virus may spread to inoculate a few hundred or thousand cells from the point of infection, but further spread is then inhibited and a necrotic local lesion forms. Generally, the evidence suggests that the necrosis and cell death is not the primary block to virus movement, but may be a secondary defense reaction (reviewed in Fraser, 1985). Virus particles can be detected outside the necrotic area in some cases, but are clearly unable to spread further. Perhaps some form of modification or gating of the plasmodesmata has become operative in these cases.

The determinant of virulence/avirulence has been shown to be located in the viral coat protein gene for at least two examples of hypersensitive defense reactions. These are the *N'*gene for TMV resistance in tobacco and the *Nx* gene for potato virus X (PVX) resistance in potato (Pfitzner and Pfitzner, 1992; Santa Cruz and Baulcombe, 1993). Presumably, the different forms of coat protein are involved in recognition events with host components which may trigger a cascade of events including eventual virus localization and the necrotic response. The coat protein gene and its product need not be directly involved in the ultimate expression of the defense mechanism.

For the *N* gene for TMV resistance in tobacco, the determinant of virulence/avirulence was shown to be located in the gene for the viral replicase (Padgett and Beachy, 1993). It is interesting that *N* and *N'* appear to recognize quite different viral functions as determinants of virulence/avirulence. It has been suggested that *N* and *N'* are allelic, but the evidence for allelism is far from complete (Dunigan *et al.,* 1987).

Further interesting observations on the potato *Rx* gene came from David Baulcombe's group. Using synthesized recombinants between PVX strains inducing different responses, they showed that the determinant of the hypersensitive response is a threonine residue at position 121 of the viral coat protein. This feature induced hypersensitivity not only in potato carrying *Rx*, but also in *Gomphrena globosa*. This suggests that there are homologous components in potato and *Gomphrena* which recognize threonine 121 in the PVX coat protein and activate the resistance response (Goulden and Baulcombe, 1993). Further experiments showed that the resistance induced by PVX in *Rx* plants was also effective against the unrelated virus CMV (Köhm *et al.*, 1993).

In *N* gene tobacco, Loebenstein and colleagues have produced evidence for an inhibitor of virus replication (IVR) which may be involved in the resistance response. This protein does inhibit the multiplication of several plant viruses in test systems (Gera *et al.*, 1990) and there is good circumstantial evidence in support of its association with *N* gene activity (Gera *et al.*, 1993). However, its mode of action remains to be established and it is not clear whether the inhibition of TMV multiplication is sufficient to explain the complete cessation of virus spread in *N* gene tobacco. In an alternative model, Moser *et al.* (1988) have shown that the 30kDa movement protein of TMV in the cell wall fraction decreased sharply in amount as the virus became localized and necrosis began.

Resistance mechanisms involving gene-dosage dependent alleles seem to be broadly associated with mechanisms which permit some spread of the virus through the plant, but which tend to reduce multiplication. Thus, the *Tm-1*gene in tomato permits full systemic spread of ToMV but inhibits its multiplication generally by about 60% in heterozygous plants and up to 95% in homozygotes (Fraser and Loughlin, 1980). Interestingly, TMV strains which overcome the *Tm1* gene have an alteration in the viral replicase gene and again artificial recombination has shown that this is responsible for resistance breaking behaviour (Meshi *et al.*, 1988). The suggestion in this case is that the product of the resistance gene is an inhibitor of viral replicase activity. Clearly, this would reduce virus multiplication but would not necessarily interfere with systemic spread. Clearly, also, homozygous plants with twice the level of expression of inhibitor would be expected to be more resistant than heterozygotes. Presumably, the replicase of resistance breaking isolates is not recognized or inhibited by the *Tm-1* gene product.

Little seems to be known about mechanisms of resistance associated with recessive genes, although any associated with negative mechanisms involving the lack of susceptibility factors would, by definition, be difficult to study. These mechanisms should tend to be associated with complete immunity types of response and there does seem to be a little evidence from studies of recessive mechanisms that this may occur. However, more work is clearly needed.

Pokeweed antiviral protein (PAP) reduces the infectivity of a number of plant viruses when co-inoculated to leaves of susceptible species (Chen *et al.*, 1991). PAP is a member of a larger class of ribosome-inactivating proteins. These toxins operate at very low concentrations to cause complete ribosome inactivation in cells by site-specific removal of a single adenine base from the larger ribosomal RNA, thus interfering with elongation factor binding (Endo *et al.*, 1987). PAP is a cell wall protein. A model for its antiviral mechanism was that when damage occurred during virus infection or vector transmission, the PAP selectively entered infected cells, disrupted protein synthesis, and caused local suicide with a restriction of virus multiplication. A problem with this model was that pokeweed ribosomes appeared unaffected by PAP. However, recent experiments with ribosomes prepared from protoplasts of pokeweed (hence not exposed to PAP during homogenization of leaf material) have clearly shown that pokeweed ribosomes are indeed inactivated by PAP. The local suicide model of virus resistance in plants which accumulate RIPs in their cell walls, therefore, seems to be validated. Genes for RIPs (including PAP) have been cloned (Lin *et al.,* 1991) and will be transferred to crop plants for tests of antiviral activity. In this case, however, the human and animal side effects will need careful evaluation.

ACQUIRED RESISTANCE

This covers a diversity of mechanisms. Some provide very effective control of virus infections; others have been much more difficult to substantiate as meaningful resistance mechanisms against viruses although they may well have effects against fungal and bacterial invaders.

Systemic Acquired Resistance and the Pathogenesis Related Proteins

When plants are inoculated by viruses to which they respond by formation of necrotic local lesions, they often respond to a second inoculation at a later date by forming lesions which are smaller or less numerous than those formed on a previously uninoculated control plant (Ross, 1961). This form of apparent resistance is associated with the accumulation of the so-called pathogenesis related (PR) proteins which tend to share common properties (van Loon, 1983). Thus, one group in tobacco is soluble in low pH buffers, resistant to digestion by trypsin, and located in the apoplast. Salicylic acid appears to be the endogenous systemic signal responsible for the induction of PR proteins and the apparent resistance (Yalpani *et al.,* 1993). An antiviral function was long argued for PR proteins (Antoniw and White, 1986) but there appears to be little direct evidence for this. Transgenic plants expressing cloned PR protein genes do not appear to have enhanced virus resistance (Linthorst *et al.,* 1989). Many PR proteins have now been found to have functions such as chitinase and glucanase activities, which could involve them in resistance to fungi and bacterial pathogens (Fritig *et al.,* 1987; Van de Rhee *et al.,* 1994). Moreover, it has been questioned whether the smaller lesions formed on plants showing acquired resistance are actually associated with any reduction of virus multiplication (Fraser and Clay, 1983).

Cross Protection

A valuable form of resistance can be induced in susceptible plants by deliberate infection with mild strains of a virus normally causing severe effects on the plant. Cross protection is widely used against ToMV in tomato lines which do not carry ToMV resistance, in citrus against citrus tristeza virus, and in zucchini against

ZYMV (Sherwood, 1990). The commercial benefits can be considerable. Interestingly, use of mild strain cross protecting isolates of virus in some countries requires that they be registered for use as a pesticide in the same way as fungicides and insecticides. This is despite the fact that severe strains of the virus will already be widely prevalent.

There has been considerable debate about the mechanism or mechanisms of cross protection (De Zoeten and Fulton, 1975; Zaitlin, 1976), but evidence suggests that the protecting strain coat protein may prevent uncoating and gene expression by an invading severe strain (Wilson and Watkins, 1986). The total mechanism is probably more complex than this, but evidence from transgenic plants expressing the coat protein gene of a virus and showing resistance to infection by that virus, similar to cross protection, does emphasize the central role of coat protein (Nelson *et al.*, 1987; Tumer *et al.*, 1987).

Transgenic Plants

In recent years, plants have been transformed with and made to express numerous types of virus-derived genetic elements. A surprising number of these have conferred resistance to subsequent challenge inoculation with the parent or related viruses (reviewed in Harms, 1992; Grumet, 1994; Pappu *et al.*, 1995). The most basic form is transformation with the complete genome of a cross protecting mild isolate. A more refined version has been transformation with the coat protein gene only and this has now been achieved for more than 20 different viruses in a number of different crops, some of which are now undergoing field trials (Harms, 1992; Register and Nelson, 1992). There is no doubt that impressive levels of resistance can be obtained. Fears have been expressed that the presence of excess virus coat protein in released transgenic plants might permit transcapsidation and spread of potentially damaging or pathogenic nucleic acids. These concerns need to be addressed by appropriate risk assessment experiments to inform the protocols for testing the release and commercial exploitation of coat protein and other virus-derived resistance genes (Hull, 1990).

Further effective resistances have been demonstrated in plants transgenically expressing viral satellite RNAs which modify multiplication and symptom expression of those few viruses subject to satellites, such as CMV (Harrison *et al.*, 1987; Baulcombe, 1989). Other viral functions, such as replicase gene components, have also been shown to have potential to inhibit multiplication of the parent virus in a highly effective manner (Golemboski *et al.*, 1990; MacFarlane and Davies, 1992). The mode of action appears to be an inhibition of replicase function or assembly (Carr and Zaitlin, 1991; Carr *et al.*, 1994). There are further possibilities to interfere with virus multiplication or symptom formation by using antisense technology but with a rare exception (Kawchuk *et al.*, 1991) the level of protection has not been very effective (Rezaian *et al.*, 1988; Powell *et al.*, 1989). Ribozymes, catalytically active RNAs designed to cut viral RNA molecules at specific sequences, and thus disable them, have shown potential as antiviral devices.

At present most of the evidence is from *in vitro* and protoplast studies (Edington and Nelson, 1992; Edington *et al.*, 1992). In principle, as soon as any viral function is understood at the level of its molecular mechanism and the responsible nucleic acid sequenced, it should be possible to think of ways of interfering with this by introduction of mimic defective interfering or other blocking mechanisms.

CONCLUSIONS

Viruses are likely to continue to present one of the most important biotic stress challenges to successful production of plants for food, ornamentals, fibers, and wood products. Control by classical resistance breeding has scored notable successes but has probably tackled only a small proportion of problems world wide. In many cases, no suitable sources of resistance are known and the erosion of the genetic base of some crop species may pose hazards for future crop protection. On the other hand, the ability to mark resistance genes with linked molecular markers and, in future, the increasing ability to shift genes between species by molecular techniques, should enhance the ability of the plant breeder to develop resistant varieties. Great attention needs to be given to the pathogenic race structure of the viral population so that resistance breaking behaviour can be carefully managed in the plant breeding context.

Production of transgenic plants with novel types of virus resistance gene must offer a quantum leap in ability to control the most important viruses and gives considerable optimism for the future. Successes with several different components derived from virus genomes would suggest that theoretically this route may be open for large numbers of different types of plant viruses. The limiting factor is likely to be the ease with which facile transformation systems can be developed for a large number of crop species.

REFERENCES

Antoniw J.F. and White R.F. (1986) Changes with time in the distribution of virus and PR protein around single local lesions of TMV-infected tobacco. *Plant Molecular Biology*, **6**, 145–150.

Bald J.G. and Tinsley T.W. (1967a) A quasi-genetic model for plant virus host ranges. I. Group reactions within taxonomic boundaries. *Virology*, **31**, 616–624.

Bald J.G. and Tinsley T.W. (1967b) A quasi-genetic model for plant virus host ranges. II. Differentiation between host ranges. *Virology*, **32**, 321–327.

Bald J.G. and Tinsley T.W. (1967c) A quasi-genetic model for plant virus host ranges. III. Congruence and relatedness. *Virology*, **32**, 328–336.

Baulcombe D. (1989) Strategies for virus resistance in plants. *Trends in Genetics*, **5**, 56-60.

Bonness M.S., Ready M.P., Irvin J.D. and Mabry T.J. (1994) Pokeweed antiviral protein inactivates pokeweed ribosomes; implications for the antiviral mechanism. *The Plant Journal*, **5**, 173–183.

Bos L. (1971) Bean common mosaic virus. *CMI/AAB Descriptions* of *Plant Viruses*, **73**.

Calder V.L. and Palukaitis P. (1992) Nucleotide sequence analysis of the movement genes of resistance breaking strains of tomato mosaic virus. *Journal of General Virology*, **73**, 165–168.

Carr J.P., Gal-On A., Palukaitis P. and Zaitlin M. (1994) Replicase-mediated resistance to cucumber mosaic virus in transgenic plants involves suppression of both virus replication in the inoculated leaves and long distance movement. *Virology*, **199**, 439–447.

Carr J.P. and Zaitlin M. (1991) Resistance in transgenic tobacco plants expressing a non-structural gene sequence of tobacco mosaic virus is a consequence of markedly reduced virus replication. *Molecular Plant-Microbe Interactions*, **4**, 579–585.

Cassells A.C. (1983) Chemical control of virus diseases of plants. In: *Progress in Medicinal Chemistry*, Vol. **20**, edited by G.P. Ellis, and G.B. West, pp.119–155. Amsterdam: Elsevier.

Chen Z.C., White R.F., Antoniw J.F. and Lin Q. (1991) Effect of pokeweed antiviral protein (PAP) on the infection of plant viruses. *Plant Pathology*, **40**, 612–620.

Cheo P.C. and Garard J.S. (1971) Differences in virus-replicating capacity among species inoculated with tobacco mosaic virus. *Phytopathology*, **61**, 1010–1012.

Csillery G., Tobias I. and Rusko J. (1983) A new pepper strain of tomato mosaic virus. *Acta Phytopathologia Academia Scientiarum Hungaricae*, **18**, 195–200.

Dempsey D.A., Wobbe K.K. and Klessig D.F. (1993) Resistance and susceptible responses of *Arabidopsis thaliana* to turnip crinkle virus. *Phytopathology*, **83**, 1021–1029.

De Zoeten G.A. and Fulton R.W. (1975) Understanding generates possibilities. *Phytopathology*, **65**, 221.

Drijfhout E. (1978) Genetic interaction between *Phaseolus vulgaris* and bean common mosaic virus with implications for strain identification and breeding for resistance. *Agricultural Research Reports, Wageningen*, **872**, 1–98.

Edington B.V., Dixon R.A. and Nelson R.S. (1992) Ribozymes: Descriptions and uses. In *Transgenic Plants*, edited by A. Hiatt, pp. 301–323. New York: Marcel Dekker.

Edington B.V. and Nelson R.S. (1992) Utilization of ribozymes in plants - Plant viral resistance. In *Gene Regulation: Biology of Antisense RNA and DNA*, edited by R.P. Erickson and J.G. Izant, pp.209–221. New York: Raven Press Ltd.

Endo Y., Mitsui K., Motizuki M. and Tsurugi K. (1987) The mechanism of action of ricin and related toxin lectins on eukaryotic ribosomes. The site and the characteristics of the modification in 28S ribosomal RNA caused by the toxins. *Journal of Biological Chemistry*, **262**, 5908–5912.

Fletcher J.T. (1992) Disease resistance in protected crops and mushrooms. *Euphytica*, **63**, 33–49.

Fraser R.S.S. (1985) Mechanisms of genetically controlled resistance and virulence: virus diseases. In *Mechanisms of Resistance to Plant Diseases*, edited by R.S.S. Fraser, pp.143–196. Dordrecht: Martinus Nijhoff/Dr W Junk.

Fraser R.S.S. (1986) Genes for resistance to plant viruses. *CRC Critical Reviews in Plant Sciences*, **3**, 257–294.

Fraser R.S.S. (1990a) The genetics of resistance to plant viruses. *Annual Review of Phytopathology*, **28**, 179–200.

Fraser R.S.S. (1990b) The genetics of plant-virus interactions: mechanisms controlling host range, resistance and virulence. In *Recognition and Response in Plant-Virus Interactions*, edited by R.S.S. Fraser, pp.71–91. Heidelberg: Springer- Verlag.

Fraser R.S.S. (1992a) Plant viruses as agents to modify the plant phenotype for good or evil. In *Genetic Engineering with Plant Viruses*, edited by T.M.A. Wilson and J.W. Davies, pp.1-24. Boca Raton: CRC Press.

Fraser R.S.S. (1992b) The genetics of plant-virus interactions: implications for plant breeding. *Euphytica*, **63**, 175–185.

Fraser R.S.S. and Clay C.M. (1983) Pathogenesis-related proteins and acquired systemic resistance: causal relationship or separate effects? *Netherlands Journal of Plant Pathology*, **89**, 283–292.

Fraser R.S.S. and Loughlin S.A.R. (1980) Resistance to tobacco mosaic virus in tomato: effects of the *Tm-1* gene on virus multiplication. *Journal of General Virology*, **48**, 87–96.

Fritig B., Kauffmann S., Dumas B., Geoffroy P., Kopp M. and Legrand M. (1987) Mechanism of the hypersensitivity reaction of plants. In *Plant Resistance to Viruses*, Ciba Foundation Symposium 133, edited by D. Evered and S. Harnett, pp.92–102. Chichester: John Wiley & Sons.

Furusawa I. and Okuno T. (1978) Infection with BMV of protoplasts derived from five plant species. *Journal of General Virology*, **40**, 489–491.

Gera A., Loebenstein G., Salomon R. and Frank A. (1990) An inhibitor of virus replication (IVR) from protoplasts of a hypersensitive tobacco cultivar infected with tobacco mosaic virus, is associated with a 23K protein species. *Phytopathology*, **80**, 78–81.

Gera A., Tam Y., Teverovsky E. and Loebenstein G. (1993) Enhanced tobacco mosaic virus production and suppressed synthesis of the inhibitor of virus replication in protoplasts and plants of local lesion responding cultivars exposed to 35°C. *Physiological and Molecular Plant Pathology*, **43**, 299–306.

Gibson R.W., Pickett J.A., Dawson G.W., Rice A.D., and Stribley M.F. (1984) Effects of aphid alarm pheromone derivatives and related compounds on non- and semi-persistent plant virus transmission by *Myzus persicae*. *Annals of Applied Biology*, **104**, 203–209.

Goldbach R., Eggen R., de Jager C., van Kammen A., van Lent J., Rezelman G. and Wellink, J. (1990) Genetic organization, evolution and expression of plant viral RNA genomes. In *Recognition and Response in Plant-Virus Interactions*, edited by R.S.S. Fraser, pp.147–162. Heidelberg: Springer-Verlag.

Golemboski D.B., Lomonossoff G.P. and Zaitlin M. (1990) Plants transformed with a tobacco mosaic virus nonstructural gene sequence are resistant to the virus. *Proceedings of the National Academy of Sciences, USA*, **87**, 6311–6315.

Goulden M.G. and Baulcombe D.C. (1993) Functionally homologous host components recognize potato virus X in *Gomphrena globosa* and potato. *The Plant Cell*, **5**, 921–930.

Grumet R. (1994) Development of virus resistant plants via genetic engineering. *Plant Breeding Reviews*, **12**, 47–79.

Haley S.D., Afanador L. and Kelly J.D. (1994) Identification and application of a random amplified polymorphic DNA marker for the *I* gene (potyvirus resistance) in common bean. *Phytopathology*, **84**, 157–160.

Hall T.J. (1980) Resistance at the *Tm-2* locus in the tomato to tomato mosaic virus. *Euphytica*, **29**, 189–197.

Harms C.T. (1992) Engineering genetic disease resistance into crops: biotechnological approaches to crop protection. *Crop Protection*, **11**, 291–306.

Harrison B.D., Mayo M.A. and Baulcombe D.C. (1987) Virus resistance in transgenic plants that express cucumber mosaic virus satellite RNA. *Nature*, **328**, 799–802.

Hayes R.J. and Buck K.W. (1990) Complete replication of a eukaryotic virus RNA *in vitro* by a purified RNA-dependent RNA polymerase. *Cell*, **63**, 363–368.

Holmes F.O. (1938) Inheritance of resistance to tobacco mosaic virus in tobacco. *Phytopathology*, **28**, 553–561.

Holmes F.O. (1955) Additive resistances to specific viral diseases in plants. *Annals of Applied Biology*, **42**, 129–139.

Hull R. (1990) The use and misuse of viruses in cloning and expression. In *Recognition and Response in Plant-Virus Interactions*, edited by R.S.S. Fraser, pp.443–457. Heidelberg: Springer-Verlag.

Jellis G.J. (1992) Multiple resistance to diseases and pests in potatoes. *Euphytica*, **63**, 51–58.

Jørgensen J.H. (1992) Discovery, characterization and exploitation of Mlo powdery mildew resistance in barley. *Euphytica*, **63**, 141–152.

Kawchuk L.M., Martin R.R. and McPherson J. (1991) Sense and antisense RNA-mediated resistance to potato leafroll virus in Russet Burbank potato plants. *Molecular Plant-Microbe Interactions*, **4**, 247–253.

Köhm B.A., Goulden M.G., Gilbert J.E., Kavanagh T.A. and Baulcombe D.C. (1993) A potato virus X resistance gene mediates an induced, nonspecific resistance in protoplasts. *The Plant Cell*, **5**, 913–920.

Lin Q., Chen S.C., Antoniw J.F. and White R.F. (1991) Isolation and characterization of a cDNA clone encoding the antiviral protein from *Phytolacca americana*. *Plant Molecular Biology*, **17**, 609–614.

Linthorst H.J.M., Meuwissen R.L.J., Kauffmann S. and Bol J.F.(1989) Constitutive expression of pathogenesis-related proteins PR-1, GRP, and PR-S in tobacco has no effect on virus infection. *The Plant Cell*, **1**, 285–291.

MacFarlane S.A. and Davies J.W. (1992) Plants transformed with a region of the 201-kilodalton replicase gene from pea early browning virus RNA1 are resistant to virus infection. *Proceedings of the National Academy of Sciences, USA*, **89**, 5829–5833.

Martin T.J., Harvey T.L., Bender C.G. and Seifers D.L. (1984) Control of wheat streak mosaic virus with vector resistance in wheat. *Phytopathology*, **74**, 963–964.

Martin G.B., Brommonschenkel S.H., Chunwongse J., Frary A., Ganal M.W. and Spivey R. (1994) Map-based cloning of a protein kinase gene conferring disease resistance in tomato. *Science* (in press).

Meshi T., Motoyoshi F., Adachi A., Watanabe Y., Takamatsu N. and Okada Y. (1988) Two concomitant base substitutions in the putative replicase genes of tobacco mosaic virus confer the ability to overcome the effects of a tomato resistance gene, *Tm-1*. *The EMBO Journal*, **7**, 1575–1522.

Meshi T., Motoyoshi F., Maeda T., Yoshiwoka S., Watanabe H. and Okada Y. (1989) Mutations in the tobacco mosaic virus 30-kD protein gene overcome *Tm-2* resistance in tomato. *The Plant Cell*, **1**, 515–522.

Moser O., Gagey M.J., Godefroy-Colburn T., Stussi-Garaud C., Ellwart-Tschurtz M. and Nitschko H. (1988) The fate of the transport protein of tobacco mosaic virus in systemic and hypersensitive tobacco hosts. *Journal of General Virology*, **69**, 1367–1378.

Nelson R.S., Abel P.P. and Beachy R.N.(1987) Lesions and virus accumulation in inoculated transgenic tobacco plants expressing the coat protein gene of tobacco mosaic virus. *Virology*, **158**, 126–132.

Nishiguchi M. and Motoyoshi F. (1987) Resistance mechanisms of tobacco mosaic virus strains in tomato and tobacco. In *Plant Resistance to Viruses*, Ciba Foundation Symposium 133, edited by D. Evered and S. Harnett, pp. 38–46. Chichester: John Wiley & Sons.

Padgett H.S. and Beachy R.N. (1993) Analysis of a tobacco mosaic virus strain capable of overcoming *N* gene-mediated resistance. *The Plant Cell*, **5**, 577–586.

Pappu H.R., Nilbett C.L. and Lee R.F. (1995) Application of recombinant DNA technology to plant protection: molecular approaches to engineering virus resistance in crop plants. *World Journal of Microbiology and Biotechnology*, **11**, 426–434.

Pfitzner U.M. and Pfitzner A.J. (1992) Expression of a viral avirulence gene in transgenic plants is sufficient to induce the hypersensitive defense reaction. *Molecular Plant-Microbe Interactions*, **6**, 318–321.

Powell P.A., Stark D.M., Sanders P.R. and Beachy R.N. (1989) Protection against tobacco mosaic virus in transgenic plants that express tobacco mosaic virus antisense RNA. *Proceedings of the National Academy of Sciences, USA*, **86**, 6949–6952.

Quadt R., Kao C.C., Browning K.C., Hershberger R.P. and Ahlquist P. (1993) Characterisation of a host protein associated with brome mosaic virus RNA-dependent RNA polymerase. *Proceedings of the National Academy of Sciences, USA*, **90**, 1498–1502.

Register J.C. III and Nelson R.S. (1992) Early events in plant virus infection: relationships with genetically engineered protection and host gene resistance. *Virology*, **3**, 441–451.

Ross A.F. (1961) Systemic resistance induced by localized virus infections in plants. *Virology*, **14**, 340–358.

Santa Cruz S. and Baulcombe D.C. (1993) Molecular analysis of potato virus X isolates in relation to the potato hypersensitivity gene *Nx*. *Molecular Plant-Microbe Interactions*, **6**, 707–714.

Stace-Smith R. (1990) Tissue Culture. In *Plant Viruses, Vol. II*, edited by C.L. Mandahar, pp. 295–320. Boca Raton : CRC Press.

Sulzinski M.A. and Zaitlin M. (1982) Tobacco mosaic virus replication in resistant and susceptible plants: in some resistant species virus is confined to a small number of initially infected cells. *Virology*, **121**, 12–19.

Tumer N.E., O'Connell K.M., Nelson R.S., Sanders P.R., Beachy R.N. and Fraley R.T. (1987) Expression of alfalfa mosaic virus coat protein gene confers cross-protection in transgenic tobacco and tomato plants. *The EMBO Journal*, **6**, 1181–1188.

Urban L.A., Sherwood J.L., Rezende J.A.M. and Melcher U. (1990) Examination of mechanisms of cross protection with non-transgenic plants. In *Recognition and Response in Plant-Virus Interactions*, edited by R.S.S. Fraser, pp. 415–426. Heidelberg: Springer-Verlag.

Van de Rhee M., Linthorst H.J.M. and Bol J.F. (1994) Pathogen- induced gene expression In *Stress-Induced Gene Expression in Plants*, edited by A.S.Basra, pp.249–284. Chur, Switzerland: Harwood Academic Publishers.

Van Loon L.C. (1983) The induction of pathogenesis-related proteins by pathogens and specific chemicals. *Netherlands Journal of Plant Pathology*, **88**, 265–273.

Van Loon L.C. (1987). Disease induction by plant viruses. *Advances in Virus Research*, **33**, 205–256.

Wilson T.M.A. and Watkins P.A.C. (1986) Influence of exogenous virus coat protein on the cotranslational disassembly of tobacco mosaic virus (TMV) particles, *in vitro*. *Virology*, **149**, 132–135.

Yalpani N., Shulaev V. and Raskin I. (1993) Endogenous salicylic acid levels correlate with accumulation of pathogenesis-related proteins and virus resistance in tobacco. *Phytopathology*, **83**, 702–708.

Yu Y.G., Saghai Maroof M.A., Buss G.R., Maughan P.J. and Tolin S.A. (1994) RFLP and microsatellite mapping of a gene for soybean mosaic virus resistance. *Phytopathology*, **84**, 60–64.

Zaitlin M. (1976) Virus cross protection: more understanding is needed. *Phytopathology,* **66**, 382.

Zheng Y. and Edwards M.C. (1990) Expression of resistance to barley stripe mosaic virus in barley and oat protoplasts. *Journal of General Virology*, **71**, 1865–1868.

14. RESISTANCE AGAINST FUNGAL PATHOGENS: ITS NATURE AND REGULATION

DAVID B. COLLINGE[1], TOMAS BRYNGELSSON[2], PER L. GREGERSEN[1], V. SMEDEGAARD-PETERSEN[1] and HANS THORDAL-CHRISTENSEN[1].

[1]*Section for Plant Pathology, Department of Plant Biology, Royal Agricultural and Veterinary University, Thorvaldsensvej 40, 1871 Frederiksberg C Copenhagen, Denmark.*

[2]*Department of Plant Breeding Research, The Swedish University of Agricultural Sciences, S-268 31 Svalöv, Sweden.*

INTRODUCTION

As static organisms, plants require mechanisms by which they can adapt to adverse environmental conditions. Physiological adaption primarily occurs through biochemical changes brought about by *de novo* transcription and resultant protein accumulation. The application of molecular technologies has provided much new information about the transcripts and the encoded proteins which accumulate as a response to different forms of environmental stress (see Wray, 1992; Basra, 1994). Nevertheless, relatively little is known of the physiological mechanisms through which they act to protect the plant. Likewise, although the stimuli which induce the induction of defenses are described, our knowledge of the mechanisms through which these signals are perceived by the plant is at best fragmentary.

Of the natural mechanisms imparting stress tolerance, the nature of resistance against pathogens has perhaps received most attention. In some respects, the interactions between plant and pathogen are the most fascinating, and by their very nature the most complex. Each host plant has to contend with a myriad of more or less specialized pathogens. The interaction between plant and pathogen is the result of co-evolution, governed by the rule of natural selection: it is to the host's advantage to evolve mechanisms to inhibit pathogen growth (resistance mechanisms — see below) and to the pathogen's advantage to evolve mechanisms which overcome these defenses (pathogenicity factors — see Schäfer, 1994; VanEtten, 1994b). It is then to the host's advantage to develop an improved defense and so *ad infinitum*. Analysis of the defense mechanisms has to take this process of co-evolution into account (see Newton and Crute, 1989).

PLANT PATHOGEN INTERACTIONS

Two facets of resistance can be considered: the nature of the mechanisms which actually arrest pathogen development, and the means by which the implementation of these mechanisms are regulated. The defense mechanisms which actually arrest pathogen penetration or development include the hypersensitive response, physical barriers, antimicrobial proteins and metabolites (see Collinge *et al.*, 1994; Dixon *et al.*, 1994). The hypersensitive response is a form of induced programmed cell death which is induced in certain interactions (see Baker and Orlandi, 1994; Collinge *et al.*, 1994; Dietrich *et al.*, 1994; Mehdy, 1994). The other categories include both constitutive and induced forms. The physical barriers include the pre-formed cuticle and cell wall (Smart, 1991), and induced lignification of the cell wall (Boudet *et al.*, 1995), oxidative cross-linking of proteins (Bradley *et al.*, 1992) and formation of callose (Köhle *et al.*, 1985). Lignification requires *de novo* transcription of genes, whereas callose formation (Köhle *et al.*, 1985) and oxidative cross-linking of proteins (Bradley *et al.*, 1992) occur through the rapid activation of post translational events. Likewise, some antimicrobial metabolites are constitutively present, while others are synthesized *de novo* (see Bennett and Wallsgrove, 1994); the former are known as phytoanticipins, the latter as phytoalexins (VanEtten *et al.*, 1994a).

Extracellular Stimuli Inducing Resistance

The activation of defense responses by pathogens apparently utilizes essentially the same types of intracellular signal transduction pathway as used in other perception-response systems described for eukaryotes. The primary signal perceived by the plant is termed as elicitor. A diverse range of pathogen cell wall fragments and exudates can function as elicitors (see Ebel and Cosio, 1994; Kessmann *et al.*, 1994). Elicitor-active molecules can be released from plant tissues by the action of hydrolytic enzymes produced by the pathogen e.g. microbial pectinases can release elicitor active oligogalacturans from the cell wall. Other elicitor-active molecules are constitutively present in plants (e.g. Yokoyama *et al.*, 1991; Hiramoto *et al.*, 1992). Furthermore, exogenous application of various endogenous molecules such as active oxygen species, jasmonates and salicylates is known to induce expression of defense-related genes; these molecules are implicated in intercellular signalling related to local and systemic induced resistance (Enyedi *et al.*, 1992; Farmer and Ryan, 1992; Kessmann *et al.*, 1994; Mehdy, 1994; Lyon *et al.*, 1995).

The significance of the induction of resistance by various chemicals, such as heavy metal ions, and physical stimuli, such as UV light, is less clear. The elicitors are perceived fortuitously by receptors coupled to intracellular signal transduction systems. That a particular molecule can be isolated from a particular pathogen does not mean that the molecule has a role in the intact interaction, merely that it can function as an elicitor in semi-*in vitro* systems. The ultimate response to different elicitors is generally the same, i.e. the response to a number of individual signals, unwittingly produced by different pathogen species (or races of the same pathogen), is the same, involving essentially a single set of defense-related genes. Nevertheless, not all elicitors induce the whole battery of defenses; this can be interpreted to indicate branch points in signal transduction pathways (see Dixon and Lamb, 1990; Davis *et al.*, 1993; Ebel and Cosio, 1994). Under the simplest model implied, a single common signal transduction pathway is responsible for the coordinated induction of diverse defenses (e.g. Gabriel and Rolfe, 1990). However, studies with elicitors and the endogenous signals are complicated by the difficulty in distinguishing direct and pleiotropic effects. This is exemplified by the controversy surrounding the role of salicylic acid (and related molecules) in the systemic induction of resistance (see Chasan, 1995; Dong, 1995 and below). Furthermore, individual defense-related genes can be subject to developmental regulation, being expressed during the normal development of the plant. They can also be induced separately for different purposes by independent stimuli. Examples include the constitutive production of many defense-related proteins in roots and flower parts, and the role of enzymes of phenylpropanoid metabolism in defense and stem-related lignification, signalling and pigmentation. Alternatively, different gene family members can represent the organ/developmental specific expression.

The specific components of the induced defenses do not all accumulate in the same tissues relative to the attacking microorganism. Thus a hypersensitive response can be restricted to the cell or cells in direct contact with the pathogen, the phytoalexins can accumulate in adjacent cells and the antimicrobial "pathogenesis-related" proteins in other tissues or elsewhere in the plant (e.g. Schmelzer *et al.*, 1989). The nature of the signals responsible for local and systemic induction of resistance remains unknown, although several endogenous compounds, which can induce different facets of the defenses, accumulate in tissues exhibiting a defense response. There is good evidence that active oxygen species (such as hydrogen peroxide) can act between adjacent cells and perhaps salicylate acts over longer distances (see Chasan, 1995; Dong, 1995).

Race Specificity

The activity of a molecule as an elicitor is a function defined by the plant rather than by the pathogen producing the molecule. Thus, there is substantial selective pressure on the pathogen to mask production, or at least the secretion, of elicitor active molecules. One consequence is that mutations which result in the non-production of elicitors will be of selective advantage in allowing the parasitization of hosts which cannot detect them, even if a fitness cost is entailed (*c.f.* sickle-cell anaemia). Likewise, new mutations in the host which facilitate the recognition of other molecules produced by the pathogen will be advantageous to the host. The result is polymorphic populations of both host and pathogen with respect to potential host range: this is termed race specificity (see Pryor 1987; Keen, 1992). Many specific resistance genes are described, including some 70 (see Dangl, 1992) in barley acting against different pathogens. These include some 28 alleles atthe complex *Mla* locus alone, and a further 16 genes conferring resistance against powdery mildew attack (Jørgensen, 1992a, 1994). The pathogen genes responsible for detection by the host are termed avirulence (*avr*) genes. At the molecular level, a large number of *avr* genes have been cloned and sequenced, particularly from bacteria (Keen, 1992, De Wit, 1995). The genetic function of an *avr* gene is confirmed by genetic complementation strategies: transfer of an *avr* gene to a virulent race of the pathogen will make it avirulent on a host carrying the corresponding specific resistance gene since the *avr* gene is functionally dominant. Remarkably little, however, has been gained from this enormous effort: few specific elicitor molecules have been identified which can be attributed directly to any specific bacterial *avr* gene product. The best documented example is the *avrD* gene of *Pseudomonas syringae*pv *tomato*(see Keen *et al.*, 1994). In contrast, several specific elicitor molecules have been identified in fungal interactions, and the corresponding *avr* genes cloned by reverse genetic approaches; other fungal *avr* genes have been cloned by map-based methods (see De Wit, 1995).

The cloning and characterization of race-specific resistance genes has had high priority in plant pathology research for a number of years since race specificity represents a simple difference between the states of health and disease. Molecular genetic approaches have led to the recent cloning of specific resistance genes from several species representing interactions with bacterial, fungal and viral pathogens (see Dangl, 1995; De Wit 1995; Staskawitz *et al.*, 1995). Comparison of their sequences to others in the data bases implies that the products are likely to be components of signal transduction pathways responsible for activating defense responses. On the basis of sequence homologies, the products of the L^6, RPS2 and N genes of flax, Arabidopsis and tobacco, respectively, are thought to recognize intracellular ligands. The former would appear to be membrane bound, the latter two cytoplasmic. In contrast, the *Cf9* gene of tomato and *Xa21* gene of rice, which confer resistance to the fungal pathogen *Cladosporium fulvum* and bacterial pathogen *Xanthomonas oryzae*, respectively, show characteristics of extra-cytoplasmic receptors (Jones *et al.*, 1994; Song *et al.*, 1995). Interestingly, both the physical mapping data obtained, and earlier genetic data show that a number of specific resistance genes are clustered (see Dangl, 1992; Jørgensen 1992a; De Wit, 1995). This may suggest that novel specificities arise through uneven crossing over during meiosis.

The cloning of specific resistance genes is a substantial contribution towards comprehending the signal transduction processes regulating the defense response. However, it is necessary to identify the remaining components of these pathways in order to understand how these specific resistances work. A combination of genetic and molecular approaches is being used to identify the other components in the pathways. In animal systems, the transmission of the signal usually involves a physical interaction with the next component. One approach being taken is to exploit this through a genetic screening technique called the "yeast two hybrid system" (Fields and Song, 1991), which facilitates the identification of genes whose products interact with a known component of the pathway, for example the product of a specific resistance gene. A second approach is to identify mutants in which the other genes necessary for the function of a specific resistance gene are inactivated. Thus, the *Rar* (formerly *Nar*) genes of barley are necessary for the function of some *Mla* genes but not other resistance genes in barley (Freialdenhoven *et al.*, 1994, 1995; Jørgensen, 1996). Similar mutations have been observed in Arabidopsis and tomato (*Yu* et al., 1993; Hammond-Kosack *et al.*, 1994; Salmeron *et al.*, 1994). Map-based cloning strategies are being used to clone these genes. A combination of these two approaches will, with considerable effort, lead to a real understanding of the regulation of plant defense responses, and signal transduction systems in general.

Much of the evidence gathered relates to studies in dicots. In this chapter, we concentrate on recent studies in monocots, particularly in relation to the biotrophic powdery mildew fungus, *Erysiphe graminis* f.sp. *hordei.* (*Egh*) (syn. *Blumeria graminis*(DC.) Speer f.sp. *hordei* Em Marchal). Much of the older literature is accessible through specific reviews (Aist and Bushnell, 1991; Bryngelsson and Collinge, 1992; Scott, 1992; Kunoh, 1995; Jørgensen, 1994).

THE INTERACTION BETWEEN BARLEY AND THE POWDERY MILDEW FUNGUS, *ERYSIPHE GRAMINIS* F.SP. *HORDEI*

The interaction between barley and the powdery mildew fungus provides an excellent model for studying defense responses to a pathogen, since it is possible to inoculate at high density while obtaining synchronized development of the fungus. There is an abundance of cytological data (see Aist and Bushnell, 1991; Bryngelsson and Collinge, 1992; Wei *et al.*, 1994) which demonstrates a remarkable robustness in this system compared to most plant-fungal interactions. This robustness, in terms of easily scored and stable phenotypes, has led to a wealth of genetic studies describing the specificity of the interaction and the nature of the host response (see Figure 14.1 and Kita *et al.*, 1981; Jørgensen, 1994). Resistance is expressed at different specific stages during the course of infection: (1) at the penetration stage, termed papilla resistance; (2) At the stage of haustorial formation, as a single cell hypersensitive response (HR); (3) During secondary hyphal elongation, as a multi-cell necrosis (Tosa and Shishiyama, 1984; Aist and Bushnell, 1991). This classification is a simplification: while individual resistances work primarily at a particular stage, a small proportion survive to a later stage. This phenotypic leakage may be influenced by the genetic background of the fungus and host, by the environment, and by inoculation density. Finally, differences in the reaction of different cell types has been observed: the long epidermal cells are penetrated more slowly than the short cells, consequently the HR develops later in these (Koga *et al.*, 1990; Andersen and Jørgensen, 1992b; Görg *et al.*, 1993).

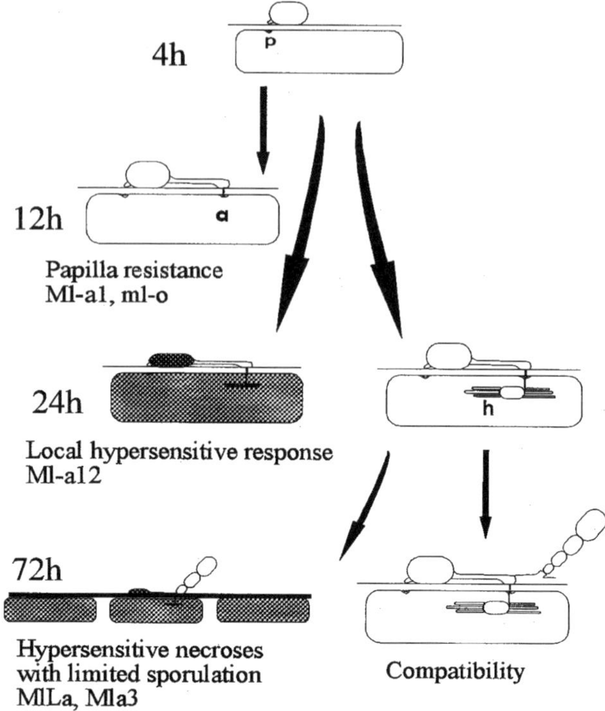

Figure 14.1 Different interaction types controlled by different specific resistance genes in which resistance is exhibited at different specific developmental stages (hours after inoculation). p – papilla; a – appressorium; h – haustorium.

Papilla Resistance

In all penetration attempts, a plant cell wall apposition, comprising a central papilla surrounded by a halo, forms subjacent to the appressorium. The whole structure is generally, though strictly inaccurately, referred to as the papilla for convenience. In papilla-resistant plants, the papillae are rarely penetrated. In our laboratory, penetration of *Mla* plants by avirulent isolates occurs in less than 3% of attempts (Thordal-Christensen and Smedegaard-Petersen, 1988b), though penetration occurs at a higher frequency in other laboratories (Boyd *et al.*, 1995; Bushnell and Scott, *pers. comm.*). A high rate of failure to penetrate papillae (up to 90%) is also observed in compatible interactions (see e.g. Aist and Israel, 1977; Görg *et al.*, 1993; Wei *et al.*, 1994), particularly at high levels of inoculum (Thordal-Christensen and Smedegaard-Petersen, 1988a), or when the appressorial germ tube interacts with the same epidermal cell as the primary germ tube (Woolacott and Archer, 1984). The effect of papilla resistance is also observed in non-host-specific interactions. The papillae of plants homozygous for the *mlo* mutation are larger than those of *Mlo*(Skou *et al.,* 1984), and are penetrated by less than 1% of the conidia (Aist *et al.,* 1988), except for the aggressive isolate HL 3/5, which infects 3-8% of cells attacked (Andersen and Jørgensen, 1992). Likewise, local induced resistance results in the formation of larger papillae in challenged plants (Thordal-Christensen and Smedegaard-Petersen, 1988b). Papilla formation is also observed in penetration attempts by *E. graminis* f. sp. *tritici*(*Egt*), necrotrophic fungi such as *Dreschlera teres* and *Rhyncosporium secalis*(Keon and Hargreaves 1983; Thordal-Christensen and Smedegaard-Petersen, 1988b; Jørgensen *et al.,* 1993), and in unsuccessful penetration attempts by saprophytes (Gregersen and Smedegaard, 1989; Christiansen and Smedegaard-Petersen, 1990). Thus, it is clear that this phenomenon is an important factor in resistance, the study of which, at the molecular level, has been relatively neglected in comparison to the hypersensitive response.

As the papillae appear to prevent many penetration attempts in both compatible and incompatible interactions, they have been investigated thoroughly with histological stains. Papillae contain callose, phenolic compounds, guanidine-containing compounds, proteins including peroxidase and thionins, hydrogen peroxide or other active oxygen species and silica, and there is evidence that many of these components are oxidatively cross-linked (see Kita *et al.*, 1980; Zeyen *et al.*, 1983; Koga *et al.,* 1988; Aist and Bushnell, 1991; Bryngelsson and Collinge, 1992; Ebrahim-Nesbat *et al.,* 1993; Scott-Craig *et al.,* 1993; Wei *et al.,* 1994; Zhang, 1995). Papillae of wheat stain strongly with phloroglucinol indicating lignin, whereas those of barley exhibit, at best, weak staining with this reagent (Smart *et al.,* 1986; Wei *et al.,* 1994). Nevertheless, the presence of phenolic compounds in barley papillae is suggested by epifluoresence (Mayama and Shishiyama, 1978) and weak green staining with toluidine blue (Smart *et al.,* 1986).

Single Cell Hypersensitive Response

Following penetration, the penetration peg differentiates to form the multilobed haustorium, which invaginates the plasma membrane of the epidermal cell. In some interactions, e.g. *Mla12,* (and often *Mla*, e.g. *Boyd et al.,* 1995), resistance is expressed at this stage as HR, limited to the penetrated cell (e.g. Görg *et al.*, 1993; Freialdenhoven *et al.,* 1994). This HR is associated with electrolyte leakage (Lee-Stadelmann *et al.,* 1992), epifluoresence, polyphenols, guanidine compounds, peroxidase and hydrogen peroxide (Aist and Bushnell, 1991; Wei *et al.,* 1994; Zhang, 1995). The *Rar1*and *Rar2* genes (previously known as *Nar1* and *Nar2*, respectively) are necessary for the development of this response in plants possessing *Mla*genes (Freialdenhoven *et al.,* 1994; Jørgensen, 1996). Studies with the inhibitor of mRNA synthesis, cordycepin, indicate that the development of HR requires the transcription of host genes (Bushnell and Liu, 1994).

Necrotic Resistance

In some interactions, e.g. *Mla3 and Ml(La)*, resistance is exhibited at a later stage as necroses extending over several cells. This form of resistance is associated with epifluoresence (Zhang, 1995). These necroses do not always arrest fungal development (Tosa and Shishiyama, 1984), and it can be debated, therefore, whether they represent HR or not. Thus, sporulation occurs, albeit at a reduced rate; these can be considered a form of race-specific necrotic partial resistance.

Though little understood, partial resistance to *Egh*is well documented (see Geiger and Heun, 1989), and, in many respects, of greater practical importance in the control of disease: this form of resistance is typically controlled by a number of genes (the so-called quantitative trait loci, or QTLs) which makes it more difficult for the pathogen to adapt (Asher and Thomas, 1987). Some progress has been made in mapping QTLs in barley using restriction fragment length polymorphisms (RFLP). Partial resistance is known to be affected by environment: thus higher nitrogen supply (Oerke *et al.,* 1989) and decreased day length (Carver and Williams, 1980) increases susceptibility to powdery mildew. The physiological basis of partial resistance remains unknown, its multigene nature and susceptibility to environmental conditions makes the design of critical studies difficult. However, the expression of partial resistance seems to involve quantitative modulation of the general resistance response described in this review, e.g. HR responses (Carver, 1986, Asher and Thomas, 1987) and accumulation of defense-related gene transcripts (Boyd *et al.,* 1994). It is likely that progress will come through the genetic approach, where the mapping of defense-related genes (described below) and QTLs may indicate associations which can be tested by molecular genetic approaches.

EXPRESSION OF THE DEFENSE RESPONSE

Infection or inoculation with *E. graminis*is associated with a metabolic cost to the host plant (Smedegaard-Petersen and Tolstrup, 1985). Inoculation causes the local induction of resistance against subsequent infection attempts (Cho *et al.,* 1986; Thordal-Christensen and Smedegaard-Petersen, 1988a). Some of the physiological changes caused are cytological (listed above), others are invisible. Evidence of alterations in transcription following inoculation initially came from two-dimensional gel electrophoretic (2D-PAGE) analyses of *in vitro*translation products obtained using mRNA populations from plants inoculated with either *Egh*(Manners *et al.,* 1985; Bryngelsson and Collinge, 1992) or *Egt*(Gregersen *et al.,* 1990). Shotgun cloning strategies (differential hybridization and subtractive hybridization) have subsequently led to the isolation of at least 25 cDNA clone species to date in our laboratory (Thordal-Christensen *et al.,* 1992; Gregersen *et al.,* 1993; Bryngelsson *et al.,* 1994; Wei *et al.,* 1996). These represent about 15 distinct transcript families which accumulate in barley leaves following inoculation with *E. graminis*. The majority of these have been identified by sequence comparisons, confirmed by enzyme activities where possible (see Table 14.1). Several species have been isolated for which no homologues have been identified in the data bases. Additional barley response sequences have been isolated using similar strategies in the laboratories of K.J. Scott (Davidson *et al.*, 1987) and J.K.M. Brown (Green, 1991; Boyd *et al.*, 1995), and from developing grains in other laboratories (Høj *et al.*, 1989; Leah *et al.*, 1991).

Application of the shotgun strategy to wheat inoculated with the non-host pathogen, *Egh*, yielded clones representing six classes of transcript encoding, e.g. a peroxidase highly homologous to *Prx8* (Schweizer *et al.*, 1989; Rebmann *et al.*, 1991a), a putative glutathione-S-transferase (Dudler *et al.*, 1991), a putative transmembrane protein (Bull *et al.*, 1992) and a thaumatin-like protein (PR-5-highly homologous to Hv-1 of barley) (Bryngelsson and Gréen, 1989; Rebmann *et al.*, 1991b). These wheat cDNAs have been used subsequently in a number of studies involving barley (Hahn *et al.*, 1993; Kogel *et al.*, 1994, 1995; Valè *et al.*, 1994). Finally, certain leaf thionins, isolated using barley seed probes, have been shown to accumulate in response to powdery mildew (Bohlmann *et al.*, 1988). The induced defenses of cereals thus represent diverse proteins implicated in all aspects of defense, including potential regulators of the defense response, components with proven direct antimicrobial activities, and enzymes involved in the biosynthesis of secondary metabolites (see Table 14.1).

Before describing the individual groups of defense related transcripts and discussing their possible roles, we describe the dynamics of the response and the localization of specific components. Analyses of the accumulation of defense related transcripts in barley leaves interacting with different powdery mildew isolates indicate several response patterns (Figure 14.2). The availability of near-isogenic lines of cultivar "Pallas" carrying different individual resistance genes (Kølster *et al.*, 1986) has proved an invaluable resource for these studies. The differences in response patterns reflect different interaction types. The most common expression pattern observed, dubbed the "PR response", is typified by the responses of the transcripts for the PR proteins, the peroxidase, *Prx*8, and the GRP94 transcripts. In all combinations of host and pathogen chosen, accumulation can be detected as early as 4 hours after inoculation (hai) at high inoculation densities (e.g. 400 spores mm^{-2} (spm)), concomitant with the production of the primary germ tube (see Brandt *et al.*, 1992; Thordal-Christensen *et al.*, 1992; Gregersen *et al.*, 1993; Walther-Larsen *et al.*, 1993 and unpublished data). This peak is not readily observed at lower inoculation densities (<40 spm). Lower levels of transcript are consistently recorded at 9 hai. A second and higher peak of transcript amount is observed at about 12-15 hai during appressorium development (Figure 14.2b). These observations suggest that an elicitor is produced by the germinating hypha. In barley plants exhibiting a localized necrotic response (often termed hypersensitive resistance in this interaction), high response gene transcript levels continued up to 120 hai (see Collinge *et al.*, 1993). In compatible interactions, a third, marked accumulation peak is observed between 72 and 96 hai. Similar results have been observed with these and other transcripts by Clark *et al.* (1994) and Boy *et al.* (1993) studying *Mla*, as well as Davidson *et al.* (1987) studying *Mlp*. The pronounced accumulation of a range of response gene transcripts in compatible interactions at late time points after inoculation has been corroborated by 2D- PAGE of *in vitro* translation products of RNA isolated 72 hai (unpublished).

Table 14.1. cDNA clones representing gene transcripts from barley which accumulate following inoculation with *Erysiphe graminis* f.sp. *hordei*.

Sequence homology (reference)	Related clones (nos.)	Induction by early interactions	*E. graminis*: late compatible interaction	location	Data base accession number (EMBO)
14-3-3 protein from mammals. Putative regulatory protein of signal transduction [1,3].	3	+	–	epidermis	X62388 X93179
GRP94, ER-resident member of the HSP90 family [9].	1	+++	+++	mes. & epid.	X67960
PR-1 protein (tobacco) [4,8] (protein: Hv-4 & 8)	2	++(++)	++++	mesophyll	X74939, X74940 Z21494
PR-2 (β-1,3-glucanase) [7,11] (Hv-7)	1	++(++)	++++	mesophyll	X16274
PR-3 (type II chitinase) [11] (Hv-5)	2	+++(+)	++++	mesophyll	X78671 X78672
PR-4 protein ("barwin") [11]	2	++(++)	++++	mesophyll	
PR-5, thaumatin-like protein [6,11] (Hv-1; acidic, 6, 9 & 10: all basic)	4	++(++)	++++	mesophyll	X58564
Novel PR-proteins [11] (BH6-12 & 17)	2	++(++)	+++(+)	mes. & epid.	
Phenylalanine ammonia lyase [2]	1	+	+	–	
Chalcone synthase (*Chs3*) [11]	1	–	+	–	
Flavonoid *O*-methyltransferases [5]	1	+	+++	–	X77467
Peroxidase, *Prx7** [8]	1	+	++++	mes. & epid.	X62438 L36093
Peroxidase, *Prx8* [8]	1	++++	++++	mesophyll	X58396
Oxalate oxidase-like ("903") [11]	1	++++	+	epidermal	
Oxalate oxidase# [11]	1	+++	+++	mesophyll	
Sucrose synthetase [11]	1	+	+		X66728

+ – ++++: relative accumulation (to each other) of gene transcripts observed in northern blot hybridizations to RNA from different host–pathogen interactions. Transcript localization determined using probe prepared from mRNA isolated from epidermal strips, and *in situ* hybridization data. Hv–1 to 10 are the corresponding barley PR proteins. All cDNA clones isolated using differential or subtractive hybridization proceedures from cDNA libraries prepared using inoculated barley leaf (see 1 & 4) except * and # which were isolated by cross hybridization with seed cDNAs. # preliminary data. References: [1] Aitken *et al.*, 1992; [2] Boyd *et al.*, 1995; [3] Brandt *et al.*, 1992; [4] Bryngelsson e*t al.*, 1994; [5] Gregersen *et al.*, 1994; [6] Hahn *et al.*, 1993; [7] Jutidamrongphan *et al.*, 1991; [8] Muradov *et al.*, 1993; [9] Thordal–Christensen *et al.*, 1992; [10] Walther–Larsen *et al.*, 1993; [11] ; This group unpublished results.

The activated defense against pathogens in dicots, such as tobacco, appears to involve a set of common mechanisms: as a generalization, the majority of defenses are activated against the majority of pathogens. However, in cereals, there is not nearly as much data available for other pathogens as for Egh. Nevertheless, the studies available indicate that the majority of transcripts and proteins tested also accumulate in response to necrotrophic pathogens (e.g. Scott *et al.,* 1990; Hahn *et al.,* 1993). Our own data indicate accumulation of one or more of the transcripts HvPR-1, *Prx8,* HvGRP94, F1-OMT and BH6-12, in response to leaf and root pathogens including *Cochliobolus sativus*, *Pyrenophora teres (Syn: Dreschlera teres*), *Rhyncosporium secalis* and *Septoria nodorum* (Smedegaard-Petersen *et al.,* 1992, Gregersen *et al.*, 1994; and unpublished data and figure 14.2a). Antisera raised against native PR proteins from barley demonstrate accumulation of chitinase, β-1,3-glucanase and PR-5 in response to *P. teres* and *Puccinia hordei*(Reiss and Bryngelsson, submitted).

Although the leaf comprises a number of tissues, the powdery mildew fungus is confined to one, the epidermis. It is, therefore, likely that the events determining the fate of interactions are themselves restricted to the epidermis. The distribution of the defense related molecules relative to the site of infection will therefore give an indication as to whether a particular response is likely to have an important role in resistance. Accumulation of a particular protein in contact with the pathogen indicates a direct role in arresting fungal development; accumulation in adjacent tissues indicates, at best, a role in limiting spread into those tissues. Two types of technique can be used to study the tissue location of particular molecules: histological techniques and tissue-specific biochemical studies. Histological techniques have been used to demonstrate the occurrence of various classes of molecules in papillae and cells undergoing necrotic responses (see above). Immunocytological localization has localized a peroxidase (PRX8/P8.5) to papillae (Scott-Craig *et al.*, 1995). Preliminary immunogold localization data from our laboratory (Hansen *et al.*,unpublished) indicates that the apoplastic BH6-12 protein (see below) accumulates in papillae and round the haustorial primaries.

Comparison of fluorographic/2D-PAGE analyses of *in vitro* translation products obtained using epidermal mRNA indicates that many of the same products accumulate in the epidermis and mesophyll of epidermal tissues 5 hai with the wheat pathogen, *Egt* (Gregersen *et al.,* 1990). mRNA prepared from adaxial epidermis, residual leaf and whole leaf has been used to prepare radioactive first strand cDNA probes, and these probes were hybridized to Southern blots prepared with a range of defense-related cDNAs from barley. The results obtained confirm that the majority of transcripts encoding PR proteins (see below) accumulate in the mesophyll rather than inoculated epidermal tissues. The BH6-903 and *Prx7* transcripts, however, accumulate primarily in the inoculated epidermis. The BH6-12 and HvGRP94 transcripts accumulate in both tissues (Thordal-Christensen, unpublished results). The BH6-903 transcript accumulates in the epidermis in both compatible and incompatible interactions (Schmelzer and

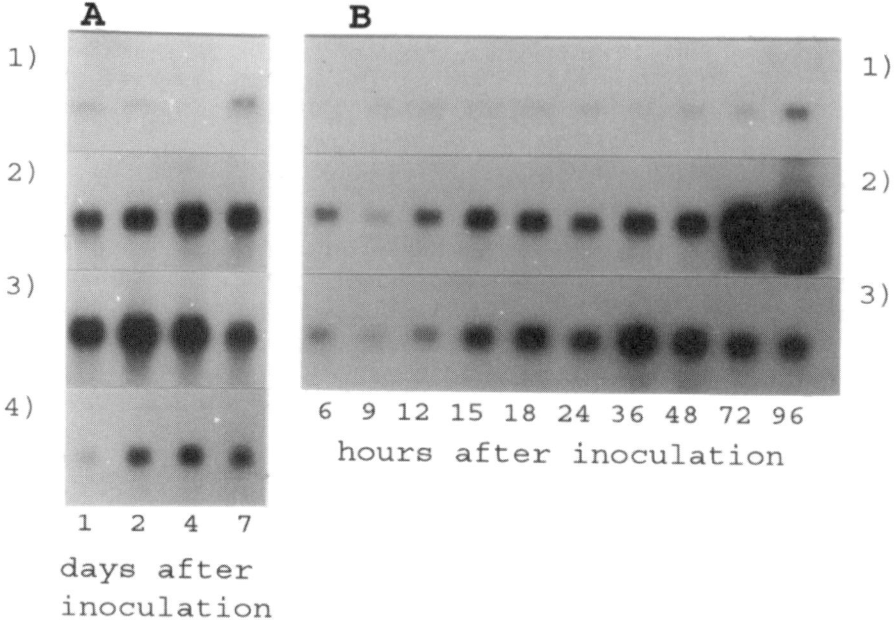

Figure 14.2 Hybridizations with a ^{32}P–labelled probe of a PR–1 cDNA clone to northern blots of total RNA (15 μg/lane) isolated from uninoculated and inoculated leaves of near–isogenic barley line of cultivar Pallas P–01 (Kølster *et al.*, 1986) **A:** 1) uninoculated controls, 2) *Drechslera teres*, 3) *Bipolaris sorokiniana*, 4) *Rhyncosporium secalis*. **B:** 1) uninoculated controls, 2) isolate A6 of *E. graminis* f.sp. *hordei* (compatible interaction), 3) isolate C15 of *E. graminis* f.sp. *hordei* (incompatible interaction)

Gregersen, unpublished). Thus, the PR-proteins do not appear to be involved directly in resistance against this fungus.

INTERCELLULAR SIGNALLING IN THE INDUCTION OF BARLEY RESPONSES

Relatively little is known of the molecules responsible for the activation of defenses in barley. The production of soluble elicitors from conidia (Kristensen, 1994) and endogenous elicitors (Inoue *et al.*, 1994) have been demonstrated; the effect of exogenous application of molecules such as chitin (in wheat-Barber *et al.*, 1989), salicylates and jasmonates (see below) have been studied. It is important to bear in mind that an effect following the application of such substances is not a demonstration that such a substance has a role in the intact biological system.

Exogenously supplied salicylic acid, and a number of molecules of related structure, induce resistance in many plant species against pathogens (see Malamy and Klessig, 1992; Raskin, 1992); 2,6 dichloroisonicotinic acid (DCINA), an analogue of salicylic acid, is also effective in barley, inducing resistance against powdery mildew and inducing accumulation of thionins concomitantly with resistance (Kogel *et al.*, 1995). Endogenous salicylate accumulates in tobacco tissues in which resistance is being induced at levels capable of inducing resistance (see Enyedi *et al.*, 1992). Thus, salicylate has been considered a likely candidate as the transported molecule responsible for systemic induction of resistance. A salicylate-binding protein purified from tobacco proved to possess catalase activity. As binding of salicylate inhibits the catalase, it was proposed that the consequent accumulation of hydrogen peroxide induced the accumulation of defense-related proteins (Chen *et al.*, 1993). However, this seems unlikely as the levels of endogenous salicylate reached in the SAR induced tissues of tobacco are not thought to be sufficient to inhibit the catalase (see Bi *et al.*, 1995; Chasan *et al.*, 1995). Mutants affected in the synthesis of salicylate (or antisense plants) will be useful in clarifying this complex area.

Jasmonate, its methyl ester and related conjugates, have diverse pleiotropic effects on plant metabolism and development. Jasmonates are implicated in regulating the onset of leaf senescence in many species, tuber formation in potato, induction of flavonoid biosynthesis in several dicots and as translocated signals implicated in wound induction of protease inhibitors in the Solanaceae and Fabaceae (see Farmer and Ryan, 1992). These physiological effects are associated with the accumulation of a number of transcripts and proteins known as JIPs (Jasmonate-induced proteins). In contrast to the effects of other growth regulators, there is essentially no homology among jasmonate-induced transcripts among the different taxa (Wasternack, 1994). JIPs include protease inhibitors in solanaceous plants and *Medicago* (Farmer and Ryan, 1990), lipoxygenases in soybean (Grimes *et al.*, 1992) and in the chloroplasts of barley (Feussner *et al.*, 1995), the phenylpropanoid biosynthetic enzyme phenylalanine ammonia lyase (PAL) in soybean (Gundlach *et al.*, 1992), water stress-related transcripts in barley (e.g. Leopold *et al.*, 1996), a ribosome inhibiting protein and at least one thionin in barley (Andresen *et al.*, 1992; Kogel *et al.*, 1995). Osmotic stress also induces JIPs in barley (Lehmann *et al.*, 1995). Jasmonate is also involved in the inhibition of the initiation of mRNA translation (Reinbothe *et al.*, 1993). The induction of so many classes of defense-related transcripts by jasmonate suggests that jasmonate may have an important role in resistance against pathogens. Indeed, jasmonate induces resistance against *Phytophthora infestans* in potato and tomato (Cohen *et al.*, 1993). However, unlike salicylate and related molecules, jasmonate does not induce durable resistance in barley against powdery mildew, but possesses, rather, a direct toxic effect (Schweizer *et al.*, 1993; Kogel *et al.*, 1995). The accumulation of PR proteins is not induced (Schweizer *et al.*, 1993).

SIGNAL TRANSDUCTION PATHWAYS IN BARLEY

There is an accumulating body of data indicating that the intracellular signal transduction pathways regulating the response of plant cells to external stimuli share many features with those of animal cells (Bowler and Chua, 1994). This is also true for the response to pathogens, but the data here are more limited. The structures of some of the proteins encoded by cloned specific resistance genes show features typical for components of intracellular signal transduction: protein kinase domains and leucine rich repeats indicative of protein-protein interactions important for receptor selectivity (Dangl, 1995; Staskawitz *et al.*, 1995). There is also evidence that protein kinases (e.g. Després *et al.*, 1995; Suzuki and Shinshi, 1995), phospholipase C (Legrendre *et al.*, 1993) and G-proteins (Vera-Estrella *et al.*, 1994b) are involved in the intracellular transduction of signals with elicitors.

The only evidence for a specific component of signal transduction in interactions of barley with pathogens concerns a 14-3-3 protein homologue: the transcript of a homologue, designated Hv1433a, accumulates in barley leaves inoculated with the *Egh* (Brandt *et al.*, 1992). The enigmatic 14-3-3 proteins are involved in the regulation of several aspects of signal transduction in eukaryotes. Biochemical activities described include interactions with protein kinases, G-proteins, Ca^{2+} and transcription factors (see Aitken *et al.*, 1992; Aitken 1995). Whereas our data indicate that the majority of the defense related proteins accumulate in the mesophyll, the accumulation of barley 14-3-3 protein is found in the epidermis during early stages of the interaction with *Egh* (Andersen *et al.*, unpubl.). This is of particular interest as this suggests a role in the regulation of the defense response. In maize, *Arabidopsis* and tobacco, 14-3-3 proteins are associated with transcription regulator complexes associated with the G-box (de Vetten *et al.*, 1992; de Vetten and Ferl, 1994; Oecking *et al.*, 1994; Bressan, 1995; Chen *et al.*, 1995), suggesting that 14-3-3 can possess a role as a regulator of transcription. We have identified two further transcript species in barley leaves (Hv1433b and Hv1433c). It can be envisaged that the epidermal 14-3-3 proteins regulate (1) papilla formation, (2) the post-translationally activated defenses or (3) signalling to the mesophyll for the induction of the PR protein response. Interestingly, 14-3-3 proteins are also implicated as receptors of the fungal toxin fusicoccin (Korthout and De Boer, 1994; Korthout *et al.*, 1994; Marra *et al.*, 1994). Fusicoccin, like the NIPs (necrosis-inducing proteins) of the barley pathogen *Rhyncosporium secalis*, stimulate H^+ATPase (Wevelsiep *et al.*, 1993) (NIP1 also acts as a race-specific elicitor-Hahn *et al.*, 1993). Fusicoccin treatment of barley results in association of 14-3-3 protein with the plasma membrane, and concomitant induction of the fusicoccin-binding activity. Interestingly, the AVR5 elicitor of *Cladosporium fulvum* resulted in stimulation of the H^+ATPase in resistant (Cf5) cells due to dephosphorylation (Vera-Estrella *et al.*, 1994 a, b), and it is thought that enhanced H^+ATPase activity may stimulate callose formation (De Wit, 1995). Furthermore, the constitutive expression of a bacterial proton pump in tobacco led to resistance against TMV and *Pseudomonas syringae* pv *tabaci* (Mittler *et al.*, 1995). We have, in collaboration with Preben Bach Holm and Søren Knudsen at the Carlsberg Laboratory, prepared a number of transgenic barley lines expressing Hv1433a as sense and antisense transcripts. Furthermore, we have prepared tobacco and potato (in collaboration with Jørn D. Mikkelsen at Danisco Biotechnology), and *Arabidopsis* lines which express the same transcript constitutively. We hope that this material will enable us to determine the role of this fascinating protein in interactions with pathogens.

The *mlo* mutant of barley is remarkable as one of the few examples of mutation of a plant species towards resistance. This resistance is essentially independent of race, being effective against virtually all isolates (see Jørgensen, 1992b). Cytological studies have demonstrated that *mlo* resistance is associated with larger than normal papillae, and apparently contain higher amounts of callose and phenylpropanoid material (Skou, 1982; Skou *et al.*, 1984; Gold *et al.*, 1986). Interestingly, inhibition of PAL does not affect the outcome of interactions involving *mlo*, whereas those involving *Mla* are inhibited (Carver and Zeyen, 1993). Ca^{2+} concentrations apparently play a role in mediating *mlo* resistance (Bayles and Aist, 1987). In some *mlo*mutants, the papillae form spontaneously, even in the absence of the fungus (Wolter *et al.*, 1993).

In the *mlo* Pallas line P-22, relatively high constitutive levels of almost all of the defense-related transcripts tested are detected in uninoculated plants (Collinge *et al.*, unpublished results). This can be interpreted to mean that the recessive *mlo* is a mutation in a factor normally down-regulating the defense response. It is not possible, on the basis of existing data, to determine whether *mlo* is a lesion in a factor which regulates (1) papilla formation directly, (2) the transcriptional regulation of the defense related genes, or whether its role in one of these has a pleiotropic effect on the other. The order of these events cannot be distinguished. The results nevertheless imply a situation where the functional product of the wild type allele represses (in the broadest sense) the transcription of the defense-related gene products. An analogous situation is observed with the catalase associated systemic induction of resistance in tobacco (Chen *et al.*, 1993). Inhibition of the catalase by salicylic acid results in the potential accumulation of hydrogen peroxide, and hydrogen peroxide is capable of inducing the accumulation of PR proteins. A further analogy can be drawn to this system, since constitutive expression of salicylic acid hydroxylase gene from *Pseudomonas putida* (*NahG*) results in two phenomena: an inhibition in the ability of tobacco plants to systemically acquire resistance, and increased local susceptibility to pathogens (Gaffney *et al.*, 1993; Delaney *et al.*, 1994). Mutants have been identified which demonstrate two genes (*Ror1* and *Ror2*) required for *mlo* resistance (Freialdenhoven *et al.*, 1996). These mutations are likely to affect the regulation of the *mlo*-response since the structure of the papillae formed appear unaltered.

ACTIVE OXYGEN SPECIES

The term "active oxygen species" or AOS covers molecules such as the superoxide anion radical (O_2^-), hydrogen peroxide (H_2O_2), and the hydroxyl radical

(OH^-) which are produced by successive reduction of oxygen. Various AOS accumulate rapidly in tissues undergoing a hypersensitive response, a phenomenon known as the oxidative burst. These act in several ways in interactions with pathogens (see Mehdy 1994; Baker and Orlandi, 1995). Their most direct effect, at high concentrations, is as a potential antimicrobial agent. Levels as high as 1.2 mM H_2O_2 have been detected transiently following elicitor treatment of soybean cell suspension cultures (Legrende *et al.*, 1993). They are also implicated, at lower concentrations, as the molecules responsible for the generation of the hypersensitive response (Levine *et al.*, 1994). A third role proposed is as messengers involved in the induction of PR-protein gene expression (but see Chasan, 1995; Dong, 1995). Finally, they are necessary for peroxidase-mediated cross-linking of molecules including sugars, lignin precursors, lipids and proteins in secondary cell walls (Everse *et al.*, 1991; Welinder *et al.*, 1993). The latter can occur rapidly, in the absence of *de novo* gene expression (Bradley *et al.*, 1992; Brisson *et al.*, 1994). Convincing evidence of their importance in these interactions comes from experiments using scavangers and enzyme inhibitors (e.g. Levine *et al.*, 1994) and by producing transgenic plants with an constitutively elevated level of H_2O_2 (Wu *et al.*, 1995).

An accumulating body of data indicates that AOS metabolism is important in interactions between barley and the powdery mildew fungus. As mentioned above, we have demonstrated the accumulation of two peroxidase transcripts in barley leaves following inoculation with *Egh* (Thordal-Christensen *et al.*, 1992). We have also observed oxidative cross-linking of substances in the papillae and of peroxidase activity in papillae. A cDNA library prepared from barley leaves exhibiting an incompatible interaction with *Egh* six hai (Thordal-Christensen *et al.*, 1992) was screened by differential hybridization using probes prepared from inoculated and non-inoculated barley epidermal tissues (Wei *et al.*, 1996). A barley cDNA clone (pBH6-903) isolated by this means shows approx. 45% sequence identity to oxalate oxidases from barley and wheat. Oxalate oxidase is of interest in the context of active oxygen species as it releases hydrogen peroxide from oxalic acid. The transcript and protein corresponding to this cDNA accumulate early in the epidermis after inoculation with *Egh*, concomitantly with papilla formation and prior to PR protein and peroxidase accumulation (Wei, 1994; Zhang, 1995). While the "903" protein appears not to have oxalate oxidase activity, this activity, conferred by other proteins, also increases in barley leaves following inoculation with *Egh* (Zhang *et al.*, 1995). The extractable protein and enzyme activity accumulate in the mesophyll, concomitantly with the PR proteins and PRX8. The sequence of a novel oxalate oxidase cDNA clone isolated from a barley leaf cDNA library matches N-terminal amino acid sequence data obtained from the purified oxalate oxidase protein (Zhou, unpublished).

PHENYLPROPANOID AND OTHER SECONDARY METABOLIC PATHWAYS ASSOCIATED WITH DEFENSE IN BARLEY

Three groups of phenolic compounds are implicated in defense responses: (1) the signal molecule salicylic acid (see Enyedi *et al.*, 1992; Ryals *et al.*, 1992), (2) phenylpropanoid phytoalexins (see Dixon, 1986; Kuc, 1995) and (3) lignin and related polyphenolics (see Boudet *et al.*, 1995; Whettan and Sederoff, 1995). Some central features of the pathway for phenylpropanoid metabolism are illustrated in Figure 14.3.

The presence of phenolic compounds in the papillae which form subjacent to the appressorium in barley (Aist and Israel, 1986; Shiraishi *et al.*, 1989) suggests a role for at least that branch of the phenylpropanoid pathway which is involved in lignification, although histochemical tests for lignin give negative results in barley (Smart *et al.*, 1986; Wei *et al.*, 1994). There is also evidence that the lignins of grasses include cross-linked feruloyl esters (Ralph *et al.*, 1992), though these are not implicated in defense. The activities of the enzymes PAL, C4H, 4CL and COMT (see Figure 14.3) increase round the site of lignification in wheat associated with *Botrytis cinerea*(Maule and Ride, 1976, 1983). Evidence of the importance of this lignification comes from studies with inhibitors. Infusion of inhibitory concentrations of the PAL inhibitors α-aminooxy-β-phenylpropionic acid or α-aminooxyacetic acid into oat leaves suppressed the induction of epifluorescent material in the papillae (Carver *et al.*, 1992a,b). Furthermore, inhibition of PAL resulted in a higher frequency of penetration in barley (Carver *et al.*, 1994).

The final step of lignification is polymerization, which can be catalyzed by peroxidase or laccase (polyphenol oxidase). Peroxidases accumulate in the apoplast of barley leaves infected with powdery mildew (Hislop and Stahmann, 1971; Kerby and Somerville, 1989, 1993), and peroxidase activity can be detected in papillae (Wei, 1994; Zhang, 1995), the latter observation indicating the occurrence of cross-linking reactions in papillae. At least two peroxidase transcripts accumulate in barley leaves inoculated with *Egh*(Thordal-Christensen *et al.*, 1992), and the accumulation of one of these is concomitant with the accumulation of transcripts cross-hybridizing with a clone encoding PAL and COMT from barley and maize, respectively (Collazo *et al.*, 1992; Gregersen *et al.*, 1994). The peroxidases have been designated *Prx7*and *Prx8* (S.K. Rasmussen pers. comm.). Confirmation that the pBH6-301 (*Prx8*) cDNA indeed encodes a peroxidase was obtained following expression of the cDNA in transgenic potato and tobacco plants. The transgenics accumulate an extra apoplastic peroxidase which exhibits the same isoelectric point of pI 8.5 as the apoplastic barley peroxidase, and it cross-reacts with antiserum raised against this peroxidase (Scott-Craig *et al.*, 1995; Wyke *et al.*, unpublished results).

In addition to yielding clones encoding five classes of PR protein homologues, shotgun screening of a cDNA library prepared from a compatible interaction

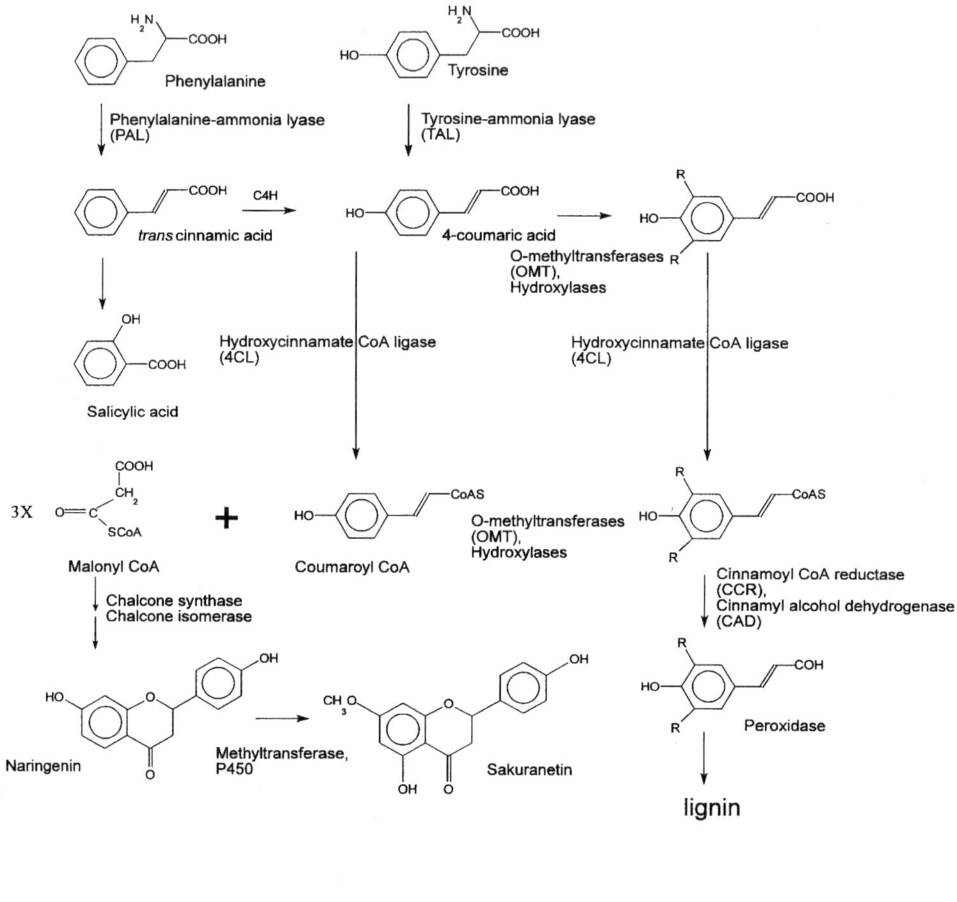

Figure 14.3. Central phenylpropanoid pathway. See Matsuyama and Dimond, 1974; Dixon, 1986; Enyedi *et al.*, 1992; Kodama *et al.*, 1992; Ryals *et al.*, 1992; Boudet *et al.*, 1995; Kuc , 1995; Whettan and Sederoff, 1995 for details.

three days after inoculation yielded two cDNAs which represented transcripts exhibiting a markedly later accumulation pattern in compatible interactions, relative to transcripts of PAL, a COMT and the peroxidase, *Prx8* (Gregersen *et al.*, 1994). Comparison of these sequences to the data bases indicates significant sequence identity to *O*-methyltransferases COMT, and chalcone synthases. The OMT sequence is most closely related to a putative methyltransferase from maize, different from COMT. Both cDNAs have been expressed functionally in *E.coli*. The chalcone synthase uses both 4-coumaroyl-CoA and caffeoyl-CoA precursors, leading to production of naringenin and eriodictyol, respectively.

In addition to induction by pathogens, the OMT and CHS transripts also accumulate following UV treatment. There are several reports on UV-and pathogen-induced flavonoids in cereals: two glucosylated flavonoids, saponarin and lutonarin, are expressed in barley leaves in response to UV irradiation (Liu *et al.*, 1995); sorghum expresses the flavonoid phytoalexins apigeninidin and luteolinidin in response to *Colletotrichum graminicola* inoculation and also a caffeic acid ester of apigeninidin (Nicholson, 1992); and sakuranetin, a methylated derivative of naringenin, has been shown to be one of the major phytoalexins of rice also induced by UV light (Kodama *et al.*, 1992). The antifungal activity of sakuranetin is much higher towards the rice blast fungus, than the non-methylated precursor. The methylation step may additionally have a role in retaining the putative phytoalexin in the apoplast in the presence of peroxidase by preventing cross linking. We are currently studying the flavonoids which accumulate in barley leaves interacting with the powdery mildew fungus in order to identify the natural substrates of the CHS and OMT enzymes.

It has been proposed that cyanogenic glucosides have a function as antimicrobial agents (See Fry and Myers, 1981; Bennett and Wallsgrove, 1994). Cyanogenic glucosides and their specific β-glucosidases are compartmented separately in plant tissues. Tissue damage caused by frost, herbivores or pathogens brings the glucosides in contact with the glucosidases. Spontaneous and enzymatic degradation of the aglycone results in the release of toxic hydrogen cyanide. Epidermal cells of barley contain considerable amounts of soluble cyanogenic glucosides, however, the cyanogenic glucoside content of barley leaves is proportional to the susceptibility of the leaves to the powdery mildew fungus (Pourmohseni and Ibenthal, 1991). Pathogens of cyanogenic species are generally adapted to cyanogenesis by the host as they are capable of assimilating the cyanide released by various mechanisms (see Fry and Myers, 1981; Bennett and Wallsgrove, 1994). Formamide hydrolyase releases formamide from hydrogen cyanide. In a survey of 30 species of fungi, it was found that all the pathogens of cyanogenic plants accumulated formamide hydrolyase, whereas none of the non-pathogens tested could do so (see Fry and Evans, 1977). The powdery mildew fungus does not cause necrosis but invaginates the plasma membrane. The enhanced susceptibility of the barley leaves producing greater levels of cyanogenic glucosides may therefore reflect the higher levels of carbohydrate available to the fungus in these tissues.

PATHOGENESIS-RELATED (PR) AND OTHER ANTIMICROBIAL PROTEINS

The "pathogenesis-related" or PR proteins were originally described in tobacco, where their accumulation in leaves during infection with tobacco mosaic virus was interpreted to indicate a role in virulence (see Van Loon, 1985; Collinge *et al.*, 1994). Subsequent studies have demonstrated that these possess antimicrobial activities (see Collinge *et al.*, 1994), suggesting a role in resistance. The current consensus on their nomenclature is to recognize five families with reference to tobacco (designated PR-1 to PR-5), and additional families on the basis of sequences originally identified from other plant species (currently PR-6 to PR-11) (Van Loon *et al.*, 1994).

PR proteins were described as accumulating in barley leaves exhibiting a hypersensitive response to *Egh* (White *et al.*, 1987; Bryngelsson *et al.*, 1988), and have been purified subsequently (see Table 14.2). cDNA clones encoding many of these were isolated independently by differential hybridization from libraries prepared from leaves infected with *Egh* (Jutidamrongphan *et al.*, 1989; Muradov *et al.*, 1993; Bryngelsson *et al.*, 1994). At least the PR-1 (Bryngelsson *et al.*, 1994), β-1,3-glucanase and chitinase (type II, Collinge *et al.*,unpublished) proteins are secreted into the apoplast of infected leaves (see Figure 14.4), and of leaves exhibiting a necrotic resistance form (*MlLa*). Three chitinases, one of which corresponds to the type II chitinase, and a β-1,3-glucanase accumulate in the media from embryonic barley suspension cultures (Kragh *et al.*, 1991). The possession of similar putative signal peptides in translated open reading frames for the leaf PR-2 (β-1,3-glucanase), PR-4 ("barwin") and PR-5 ("thaumatin-like") cDNAs suggests that these too are secreted. Some of the same (or very closely related) proteins apparently accumulate during normal seed development (Hejgaard *et al.*, 1991, 1992; Svensson *et al.*, 1992). Sequence analyses indicate that, with the exception of PR-3 (chitinase), the developmentally regulated seed and induced leaf proteins are encoded by the same or closely related genes (unpublished data). The antimicrobial activities of the seed PR-2 to PR-5 proteins have been studied *in vitro*: all possess antimicrobial activities when tested in combination, some mixtures exhibit synergistic effects (Hejgaard *et al.*, 1991, 1992; Leah *et al.*, 1991). β-1,3-glucanase and type I and II chitinases from barley seeds exhibit *in vitro* antifungal activities against *Trichoderma and Fusarium* species (Jacobsen *et al.*, 1990; Leah *et al.*, 1991; Swegle *et al.*, 1992). Synergistic antifungal effects have been obtained between the seed β-1,3-glucanase and type II chitinases *in vitro* (Leah *et al.*, 1991) and *in vivo* following expression in tobacco (Jach *et al.*, 1995) in the presence of another class of protein which is constitutive in the seed, namely the ribosomal inactivating proteins (RIP). RIP60 accumulates in jasmonate treated, but not inoculated leaves (Reinbothe *et al.*, 1994; Kogel *et al.*, 1995).

Table 14.2 Purified barley PR proteins

Protein	MW	pI	Homology to	antiserum	peptide	cDNA# sequence
HvPR-1a	15	10.5	PR 1 [1]	x	x	x
HvPR-b	15	>10.5	PR 1 [1]	x	x	x
HvPR-2a	31	10.5	β-1,3-glucanase*	x	x	x
HvPR-2b	36	<3.2	β-1,3-glucanase*		x	
HvPR-2c	33	3.2	β-1,3-glucanase*			
HvPR-3	26	>10.5	chitinase*	x	x	x
HvPR-5a	19	3.4	thaumatin [2]	x	x	x
HvPR-5b	27	9.3	thaumatin		x	
HvPR-5c	26	10.2	thaumatin		x	x?
HvPR-5d	25	>10.5	thaumatin		x	x?
HvPR-5e	22	>10.5	thaumatin		x	

*identification confirmed by enzyme activity determinations. # see Table 14.1.
[1] Bryngelsson and Gréen, 1989, [2] Bryngelsson *et al.*, 1994.

Thionins are cysteine rich 5 kDa polypeptides which possess antimicrobial activities *in vitro* against *Clavibacter michiganensis*, *Egh*, *Pseudomonas syringae*, *Puccinia graminis f.* sp. *tritici*, *Pyrenothora teres* and *Thieraliopsis paradoxa* (Bohlmann *et al.*, 1988; Carmona *et al.*, 1993; Kogel *et al.*, 1995). Some thionins are constitutively present in barley seeds, others accumulate in papillae in inoculated leaves (Bohlmann *et al.*, 1988; Ebrahim-Nesbat *et al.*, 1993), and following treatment with jasmonate, salicylate and related molecules (Andresen *et al.*, 1992; Kogel *et al.*, 1995; Wasternack *et al.*, 1995—see above). Expression in transgenic tobacco of an α-thionin gene from barley grain under the regulation of the 35S promoter of cauliflower mosaic virus conferred enhanced resistance against the bacterial pathogens *Pseudomonas syringae* pv. *tabacini* and *P. s.*pv. *syringae* (Carmona *et al.*, 1993). Non-specific lipid transfer proteins are present in uninoculated, etiolated barley and maize leaves, and accumulate following induction with powdery mildew (Molina *et al.*, 1993). These have antibacterial (against *Clavibacter michiganensiss* sp. *sepadonicus* and *Pseudomonas solanacearum*) and antifungal (against *Fusarium solani*) activities *in vitro*. A synergistic effect was observed in combination with wheat α and β thionin against *F. solani*.

Several sequences have been identified which apparently represent novel PR proteins; no sequence identity is shown to any proteins in the data bases. For

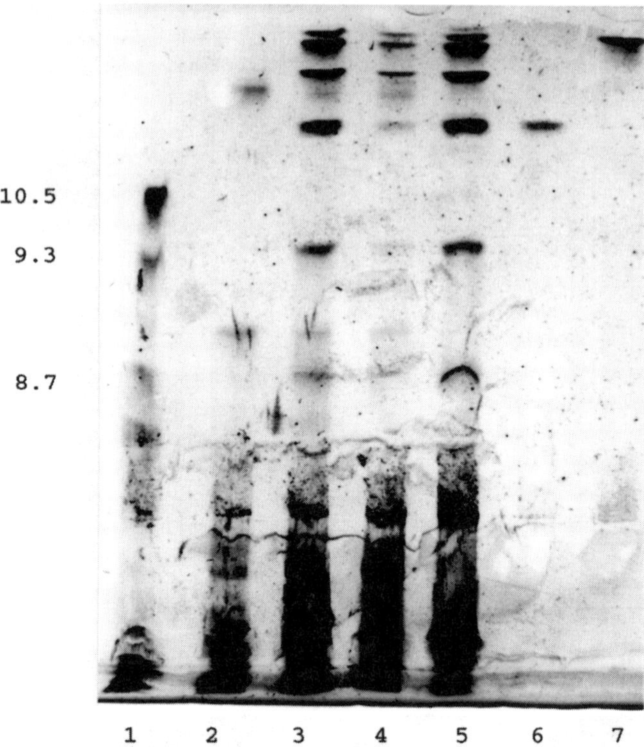

Figure 14.4 Isoelectric focussing gel, approx pI 8 to 11, stained with silver. Lane 1: marker; Lane 2: uninoculated leaves; Lanes 3, 4 and 5: infected leaves, compatible interaction, 6 days after inoculation. Lanes 2 and 3 total extracted proteins (homogenized in citrate–phosphate buffer pH 2.8); Lane 4: intracellular after intercellular wash; Lane 5: apoplastic fraction; Lane 6: HvPr1a; Lane 7: HvPr1a (see Bryngelsson *et al.*, 1994).

example, the cDNAs pBH6-12 and pBH6-17 encode proteins with typical signal peptides, a predicted M_r values of 23 and 25 kDa. We have raised antisera against the corresponding proteins following expression in *E.coli*. Western blot analysis demonstrates that the corresponding BH6-1 2 and BH6-17 proteins accumulate in the apoplast of inoculated barley. In collaboration with Jørn Mikkelsen, Danisco, we have prepared transgenic potato and tobacco plants in which the pBH6-12 cDNA is fused to the enhanced 35S promoter of CaMV. The BH6-12 protein is secreted into the apoplast, but no significant differences in the susceptibility of these potato plants to the pathogen *Phytophthora infestans*, have been detected. The apoplastic fluid of the uninoculated transgenic tobacco plants does not exhibit antifungal activity against yeast, *Trichodema reesei* or *Egh*.

The secretion to the apoplast of barley PR proteins and peroxidases implicates the activation of the endoplasmic secretory mechanism. We were the first to clone a plant homologue of the vertebrate chaperon, GRP94 (Walther-Larsen *et al.*, 1993) which is implicated in the secretory pathway (see Vitale *et al.*, 1993). The transcript accumulates in barley leaves concomitantly with the PR-protein and *Prx8* transcripts following inoculation with *Egh*.

MOLECULAR GENETICS OF THE PATHOGEN

Most information about the interaction between barley and the powdery mildew fungus originates from the host. Knowledge of the pathogen, apart from the intrinsic interest in the biology of an organism superbly adapted to its host, might lead to the development of original and specific strategies for its control. The obligate nature of the pathogen makes pathogenicity studies involving gene disruption unrealistic with current technologies. Nevertheless, knowledge of the molecular biology of this organism is emerging. Virtually the only conventional genetic information available concerns race specificities and sensitivity to fungicides (Brown *et al.*, 1992 and see below). The development of RFLP (restriction fragment length polymorphism) has facilitated the development of a partial genetic map of the fungus comprising seven linkage groups (Christiansen and Giese, 1990; Brown and Simpson, 1994). Three *avr* genes are localized on this map and the closest linkage observed was 9 centimorgans (cM). More recent maps incorporate 7 *avr* genes (Brown and Simpson, 1994; Jensen *et al.*, 1995). Cytological studies indicate five small and at least two larger chromosomes, and field inversion gel electrophoresis studies separate several bands including a mitochondrial band. The exact genome size could not be estimated since the larger chromosomes could not be resolved (Borbye *et al.*, 1992), but is estimated to 35 Mb (Giese, pers. comm.). The frequencies of genetic recombination (i.e. the size of the genetic map in cM) relative to genome size appear high compared to many other organisms. Speculatively, this may reflect an ability to evolve rapidly. The genome has a high content of repetitive DNA compared to other filamentous fungi (Rasmussen *et al.*, 1993). Part of the repetitive DNA is present as a SINE-like element (Rasmussen *et al.*, 1993). This may provide a mechanism for frequent illegitimate recombination through uneven crossing-over. A second retroposon element is also present, and the probable master element has been identified from a cDNA library prepared from inoculated epidermal tissues (Wei *et al.*, 1996). Another feature is an episome in the mitochondria of some isolates (Giese *et al.*, 1990). The biological significance of these elements and episome remain unresolved. However, medium copy number repetitive DNA has proved useful for finger-printing populations in epidemiological studies (e.g. O'Dell *et al.*, 1989; Brown *et al.*, 1990; 1993; Wolfe and McDermott, 1994).

Very little is known of the physiology of pathogenesis in the pathogen: esterase and cutinase are secreted by the fungus during spore germination (Staub *et al.*, 1974; Nicholson *et al.*, 1988; Kunoh *et al.*, 1990). Genetic and cytological phenotypic studies suggest that expression of some *avr* genes is associated with specific developmental stages, such as haustorial initiation (Boyd *et al.*, 1995). By cloning transcripts expressed at these stages, it may be possible to identify some *avr* genes directly. A cDNA library prepared from germinating conidia was screened by subtractive hybridization in order to identify transcripts associated with specific stages in development. Two cDNAs corresponding to transcripts expressed in germinating conidia have been sequenced, but no identification is suggested from the data bases (Justesen *et al.*, 1995). One of these transcripts is associated with hyphal growth. A cDNA library from infected barley leaf tissue screened by differential hybridization (Bryngelsson *et al.*, 1994) also yielded several powdery mildew sequences in addition to many of the host defense related sequences listed in Table 14.1. The *Egh* sequences identified encoded putative ribosomal proteins and a fragment of a putative ATPase (Gregersen, unpublished). The β-tubulin gene of *Egh* has also been sequenced (Sherwood and Somerville, 1990). The development of genetic maps incorporating *avr* genes (Brown and Simpson, 1994; Jensen *et al.*, 1995), cosmid (Rasmussen *et al.*, 1993) and YAC (yeast artificial chromosome) libraries, and transformation technologies (Christiansen *et al.*, 1996) provide the tools for map-based cloning of *avr* genes from *Egh*.

PERSPECTIVES

All gene transcripts identified in interactions between barley and isolates of *Egh* accumulate at some time-point in all interactions tested. One interpretation of this is that the defenses induced are ineffective. However, the more rapid accumulation in interactions involving a hypersensitive response provides correlative evidence that at least some of these are effective under certain circumstances. Given that compatibility is the exception rather than the rule, it seems more than likely that the induced defense responses have a role in limiting the spread and hence the damage caused by pathogens, in addition to limiting the range of pathogens causing infection. Formal proof for these suppositions will require an enormous effort involving, for example, transgenic antisense plants in which production of specific proteins is suppressed. Important evidence has been obtained through the use of inhibitors: inhibition of PAL results in penetrable papillae in barley and oat (Carver *et al.*, 1992a,b, 1994). Biochemical functions have been identified about half of the transcripts isolated. The identification of the remainder is a high priority. Thus, the characterization of the response comprises both identification of the actual response components, the biological role, and the identification of the mechanisms involved in the regulation of its induction, e.g. the specific resistance genes, and genes involved in cell-cell signalling.

No specific resistance genes have been cloned from monocots to date, except for a single rice gene (Song *et al.*, 1995). Many independent specific resistance gene loci for powdery mildew have been incorporated in the barley genomic maps (see Jørgensen, 1992b; 1994; Maroof *et al.*, 1994), which are well supplemented with RFLP and RAPD markers. The information obtained has demonstrated that the cereals exhibit considerable collinearity among genomes. This means that markers mapped in one species (e.g. rice) can be used in others, e.g. barley and wheat. Thus, the available molecular tools make barley an attractive plant for map based cloning studies of the type which yielded the *Pto* gene from tomato (Martin *et al.*, 1993) and *Xa21*gene from rice (Song *et al.*, 1995). Indeed, RFLP markers have been identified which are closely linked to e.g. *Mla* (DeScenzo *et al.*, 1994), MlLa (Hilbers *et al.*, 1992; Giese *et al.*, 1993) and *Mlg* (Görg *et al.*, 1993), as well as to one of the genes necessary for *Mla* resistance (Freialdenhoven *et al.*, 1994). A second strategy will be to use the cloned resistance genes as heterologous probes. This will undoubtedly yield homologues genes, but it will take a lot of effort to establish the specificity, if any, associated with a candidate resistance gene.

One of the ways to test critically whether individual components of the defense response have a role in resistance is to express the individual components constitutively in a plant and determine whether the transgenic plants have become more resistant to their pathogens. A positive result will indicate that the coding sequence can contribute to the improvement of the resistance of crops. This strategy cannot, however, prove a role for the defense in question in the original interaction, but has the potential application of providing disease resistant plants. In view of the propensity of pathogens to evolve to overcome the defenses of plants, the most durable results will be obtained when individual factors are combined to express at least two independent mechanisms. It is predicted that the pathogens of a particular plant host are adapted to overcome the defenses employed by its host. The mechanisms of adaptation include the ability of the pathogen to neutralize the defenses—examples include phytoalexin detoxification mechanisms (see Van Etten *et al.*, 1989). Transfer of the genes encoding defense mechanisms from one plant species to another will, in some cases, overcome the adaptation problem (e.g. Hain *et al.*, 1993). Indeed barley genes have been expressed in dicots with the development of transgenic lines with enhanced resistance to both fungal (Jach *et al.*, 1995) and bacterial (Carmona *et al.*, 1993) pathogens. The development of a relatively efficient biolistic transformation method for barley (Wan and Lemaux, 1994) and other cereals (see Scott, 1994) make the reciprocal approach an achievable goal in the foreseeable future. Though technically the resistance achieved by this means represents a single factor resistance, it differs in principal from that provided by transfer of resistance genes. The induction of specific resistance genes using single factors (the *avr*gene product—see above), and their "breakdown" only requires inactivation of the *avr*gene. "Breakdown" of resistance achieved through transfer of novel defense mechanisms between species requires the development a positive character (e.g. an inhibition mechanism).

There is a risk that the constitutive expression of defense-related genes can have metabolic costs for the transgenic plants. It may also be considered undesirable to add antifungal proteins or metabolites to food, although these are present naturally in many foodstuffs, such as cereal grains. The expression of defense-related genes can be regulated to ensure that the gene products are produced only where and when the pathogen attacks through the use of appropriate promoters (see Dangl, 1991). That this, of course, happens already with existing induced defenses is not necessarily a hindrance, since the antimicrobial products are not necessarily produced in the right tissues, or rapidly enough relative to the invading pathogen. One possible explanation is suppression; indeed there is good evidence that the pathogenicity mechanisms used by some pathogens include the ability to suppress the activation of host defenses (Kunoh *et al.*, 1985; Yamada *et al.*, 1989). The corollary to this is that it may be possible to obtain resistant plants through engineering of factors associated with, or circumventing, signal transduction. *mlo* resistance may represent a natural mutation in such a regulator. Indeed, the engineered resistance obtained by expressing a hydrogen peroxide generating glucose oxidase in tobacco (Wu *et al.*, 1995) and proton pump (Mittler *et al.*, 1995) may be examples of circumventing the natural regulatory mechanisms since AOS and changes in proton balances are implicated as local intercellular signals inducing defenses (however, see Chasan, 1995; Dong, 1995).

Much is now known of defenses in cereals through studies of the barley-powdery mildew system: the defenses in barley, though essentially similar, apparently exhibit distinct differences to dicot systems (e.g. different effects of jasmonate, *mlo*). A combination of mutational and functional studies will clarify the basis of resistance. Molecular and genetic studies on the interaction of barley and the powdery mildew fungus in particular have led to significant contributions to our knowledge of the mechanisms underlying disease resistance in plants.

ACKNOWLEDGEMENTS

Without Claus Andersen, Jakob Brandt, Anders B. Christensen, Alice B. Jørgensen, Kenneth Madriz-Ordeñana, Michael Næsby, Yangdou Wei, Lene Wyke, Fasong Zhou and Zhang Ziguo, this review would not have been possible. We are grateful to Jeanett Christiansen, Tina Haar and Britt Gréen for their invaluable technical support. We are indebted to Drs Bert De Boer, Kirsten Bojsen, Kirsten Brandt, Søren Knudsen, Preben Bach Holm, Jørn D. Mikkelsen, Elmon Schmelzer and Joachim Schröder for permission to cite collaborative results prior to publication. Drs James Brown, Andreas Freialdenhoven, Henriette Giese, Jørgen Helms Jørgensen, Lone Rossen, Ken J. Scott and Claus Wasternack, are gratefully acknowledged for providing copies of manuscripts prior to publication, and comments on parts of the manuscripts. The Centre for Plant Biotechnology (Framework Programme for Disease Resistance), Nordic Council of Ministers (via the Nordic Contact Organ for Agriculture), the Danish Research Council for Agricultural Research (SJVF), and the Royal Veterinary and Agricultural University (for a fellowship to PLG and Ph.D. studentships to CHA, JB and ABC) are acknowledged for financial support.

REFERENCES

Aist J.R. and Bushnell W.R. (1991) Invasion of plants by powdery mildew fungi, and cellular mechanisms of resistance. In *The Fungal Spore and Disease Initiation in Plants and Animals*, edited by G.T. Cole and H.C. Hoch, p.p. 321–343, New York: Plenum Press.

Aist J.R., Gold R.E, Bayles C.J., Morrison G.H., Chandra S. and Israel H.W. (1988) Evidence that molecular components of papillae may be involved in ml-o resistance to barley powdery mildew. *Physiological and Molecular Plant Pathology*, **33**, 17–32.

Aist J.R. and Israel H.W. (1977) Papilla formation: timing and significance during penetration of barley coleoptiles by *Erysiphe graminis hordei*. *Phytopathology*, **67**, 455–461.

Aist J.R. and Israel H.W. (1986) Autofluorescent and ultraviolet-absorbing components in cell walls and papillae of barley coleoptiles and their relationship to disease. *Canadian Journal of Botany*, **64,** 266–272.

Aitken A. (1995) 14-3-3 proteins on the MAP. *Trends in Biochemical Sciences*, **20,** 95–97.

Aitken A., Collinge D.B., van Heusden G.P.H., Isobe T., Roseboom P.H., Rosenfeld G. (1992) 14-3-3 proteins: A highly conserved widespread family of eukaryotic proteins. *Trends in Biochemical Sciences*, **17,** 498–500.

Andresen I., Becker W., Schlüter K., Burges J., Parthier B. and Apel K. (1992) The identification of a leaf thionin as one of the main jasmonate-induced proteins of barley (*Hordeum vulgare*). *Plant Molecular Biology*, **19,** 193–204.

Andersen L. and Jørgensen J.H. (1992) Mlo aggressiveness of barley powdery mildew. *Norwegian Journal of Agricultural Sciences*,. Supplement No. **7**, 77–87.

Asher M.J.C. and Thomas C.E. (1987) The inheritance of mechanisms of partial resistance to *Erysiphe graminis* f. sp. *hordei* in spring barley. *Plant Pathology*, **36,** 66–72.

Baker C.J. and Orlandt E.W. (1995) Active oxygen species in plant pathogenesis. *Annual Review of Phytopathology*, **33,** 299–321.

Barber M.S., Bertram R.E. and Ride J.P. (1989) Chitin oligosaccharides elicit lignification in wounded wheat leaves. *Physiological and Molecular Plant Pathology*, **34,** 3–12.

Basra A.S. (1994) *Stress-induced Gene Expression in Plants*. Chur, Switzerland: Harwood Academic Publishers.

Bayles C. J. and Aist J. R. (1987) Apparent calcium mediation of resistance of an ml-o barley mutant to powdery mildew. *Physiological and Molecular Plant Pathology*, **30,** 337–345.

Bennett R.N. and Wallsgrove R.M. (1994) Secondary metabolites in plant defense mechanisms. *New Phytologist*, **127,** 617–633.

Bi Y.M., Kenton P., Mur L., Darby R. and Draper J. (1995) Hydrogen peroxide does not function downstream of salicylic acid in the induction of PR protein expression. *The Plant Journal*, **8,** 235–245.

Bohlmann H., Clausen S., Behnke S., Giese H., Hiller C., Reimann-Philipp U. *et al.* (1988) Leaf-specific thionins of barley -a novel class of cell wall proteins toxic to plant-pathogenic fungi and possibly involved in the defense mechanism of plants. *The EMBO Journal*, **7**, 1559–1565.

Borbye L., Linde-Laursen I., Christiansen S.K. and Giese H. (1992) The chromosome complement of *Erysiphe graminis* f. sp. *hordei* analysed by light microscopy and field inversion gel electrophoresis. *Theoretical and Applied Genetics*, **79,** 705–712.

Boudet A.M., Lapierre C. and Grima-Pettenati J. (1995) Biochemistry and molecular biology of lignin. *New Phytologist*, **129**, 203–236.

Bowler C. and Chua N.H. (1994) Emerging themes of plant signal transduction. *The Plant Cell*, **6,** 1529–1541.

Boyd L.A., Smith P.A., Foster E.M. and Brown J.K.M. (1995) The effects of allelic variation at the Mla locus in barley on the early devlopment of *Erysiphe graminis* f. sp. *hordei* and host defenses. *The Plant Journal*, **7,** 959–968.

Boyd L.A., Smith P.A., Green R.M. and Brown J.K.M. (1994) The relationship between the expression of defense-related genes and mildew devlopment in barley. *Molecular Plant-Microbe Interactions*, **7**, 401–410.

Bradley D.J., Kjellbom P. and Lamb C.J. (1992) Elicitor- and wound-induced oxidative cross-linking of a proline-rich plant cell wall protein: a novel, rapid defense response. *Cell*, **70**, 21–30.

Brandt J., Thordal-Christensen H., Vad K., Gregersen P.L. and Collinge D.B. (1992) A pathogen-induced gene of barley encodes a protein showing high similarity to a protein kinase regulator. *The Plant Journal*, **2**, 815–820.

Bressan R.A., Pardo J., Yun D.J., Zhao Y., Todd L., Paino D'Urzo M. *et al.* (1995) Structure and function of family 5–PR–genes. In *"The 4th International Workshop on Pathogeneis–Related Proteins in Plants: Biology and Biotechnological Potential"*, Irsee, Germany, Sept. 3–7.

Brisson L.F., Tenhaken R. and Lamb C. (1994) Function of oxidative cross–linking of cell wall structural proteins in plant disease resistance. *The Plant Cell*, **6**, 1703–1712.

Brown J.K.M., Jessop, A.C., Thomas S. and Rezanoor H.N. (1992) Genetic control of the response of *Erysiphe graminis* f. sp. *hordei* to ethirimol and tridimenol. *Plant Pathology*, **41,** 126–135.

Brown J.K.M. and Simpson C.G. (1994) Genetic analysis of DNA fingerprints and virulences in *Erysiphe graminis* f. sp. *hordei. Current Genetics*, **26,** 172–178.

Brown J.K.M., Simpson C.G., O'Dell M. and Wolfe M.S. (1990) The use of DNA polymorphisms to test hypotheses about a population of *Erysiphe graminis* f. sp. *hordei. Plant Pathology*, **39,** 391–401

Brown J.K.M., Simpson C.G. and Wolfe M.S. (1993) Adaptation of barley powdery mildew populations in England to varieties with two resistance genes. *Plant Pathology*, **42,** 108–115

Bryngelsson T. and Collinge D.B. (1992) Biochemical and molecular analyses of the response of barley to infection by powdery mildew. In *Barley: Genetics, Molecular Biology, Biochemistry and Biotechnology*, edited by P.R. Shewry, pp.459–480, Wallingford: C.A.B. International.

Bryngelsson T. and Gréen B. (1989) Characterization of a pathogenesis-related, thaumatin-like protein isolated from barley challenged with an incompatible race of mildew. *Physiological and Molecular Plant Pathology*, **35**, 45–52.

Bryngelsson T., Gustafsson M., Ramos Leal M. and Bartonek E. (1988) Induction of pathogenesis-related proteins in barley during the resistance reaction to mildew. *Journal of Phytopathology*, **123**, 193–198.

Bryngelsson T., Sommer–Knudsen J., Gregersen P.L., Collinge D.B., Ek B. and Thordal–Christensen H. (1994) Purification, characterization and molecular cloning of basic PR-1-type pathogenesis related proteins form barley. *Molecular Plant-Microbe Interactions*, **7**, 267–275.

Bull J. Mauch F., Hertig C., Rebmann G. and Dudler R. (1992) Sequence and expression of a wheat gene that encodes a novel protein associated with pathogen defense. *Molecular Plant-Microbe Interactions*, **5**, 516–519.

Bushnell W.R. and Liu Z. (1994) Incompatibility conditioned by the Mla gene in powdery mildew of barley: timing of the effect of cordycepin on hypersensitive cell death. *Physiological and Molecular Plant Pathology*, **44**, 389–402.

Carmona M.J., Molina A., Fernandez J.A., López–Fando J.J., García–Olmedo F. (1993) Expression of the α-thionin gene from barley in tobacco confers enhanced resistance to bacterial pathogens. *The Plant Journal*, **3**, 457–462.

Carver, T.L.W. (1986) Histology of infection by *Erysiphe graminis* f. sp. *hordei* in spring barley lines with various levels of partial resistance. *Plant Pathology*, **35**, 232–240.

Carver T.L.W., Robbins M.P. and Zeyen, R.J. (1992a) Effects of two PAL inhibitors on the susceptibility and localized autofluorescent host cell responses of oat leaves attacked by *Erysiphe graminis* D.C. *Physiological and Molecular Plant Pathology*, **39**, 269–287.

Carver T.L.W., Robbins M.P., Zeyen, R.J., and Dearne G.A. (1992b) Effects of PAL–specific inhibition on suppression of activated defense and quantitative susceptibility of oats to *Erysiphe graminis*. *Physiological and Molecular Plant Pathology*, **41**, 149–163

Carver T.L.W. and Williams O. (1980) The influence of photoperiod on growth patterns of *Erysiphe graminis* f. sp. hordei. *Annals of Applied Biology*, **94,** 405–414.

Carver T.L.W., and Zeyen R.J. (1993) Effects of two PAL and CAD inhibitors on powdery mildew resistance phenomena in cereals. In *Mechanisms of Plant Defense Responses*, edited by B. Fritig and M. Legrand, pp.324–327. Dordrecht: Kluwer Academic Publishers.

Carver T.L.W., Zeyen, R.J., Bushnell W.R. and Robbins M.P. (1994) Inhibition of phenylalanine ammonia lyase and cinnamic acid dehydrogenase increases quantitative susceptibility of barley to powdery mildew (*Erysiphe graminis* D.C.). *Physiological and Molecular Plant Pathology*, **44,** 261–272.

Chasan R. (1995) SA: Source or signal for SAR? *The Plant Cell*, **7,** 1519–1521.

Chen Z., Fu H., Liu D., Chang P.F.L., Narasimhan M., Ferl R., (1995) A NaCl–regulated plant gene encoding a brain protein homolog that activates ADP ribosyltransferase and inhibits protein kinase C. *The Plant Journal*, **6**, 729–740.

Chen Z., Silva H. and Klessig D.F. (1993) Active oxygen species in the induction of plant systemic acquired resistance by salicylic acid. *Science*, **262**, 1883–1886.

Cho B.K. and Smedegaard–Petersen V. (1986) Induction of resistance to *Erysiphe graminis* f. sp. *hordei* in near isogenic barley lines. *Phytopathology*, **76,** 301–305.

Christiansen S.K. and Giese H. (1990) Genetic analysis of the obligate parastic barley powdery mildew fungus based on RFLP and virulence loci. *Theoretical and Applied Genetics*, **79,** 705–712.

Christiansen S.K., Knusden S. and Giese H. (1995) Biolistic transformation of the obligate plant pathogenic fungus *Erysiphe graminis* f. sp. *hordei*. *Current Genetics*, **29,** 100–102

Christiansen S. K. and Smedegaard V. (1990) Microscopic studies of the interaction between barley and the saprophytic fungus *Cladosporium macrocarpum*. *Journal of Phytopathology*, **128**, 209–219.

Clark T.A., Zeyen R.J., Smith A.G., Bushnell W.R., Szabo L.J. and Vance C.P. (1993) Host response gene transcript accumulation in relation to visible cytological events during *Erysiphe graminis* attack in isogenic barley lines differing at the Ml–a locus. *Physiological and Molecular Plant Pathology*, **43,** 283–298.

Cohen Y., Gisi U. and Niderman T. (1993). Local and systemic protection against *Phytophthora infestans* induced by unsaturated fatty acids. *Phytopathology*, **83,** 1054–1062

Collazo P., Montoliu L., Puigdomènech P. and Rigau J. (1992) Structure and expression of the lignin O–methyltransferase gene from *Zea mays* L. *Plant Molecular Biology*, **20,** 857–867.

Collinge D.B., Bryngelsson T., Gregersen P.L., Thordal–Christensen H. and Tronsmo A.M. (1993) The molecular and biochemical basis of plant disease resistance. *Växtskydsnotiser*, **57,** 102–107.

Collinge D.B., Gregersen P.L. and Thordal–Christensen H. (1994). The induction of gene expression in response to pathogenic microbes. In *Mechanisms of Plant Growth and Improved Productivity: Modern Approaches*, edited by A.S. Basra, pp. 391–433. New York: Marcel Dekker.

Dangl J.L. (1991) Regulatory elements controlling developmental and stress–induced expression of phenylpropanoid genes. In *Genes Involved in Plant Defense*, edited by T. Boller and F. Meins, pp 303–326. Wein and New York: Springer–Verlag.

Dangl J.L. (1992) The major histocompatibility complex à la carte: are there analogies to plant disease resistance genes on the menu? *The Plant Journal*, **2,** 3–11.

Dangl J.L. (1995) Pièce de Résistance: novel classes of plant disease resistance genes. *Cell*, **80,** 363–366.

Davidson A. D., Manners J. M., Simpson R. S. and Scott K. J. (1987) cDNA cloning of mRNAs induced in resistant barley during infection by *Erysiphe graminis* f. sp *hordei*. *Plant Molecular Biology*, **8,** 77–85.

Davis D., Merida J., Legrendre L., Low P. and Heinstein P. (1993) Independent elicitation of the oxidative burst and phytoalexin formation in cultured plant cells. *Phytochemistry*, **32,** 607–611.

de Vetten N. and Ferl R.J. (1992). Two genes encoding GF14 (14–3–3) proteins in *Zea mays*. *Plant Physiology*, **106**, 1593–1604.

de Vetten N.C., Lu G. and Ferl R.J. (1994) A maize protein associated with the G–box binding complex has homology to brain regulatory proteins. *The Plant Cell*, **4,** 1295–1307.

de Wit P.J.G.M. (1995) Fungal avirulence genes and plant resistance genes: unravelling the molecular basis of gene–for–gene interactions. *Advances in Botanical Research*, **21,** 147–185.

Delaney T.P., Uknes S., Vernooij B., Friedrich L., Weymann K., Negrotto D. *et al.* (1994) A central role of salicylic acid in plant disease resistance. *Science*, **266**, 1247–1250.

DeScenzo R.A., Wise R.P. and Mahadevappa M. (1994) High-resolution mapping of the Hor1/Mla/Hor2 region on chromosome 5S in barley. *Molecular Plant–Microbe Interactions*, **47**, 657–666.

Després C., Subramaniam R., Matton D.P. and Brisson N. (1995) The activation of the potato PR–10a gene requires the phosphorylation of the nuclear factor PBF–1. *The Plant Cell*, **67**, 589–598.

Dietrich R.A., Delaney T.P., Uknes S.J., Ward E.R., Ryals J.A. and Dangl J.L. (1994) *Arabidopsis* mutants simulating disease resistance. *Cell*, **77**, 565–577.

Dixon R.A. (1986) The phytoalexin response: elicitation, signaling and the control of host gene expression. *Biological Reviews*, **61**, 239–291.

Dixon R.A., Harrison M. and Lamb C.J. (1994) Early events in the activation of plant defense responses. *Annual Review of Phytopathology*, **32**, 479–501.

Dixon R.A. and Lamb C.J. (1990) Molecular communication in interactions between plants and microbial pathogens. *Annual Review of Plant Physiology and Plant Molecular Biology*, **41**, 339–367.

Dong X. (1995) Finding the missing pieces in the puzzle of plant disease resistance. *Proceedings of the National Academy of Sciences, USA*, **92**, 7137–7139.

Dudler R., Hertig C., Rebmann G., Bull J. and Mauch F. (1991) A pathogen–induced wheat gene encodes a protein homologous to Glutathione–S–transferase. *Molecular Plant–Microbe Interactions*, **4**, 14–18.

Ebel J. and Cosio E.G. (1994) Elicitors of plant defense responses. *International Review of Cytology*, **148,** 1–36.

Ebrahim–Nesbat F., Bohl S., Heitefuss R. and Apel K. (1993) Thionin in cell walls and papillae of barley in compatible interactions with *Erysiphe graminis* f. sp. *hordei*. *Physiological and Molecular Plant Pathology*, **43**, 343–352.

Enyedi A.J., Yalpani N., Silverman P. and Raskin I. (1992) Signal molecules in systemic resistance to pathogens and pests. *Cell*, **70,** 879–886.

Everse J., Everse K.E. and Brisham, M.B. (1991) *Peroxidases in Chemistry and Biology. Vol. 2.* Boca Raton: CRC Press.

Farmer E.E. and Ryan C.A. (1990) Interplant communication: Airborne methyl jasmonate induces synthesis of proteinase inhibitors in plant leaves. *Proceedings of the National Academy of Science, USA*, **87**, 7713–7716.

Farmer E.E. and Ryan C.A. (1992) Octadecanoid precursors of jasmonic acid activate the synthesis of wound–inducible proteinas inhibitors. *The Plant Journal,* **4,** 129–134.

Feussner I., Hause B., Vörös K., Parthier B. and Wasternack C. (1995) Jasmonate–induced lipoxygenase forms are localized in chloroplasts of barley leaves. *The Plant Journal*, **7,** 949–957.

Fields S. and Song O.K. (1991) A novel genetic system to detect protein–protein interactions. *Nature*, **340,** 245–246.

Freialdenhoven A., Peterkänsel C., Kurth J., Kreuzaler F. and Schulze–Lefert P. (1996) Identification of genes required for the function of non-race-specific mlo resistance to powdery mildew in barley. *The Plant Cell*, **8,** 5–14.

Freialdenhoven A., Scherag B., Hollricher K., Collinge D. B., Thordal–Christensen H. and Schulze–Lefert P. (1994) Nmr–1 and Nmr–2, two loci required for Mla_{12} specified race–specific resistance to powdery mildew in barley. *The Plant Cell*, **6,** 983–994.

Fry W.E. and Evans P.H. (1977) Association of formamide hydrolyase with fungal pathogenicity to cyanogenic plants. *Phytopathology*, **67,** 1001–1006.

Fry W.E. and Myers D.F. (1981) Hydrogen cyanide metabolism by fungal pathogens of cyanogenic plants. In *Cyanide in Biology*, edited by B. Vennesland, E.E. Conn, C.J. Knowles, J. Westley and F. Wissing, pp.321–334. New York, Academic Press.

Gabriel D.W. and Rolfe B.G. (1990) Working models of specific recognition in plant–microbe interactions. *Annual Review of Phytopathology*, **28,** 365–391.

Gaffney T., Friedrich L., Vernooij B., Negrotto D., Nye G., Uknes S., *et al.* (1993) Requirement of salicylic acid for the induction of systemic acquired resistance. *Science*, **261**, 754–756.

Geiger H.H. and Heun M. (1989), Genetics of quantitative resistance to fungal diseases. *Annual Review of Phytopathology*, **27**, 317–341.

Giese H., Christiansen S.K. and Jensen H.P. (1990) Extrachromosomal plasmid–like DNA in the obligate parasitic fungus *Erysiphe graminis* f.sp. *hordei*. *Theoretical and Applied Genetics*, **79,** 56–64.

Giese H., Holm–Jensen A.G., Jensen H.P. and Jensen J. (1993) Localization of the Laevigatum powdery mildew resistance gene to barley chromosome 2 by the use of RFLP markers. *Theoretical and Applied Genetics*, **85,** 897–900.

Gold R.E, Aist J.R., Hazen B.E., Stolzenburg M.C., Marshall M.R. and Israel H.W. (1986) Effects of calcium nitrate and chlortetracycline on papilla formation, ml-o resistance and susceptibility of barley to powdery mildew. *Physiological and Molecular Plant Pathology*, **29**, 115–129.

Görg R., Hollricher K. and Schulze–Lefert P. (1993), Functional analysis and RFLP mapping of the Mlg resistance gene in barley. *The Plant Journal*, **3,** 857–866.

Green R.M. (1991) Isolation and characterization of genes induced in barley during powdery mildew infection. *Ph.D. Thesis*, Cambridge University, UK.

Gregersen P. L. and Smedegaard V. (1989) Induction of resistance in barley against *Erysiphe graminis* f. sp. *hordei* after preinoculation with the saprophytic fungus, *Cladosporium macrocarpum*. *Journal of Phytopathology*, **124**, 128–36.

Gregersen P.L., Collinge D.B. and Smedegaard V. (1990) Early induction of new mRNAs accompanies the resistance reactions of barley to the wheat pathogen *Erysiphe graminis* f. sp. *tritici*. *Physiological and Molecular Plant Pathology*, **36**, 471–481.

Gregersen P.L., Brandt, J. Thordal–Christensen H., and Collinge D.B. (1993) cDNA cloning and characterization of mRNAs induced in barley by the fungal pathogen *Erysiphe graminis*. In *Mechanisms of Plant Defense Responses*, edited by B. Fritig and M. Legrand, pp.304–307. Dordrecht: Kluwer Academic Publishers.

Gregersen P.L., Christensen A.B., Sommer–Knudsen J. and Collinge D.B. (1994) A putative O–methyltransferase from barley is induced by fungal pathogens and UV light. *Plant Molecular Biology*, **26**, 1797–1806.

Grimes H.D., Koetje D..S. and Franceschi V.R.(1992) Expression, activity, and cellular accumulation of methyl jasmonate–responsive lipoxygenase in soybean seedlings. *Plant Physiology*, **100,** 433–443.

Gundlach H., Müller M.J., Kutchan T.M. and Zenk M.H. (1992) Jasmonic acid is a signal transducer in elicitor–induced plant cell cultures. *Proceedings of the National Academy of Science, USA*, **89,** 2389–2393.

Hahn M., Jüngling S. and Knogge W. (1993) Cultivar–specific elicitation of barley defense reactions by the phytotoxic peptide NIP1 from *Rhyncosporium secalis*. *Molecular Plant–Microbe Interactions*, **6,** 745–754.

Hain R., Reif, H.J., Karuse E., Langbartels R., Kindl H., Vornam B. *et al* (1993) Disease resistance results from foreign phytoalexin expression in a novel plant. *Nature*, **361,** 153–156.

Hammond–Kosack K.E., Jones D.A. and Jones J.D.G. (1994) Identification of two genes required in tomato for full Cf–9 dependent resistance to *Cladosporium fulvum*. *The Plant Cell*, **6,** 361–374.

Hejgaard J., Jacobsen S. and Svendsen I. (1991) Two antifungal thaumatin–like proteins from barley grain. *FEBS Letters*, **291,** 127–131.

Hejgaard J., Jacobsen S., Bjørn S.E. and Kragh K.M. (1992) Antifungal activity of chitin–binding PR–4 type proteins from barley grain and stressed leaf. *FEBS Letters*, **307**, 389–392.

Heun M. (1989) Mapping quantitative powdery mildew resistance of barley using a restriction fragment polymorphism map. *Genome*, **35,** 1019–1025.

Hilbers S., Fischbeck G. and Jahoor A. (1992) Localization of the *lavigatum* resistance gence *Mlla* against powdery mildew in the barley genome by the use of RFLP markers. *Plant Breeding*, 109, 335–338.

Hiramoto T., Tobimatsu R., Shiraishi T., Yamada T., Ichinose Y. and Oku H. (1992) Endogenous elicitor present in barley seeds II. Mechanisms of the induction of resistance in barley leaves to *Erysiphe graminis*. *Journal of Phytopathology*, **135**, 167–176.

Hislop E. C. and Stahmann M. A. (1971) Peroxidase and ethylene production by barley leaves infected with *Erysiphe graminis* f. sp. *hordei*. *Physiological Plant Pathology*, **1,** 297–312.

Høj P.B., Hartman D.J., Morrice N.A., Doan D.N.P. and Fincher G.B. (1989) Purification of (1–>3)–β–glucan endohydorlase isoenzyme II from germinating barley and determination of its primary structure from a cDNA clone. *Plant Molecular Biology*, **13**, 31–42.

Inoue S., Macko V. and Aist J.R. (1994) Identification of the active component in the papilla–regulating extract from barley leaves. *Physiological and Molecular Plant Pathology*, **44**, 441–453.

Jach G., Görnhardt B., Mundy J., Logemann J., Pinsdorf E., Leah R. *et al.* (1995) Enhanced quantitative resistance against fungal disease by combinatorial expresion of different barley antifungal proteins in transgenic tobacco. *The Plant Journal*, **8,** 97–109.

Jacobsen S., Mikkelsen J.D. and Hejgaard J. (1990) Characterization of two antifungal chitinases from barley grain. *Physiologia Plantarum*, **79,** 554–562.

Jensen J., Jensen HP. and Jørgensen J.H. (1995) Linkage studies of barley powdery mildew virulence loci. *Hereditas*, **122,** 197–209.

Jones D.A., Thomas C.M., Hammond-Kosack K.E., Balint-Kurti P.J. and Jones J.D. G. (1994) Isolation of the tomato Cf-9 gene for resistance to *Cladosporium fulvum* by transposon tagging. *Science*, **266,** 789 –793.

Jørgensen H.J.L., Neergaard E. de and Smedegaard–Petersen V. (1993) Histological examination of the interaction between *Rhyncosporium secalis* and susceptible and resistance cultivars of barley. *Physiological and Molecular Plant Pathology*, **42**, 345–358.

Jørgensen J.H. (1992a) Discovery, characterization and exploitation of Mlo powdery mildew resistance in barley. *Euphytica*, **63,** 141–152.

Jørgensen J.H. (1992b) Multigene families of powdery mildew resistance genes in locus Mla on barley chromosome 5. *Plant Breeding*, **108,** 53–59.

Jørgensen J.H. (1994) Genetics of powdery mildew resistance in barley. *CRC Critical Reviews of Plant Sciences*, **13,** 97–119.

Jørgensen J.H. (1996) Effect of three suppressor mutant genes on the expression of 17 powdery mildew resistance genes in barley. *Genome* **39**, 492–98.

Justesen A., Somerville S.C., Christiansen S.K. and Giese H. (1995) Isolation and characterization of two novel genes expressed in germinating conidia of the obligate biotroph *Erysiphe graminis* f.sp. *hordei*. *Gene* **170**, 131–135.

Jutidamrongphan, W., Andersen, J.B., Mackinnon, G., Manners, J.M., Simpson, R.S. and Scott, K.J. (1991) Induction of a β-*1, 3-glucanase* in barley ion response to infection by fungal pathogens. *Molecular Plant-Microbe Interactions,* **4**, 234–238.

Jutidamrongphan W., Mackinnan G., Manners J.M. and Scot K.J. (1989) Sequence of a near–full length cDNA clone for a mRNA of barley induced by fungal infection. *Nucleic Acids Research*, **17**, 9478.

Keen N.T. (1992) The molecular biology of disease resistance. *Plant Molecular Biology*, **19,** 109–122.

Keen N.T., Midland S.L., Boyd C., Yucael I., Tsurushima T., Lorang J. *et al* (1994) Syringolide elicitors specified by avirulence gene D alleles in *Pseudomonas syringae*. In *Advances in Molecular Genectics of Plant–Microbe Interactions*, edited by M.J. Daniels, J.A. Downie and A.E. Osbourn, pp. 41–48. Dordrecht: Kluwer Academic Publishers.

Keon, J.P.R. and Hargreaves, J.A. (1983) A cytological study of the net blotch disease of barley caused by *Pyrenophora teres*. *Physiological Plant Pathology*, **22**, 321–329.

Kerby K. and Somerville S.C. (1989) Enhancement of specific intercellular peroxidases following inoculation of barley with *Erysiphe graminis* f. sp. *hordei*. *Physiological and Molecular Plant Pathology*, **35,** 323–337.

Kerby K. and Somerville S.C. (1993) Purification of an infection–related, extracellular peroxidase from barley. *Plant Physiology*, **100**, 397–402.

Kesemann H., Staub T., Hofmann C., Maetzke T., Herzog J., Ward E.R. *et al.*, (1994). Induction of systemic acquired resistance in plants by chemicals. *Annual Review of Phytopathology*, **32,** 439–459.

Kita N., Toyoda H. and Shishiyama J. (1980) Histological reactions of papilla and cytoplasmic aggregate in epidermal cells of barley leaves infected by *Erysiphe graminis hordei*. *Annals of the Phytopathology Society, Japan*, **46,** 263–65.

Kita N., Toyoda H. and Shishiyama J. (1981) Chronological analysis of cytological responses in powdery–mildewed barley leaves. *Canadian Journal of Botany*, **59,** 1761–1768.

Kodama O., Miyakawa J., Akatsuka T. and Kiyosawa S. (1992) Sakuranetin, a flavanone phytoalexin from ultravioloet–irradiated rice leaves. *Phytochemistry*, **31,** 3897–3909.

Kogel K.H., Beckhove U., Dreschers J., Münch S. and Rommé Y. (1994) The resistance mechanism induced by 2,6–dichloroisonicotinic acid is a phenocopy of genetically based mechanism governing race–specific powdery mildew resistance. *Plant Physiology*, **106,** 1269–1277.

Koga H., Bushnell W.R. and Zeyen R.J. (1990), Specificity of cell type and timing of events associated with papilla fromation and the hypersensitive reaction in leaves of *Hordeum vulgare* attacked by *Erysiphe graminis* f. sp. *hordei*. *Canadian Journal of Botany*, **68,** 2344–2352.

Kogel K.H., Ortel B., Jarosch B., Atzorn R, Schiffer R and Wasternack C. (1995) Resistance in barley against the powdery mildew fungus *(Erysiphe graminis* f. sp. *hordei)* is not associated with enhanced levels of endogenous jasmonates. *European Journal of Plant Pathology*, **101,** 319–332.

Koga H., Zeyen R.J., Bushnell W.R. and Ahlstrand G.G. (1988). Hypersensitive cell death, autofluorescence, and insoluble silicon accumulation in barley leaf epidermal cells under attack by *Erysiphe graminis* f. sp. *hordei*. *Physiological and Molecular Plant Pathology*, **32,** 395–409.

Köhle H., Jeblick W., Poten F., Blaschek W. and Kauss H. (1985) Chitosan–elicited callose synthesis in soybean cells as a Ca^{2+} –dependent process. *Plant Physiology*, **77,** 544–551.

Kølster P., Munk L., Stølen O., and Løhde J. (1986) Near–isogenic barley lines with genes for resistance to powdery mildew. *Crop Science*, **26,** 903–907.

Korthout H.A.A.J.. and De Boer A.H. (1994) A fusicoccin binding protein belongs to the family of 14–3–3 brain protein homologues. *The Plant Cell*, **6,** 1681–1692.

Korthout H.A.A.J., van der Hoeven P.C.J., Wagner M.J., Van Hunnik E. and De Boer A.H. (1994) Purification of the fusicoccin–binding protein from oat root plasma membrane by affinity chromatography with biotinylated fusicoccin. *Plant Physiology*, **105,** 1281–1288.

Kragh K.M., Jacobsen S., Mikkelsen J.D. and Nielsen K.A. (1991) Purification and characterization of three chitinases and one β–1,3–glucanase accumulating in the medium of cell suspension cultures of barley *(Hordeum vulgare* L.). *Plant Science*, **76,** 65–77.

Kristensen H.J. (1994) Activation of plant defense responses with emphasis on barley powdery mildew. *Ph.D. Thesis*, Royal Veterinary and Agricultural University, Frederiksberg C, Denmark.

Ku J. (1995) Phytoalexins, stress metabolism, and disease resistance in plants. *Annual Review of Phytopathology*, **33,** 275–297.

Kunoh H. (1995) Host–parasite specificity in powdery mildews . In *Pathogeneis and Host Specificity in Plant Diseases*. Volume II. Eukaryotes, edited by K. Kohmoto, U.S. Singh, and R.P. Singh, pp 239–250. Oxford: Elsevier.

Kunoh H., Hayashimotot A., Harui M. and Ishizaki H. (1985) Induced susceptibility and enhanced resistance at the cellular level in barley coleoptiles. I. The significance of timing of fungal invasion. *Physiological Plant Pathology*, **27,** 43–54.

Kunoh H., Nicholson R.L., Yosioka H., Yamaoka N. and Kobayashi I. (1990) Preparation of the infection court by *Erysiphe graminis*: degradation of the host cuticle. *Physiological and Molecular Plant Pathology*, **36,** 397–407.

Leah R., Tommerup H., Svendsen I. and Mundy J. (1991) Biochemical and molecular characterization of three barley seed proteins with antifungal properties. *Journal of Biological Chemistry*, **226,** 1564–1573.

Lee–Stadelmann O.Y., Curran C.M., and Bushnell W.R. (1992) Incompatibility conditioned by the Mla gene in powdery mildew of barley: change in permeability to non–electrolytes. *Physiological and Molecular Plant Pathology*, **41,** 165–177.

Legrendre L., Yueh Y.G., Crain R., Haddock N., Heinstein P.F. and Low P. (1993) Phospholipase C activation during elicitation of the oxidative burst in cultured plant cells. *Journal of Biological Chemistry*, **268,** 24559–24563.

Lehmann J., Atzorn R., Brückner C., Reinbothe S., Leopold J., Wasternack C. *et al.* (1995) Accumulation of jasmonate, abscisic acid, specific transcripts and proteins in osmotically stressed barley leaf segments. *Planta*, **197,** 156–162.

Leopold J., Hause B., Lehmann J., Graner A., Parthier B. and Wasternack C. (1996) Isolation, characterization and expression of a cDNA coding for a jasmonate inducible protein of 37 kDA in barley leaves. *Plant, Cell and Environment* (in press).

Levine A., Tenhaken R., Dixon R. and Lamb C. (1994) H_2O_2 from the oxidative burst orchestrates the plant hypersensitive disease resistance response. *Cell*, **79**, 583–593.

Liu L., Gitz D.C. and McClure J.W. (1995) Effects of UV–B on flavonoids, ferulic acid, growth and photosynthesis in barley primary leaves. *Physiologia Plantarum*, **93**, 725–733.

Lyon G.D., Reglinski T. and Newton A.C. (1995) Novel disease controlling compounds: the potential to "immunize" plants against infection. *Plant Pathology*, **44,** 407–427.

Malamy J. and Klessig D.F. (1992) Salicylic acid and plant disease resistance. *The Plant Journal*, **2**, 643–654.

Manners J.M., Davidson A.D. and Scott K.J. (1985). Patterns of post–infectional protein synthesis in barley carrying different genes for resistance to the powdery mildew fungus. *Plant Molecular Biology*, **4**, 275–283.

Maroof M.A.S., Zhang Q. and Biyashev R.M. (1994) Molecular marker analyses of powdery mildew resistence in barley. *Theoretical and Applied Genetics*, **88**, 773–740.

Marra M., Fullone M.R., Fogliano V., Pen J., Mattei M., Masi S. *et al.* (1994) The 30–kilodalton protein present in purified fusicoccin receptor preparations is a 14–3–3–like protein. *Plant Physiology*, **106**, 1497–1501.

Martin G.B., Brommonschenkel S.H., Chunwongse J., Frary A., Ganal M.W., Spivey R. *et al.* (1993) Map–based cloning of a protein kinase gene conferring disease resistance in tomato. *Science*, **262**, 1432–1436.

Maule A.J. and Ride J.P. (1976) Ammonia–lyase and O–methyl transferase activities related to lignification in wheat leaves infected with *Botrytis*. *Phytochemistry*, **15**, 1661–1664.

Maule A.J. and Ride J.P. (1983) Cinnamate 4–hydroxylase and hydroxycinnamate: CoA ligase in wheat leaves infected with *Botrytis cinerea*. *Phytochemistry*, **22**, 1113–1116.

Mayama S. and Shishiyama J. (1978) Localized accumulation of fluorescent and u.v.-absorbing compounds at penetration sites in barley leaves infected with *Erysiphe graminis hordei*. *Physiological Plant Pathology*, **13**, 347–354.

Mehdy M.C. (1994) Active oxygen species in plant defense against pathogens. *Plant Physiology*, **105**, 467–472.

Mittler R., Shulaev V. and Lam E. (1995) Coordinated activation of programmed cell death and defense mechaisms in transgenic tobacco plants expressing a bacterial proton pump. *The Plant Cell*, **7**, 29–42.

Molina A., Segura A. and García–Olmedo F. (1993) Lipid transfer proteins (nsLTPs) from barley and maize leaves are potent inhibitors of bacterial and fungal plant pathogens. *FEBS Letters*, **316**, 119–122.

Muradov A., Petrasovits L., Davidson A. and Scott, K. J. (1993) A cDNA clone for a pathogenesis–related proteins 1 from barley. *Plant Molecular Biology*, **23**, 439–442.

Newton A.C. and Crute I.R. (1989) A consideration of the genetic control of species specificity in fungal plant pathogens and its relevance to a comprehension of the underlying mechanisms. *Biological Reviews*, **64**, 35–50.

Nicholson R.L. (1992) *Colletotrichum graminicola* and the antrachnose diseases of maize and sorghum. In *Colletotrichum: Biology, Pathology, and Control*, edited by J.A. Bailey and M.J. Jeger, pp. 186–202. Wallingford: CAB International.

Nicholson R.L., Yosioka H., Yamaoka N. and Kunoh H. (1988) Preparation of the infection court by *Erysiphe graminis*: II Release of esterase enzyme from conidia in response to a contact stimulus. *Experimental Mycology*, **12**, 336–349.

O'Dell M., Wolfe M.S., Flavell R.B., Simpson C.G. and Summers R.W. (1989) Molecular variation in populations of *Erysiphe graminis* on barley, oats and rye. *Plant Pathology*, **38**, 340–351.

Oecking C., Eckerskorn C. and Weiler E.W. (1994) The fusicoccin receptor of plants is a member of the 14–3–3 superfamily of eukaryotic regulatory proteins. *FEBS Letters*, **352**, 163–166.

Oecke E.C., Steiner U. and Schönbeck F. (1989) Zur Wirksamkeit der induzierten Resistenz unter praktischen Anbaubedingungen V. Mehltaubefall und Ertag von Winter– und Sommergerste in Abhängigkeit von der Stickstoffdüngung. *Zietschrift für Pflanzenkrankheiten und Pflanzenschutz*, **96**, 140–153.

Pourmohseni H. and Ibenthal W.D. (1991) Novel β–cyanoglucosides in the epidermal tissue of barley and their possible role in the barley–powdery mildew interaction. *Angewandte Botanik*, **65**, 341–350.

Pryor A. (1987) The origin and structure of fungal disease resistance genes in plants. *Trends in Genetics*, **3**, 157–161.

Ralph J., Helm R., Quideau S. and Hatfield R.D. (1992) Lignin–feruloyl ester cross–links in grasses. Part 1. Incorporation of feruloyl esters into coniferyl alcohol dehydrogenation polymers. *Journal Chemical Society Transactions*, **1**, 2961–2969.

Raskin I. (1992) Role of salicylic acid in plants. *Annual Review of Plant Physiology and Plant Molecular Biology*, **43**, 439–463.

Rasmussen M., Rossen L. and Giese H. (1993) SINE–like properties of a highly repetitive element in the genome of the obligate parasitic fungus *Erysiphe graminis* f.sp. *hordei*. *Molecular and General Genetics*, **239**, 298–303.

Rebmann G., Hertig C., Bull J., Mauch F. and Dudler R. (1991a) Cloning and sequencing of cDNAs encoding a pathogen–induced putitive peroxidase of wheat (*Triticum aestivum*). *Plant Molecular Biology*, **16**, 329–331.

Rebmann G., Mauch F. and Dudler R. (1991b) Sequence of a wheat cDNA encoding a pathogen–induced thaumatin–like protein. *Plant Molecular Biology*, **17**, 283–285.

Reinbothe S., Reinbothe C., Lehmann J., Becker W. Apel K. and Parthier B. (1994) JIP60, a methyl jasmonate–induced ribosome–inactivating protein involved in plant stress reactions. *Proceeedings of the National Academy of Sciences, U.S.A.*, **91**, 7012–7016.

Reinbothe S., Reinbothe C. and Parthier B. (1993) Methyl jasmonate represses translation intiation of a specific set of mRNAs in barley. *The Plant Journal*, **4**, 459–467.

Ryals J., Ward E., Ahl–Goy P. and Métreaux J.P. (1992). Systemic acquired resistance: an inducible defense mechanism in plants. In *Inducible Plant Proteins*, edited by J. L. Wray, pp. 205–229. *Society of Experimental Biology Seminar Series*, **49**, Cambridge: Cambridge University Press.

Salmeron J.M., Barker S.J., Carland F.M., Mehta A.Y. and Staskawitz B.J. (1994) Tomato mutants altered in bacterial disease resistance provide evidence for a new locus controlling pathogen recognition. *The Plant Cell*, **76**, 511–520.

Schäfer W. (1994) Molecular mechanisms of fungal pathogenicity to plants. *Annual Review of Phytopathology*, **32**, 461–477.

Schmelzer E., Krüger–Lebus S. and Hahlbrock K. (1989) Temporal and spacial patterns of gene expression around sites of attemped fungal infection in parsley leaves. *The Plant Cell*, **1**, 993–1001.

Schweizer P., Gees R. and Mösinger E. (1993) Effect of jasmonic acid on the interaction of barley (*Hordeum vulgare*) with the powdery mildew *Erysiphe graminis* f. sp. *hordei*. *Plant Physiology*, **102**, 503–511.

Schweizer P., Hunziker W. and Mösinger E. (1989). cDNA cloning, *in vitro* transcription and partial sequence analysis of mRNAs from winter wheat (*Triticum aestivum* L.) with induced resistance to *Erysiphe graminis* f. sp. *tritici*. *Plant Molecular Biology*, **12**, 643–654.

Scott K.J. (1992) The molecular analysis of barley resistance to powdery mildew. In *Barley: Genetics, Molecular Biology, Biochemistry and Biotechnology*, edited by P.R. Shewry, pp.481–495. Wallingford: C.A.B. International.

Scott K. J. (1994) Genetic engineering of cereals for resistance to phytopathogens. *Australian Plant Pathology*, **23**, 154–162.

Scott K. J., Davidson A. D., Jutidamrongphan W, Mackinnon G. and Manners J. M. (1990). The activation of genes of wheat and barley by fungal pathogens. *Australian Journal of Plant Physiology*, **17**, 229–238.

Scott–Craig J.S., Kerby K., Stein B.D. and Somerville S.C. (1995) Expression of an extracellular peroxidase that is induced in barley (*Hordeum vulgare*) by the powdery mildew pathogen (*Erysiphe graminis* f. sp. *hordei*). *Physiological and Molecular Plant Pathology*, **47**, 407–418.

Sherwood J.E. and Somerville S.C. (1990) Sequence of the *Erysiphe graminis* f. sp. *hordei* gene encoding β–tubulin. *Nucleic Acids Research*, **18**, 1052.

Shiraishi T., Yamaoka N. and Kunoh H. (1989) Association between increased phenylalanine ammonia lyase activity and cinnamic acid synthesis and the induction of temporary inaccessibility caused by *Erysiphe graminis* primary germ tube penetration of the barley leaf. *Physiological and Molecular Plant Pathology*, **34**, 75–83.

Skou J.P. (1982) Callose formation responsible for the powdery mildew resistance in barley with genes in the ml-o locus. *Phytopathologische Zeitschrift*, **104**, 90–95.

Skou J.P., Jϕrgensen J.H. and Lilholt U. (1984) Comparative studies on callose formation in powdery mildew compatible and incompatible barley. *Phytopathologische Zeitschrift*, **109**, 147–168.

Smart M.G. (1991). The plant cell wall as a barrier to fungal invasion. In *The Fungal Spore and Disease Initiation in Plants and Animals* edited by G.T. Cole and H.C. Hoch, pp. 47–66. New York: Plenum Press.

Smart, M.G., Aist, J.R. and Israel, H.W. (1986) Structure and function of wall appositions. 1. General histochecistry of papillae in barley coleoptiles attacked by *Erysiphe graminis* f. sp. *hordei*. *Canadian Journal of Botany*, **64**, 793–801.

Smedegaard–Petersen V., Collinge D.B., Thordal–Christensen H., Brandt J., Gregersen P.L., Cho B.H. *et al.* (1991). Induction and molecular analyses of resistance to barley powdery mildew. In *Biological Control of Plant Diseases: Progress and Challenges for the Future,* edited by E.C. Tjamos, G. Papavizas, R.J. Cook, pp. 321–326. New York: NATO–ASI Plenum Press.

Smedegaard–Petersen V. and Tolstrup K. (1985) The limiting effect of disease resistance on yield. *Annual Review of Phytopathology*, **23**, 475–490.

Song W.Y., Wang G.L., Chen L.L, Kim H.S., Pi L.Y., Holsten T. et al. (1995) A receptor kinase–like protein encoded by the rice disease resistance gene, Xa21. *Science*, **270**, 1804–1806.

Staskawitz B.J., Ausubel F.M., Baker B.J., Ellis J.G. and Jones J.D.G. (1995) Molecular gentics of plant disease resistance. *Science*, **268**, 661–667.

Staub T.H., Dahmen H. and Schwinn F.J. (1974) Light and scanning microsopy of cucumber and barley powdery mildew on host and nonhost plants. *Phytopathology*, **64**, 364–372.

Suzuki K. and Shinshi H. (1995) Transient activation and tyrosine phosphorylation of a protein kinase in tobacco cells treated with a fungal elicitor. *The Plant Cell*, **7**, 639–647.

Svensson B., Svendsen I., Højrup P., Roepstorff P., Ludvigsen S. and Poulsen F.M. (1992) Primary structure of barwin: a barley seed protein closely related to the C–terminal domain of proteins encoded by wound–induced plant genes. *Biochemistry*, **31**, 8767–8770.

Swegle M., Kramer K.J. and Muthukrishnan S. (1992) Properties of barley seed chitinases and release of embryo–associated isofroms during early stages of imbibition. *Plant Physiology*, **99**, 1009–1014.

Thordal–Christensen H., Brandt J., Cho B.H., Gregersen P.L., Rasmussen S.K., Smedegaard–Petersen V. *et al.* (1992) cDNAs of barley messengers induced early in response to powdery mildew include a peroxidase sequence. *Physiological and Molecular Plant Pathology*, **40**, 395–409.

Thordal–Christensen H. and Smedegaard–Petersen V. (1988a) Comparison of resistance–inducing abilities of virulent and avirulent races of *Erysiphe graminis* f. sp. *hordei* and a race of *Erysiphe graminis* f. sp. *tritici*. *Plant Pathology*, **37**, 20–27.

Thordal–Christensen H. and Smedegaard–Petersen V. (1988b) Correlation between induced resistance and host fluorescence in barley inoculated with *Erysiphe graminis*. *Journal of Phytopathology*, **123**, 34–46.

Tosa Y. and Shishiyma J. (1984) Cytological aspects of events occurring after the formation of primary haustoria in barley leaves infected with powdery mildew. *Canadian Journal of Botany*, **62**, 795–798.

Valè G.P., Torrigiani E., Gatti A., Delogu G., Porta–Puglia A., Vannacci G. *et al.* (1994) Activation of genes in barley roots in response to infection by two *Drechslera graminea* isolates. *Physiological and Molecular Plant Pathology*, **44**, 207–215.

van Loon L.C. (1985) Pathogenesis–related proteins. *Plant Molecular Biology*, **4**, 111–116.

van Loon L.C., Pierpoint W.S., Boller T. and Conejero V. (1994) Recommendations for naming plant pathogenesis–related proteins. *Plant Molecular Biology Reporter*, **12**, 245–264.

VanEtten H., Mansfield J.W., Bailey J.A. and Farmer E.E. (1994a) Two classes of plant antibiotics: phytoalexins versus "Phytoanticipins". *The Plant Cell*, **6, 1191–1192.**

VanEtten H., Matthews D.E. and Matthews P.S. (1989) Phytoalexin detoxification: importance for pathogenicity and practical implications. *Annual Review of Phytopathology*, **27**, 143–164.

VanEtten H., Soby S., Wasmann C. and McCluskey K. (1994b) Pathogenicity genes in fungi. In *Advances in Molecular Genectics of Plant–Microbe Interactions*, edited by M.J. Daniels, J.A. Downie and A.E. Osbourn, pp. 163–170. Dordrecht: Kluwer Academic Publishers.

Vera–Estrella R., Barkia B.J., Higgins V. and Blumwald E. (1994a) Plant defense response to fungal pathogens. Activation of host–plasma membrane H^+ATPase by elicitor–induced enzyme dephosphorylation. *Plant Physiology*, **104**, 209–215.

Vera–Estrella R. Higgins V. and Blumwald E. (1994b) Plant defense response to fungal pathogens. II. G – protein–mediated changes in host–plasma membrane redox reactions. *Plant Physiology*, **106**, 97–102.

Vitale A., Cerotti A. and Denecke J. (1994) The role of the endoplasmic reticulum in protein synthesis, modification and intracellular transport. *Journal of Experimental Botany*, **44**, 1417–1444.

Walther–Larsen H., Brandt J., Collinge D.B., and Thordal–Christensen H. (1993) A pathogen–induced gene of barley encodes a HSP90 homologue showing striking similarity to vertebrate forms resident in the endoplasmic reticulum. *Plant Molecular Biology*, **21**, 1097–1108.

Wan Y. and Lemaux P.G. (1994) Generation of large numbers of independently transformed fertile barley plants. *Plant Physiology*, **104**, 37–48.

Wasternack C. (1994) Jasmonate–induced alteration of gene expression in barley. *Trends in Comparative Biochemistry and Physiology*, **1**, 1255–1268.

Wasternack C., Atzorn R., Jarosch B. and Kogel K.H. (1995) Induction of a thionin, the jasmonate–induced 6 kDa protein of barley by 2,6–dichloroisonicotinic acid. *Journal of Phytopathology*, **140**, 280–284.

Wei Y.D. (1994) The biochemical and molecular basis of interactions between barley and the powdery mildew fungus. *Ph.D. Thesis*, Royal Veterinary and Agricultural University, Frederiksberg C, Denmark.

Wei Y.D., Collinge D.B., Smedegaard–Petersen V., and Thordal–Christensen H. (1996) Transcript of a new class of retroposon–type repetitive element cloned from the powdery mildew fungus, *Erysiphe graminis*. *Molecular and General Genetics*, **250**, 477–482.

Wei Y.D., Neergaard E. de, Thordal–Christensen H., Collinge D.B. and Smedegaard–Petersen V. (1994) Accumulation of a putative guanidine compound in relation to other early defense reactions in epidermal cells of barley and wheat exhibiting resistance to *Erysiphe graminis* f. sp. *hordei*. *Physiological and Molecular Plant Pathology*, **45**, 469–484.

Welinder, K.G., Rasmussen, S.K., Penel C., Gaspar H. (1993) *Plant Peroxidases: Biochemistry and Physiology*. Proceedings, University of Geneva Press, Geneva, Switzerland.

Wevelsiep L., Rüpping E., and Knogge W. (1993) Stimulation of barley plasmalemma H^+ATPase by phytotoxic peptides from the fungal pathogen *Rhyncosporium secalis*. *Plant Physiology*, **101**, 297–301.

Whettan R. and Sederoff R. (1995) Lignin biosynthesis. *The Plant Cell*, **7**, 959–965.

White R.F., Rybicki E.P., von Wechmar M.B., Dekker J.L. and Antoniw J.F. (1987) Detection of PR–1–type proteins in Amaranthaceae, Chenopodiaceae, Graminae and Solanaceae by immunoelectroblotting. *Journal of General Virology*, **68**, 2043–2048.

Wolfe M.S. and McDermott J.M. (1994) Population genetics of plant pathogen interactions: the example of the *Erysiphe graminis – Hordeum vulgare* pathosystem. *Annual Review of Phytopathology*, **32**, 89–113.

Wolter M., Hollricher K., Salamini F. and Schulze–Lefert P. (1993) The mlo resistance alleles to powdery mildew infection in barley trigger a developmentally controlled defense mimic phenotype. *Molecular and General Genetics*, **239**, 122–128.

Woolacott B. and Archer S.A. (1984) The influence of the primary germ tube on infection of barley by *Erysiphe graminis* f. sp. *hordei*. *Plant Pathology*, **33**, 225–231.

Wray J.L. (1992) *Inducible Plant Proteins. Their Biochemistry and Molecular Biology*. Society for Experimental Biology Seminar Series 49. Cambridge: Cambridge University Press.

Wu G., Shortt B.J., Lawrence E.B., Levine E.B., Fitzsimmons K.C. and Shah D.M. (1995) Disease resistance conferred by expression of a gene encoding H_2O_2–generating glucose oxidase in transgenic potato plants. *The Plant Cell*, **7**, 1357–1368.

Yamada T., Hashimoto H., Shiraishi T. and Oku H. (1989) Suppression of pisatin, phenylalanine ammonia–lyase mRNA, and chalcone synthase mRNA accumulation by a putative pathogenicity factor from the fungus *Mycosphaerella pinoides*. *Molececular Plant–Microbe Interactions*, **2**, 256–261.

Yokoyama K., Aist J. R. and Bayles C.J. (1991) A papilla–regulating extract that induces resistance to barley powdery mildew. *Physiological and Molecular Plant Pathology*, **39**, 71–78.

Yu G.L., Katagiri F. and Ausubel F.M. (1993) *Arabidopsis* mutations at the RPS2 locus result in loss of resistance to *Pseudomonas syringae* strains expressing the avirulence gene avrRpt2. *Molecular Plant–Microbe Interactions*, **76**, 434–443.

Zeyen R.J., Carver T.L.W. and Ahlstrand G.G. (1983) Relating cytoplasmic detail of powdery mildew infection to presence of insoluble silicon by sequential use of light microscopy, SEM, and X–ray analysis. *Physiological Plant Pathology*, **22**, 101–108.

Zhang, Z. (1996) Molecular approaches of resistance responses in plant/pathogen interactions. *Ph.D. Thesis*, Royal Veterinary and Agricultural University, Frederiksberg C, Denmark.

Zhang Z., Collinge D.B., and Thordal–Christensen H. (1995) Germin–like oxalate oxidase, a H_2O_2–producing enzyme, accumulates in barley attacked by the powdery mildew fungus. *The Plant Journal*, **8**, 139–145.

15. INSECT RESISTANCE IN PLANTS: NATURAL MECHANISMS AND IMPROVEMENT THROUGH BIOTECHNOLOGY

RUSSELL R. JOHNSON

Department of Biology, Colby College, Waterville ME 04901 USA

INTRODUCTION

Plants, by necessity, have evolved a fascinating array of structural and chemical defenses against insect attack. Insect resistance is an essential characteristic that all plants must possess to a certain extent. The degree of resistance and the mechanisms used to achieve this resistance, however, may vary tremendously between species and even between cultivars.

Plant insect resistance is particularly important to agriculture and a great deal of research has focused on this area as plants in an agricultural setting are especially vulnerable to insect attacks. Cultivars bred for other desirable agronomic characteristics may have lost defense systems present in wild progenitors, and crops grown in monoculture provide a much more fertile environment for foraging and breeding of monophagous insects than a wild plant community. In the recent past, chemical insecticides have been the major response to these problems, however growing difficulties with resistance to chemical insecticides in many insect species as well as concerns about environmental pollution have led to an increased interest in the development of more insect resistant crop plants. The cloning of defense genes and the development of techniques for expressing foreign genes in plants have led to a great deal of research over the last few years directed toward the goal of producing resistant varieties of crop plants through biotechnology.

After a general treatment of the types and mechanisms of insect resistance, I will focus on the various insect defense compounds produced in plants, on the applicability of defense genes to biotechnology efforts, and on efforts undertaken to develop transgenic plants with increased resistance.

Insect resistance in plants has been divided into three general categories (reviewed by Smith, 1989). Non preference or antixenosis resistance, in which the plant is unable to serve as a host, forces the insect herbivore to select a different plant. Antixenosis resistance may result from the presence of structural characteristics such as thorns, spikes, and trichomes, or from physical barriers such as thickened epidermal layers or waxy coatings. Chemicals produced by the plant which repel insects or deter feeding or oviposition can also result in antixenosis. Resistance in which feeding on a plant results in negative biological effects on the insect is referred to as antibiosis. This can result from the presence in plant tissues of toxic proteins, alkaloids, phenolic compounds, ketones, organic acids or terpenes, or from inhibitors of digestive enzymes which result in malnutrition or reduced growth. Structural factors such as trichomes which trap feeding insects may also contribute to antibiosis. A third type of resistance, in which the insects are not adversely affected and insect damage is not reduced, is referred to as tolerance. Plants exhibiting tolerance are simply able to withstand the insect infestation and produce a greater yield than susceptible plants. It is not always easy to differentiate between these three types of resistance, and two or more types may occur in combination.

Insect resistance can be due to structural characteristics of the plant or can result from chemical defenses which are present constitutively or which are induced by the onset of insect attack. An extensive bibliography of the literature relating to the screening of plants for insect resistance, the mechanisms of resistance employed, and plant characteristics related to resistance has been compiled by Stoner (1992). Many of these structural and chemical traits have been utilized in traditional plant breeding programs to develop resistant crop strains (Smith, 1989). Biochemists and molecular biologists have focused chiefly on chemical resistance, which is more amenable to improvement through biotechnology (Ryan, 1989) than are morphological traits which are generally controlled by a large number of, usually uncharacterized, genes. For the same reason, protein toxins and inhibitors have been the subject of most research aimed at producing transgenic plants with improved insect resistance. In the future it may be possible, however, to produce transgenic plants containing altered metabolic pathways which would be able to produce compounds, such as alkaloids or terpenes, that are not normally present in a particular crop plant.

PLANT DEFENSE PROTEINS

Several groups of defense related proteins have been characterized in plants (reviewed by Bowles, 1990). These proteins accumulate in plant tissues after wounding or in response to microbial invasion. Some of these defense proteins are produced only locally at the site of injury, while other responses are systemic in nature resulting in protein accumulation throughout the plant. Plant proteins which are inhibitors of proteinases and amylases, as well as plant lectins which are toxic to gut epithelial cells have been shown to have the ability to protect plants against insect attack. Other defense proteins such as thionins, extensins, hydroxyproline-rich glycoproteins, and pathogenesis-related proteins are thought to be principally involved in defense against microbial attack.

PROTEINASE INHIBITORS

Plant proteinase inhibitors, particularly those which are wound inducible, have received a great deal of attention, both from biochemists and molecular biologists interested in fundamental questions of plant-insect interactions, plant signal transduction, and gene activation (Ryan, 1992) as well as from those interested in the practical application of utilizing inhibitor genes to create insect resistant transgenic plants (Ryan, 1989). These small proteinaceous inhibitors (Mr <20,000 kDa) are found in the leaves and storage organs of a wide variety of plants, the most thoroughly studied inhibitors being those from Solanaceae and Fabaceae. Plants contain inhibitors of serine, cysteine, metallo and aspartyl proteases (Table 15.1). In seeds and vegetative storage organs of some species, these inhibitors may account for over 10% of the total protein (Laskowski and Kato, 1980).

TABLE 15.1. Inhibitor specificity of selected plant proteinase

Inhibitor	Specificity	Reference
Kunitz Inhibitor family		
soybean trypsin inhibitor	trypsin	Kunitz, 1945
winged bean inhibitor	chymotrypsin	Habu *et al.*, 1992
wheat bifunctional inhibitor	subtilisin/elastase	Mundy *et al., 1984*
potato 22 kDa inhibitor	trypsin/chymotrypsin	*Suh et al.*, 1991
potato 20 kDa inhibitor	trypsin/cathepsin D	Mares *et al.*, 1989
Bowman-Birk family		
soybean Bowman-Birk inhibitor	trypsin/chymotrypsin	Baek and kim, 1993
cowpea Bowman-Birk inhibitor	trypsin	Hilder *et al.*, 1989
alfalfa trypsin inhibitor	trypsin	Brown *et al.*, 1985
Inhibitor I family		
potato inhibitor I	chymotrypsin	Cleveland *et al.*, 1987
tomato inhibitor I	chymotrypsin	Graham *et al.*, 1985
tomato ethylene inducible inhibitor I	V8 protease	Margossian *et al.*, 1988
L. peruvianum fruit inhibitor I	trypsin	Wingate *et al.*, 1989
tobacco tumor inhibitor I	cleavage at Gln	Fujita *et al.*, 1993.
Inhibitor II family		
potato inhibitor II	trypsin/chymotrypsin	Thornburg *et al.*, 1987
tomato inhibitor II	trypsin/chymotrypsin	Lee *et al.*, 1986.
Cysteine proteinase inhibitors		
potato multicystatin	papain	Walsh and Strickland, 1993
orryzacystatin I	papain	Kondo *et al.*, 1989
oryzacystatin II	cathepsin H	Kondo *et al.*, 1990.
Metalloproteinase inhibitors		
potato carboxypeptidase inhibitor	carboxypeptidase A carboxypeptidase B	Hass and Ryan, 1981

Role of proteinase inhibitors in plant protection

Proteinase inhibitors have been assumed to be important defense proteins based on the fact that they are induced by insect attack and on the known anti-nutritive effects of these inhibitors when included in the diets of insects (Broadway and Duffey, 1986; Broadway *et al.*, 1986). A number of feeding trials have been conducted to measure the effects of proteinase inhibitors on insect larvae (reviewed by Wolfson, 1991).

The degree of antibiosis found has varied from none to quite significant depending on the insect species tested and the inhibitor(s) present in the diet. In many cases, however, high levels of proteinase inhibitors did cause a significant reduction in the growth of larvae. Many plant proteinase inhibitors are active mainly against animal and microbial proteinases rather than against endogenous plant proteinases (Richardson, 1977), lending further support to the idea that they are present to play a defensive role. In addition, the creation of transgenic plants containing genes for proteinase inhibitors has provided more direct evidence for the ability of these inhibitors to protect plant tissues from insect attack. Tobacco plants expressing a cowpea trypsin inhibitor gene were more resistant to tobacco budworm damage than control plants (Hilder *et al.*, 1987). Similarly, transgenic tobacco plants containing either potato inhibitor II or tomato inhibitor II suffered less damage from tobacco hornworms (Johnson *et al.*, 1989).

In general, proteinase inhibitors are produced in parts of the plant which are essential for reproduction and which clearly must be protected from predation. In addition, their synthesis can be induced in leaves when necessary to protect against an insect outbreak. The accumulation of very high levels of inhibitor I and II in immature fruits of the wild tomato *Lycopersicon peruvianium* and their subsequent reduction to negligible levels in mature fruit suggests that these plants have a coordinated program to maintain defensive proteins to protect developing fruits from predation, and to degrade these proteins upon maturity to allow for seed dispersal by animal consumers (Pearce *et al.*, 1988).

It should be kept in mind, however, that not all plant proteinase inhibitors are always present to serve a uniquely defensive role. Some serine proteinase inhibitors are important storage proteins in seeds and tubers, and cysteine proteinase inhibitors may play important physiological roles in regulating the proteolytic activity of endogenous proteinases (Rodis and Hoff, 1984). Although a wide variety of proteinase inhibitors, and their corresponding genes, have been isolated from many different plant species, the majority of these inhibitors do not have clearly defined physiological roles.

Serine proteinase inhibitors

Six different evolutionary families of serine proteinase inhibitors are found in plants. All of these inhibitors employ essentially the same mechanism of action, in which the inhibitor is recognized by the proteinase as a substrate based on the side chain of the P_1 residue at the reactive site (reviewed by Laskowski and Kato, 1980). The inhibitor then binds with high affinity to the active site of the enzyme as a result of a large number of non covalent van der Waals and hydrogen bond interactions to form a stable complex. Water is thus excluded from the active site, preventing hydrolysis and release of the inhibitor. The proteinase thus remains bound to the inhibitor and is prevented from acting on its normal substrate.

Members of the Kunitz inhibitor family are proteins of about 20 kDa which are stabilized by two disulfide bridges and are composed mainly of β-type secondary structure with a central distorted β-barrel (Zemke *et al.*, 1991). The soybean trypsin inhibitor (Kunitz, 1945) was the first plant proteinase inhibitor to be isolated, and it is now known that several genetically different variants exist (Kim *et al.*, 1985; Jofuku and Goldberg, 1989). Kunitz-type inhibitors have also been isolated from silk tree (Odani *et al.*, 1979), acacia (Kortt and Jermyn, 1981), Erythrina bean (Joubert, 1982), and white mustard (Menegatti *et al.*, 1985). Several different Kunitz-type inhibitors have been isolated from winged bean (*Psophocarpus tetragonolobus*) (Kortt, 1980). The seeds of winged bean contain both chymotrypsin inhibitors (Habu *et al.*, 1992) and trypsin inhibitors (Yamamoto *et al.*, 1983) while another trypsin inhibitor is present in the root nodules (Manen *et al.*, 1991). The barley (Weselake *et al.*, 1983) and wheat (Mundy *et al.*, 1984) inhibitors are bifunctional inhibitors with specificity for both subtilisin and α-amylase. Two Kunitz-type inhibitors have been isolated from potato, a 20 kDa protein which inhibits both trypsin and cathepsin D (Mares *et al.*, 1989), and a 22 kDa protein which inhibits trypsin and chymotrypsin (Suh *et al.*, 1991).

Another family of inhibitors present in legume seeds are the Bowman-Birk inhibitors. These small proteins (Mr approx. 8 kDa) contain two inhibitor domains, each with its own reactive site. Soybeans contain several Bowman-Birk type proteins (Hammond *et al.*, 1984; Joudrier *et al.*, 1987; Baek and Kim, 1993). The Bowman-Birk inhibitor itself has one inhibitory site for trypsin and another for chymotrypsin. Each member of the closely related group of inhibitors, PI-I through IV, contains two trypsin inhibitory sites, while the C-II inhibitor inhibits not only trypsin and chymotrypsin, but also elastase. Inhibitors from cowpea (Hilder *et al.*, 1989) and alfalfa (Brown *et al.*, 1985) each contain two reactive sites for trypsin inhibition. Bowman-Birk type inhibitors with inhibitory activity against both trypsin and chymotrypsin have also been isolated from several varieties of beans (Wu and Whitaker, 1991).

The Inhibitor I family contains known members from potato (Cleveland *et al.*, 1987; Beuning and Christeller, 1993), tomato (Graham *et al.*, 1985), and tobacco (Kuo *et al.*, 1984) as well as from barley (Williamson *et al.*, 1987). The potato inhibitor I has a monomer molecular weight of approximately 8 kDa and is present in the tubers as a tetramer. Inhibitor I molecules from potato, tomato and barley are all strong inhibitors of chymotrypsin. Genes for two additional Inhibitor I molecules have been isolated from tomato, one with specificity for enzymes which cleave at glutamate residues such as *Staphylococcus aureus* V8 proteinase (Margossian *et al.*, 1988) and another with specificity for trypsin (Wingate *et al.*, 1989; Wingate and Ryan, 1991). An inhibitor from tobacco contains a glutamine reside at the reactive site and thus recognizes some as yet undetermined proteinase that cleaves at this amino acid (Fujita *et al.*, 1993).

Inhibitor II proteins (monomer Mr approx. 12.3 kDa) contain two domains, each with a reactive site, analogous to the Bowman-Birk inhibitors. Inhibitor II, like inhibitor I, is found in the tubers and leaves of potato and in the fruit and leaves of tomato (Pearce *et al.*, 1988; Lorberth *et al.*, 1992). Both potato and tomato inhibitor II contain one reactive site for inhibition of trypsin and another for inhibition of chymotrypsin (Lee *et al.*, 1986; Thoroburg *et al.*, 1987; Murray and Christeller, 1994).

Members of the barley trypsin inhibitor family have molecular weights of about 14 kDa and have inhibitory activity against a broad range of enzymes. The inhibitor from barley seeds (Odani *et al.*, 1983) inhibits trypsin, while a related inhibitor from ragi (*Eleusine coracana*) is a bifunctional inhibitor of both trypsin and α-amylase (Campos and Richardson, 1983). A member of this family isolated from the seeds of maize inhibits both trypsin and the blood clotting factor XIIa (Mahoney *et al.*, 1984).

The squash inhibitor family, a family of very small inhibitors (3.2 kDa), is present in the seeds of the Curcurbitaceae family. These inhibitors have strong inhibitory activity against trypsin, and are stabilized by three disulfide bonds (Hara *et al.*, 1989). Squash inhibitors have been isolated from pumpkin (*Curcurbita maxima*) (Hojima *et al.*, 1982), *Momordica repens* (Joubert, 1984), *Curcurbita pepo* and *Cucumis sativis* (Weiczorek *et al.*, 1985). X-ray crystallographic analysis indicates that these inhibitors are of ellipsoidal shape, and lack any helices or β-sheet structure (Bode *et al.*, 1989).

Cysteine proteinase inhibitors

Several members of the cystatin family of reversible competitive cysteine proteinase inhibitors (reviewed by Barret, 1987) have been found in plants. Cystatins inhibit endopeptidases belonging to the papain superfamily. The peptidyl bond of a conserved glycine near the amino terminus of the inhibitor is postulated to be the reactive site bond. In addition, an internal QVVAG sequence apparently forms a secondary contact area with the enzymes allowing the formation of an extremely tight complex.

A family of seven cysteine proteinase isoinhibitors (Mr=5600 Da) has been isolated from pineapple stem (Reddy *et al.*, 1975). Multicystatin, an 85 kDa inhibitor which forms crystals in potato tubers, consists of eight individual domains, each of which can inhibit one proteinase molecule (Walsh and Strickland, 1993). Two inhibitors, referred to as oryzacystatin I (Abe *et al.*, 1987; Kondo *et al.*, 1989) and oryzacystatin II (Kondo *et al.*, 1990) are present in rice seeds. These two inhibitors differ in their specificity toward cysteine proteinases, and are synthesized at different times during seed development. The seeds of *Wisteria floribunda* also contain several cysteine proteinase inhibitors, including a protein of about 17 kDa whose amino terminal region is very similar to oryzacystatin I (Hirashiki *et al.*, 1990). A cDNA encoding another protein very similar to oryzacystatin I has been cloned from avocado (Dopico *et al.*, 1993).

Aspartyl and metallo proteinase inhibitors

Two closely related inhibitors of aspartic proteinases have been isolated from potato tubers. These 20 kDa Kunitz-type inhibitors (Mares *et al.*, 1989; Ritonja *et al.*, 1990; Strukel *et al.*, 1990) inhibit the aspartyl proteinase cathepsin D in addition to trypsin. An inhibitor of metalloexopeptidases has also been isolated from potato tubers. The carboxypeptidase inhibitor (CPI) (Mr = 3900 Da), a strong inhibitor of both carboxypeptidase A and B, accumulates during tuber development along with inhibitor I and II (Hass and Ryan, 1981).

Expression of proteinase inhibitor genes

Proteinase inhibitor (*pin*) gene expression is regulated by a complex set of developmental and environmental signals. Inhibitors accumulate in an organ or tissue specific manner during the development of seeds, tubers and fruits. In addition, serine proteinase inhibitors such as inhibitor I and II are produced in leaves after wounding caused by chewing insects or other mechanical damage. Wounding of tomato or potato leaves results not only in local accumulation of these two inhibitors, but also in a systemic response throughout the plant in which unwounded leaves also produce the inhibitors (reviewed by Ryan, 1992). Local accumulation of inhibitors is mediated by the release of oligogalacturonides near the site of wounding (Ryan, 1992; Doares *et al.*, 1995). These signalling molecules induce rapid accumulation of a variety of defense compounds, including proteinase inhibitors. As these compounds are not mobile (Baydoun and Fry, 1985) another signal, or signals, must be responsible for the systemic accumulation of inhibitors far from the wound site. A strong candidate for such a signal is the 18 amino acid peptide systemin, which is an extremely potent elicitor of *pin* gene expression in tomato plants (Pearce *et al.*, 1991, 1993). Systemin is proteolytically cleaved from a 200 amino acid precursor, prosystemin, and accumulates in a wide variety of plant tissues. It is an essential part of the signal transduction pathway leading to inhibitor induction, as transformation of tomato plants with an antisense prosystemin gene greatly reduced wound-induced *pin* gene expression (McGurl *et al.*, 1992). Additional evidence for the importance of systemin in the production of the systemic signal comes from transgenic plants which overexpress a prosystemin gene. Not only did the overproduction of systemin result in a high constitutive level of inhibitors, but the rootstocks of these transgenic plants were also able to induce a similar overexpression of inhibitors in scions from untransformed plants (McGurl *et al.*, 1994). When radiolabeled systemin was applied to leaf wounds, it moved throughout the leaf via the apoplast and then entered the phloem in a manner similar to that of labelled sucrose (Narváez-Vásquez *et al.*, 1995). Within 90 minutes of application, the systemin had travelled to other leaves on the plant. Inhibition of phloem loading at wound sites prevented both the mobility of systemin and the systemic accumulation of inhibitors (Narváez-Vásquez *et al.*, 1994). A 50 kDa systemin-binding protein present in the plasma membrane (Schaller and Ryan, 1994) may serve as a receptor in these distal tissues which then triggers an intracellular cascade resulting in the induction of inhibitor genes.

Both local induction, mediated by oligogalacturonides, and systemic induction, mediated by systemin (or other signals), occur via the octadecanoid pathway (Farmer and Ryan, 1992; Blechert *et al.*, 1995; Doares *et al.*, 1995). According to currently accepted models, these signals, upon binding to their appropriate receptors, activate the release of linolenic acid from the plasma membrane into the cytoplasm. Linolenic acid is then converted to jasmonic acid by cytoplasmic enzymes, and jasmonic acid then activates transcription of *pin* genes by an as yet unknown mechanism (Farmer *et al.*, 1992).

In addition to oligogalacturonides and systemin, traditional plant hormones are also able to stimulate *pin* gene expression. External application of abscisic acid (ABA) induces systemic *pin*2 expression in potatoes in a manner similar to wounding (Peña-Cortés *et al.*, 1991). The involvement of ABA in wound signalling *in vivo* is not entirely clear, for while ABA levels increase upon wounding, other environmental stresses such as drought cause accumulation of ABA without a resulting increase in proteinase inhibitors. The activation of inhibitor genes by ABA apparently occurs through the octadecanoid pathway (Peña-Cortés *et al.*, 1995). Auxin has been found to decrease wound-induced *pin* expression, and upon wounding, auxin levels were seen to decline (Kernan and Thornburg, 1989). There is, however, one tomato root proteinase inhibitor which is auxin-inducible (Young *et al.*, 1994).

Evidence also exists that electrical signals can play an important role in wound signal transduction (Davies, 1987; Wildon *et al.*, 1992). Wounding of young tomato cotyledons resulted in the transmission of an electrical potential out of the cotyledon and throughout the plant. Systemic accumulation of proteinase inhibitors was correlated with the systemic electrical signal. Inhibition of the phloem transport of chemical signals by stem chilling, which did not interfere with the transmission of the electrical signal, did not prevent the systemic wound-induction of inhibitors (Wildon *et al.*, 1992). These results would appear to conflict with those of Narváez-Vásquez *et al.* (1994), discussed above, who found that inhibition of phloem loading did prevent systemic induction of inhibitors. External application of an electric current to tomato leaves resulted in the systemic accumulation of *pin*2 mRNA (Herde *et al.*, 1995). How a systemic electrical signal would result in *pin* gene activation, perhaps through interactions with other known modulators of *pin* genes, is not clear.

The fact that a number of different compounds are able to induce systemic accumulation of *pin* genes might suggest that multiple, perhaps redundant, pathways could be utilized for wound signalling. However, the fact that Lightner *et al.* (1993) were able to isolate two different tomato signalling mutants, both caused by a single recessive mutation, makes it unlikely that more than one pathway exists, at least in tomato.

Proteinase inhibitor genes in transgenic plants

The important role that some proteinase inhibitors apparently play in insect defense suggests that it should be possible to increase the resistance of important crop plants to insect predators by introducing inhibitor genes. Both serine and cysteine proteinases are found in the guts of herbivorous insects, and inhibitors of these digestive enzymes should have the potential to disrupt feeding and nutrition (Ryan, 1989). Insects which have developed the ability to tolerate the naturally occurring inhibitors in their normal plant host could be targeted by other inhibitors with a different specificity. For example, the Colorado potato beetle and the cowpea weevil are both able to feed on plant material containing high levels of serine proteinase inhibitors, since they apparently utilize cysteine proteinases as digestive enzymes (Gatehouse *et al.*, 1986; Wolfson and Murdock, 1987). Transgenic plants expressing a cysteine proteinase inhibitor gene should prove to be more resistant to these insects.

Efforts to boost insect resistance by introducing proteinase inhibitor genes into plants have yielded promising results. Transgenic tobacco plants containing the cowpea trypsin inhibitor gene under the control of the constitutive cauliflower mosaic virus (CaMV) 35S promoter produced active inhibitor in amounts up to 1% of the soluble leaf protein. Larvae of *Heliothis virescens* (tobacco budworm) grew more slowly and caused considerably less damage to leaves when growing on the transgenic plants as compared to control plants (Hilder *et al.*, 1987). An inhibitor I gene from tomato and inhibitor II genes from both potato and tomato have also been expressed in transgenic tobacco under the control of the CaMV 35S promoter. These proteins accumulated in levels up to 300 μg/g leaf tissue and were all properly processed and active in the transgenic plants. Larvae of *Manduca sexta* fed a diet of leaves from the transgenic plants containing inhibitor II from either potato or tomato grew more slowly and consumed less leaf material than larvae fed control leaves (Johnson *et al.*, 1989). For both cowpea trypsin inhibitor and inhibitor II, plants containing a higher level of the inhibitor had increased resistance over plants containing a lower level. Transgenic plants expressing the tomato *pin*1 gene did not appear to be any more resistant to *Manduca sexta* than control plants (Johnson *et al.*, 1989), perhaps because Inhibitor I may not affect the gut proteinases of tobacco hornworms. Transformation of alfalfa with a serine proteinase inhibitor gene derived from *Manduca sexta* was able to reduce thrip (*Frankiniella* spp.) predation of the transgenic plants (Thomas *et al.*, 1994).

In order to design the best strategies for introducing inhibitor genes into crop plants, it will be necessary to take into account the specificity of interactions between different inhibitors and the gut enzymes of the target insects. There can be considerable differences, for example, in the ability of a particular trypsin inhibitor to inhibit trypsin-like enzymes from different insect species (Christeller and Shaw, 1989). Surveys have been made of the midgut proteolytic activities of many insect species, and the ability of specific plant proteinase inhibitors to inhibit these enzymes (Christeller *et al.*, 1992; Purcell *et al.*, 1992). This type of work should prove very useful in determining what roles inhibitors can play in the control of particular insect pests.

Plant proteinase inhibitors with varying specificities are known (Table 15.1) and a considerable number of genes and cDNAs for these inhibitors are now available. As even more genes are isolated in the future, the ability to target particular insect digestive proteinases with transgenic plants containing inhibitors of the desired specificity will be further increased.

It is often thought that it will be more difficult for insects to rapidly develop resistance to proteinase inhibitors than to more highly toxic insecticidal compounds. This may or may not prove to be true, but there is already one documented case of adaptation of insects to transgenic plants expressing a *pin* gen. *Spodoptera exigua* larvae which had fed on transgenic tobacco plants with elevated levels of inhibitor II responded by producing elevated levels of inhibitor II insensitive gut proteases (Jongsma *et al.*, 1995.)

Lectins

Lectins are a broad class of proteins grouped on the basis of their ability to bind carbohydrates. Traditionally, only those proteins which bind reversibly to carbohydrates, without covalently altering the structure of the ligand, have been classified as lectins (Etzler, 1985). Recently, however, proteins possessing enzymatic activity have been grouped with lectins on the basis of gene homology (Moreno and Chrispeels, 1989). Lectins are involved in a variety of physiological processes, including protection from predators and pathogens, recognition signals for interaction with symbiotic bacteria, stimulation of cell proliferation, and enzymatic activities (Rudiger, 1984). Lectins that may be involved in plant defense are either cytotoxic or have enzyme inhibitory activity that could disrupt digestion in the gut of predators.

Several lectin and lectin-related proteins have severe toxic effects against animals. A variety of plant toxins which inhibit protein synthesis through ribosome-inactivation have been isolated (Gasperi-Campagni *et al.*, 1985). The best studied of these is the castor bean (*Ricinus communis*) toxin ricin (Montfort *et al.*, 1987). Ribosome inactivating proteins have been shown to protect plants against pathogenic fungi (reviewed by Lamb *et al.*, 1992), but it is unclear whether ricin plays any significant role in protection against insect attack. Ribosomes from *Trichoplusia ni* and *Spodoptera frugiperda* are sensitive to inactivation by ricin, but intact cells from these insects were resistant to levels of ricin that were lethal to mammalian cells (Maruniak *et al.*, 1990).

A number of seed lectins have been shown through feeding trials to have antibiotic effects against insects. For example, wheat germ agglutinin (Murdock *et al.*, 1990) and α-amylase inhibitor from the common bean *Phaseolus vulgaris* (Huesing *et al.*, 1991) are both quite effective in slowing the growth of the bruchid beetle *Callosobrochus maculatus*, which is a major pest of stored seeds.

A gene coding for the pea lectin (P-Lec) (Gatehouse *et al.*, 1987), which is known to have antibiotic effects on insect pests, was tested in transgenic tobacco plants. Expression of the P-Lec gene under the control of the CaMV 35S promoter resulted in increased resistance to tobacco budworm (*Heliothis virescens*) (Boulter *et al.*, 1990).

A *Phaseolus vulgaris* gene which is closely related to seed lectin genes codes for a protein with α-amylase inhibitor activity (Moreno and Chrispeels, 1989). The 28 kDa lectin-like protein (LLP) is proteolytically cleaved and glycosylated to form the α-amylase inhibitor. Transgenic pea plants were made containing the bean α-amylase inhibitor under the control of a seed-specific promoter and seeds of these plants accumulated the inhibitor at levels similar to those in bean seeds. The peas seeds were resistant to both a storage pest *(Callosobrochus maculatus)* which attacks dry mature seeds (Shade *et al.*, 1994), and to a pea weevil (*Bruchus pisorum*) which attacks the immature seeds when growing in the field (Shroeder *et al.*, 1995).

If α-amylase inhibitors are to be successfully employed in producing transgenic plants with increased insect resistance, it will be important to develop a better understanding of the specificities of α-amylase inhibitors toward the gut α-amylase activities of important pest insects. In one study, two classes of α-amylase inhibitor (the monomeric and dimeric inhibitors) from *Triticum aestivum* were tested against a variety of insect α-amylases. The monomeric inhibitor was more active than the dimeric inhibitor against the α-amylase from some insect species while the α-amylase activity from several other species was equally susceptible to both inhibitors (Gutierrez *et al.*, 1990).

OTHER PLANT DEFENSE COMPOUNDS

Plants contain a wide variety of compounds that are known or suspected to have a role in insect defense. Early agricultural insecticides were plant-derived compounds such as pyrethrins, nicotine and other related alkaloids, and rotenone. These phytochemicals have now been largely replaced in modern agriculture by synthetic insecticides, which are more economical to produce in large quantities.

The extensive literature on the wide variety of chemicals produced in plants can be discussed only very briefly here and the reader is referred to other reviews for more detailed information (Bowers, 1980; Nakanishi, 1980; and others listed below). A partial listing of plant chemicals that are involved in antibiosis plant resistance to insects and other arthropod pests has been compiled by Smith (1989).

Alkaloids

Alkaloids are aromatic nitrogen containing compounds which are produced in a wide variety of plant species. The physiological function of many alkaloids is unknown, but at least some of these compounds are known to be involved in insect defense. The well known alkaloid caffeine has an antibiotic effect on insects due to its inhibition of phosphodiesterase activity which results in elevated levels of the intracellular messenger cyclic AMP. This alkaloid is present at high levels in vulnerable, newly formed tissues of *Coffea arabica*, while concentrations are much lower in older tissues which have developed tougher mechanical protections (Frischknecht *et al.*, 1986 and references therein). Nicotine and other related pyridine-containing alkaloids are constitutively present in the leaves of tobacco plants. Insect herbivory, however, results in a dramatic increase in the level of tobacco leaf alkaloid content (Baldwin, 1988,1991). Larvae of *Manduca sexta* and *Trichoplusia ni*, two important tobacco pests, ate less and gained less weight when they were fed a diet of induced leaves containing high alkaloid concentrations (Baldwin, 1991). The antibiotic effects of nicotine are believed to be based on its interaction with cholinergic receptors, and consequent interference with animal nervous systems. Feeding trials have shown that the naphthylisquinoline alkaloid dioncophylline A strongly inhibits the growth of *Spodoptera littoralis*, a polyphagous insect pest, apparently due to antixenosis effects (Bringmann *et al.*, 1992). Evidence for the role of other alkaloids in insect defense is more circumstantial. The mediterranean shrub *Spartium junceum*, for example, produces more alkaloids (principally cytisine) when grown under conditions where they are most subject to insect attack, and alkaloid levels are highest in those tissues which are most vulnerable to the insects (Barboni *et al.*, 1994). The pyrrolizidine alkaloids, present in some species of Asteraceae, Boraginaceae, and Fabaceae, are also believed to be important in antixenosis resistance against insects (Boppre, 1990).

Terpenes

Terpene compounds consist of varying numbers of isoprenoid units condensed into ring compounds. High quantities of terpenes are produced in the resin ducts of conifer trees and are also present in the leaves of many plant species. A number of conifer species respond to bark beetle attack by producing oleoresin at the site of wounding. This oleoresin is not only toxic to beetles but when dried, forms a barrier of hardened resin to protect the tree from further damage (Steele *et al.*, 1995 and references therein). There is also evidence that the tetracyclic triterpenes from the genus *Curcurbita* known as curcurbitacins function in defense against herbivory (Tallamy and McCloud, 1991). Confirmation of the antibiotic effect of specific terpenes against particular insect species requires the ability to quantitatively incorporate volatile terpenes into artificial diets. A microencapsulation system has been utilized to test the effects of several Douglas fir (*Pseudotsuga mensiezii*) monoterpenes against the western spruce budworm (*Choristoneura occidentalis*)(Clancy *et al.*, 1992) . β-pinene, α-pinene, and limonene had little effect on larval growth, while high concentrations of camphene, β-citronellol, or linalool had deleterious effects. The extension of this type of analysis to other plant-insect systems should prove useful in determining the importance of specific terpenes in plant protection from insects.

In addition to their direct involvement in insect defense systems, terpenoid compounds are also important in attracting parasitic or predatory arthropods which are natural enemies of phytophagous insects. For example, terpenoids released by maize leaves in response to insect feeding serve to attract parasitic wasps which attack caterpillars feeding on the maize (Turlings and Tumlinson, 1992; Turlings *et al.*, 1995).

Hydroxamic Acids

Hydroxamic acids are found in the leaves of several members of the Poaceae family. The major hydroxamic acid of maize and wheat, 2,4-dihydroxy-7-methoxy-1,4-benzoxazin-3-one (DIMBOA), acts as a feeding deterrent, resulting in antixenosis resistance (Robinson *et al.*, 1978; Nicol *et al.*, 1992). When ingested, DIMBOA also exhibits antibiotic effects against the european corn borer (*Ostrinia nubilalis*) such as decreased weight gain, increased time to pupation, and increased mortality (Campos *et al.*, 1989). The antibiosis caused by DIMBOA appears to result from the inhibition of food digestion. In *in vitro* assays, DIMBOA inhibited both tryptic and chymotryptic enzymes from the midgut of *O. nubilalis* (Houseman *et al.*, 1992). Maize lines containing high concentrations of DIMBOA were found to be considerably more resistant to *Ostrinia nubilalis* than lines with lower levels (Barry *et al.*, 1994). When crosses were made between resistant and susceptible lines, corn borer resistance cosegregated with DIMBOA concentration in the progeny.

Insect Growth Regulators

Plants produce a variety of compounds which are able to specifically disrupt the developmental pattern of susceptible insects. The ecdysteroids are an important class of insect hormones which regulate molting and other developmental processes. Many plant species produce closely related molecules, referred to as phytoecdysteroids (Lafont and Horn, 1989), which have no known physiological role within the plant itself. While a role for these molecules in insect antibiosis has not been firmly established, their production in plant organs most susceptible to insect attack (Dinan, 1992) makes it a likely hypothesis.

Juvenile hormone (JH) is also extremely important in regulating the molting and metamorphosis of insects. Several compounds which exhibit JH activity have been isolated from plants. Other plant compounds, the preconenes, prevent the secretion of JH by destroying the gland which is responsible for its production. Exposure of sensitive insect species to preconenes results in improper metamorphosis and sterilization, (Bowers, 1980).

Amino Acids

A variety of analogs of the 20 amino acids normally incorporated into proteins are present in plant tissues (Rosenthal, 1982). When consumed by herbivores, and mistakenly incorporated into polypeptide chains during protein synthesis, these compounds can have strong antibiotic effects. The presence of these "incorrect" amino acids in important insect proteins can prevent the proper processing, folding, and function of these proteins, with very serious consequences for the insect.

The amino acid canavanine, for example, is present in many species of the Fabaceae family, and causes dramatic growth aberrations in lepidopteran pupae and adults when it is present in the diet of larvae (Rosenthal and Dahlman, 1986; Rosenthal, 1991). In susceptible insects, canavanine is mistakenly recognized by arginyl tRNA synthetase, resulting in the formation of abnormal, dysfunctional proteins. Interestingly, the arginyl tRNA synthetase of several insect species which feed on the seeds of legumes with high canavanine content are able to discriminate between canavinine and arginine, and thus incorporate canavanine only rarely into proteins (Rosenthal *et al.*, 1987; Rosenthal, 1991).

COMPOUNDS OF NON-PLANT ORIGIN

For the purpose of developing transgenic plants with increased insect resistance, there are a number of genes available from non-plant sources which encode proteins that are toxic to insects, and these can be used in addition to those derived from plants. Problems were initially encountered with the ability of non-plant genes to be highly expressed in plant tissues, but strategies have now been developed to overcome this difficulty. As a result, the first commercially available plants engineered for increased insect resistance were those expressing *Bacillus thuringiensis* toxin genes (Fox, 1995).

Bt Toxins

Crystal proteins of Bacillus thuringiensis

Bacillus thuringiensis (Bt) is a soil bacterium that produces protein-containing crystalline inclusions during sporulation (Aronson *et al.*, 1986). Different isolates of Bt display toxicity to lepidopteran, dipteran, and/or coleopteran larvae, and bacterial formulations have been successfully used for the control of agricultural pests (MacIntosh *et al.*, 1990b). Toxicity to insects is due to the presence of specific proteins in the parasporal crystals, many of which have been isolated and the corresponding genes cloned (Hofte and Whiteley, 1989). There is a wide diversity in the crystal protein genes present in different isolates of Bt. Many of these genes are present on conjugative plasmids and are thus quite mobile.

Mode of action

After the parasporal crystals are consumed by insect larvae, the crystals dissolve in the midgut to release the insecticidal proteins. These proteins are generally present in the form of protoxins, which must be proteolytically converted by gut enzymes into the toxic form. Toxin molecules then associate with the plasma membrane of brush border cells of the midgut epithelium by binding to specific membrane proteins which serve as receptors (Sangadala *et al.*, 1994; Knight *et al.*, 1995). The insecticidal specificities of the various Bt toxins are determined by their ability to bind to the receptors present in the midgut of different insect species. Once bound to the brush border membrane of a susceptible insect, the toxins are able to enter the lipid bilayer and form a cation-permeable channel. As a result, K^+, followed by H_2O, flows from the gut lumen into the epithelial cells, causing swelling and eventually lysis (English and Slatin, 1992).

Despite the differences in sequence between the many Bt crystal proteins, there are several regions of relatively well-conserved amino acid blocks (Hofte and Whiteley, 1989) and most of the toxins probably assume a tertiary structure similar to that determined by X-ray crystallography for CryIIIA (Li *et al.*, 1991). Domain I, located at the N terminus of the mature protein, is composed of a bundle of seven hydrophobic and amphipathic helices, and is required for membrane insertion and pore formation (Chen *et al.*, 1995). Domain II, composed of three antiparallel β-sheets, contains amino acid sequences shown to be involved in binding to midgut receptor proteins (Lee *et al.*, 1992). Domain III is required for structural stability as well as channel function and specificity (Chen *et al.*, 1993).

Classification

Bt toxin genes have been divided into four major classes and several subclasses based on insecticidal specificity and sequence similarities (Hofte and Whiteley, 1989). CryI proteins are Lepidoptera-specific, CryII proteins are toxic to both lepidopterans and dipterans, CryIII proteins are Coleoptera-specific, and CryIV proteins are Diptera-specific.

CryI genes encode 130–140 kDa proteins which are converted to 60-70 kDa active toxins in the insect gut. On the basis of sequence comparison, these genes have been divided into four subclasses: *cryIA, cryIB, cryIC, and cryID*. A survey of 29 Bt strains found CryIA to be the most common crystal protein, followed by CryIB and CryIC (Hofte *et al.*, 1988). Although all of these CryI proteins are toxic against Lepidoptera, they are specific to certain species, and this specificity differs significantly between the toxins. For example, CryIA proteins were found to be highly toxic to *Manduca sexta*, but much less toxic to *Mamestra brassicae*. CryIC, however, was more toxic to *M. brassicae* than to *M. sexta* (Hofte and Whitely, 1989).

CryII genes, which share very little sequence similarity with the other *cry* gene classes, code for proteins of about 65 kDa. Purified CryIIA protein was found to be toxic against both a lepidopteran (*M. sexta*) and a dipteran (*Aedes aegypti*), while the closely related CryIIB protein was toxic to *M. sexta* but not to *A. aegypti* (Widner and Whitely, 1989).

CryIII genes encode crystal proteins of about 73 kDa, which are toxic to coleopteran larvae. The CryIIIA protein is very toxic to larvae of the Colorado potato beetle (*Leptinotsara decemlineata*), but the southern corn root worm (*Diabrotica undecimpunctata*) was much less susceptible (Slaney *et al.*, 1992). Genes for several other closely related toxins, *cryIIIB* (Sick *et al.*, 1990), *cryIIIC* (Lambert *et al.*, 1992b), and *cryIIID* (Lambert *et al.*, 1992a), have also been cloned.

The *cryIV* gene class contains four genes, all of which were isolated from the same strain of *B. thuringiensis israelensis*. CryIVA (135 kDa), CryIVB (128 kDa),

CryIVC (78 kDa), and CryIVD (72 kDa) all exhibit toxicity toward mosquito larvae. These toxins appear to act synergistically, as intact crystals and combinations of isolated proteins were more toxic than would be expected from the additive effect of the individual proteins (Chilcott and Ellar, 1988; Delecluse *et al.*, 1988).

Bt toxin genes in transgenic plants

A truncated version of a *cry*IA(b) gene was the first insect resistance gene to be expressed in transgenic plants (Vaeck *et al.*, 1987). Transgenic tobacco plants containing this toxin gene under the control of a constitutive promoter produced sufficient Bt toxin in leaf tissues to result in high mortality of *Manduca sexta* larvae. The level of Bt toxin production, however, in these and other "first generation" transgenic plants (Barton *et al.*, 1987), was quite low (approximately 0.001% of total soluble protein). These low levels are insufficient for good field control of many important insect pests, most of which are less sensitive to Bt toxin than *M. sexta* (Delannay *et al.*, 1989).

In order to improve the expression of *cry*IA(b) and *cry*IA(c) genes in transgenic plants, modifications were made to DNA sequences predicted to inhibit efficient gene expression in plants and by altering the coding region to more closely reflect plant codon usage (Perlack *et al.*, 1990, 1991). These changes resulted in a 100-fold increase in the amount of Bt toxin production in transgenic tobacco plants. The high level accumulation of Bt toxin from this type of modified gene (up to 0.1 % of total soluble protein) greatly increases the ability to produce transgenic plants with field resistance to a wide variety of insect pests.

An alternative to the laborious and costly procedure of modifying Bt genes for higher expression in plant tissues is the incorporation of an extremely high copy number of transgenes into organelles. Transformation of tobacco chloroplasts with an unmodified CryIA(c) coding sequence under the control of a plastid promoter resulted in plants with 10,000 copies of the transgene per cell. In these plants, the toxin represented 3–5% of the total soluble leaf protein (McBride *et al.*, 1995).

However, in some cases such as in integrated pest management programs, it may be quite possible to use transgenic plants containing relatively low levels of Bt toxin. Bt toxin mediated resistance and the presence of natural enemies were found to interact synergistically to reduce populations of *Heliothis virescens* (Johnson and Gould, 1992), thus lowering the amount of toxin necessary to achieve acceptable control. Synergistic interactions have also been found between Bt toxin and low levels of serine proteinase inhibitor in insecticidal activity against several insect pests (MacIntosh *et al.*, 1990a).

Bt toxin genes have been expressed in a number of plant species including tobacco (Vaeck *et al.*, 1987), tomato (Fischoff *et al.*, 1987), potato (Peferoen *et al.*, 1990), cotton (Perlack *et al.*, 1990), maize (Koziel *et al.*, 1993) and rice (Fujimoto *et al.*, 1993). In all cases, the presence of Bt toxin conferred resistance against insect pests of the crop involved. Bt-expressing potato plants developed by Monsanto were approved by the EPA for commercial use in the United States in May of 1995. In August, similar approval was obtained by Mycogen and Ciba Seeds for their Bt-producing maize varieties (Fox, 1995).

Resistance to Bt toxin

The excitement over development of insect resistant transgenic plants expressing Bt toxin genes has been tempered by the fear that insects in the field will be able to develop resistance to these toxins, just as they have to many synthetic insecticides. It has long been known that Bt resistance could develop in insects exposed to high levels of toxin for many generations (McGaughey, 1985). Cases of resistance have also been found in the field (Kirsch and Schmutterer, 1988; Tabashnik *et al.*, 1990) where Bt insecticides have been used extensively against the diamondback moth *Plutella xylostella*. In several cases, resistance to Bt was the result of a modification in the midgut receptor which resulted in decreased toxin binding. However, in other cases resistance was not associated with reduced binding, indicating that post binding events were affected (Masson *et al.*, 1995 and references therein). In any case, resistance to Bt toxins has generally been limited to one particular group of toxin, and is usually inherited as a recessive trait (Martinez-Ramirez *et al.*, 1995). If these characteristics hold true for field-developed resistances to commercially released transgenic plants, then resistance-management strategies can be developed to combat the problem. Other toxin genes can be substituted, for example, or genes coding for new recombinant crystal proteins (Bosch *et al.*, 1994) can be developed with altered specificities. However, there is cause for concern whether this will be the case, as a strain of *Heliothis virescens* with broad spectrum Bt resistance was obtained in the laboratory after selection with only CryIA(c) toxin (Gould *et al.*, 1992). The resistance in these larvae was not correlated to a decrease in toxin binding affinity and was not inherited as a recessive trait. The mechanism(s) responsible for this broad spectrum resistance are unknown. *Plutella xylostella* exposed to only CryIA and CryII toxins developed cross-resistance to CryIF, again by an unknown mechanism (Tabashnik *et al.*, 1994).

Compounds from Arthropods and Viruses

Small peptide toxins, which have paralytic activity against insects, have been isolated from the venom of scorpions (Bontems *et al.*, 1991), mites (Tomalski and Miller, 1991), and spiders (Stapleton *et al.*, 1990). These peptides are highly toxic to insects, but appear to be safe for non-target organisms. The potential exists to generate transgenic plants producing such peptides, and if the toxins were able to survive long enough in the gut so that a sufficient quantity could be absorbed into the hemolymph, these plants would be well protected against a wide variety of insects.

An artificial gene coding for the *Buthus eupeus* insectotoxin I5A was expressed in transgenic tobacco plants, but the I5A protein produced in these plants lacked toxicity (Pang *et al.*, 1992). I5A is normally produced as part of a precursor protein which is then processed to form the mature active toxin. It will probably be necessary to express the complete toxin gene in plants before an active protein can be produced. cDNAs are now available which are able to code for active insectotoxin peptides (Stewart *et al.*, 1991; Tomalski and Miller, 1991), so it should be possible to test the effectiveness of these peptides in transgenic plants.

In addition to production of transgenic plants containing molecules that result directly in insect antibiosis, it may also be possible to engineer plants which increase the susceptibility of insects to their own natural enemies. Protein factors present in granulosis viruses (Derksen and Granados, 1988) and entomopox viruses (Xu and Hukuhara, 1992) are known to increase the susceptibility of larvae to infection by nuclear polyhedrosis viruses, a subfamily of the baculoviruses used for biological control of lepidopteran pests. These proteins disrupt the integrity of the peritrophic membrane (PM), the protective layer in the gut which is important in the prevention of microbial infection. The gene for such a viral enhancing factor has been cloned from the *Trichoplusia ni* granulosis virus, and the production of this type of protein in transgenic plants might be able to disrupt the PM of feeding larvae, rendering them more susceptible to a variety of infections (Hashimoto *et al.*, 1991).

STRATEGIES FOR DEVELOPING INSECT-RESISTANT CROP PLANTS

As discussed above, a variety of compounds are available that could be produced in transgenic plants with the goal of increasing insect resistance. Some of these compounds have already been tested in transgenic plants for their efficacy against insects, while others have not. A chronology of important landmarks along the path toward more insect resistant transgenic crop plants is presented in Table 15.2. The challenge for the future will be to use the limited genetic and financial resources that are available in the most efficient manner so that plants with durable resistance can be developed at reasonable cost, and without jeopardizing other desirable agronomic characteristics of crops.

Choice of Defense Genes

The use of transgenic plants producing Bt toxins has the advantage that these proteins are toxic to insects at very low levels. Plants can be obtained which have nearly complete resistance to pests without the necessity of producing large amounts of a foreign protein. Plant defense proteins such as proteinase inhibitors, amylase inhibitors, and other lectins, are good candidates for use in transgenic crop plants as they are naturally present in many foods and are inactivated upon cooking. These proteins have the disadvantage, however, that much higher levels are required to obtain significant insect resistance. Proteinase inhibitors may prove to be useful in combination with Bt toxins as there appears to be a synergistic effect between these two types of defense compounds when consumed together by insects (MacIntosh *et al.*, 1990a).

TABLE 15.2. A chronology of the development of insect resistant transgenic plants

Year	Plant	Gene	Result	Reference
1987	tobacco	Bt toxin	low protein levels resistant to *Maduca sexta*	Vaeck *et al.*, 1987
1987	tobacco	cowpea trypsin inhibitor	resistant to *Heliothis virescens*	Hilder *et al.*1987
1989	tobacco	potato inhibitor II tomato inhibitor II	resistant to *Manduca sexta*	Johnson *et al.*, 1989
1990	cotton	modified Bt toxin	high protein levels resistant to less sensitive insects	Perlack *et al.*, 1990
1990	tobacco	pea lectin	resistant to *Heliothis virescens*	Boulter *et al.*,, 1990
1993	maize	modified Bt toxin	high protein levels resistant to *Ostrinia nubilalis*	Koziel *et al.*,, 1993
1994	pea	bean α-amylase inhibitor	resistant to *Callosobrochus maculatus*	Shade *et al.*, 1994
1995	pea	bean α-amylase inhibitor	resistant to *Bruchus pisorum*	Schroeder *et al.*, 1995
1995	tobacco	Bt toxin	chloroplast extremely high protein levels	McBride *et al.*, 1995

1995-transgenic potato and maize plants containing Bt toxin genes Bt toxin genes approved by EPA for commercial use.

The production of non-protein insect defense compounds in transgenic crop plants, while it has not yet been reported, is a strategy that may prove useful in the future. Our understanding of the enzymatic pathways involved in the production of compounds such as alkaloids and terpenes is increasing, and genes for more of the biosynthetic enzymes involved will certainly become available. As a result, it may be possible to alter existing biosynthetic pathways or to introduce new ones into important crop plants. Alteration of biosynthetic pathways by genetic engineering has already been successfully achieved in *Arabidopsis thaliana* (Arondel *et al.*, 1992), tobacco (Tarczynski *et al.*, 1993), canola, and soybean (Falco *et al.*, 1995). It is even possible to produce transgenic plants containing a polygene which codes for the production of a series of enzymes, all under the direction of a single promoter (von Bodman *et al.*, 1995).

Expression in Transgenic Crops

Defense genes have usually been expressed in transgenic plants under the control of constitutive promoters. This strategy has the advantage of resulting in the accumulation of high levels of the defense protein in a wide variety of plant tissues. The 35S promoter from Cauliflower mosaic virus (CaMV) has usually been the promoter of choice for dicots (Sanders *et al.*, 1987). An even stronger constitutive plant promoter can be obtained if CaMV 35S gene 5' sequences are duplicated (Kay *et al.*, 1987). The level of expression of many plant genes, including those utilizing the duplicated CaMV 35S promoter, can be significantly increased by replacing the native 5' untranslated region with the leader sequences from the alfalfa mosaic virus RNA 4 or the tobacco mosaic virus omega leader sequence (Jobling and Gehrke, 1987; Warkentin *et al.*, 1992). The 35S promoter is not as highly expressed in monocots, and other sequences such as the rice actin promoter (Zhang *et al.*, 1991) and the maize ubiquitin promoter (Cornejo *et al.*, 1993) give stronger constitutive expression in these plants. In some cases, it may be advantageous to express defense genes under tissue or organ specific promoters to allow production of resistance compounds in plant parts that are especially vulnerable to attack. It may also be useful to produce defense compounds only in the non-edible portions of food crops. For example, a leaf-specific promoter might be used to express genes that would confer resistance to leaf-eating insects upon transgenic potato plants, while leaving the tuber chemistry unchanged. Wound-induced promoters, such as those of some plant proteinase inhibitors, could also be used, as this would provide protection only when needed in case of insect attack, conserving the plant's resources.

Resistance Management

Agricultural scientists today are particularly interested in biologically based insect defenses because of the resistance that many insects have developed to synthetic chemical insecticides. It is quite possible that insects will also develop resistance to Bt toxin and other defense compounds produced in genetically engineered plants. As mentioned previously, resistance to Bt in the field has already been observed. It is probably safe to assume that, under the right conditions, insects will be able to evolve resistance to any insecticidal compound that is discovered or developed. In order to delay the appearance of resistant insect strains and thus increase the utility of the limited number of defense compounds available, appropriate resistance management strategies will need to be employed (May, 1993).

Several tactics have been proposed to minimize the development of resistance, based on previous experience with pest resistant cultivars and on mathematical modelling (McGaughey and Whalon, 1992; Tabashnik, 1994). Plant varieties each containing one added defense gene could be released sequentially, and withdrawn when resistance begins to appear, to be replaced by plants with the next defense gene. This strategy would be particularly effective if susceptibility to each gene was restored upon removal of the selective pressure. This would allow the gene to be reintroduced at a later time once resistance has been lost. This should be the case if there is a fitness cost to the insects associated with the resistance. In some cases, however, insect strains that have developed resistance to high levels of Bt toxin maintained resistance even after removal of the selection (McGaughey and Beeman, 1988). Another strategy is to introduce many resistance genes together in the same transgenic plant. This would be a better imitation of the multigenic resistance strategies of wild plants. Once resistance develops to these plants, however, the usefulness of all the resistance genes present would be lost.

The dosage of defense compounds that insects are exposed to is also an important factor in determining how resistance will develop (Brunke and Meeusen, 1991). Low, sublethal levels of toxins could be used to retard the growth of insects, reducing the damage they cause and increasing their susceptibility to natural enemies. This low dose strategy should lessen the selective pressure toward development of resistance. Expression of defense genes from tissue specific, rather that constitutive, promoters might also help to reduce the selective pressure on insects to develop resistance. On the other hand, crops could be developed with very high levels of defense compounds so that marginally resistant animals would be killed and gradual development of resistance might be prevented.

Field practices will also be an important aspect of resistance management (Gould, 1988). The time required for resistance to occur can be greatly increased if a constant supply of susceptible individuals are introduced into the population each generation to dilute out the frequency of resistant loci in the gene pool. Such individuals could be provided by establishing nearby refugia of non-resistant plants, so that susceptible insects can prosper and then interbreed with insects growing on resistant crops (Brunke and Meeusen, 1991).

More field research with transgenic plants will be helpful in order to evaluate which strategies will provide the most durable resistance in agricultural settings. In order to make the best long term use of biotechnology in improving the insect resistance of crops, wisdom in the application of newly developed lines will prove to be just as important as the discovery and characterization of genes conferring resistance.

REFERENCES

Abe K., Emori Y., Kondo H., Suzuki K. and Arai S. (1987) Molecular cloning of a cysteine proteinase inhibitor of rice (oryzacystatin). *Journal of Biological Chemistry*, 262, 16793–16797.

Arondel V., Lemieux B., Hwang I., Gibson S., Goodman H.G. and Somerville C.R. (1992) Map based cloning of a gene controlling omega-3 fatty acid desaturation in *Arabidopsis*. *Science*, **258**, 1353–1355.

Aronson A.I., Beckman, W. and Dunn P. (1986) *Bacillus thuringiensis* and related insect pathogens. *Microbiological Reviews*, **50**, 1–24.

Baek J.M. and Kim S.I. (1993) Nucleotide sequence of a cDNA encoding soybean Bowman-Birk proteinase inhibitor. *Plant Physiology*, **102**, 687.

Baldwin I.T. (1988) The alkaloidal responses of wild tobacco to real and simulated herbivory. *Oecologia*, **77**, 378–381.

Baldwin I.T. (1991) Damage induced alkaloids in wild tobacco. In *Phytochemical Induction by Herbivores*, edited by D.W. Tallamy and M.J. Raupp, pp 47–69. New York: John Wiley and Sons.

Barboni L., Manzi A., Bellomaria B. and Quinto A.M. (1994) Alkaloid content in four *Spartium junceum* populations as a defensive strategy against predators. *Phytochemistry*, **37**, 1197–1200.

Barret A.J. (1987) The cystatins: A new class of peptidase inhibitors. *Trends in Biochemical Sciences*, **12**, 193–196.

Barry D., Alfaro D. and Darrah L.L. (1994) Relation of european corn borer leaf feeding resistance and DIMBOA content in maize. *Environmental Entomology*, **23**, 177–182.

Barton K.A., Whiteley H.R. and Yang N.S. (1987) *Bacillus thuringiensis* δ-endotoxin expressed in transgenic *Nicotiana tabacum* provides resistance to lepidopteran insects. *Plant Physiology*, **85**, 1103–1109.

Baydoun E.A. and Fry S.C. (1985) The immobility of pectic substances in injured tomato leaves and its bearing on the identity of the wound hormone. *Planta*, **165**, 269–276.

Beuning L.L. and Christeller J.T. (1993) Isolation of a cDNA for proteinase inhibitor I. *Plant Physiology*, **102**, 1061.

Blechert S. Brodschelm W, Holder S, nmerer L., Kutchan T., Mueller M. *et al.* (1995) The octadecanoic pathway: signal molecules for the regulation of secondary pathways. *Proceeding of the National Academy of Sciences, USA*, **92**, 4099–4105.

Bode W., Greyling H.J., Huber R., Otlewsk J. and Wilusz T. (1989) The refined 2.0Å X-ray crystal structure of the complex formed between β-trypsin and CMTI-I, a trypsin inhibitor from squash seeds (*Cucurbita maxima*). *FEBS Letters*, **242**, 285–292.

Bontems F., Roumestand C., Gilquin B., Menez A. and Toma F. (1991) Refined structure of charybdotoxin: common motifs in scorpion toxins and insect defensins. *Science*, **254**, 1521–1523.

Boppre M. (1990) Lepidoptera and pyrrolizidine alkaloids. *Journal of Chemical Ecology*, **16**, 165–185.

Bosch D., Schipper, B., van der Kleij H., de Maagal R.A. and Stiekema W.J. (1994) Recombinant *Bacillus thuringiensis* crystal proteins with new properties: possibilities for resistance management. *Bio/Technology*, **12**, 915–918.

Boulter D., Edwards G.A., Gatehouse A.M.R., Gatehouse J. and Hilder V.A. (1990) Additive protective effects of different plant-derived insect resistance genes in transgenic tobacco plants. *Crop Protection*, **9**, 351–354.

Bowers W.S. (1980) Chemistry of plant/insect interactions. In *Insect Biology in the Future*, edited by M. Locke and D.S. Smith, pp 613–633. New York: Academic Press.

Bowles D.J. (1990) Defense related proteins in higher plants. *Annual Review of Biochemistry*, **59**, 873–907.

Bringmann G., Gramatzki S., Grimm C. and Proksch P. (1992) Feeding deterrency and growth retarding activity of the naphthylisoquinoline alkaloid dioncophylline A against *Spodoptera littoralis*. *Phytochemistry*, **31**, 3821–3825.

Broadway R.M. and Duffey S.S. (1986) Plant proteinase inhibitors: mechanism of action and effect on the growth and digestive physiology of larval *Heliothis zea* and *Spodoptera exigua*. *Journal of Insect Physiology*, **32**, 827–833.

Broadway R.M., Duffey S.S., Pearce G. and Ryan C.A. (1986) Plant proteinase inhibitors: A defense against insects? *Entomologia Experimentalis et Applicata*, **41**, 33–38.

Brown W.E., Takio K., Titani K. and Ryan C.A. (1985) Wound induced trypsin inhibitor in alfalfa leaves: Identity as a member of the Bowman-Birk inhibitor family. *Biochemistry*, **24**, 2105–2108.

Brunke K. and Meeusen R.L. (1991) Insect control with genetically engineered crops. *Trends in Biotechnology*, **9**, 197–200.

Campos F. Atkinson J. Arnason J. Philogene, B., Morand P., Werstiuk N. et al., (1989) Toxicokinetics of DIMBOA in the borer *Ostrinia nubilalis*. *Journal of Chemical Ecology*, **15**, 1989–2001.

Chen X.J., Curtis A., Alcantara E. and Dean D.H. (1995) Mutations in domain I of *Bacillus thuringiensis* δ-endotoxin CryIA(b) reduce the irreversible binding of toxin to *Manduca sexta* brush border vesicles. *Journal of Biological Chemistry*, **270**, 6412–6419.

Chen X.J. Lee M.K. and Dean D.H. (1993) Site-directed mutations in a highly conserved region of *Bacillus thuringiensis* δ-endotoxin affect inhibition of short circuit across *Bombyx mori* midguts. *Proceedings of the National Academy of Sciences, USA*, **90**, 9041–9045.

Chilcott C.N. and Ellar D.J. (1988) Comparative toxicity of *Bacillus thuringiensis* var *israelensis* crystal proteins in vivo and in vitro. *Journal of General Microbiology*, **134**, 2551–2558.

Christeller J.T. Laing, W.A., Markwick, N.P. and Burgess E.P.J. (1992) Midgut protease activities in 12 phytophagous lepidopteran larvae: Dietary and protease inhibitor interactions. *Insect Biochemistry and Molecular Biology*, **22**, 735–746.

Christeller J.T. and Shaw B.D. (1989) The interaction of a range of serine proteinase inhibitors with bovine trypsin and *Costelytra zealandica trypsin*. *Insect Biochemistry*, **19**, 233–241.

Clancy K.M., Foust R.D., Huntsberger, T.G., Whitaker J.G. and Whitaker D.M. (1992) Technique for using microencapsulated terpenes in lepidopteran artificial diets. *Journal of Chemical Ecology*, **18**, 543–560.

Cleveland T.E., Thornburg R.W. and Ryan C.A. (1987) Molecular characterization of a wound-inducible inhibitor I gene from potato and the processing of its mRNA and protein. *Plant Molecular Biology*, **8**, 199–207.

Cornejo M.J., Luth D., Blankenship K.M., Anderson O.D. and Blechl A.E. (1993) Activity of a maize ubiquitin promoter in transgenic rice. *Plant Molecular Biology*, **23**, 567–581.

Davies E. (1987) Action potentials as multifunctional signals in plants: A unifying hypothesis to explain apparently disparate wound responses. *Plant, Cell, and Environment*, **10**, 623–631.

Delanny, X. LaVallee B, Proksch R., Fuchs R., Sims S., Greenplate J. *et al.* (1989) Field performance of transgenic plants expressing the *Bacillus thuringiensis* var *kurstaki* insect control protein. *Bio/Technology*, **7**, 1265–1269.

Delecluse A., Bourgouin C., Klier A. and Rapoport G. (1988) Specificity of action on mosquito larvae of *Bacillus thuringiensis* israelensis toxins encoded by two different genes. *Molecular and General Genetics*, **214**, 42–47.

Derksen A.C.G. and Granados R.R. (1988) Alteration of a lepidopteran peritrophic membrane by baculoviruses and enhancement of viral infectivity. *Virology*, **167**, 242–250 .

Dinan L. (1992) The association of phytoecdysteroids with flowering in fat hen, *Chenopodium album*, and other members of the Chenopodiaceae. *Experientia*, **48**, 305–308.

Doares S.H., Syrovets T., Weiler E.W. and Ryan C.A. (1995) Oligogalacturonides and chitosan activate plant defensive genes through the octadecanoid pathway. *Proceedings of the National Academy of Sciences, USA*, **92**, 4095–4098.

Dopico B., Lowe A.L., Wilson I.D., Merodio C. and Grierson D. (1993) Cloning and characterization of avocado fruit mRNAs and their expression during ripening and low-temperature storage. *Plant Molecular Biology*, **21**, 437–449.

English L. and Slatin S.L. (1992) Mode of action of δ-endotoxins from *Bacillus thuringiensis:* A comparison with other bacterial toxins. *Insect Biochemistry and Molecular Biology*, **22**, 1–7.

Etzler M.E. (1985) Plant lectins: Molecular and biological aspects. *Annual Review of Plant Physiology*, **36**, 209–234.

Falco S.C., Guida T., Locke M., Mauvais J., Sanders C., Ward R.T. *et al.* (1995) Transgenic canola and soybean seeds with increased lysine. *Bio/Technology*, **13**, 577–582.

Farmer E.E., Johnson R.R. and Ryan C.A. (1992) Regulation of expression of proteinase inhibitor genes by methyl jasmonate and jasmonic acid. *Plant Physiology*, **98**, 995–1002.

Farmer E.E. and Ryan, C.A. (1992) Octadecanoid precursors of jasmonic acid activate the synthesis of wound-inducible proteinase inhibitors. *The Plant Cell*, **4**, 129–134.

Fischoff, D., Bowdish K., Perlack F., Marrone P., McCormick S., Niedermeyer J. *et al.* (1987) Insect tolerant transgenic tomato plants. *Bio/Technology*, **5**, 807–813.

Fox J.L. (1995) EPA okays Bt corn: USDA eases plant testing. *Bio/Technology*, **13**, 1035–1036.

Frischknecht P.M., Ulmer-Dufek J. and Baumann T.W. (1986) Purine alkaloid formation in buds and developing leaflets of *Coffea arabica*: expresion of an optimal defense strategy? *Phytochemistry*, **25**, 613–616.

Fujimoto H., Itoh K., Yamamoto M., Kyozuka J. and Shimamoto K. (1993) Insect resistant rice generated by introduction of a modified δ-endotoxin gene of *Bacillus thuringiensis. Bio/Technology*, **11**, 1151–1155.

Fujita T., Kouchi H., Ichikawa T. and Syono K. (1993) Isolation and characterization of a cDNA that encodes a novel proteinase inhibitor I from a tobacco genetic tumor. *Plant and Cell Physiology*, **34**, 137–142.

Gasperi-Campagni A., Barberi L., Battelli M. and Stirpe F. (1985) On the distribution of ribosome-inactivating proteins amongst plants. *Journal of Natural Products*, **48**, 446–454.

Gatehouse A.M.R., Butler K.J., Fenton K.A. and Gatehouse J.A. (1986) Presence and partial characterization of a major proteolytic enzyme in the larval gut of *Callosobruchus maculatus. Entomologia Experimentalis et Applicata*, **39**, 279–286.

Gatehouse, J.A. Bown D., Evans I., Gatehouse L., Jobes D., Preston P. *et al.* (1987) Sequence of the seed lectin gene from pea *(Pisum sativum). Nucleic Acids Research*, **15**, 7642.

Gould F. (1988) Evolutionary biology and genetically engineered crops. *Bio Science*, **38**, 26–33.

Gould F., Martinez-Ramirez A., Anderson A., Ferre J., Silva F.J. and Moar W.J. (1992) Broad spectrum resistance to *Bacillus thuringiensis* toxins in *Heliothis virescens. Proceedings of the National Academy of Sciences, USA*, **89**, 7986–7990.

Graham J.S., Pearce G., Merryweather J., Titani K., Ericsson, L. and Ryan C.A. (1985) Wound induced proteinase inhibitors from tomato leaves: The cDNA-deduced primary structure of pre-inhibitor I and its post-translational processing. *Journal of Biological Chemistry*, **260**, 6555–6560.

Gutierrez C., Sanchez-Monge R., Gomez L., Ruiz-Tapiador M., Castanera P. and Salcedo G. (1990) α-amylase activities of agricultural insect pests are specifically affected by different inhibitor preparations from wheat and barley endosperms. *Plant Science*, **72**, 37–44.

Habu Y., Peyachoknagul S., Umemoto K., Sakata Y. and Ohno T. (1992) Structure and regulated expression of Kunitz chymotrypsin inhibitor genes in winged bean (*Psophocarpus tetragonolobus*). *Journal of Biochemistry*, **111**, 249–258.

Hammond R.W., Foard D.E. and Larkins B.A. (1984) Molecular cloning and analysis of a gene coding for the Bowman-Birk protease inhibitor in soybean. Journal of Biological Chemistry, **259**, 9883–9890.

Hara S., Makino J. and Ikenaka T. (1989) Amino acid sequences and disulfide bridges of serine proteinase inhibitors from bitter gourd (*Momordica charantia*) seeds. *Journal of Biochemistry*, **105**, 88–92.

Hashimoto Y., Corsaro B.G. and Granados R.R. (1991) Location and nucleotide sequence of the gene encoding the viral enhancing factor of the *Trichoplusia ni* granulosis virus. *Journal of General Virology*, **72**, 2645–2651.

Hass G.M. and Ryan C.A. (1981) Carboxypeptidase inhibitors from potatoes. *Methods in Enzymology*, **80**, 778–791.

Herde O., Fuss H., Peña-Cortés H. and Fisahn J. (1995) Proteinase inhibitor II gene expression induced by electrical stimulation and control of photosynthetic activity in tomato plants. *Plant and Cell Physiology*, **36**, 737–742.

Hilder V.A., Barker R.F., Samour R.A., Gatehouse A.M.R., Gatehouse J. and Boulter D. (1989) Protein and cDNA sequences of Bowman-Birk protease inhibitors from the cowpea (*Vigna unguiculata*). *Plant Molecular Biology*, **13**, 701–710.

Hilder V.A., Gatehouse A.M.R., Sheerman S.E., Barker R.F. and Boulter D. (1987) A novel mechanism of insect resistance engineered into tobacco. *Nature*, **330**, 160–163.

Hirashiki, I., Ogata, F., Yoshida, N., Makisumi, S. and Ito A. (1990) Purificationand complex formation analysis of a cysteine proteinase inhibitor (cystatin) from seeds of *Wisteria floribunda. Journal of Biochemistry*, **108**, 604–608.

Hofte H., Van Rie J., Jansens S., Van Outven A., Vanderbruggen H. and Vaeck M. (1988) Monoclonal antibody analysis and insecticidal spectrum of three types of lepidopteran-specific insecticidal crystal proteins of *Bacillus thuringiensis. Applied and Environmental Microbiology*, **54**, 2010–2017.

Hofte H. and Whitely H.R. (1989) Insecticidal crystal proteins of *Bacillus thuringiensis. Microbiological Reviews*, **53**, 242–255.

Hojima Y., Pierce J.V. and Pisano J.J. (1982) Pumpkin seed inhibitor of human factor XIIa (activated Hageman factor) and bovine trypsin. *Biochemistry*, **21**, 3741–3746.

Houseman J.G., Campos, F., Thie, N.M.R., Philogène, B.J.R., Atkinson, J., *et al.* (1992) Effect of the maize-derived compounds DIMBOA and MBOA on growth and digestive processes of european corn borer. *Journal of Economic Entomology*, **85**, 669–674.

Huesing J.E., Shade R.E., Chrispeels M.J., and Murdock L.L. (1991) α-amylase inhibitor, not phytohemagglutinin, explains resistance of common bean to cowpea weevil. *Plant Physiology*, **96**, 993–996.

Jobling S.A. and Gehrke L. (1987) Enhanced translation of chimeric messenger RNAs containing a plant viral untranslated leader sequence. *Nature*, **325**, 622–625.

Jofuku K.D. and Goldberg R.B. (1989) Kunitz trypsin inhibitor genes are differentially expressed during the soybean life cycle and in transformed tobacco plants. *The Plant Cell*, **1**, 1079–1093.

Jofuku K.D., Schipper R.D. and Goldberg R.B. (1989) A frameshift mutation prevents Kunitz inhibitor mRNA accumulation in soybean embryos. *The Plant Cell*, **1**, 427–435.

Johnson M.T. and Gould F. (1992) Interaction of genetically engineered host plant resistance and natural enemies of *Heliothis virescens* in tobacco. *Environmental Entomology*, **21**, 586–597.

Johnson R., Narvaez J., An G. and Ryan C.A. (1989) Expression of proteinase inhibitors I and II in transgenic tobacco plants: Effects on natural defense against *Manduca sexta* larvae. *Proceedings of the National Academy of Sciences, USA*, **86**, 9871–9875.

Jongsma M.A., Bakker P.L., Peters J., Bosch D. and Stiekema W.J. (1995) Adaptation of *Spodoptera exigua* larvae to plant proteinase inhibitors by induction of gut proteinase activity insensitive to inhibition. *Proceedings of the National Academy of Sciences, USA*, **92**, 8041–8045.

Joubert F.J. (1982) Purification and properties of the proteinase inhibitors from *Erythrina caffra* (coast erythrina) seed. *International Journal of Biochemistry*, **14**, 187–193.

Joubert F.J. (1984) Trypsin inhibitors from *Mordica repens* seeds. *Phytochemistry*, **23**, 1401–1406.

Joudrier P.E., Foard D.E., Floener L.A. and Larkins B.A. (1987) Isolation and sequence of cDNA encoding the soybean protease inhibitors PI-IV and C-II. *Plant Molecular Biology*, **10**, 35–42.

Kay R., Chan A., Daly M. and McPherson J. (1987) Duplication of CaMV 35S promoter sequences creates a strong enhancer for plant genes. *Science*, **236**, 1299–1302.

Keil M., Sanchez-Serrano J., Schell J. and Wilmitzer L. (1986) Primary structure of a proteinase inhibitor II gene from potato (*Solanum tuberosum*). *Nucleic Acids Research*, **14**, 5641–5650.

Kernan A. and Thornburg R.W. (1989) Auxin levels regulate the expression of a wound-inducible proteinase inhibitor-II chloramphenicol acetyl transferase gene fusion *in vitro* and *in vivo*. *Plant Physiology*, **91**, 73–78.

Kim S.H., Hase S., Ikenaka T., Toda H., Kitamura K. and Kaizuma N. (1985) Comparative study on amino acid sequences of Kunitz-type soybean trypsin inhibitors Ti^a, Ti^b, and Ti^c. *Journal of Biochemistry*, **98**, 435–448.

Kirsh K. and Schmutterer H. (1988) Low efficacy of a *Bacillus thuringiensis* (Berl.) formulation in controlling the diamondback moth *Plutella xylostella* in the Phillipines. *Journal of Applied Entomology*, **105**, 249–255.

Knight P.J.K., Knowles B.H. and Ellar D.J. (1995) Molecular cloning of an insect aminopeptidase N that serves as a receptor for *Bacillus thuringiensis* CryIA(c) toxin. *Journal of Biological Chemistry*, **270**, 17765–17770.

Kondo H., Abe K., Nishimura I., Watanabe H., Emori Y. and Arai S. (1990) Two distinct cystatin species in rice seeds with different specificities against cysteine proteinases. *Journal of Biological Chemistry*, **26**, 15832–15837.

Kondo H., Emori Y., Abe K., Suzuki K. and Arai S. (1989) Cloning and sequence analysis of the genomic DNA fragment encoding oryzacystatin. *Gene*, **81**, 259–265.

Kortt A.A. (1980) Isolation and properties of a chymotrypsin inhibitor from winged bean seed (*Psophocarpus tetragonolobus*). *Biochimica et Biophysica Acta*, **624**, 237–248.

Kortt A.A. and Jermyn M.A. (1981) Acacia proteinase inhibitors: Purification and properties of the trypsin inhibitors from *Acacia elata* seed. *European Journal of Biochemistry*, **115**, 551–557.

Koziel M.G. Beland G., Bowman C., Carozzi N., Crenshaw R., Crossland L. *et al.* (1993) Field performance of elite transgenic maize plants expressing an insecticidal protein derived from *Bacillus thuringiensis*. *Bio/Technology*, **11**, 194–200.

Kunitz M. (1945) Crystallization of a trypsin inhibitor from soybeans. *Science*, **101**, 668–669.

Kuo T.M., Pearce G. and Ryan C.A. (1984) Isolation and characterization of proteinase inhibitor I from etiolated tobacco leaves. *Archives of Biochemistry and Biophysics*, **230**, 504–510.

Lafont R. and Horn D.H.S. (1989) Phytoecdysteroids: Structure and occurence. In *Ecdysone: From Chemistry to Mode of Action*, edited by J. Koolman, pp 39–64. Georg Thieme Verlag.

Lamb C.J., Ryals J.A., Ward E.R. and Dixon R.A. (1992) Emerging strategies for enhancing crop resistance to microbial pathogens. *Bio/Technology*, **10**, 1436–1440.

Lambert B., Hofte H., Annys K., Jansens S., Soetaert P. and Peferoen M. (1992b) Novel *Bacillus thuringiensis* insecticidal crystal protein with a silent activity against coleopteran larvae *Applied and Environmental Microbiology*, **58**, 2536–2542.

Lambert, B., Van Audenhove K., Decock C., Thenuis W., Agouda F., Janens S. *et al.* (1992a) Nucleotide sequence of gene *cry*IIID encoding a novel coleopteran-active crystal protein from strain BTI109P of *Bacillus thuringiensis*. *Gene*, **110**, 131–132.

Laskowski M., and Kato I. (1980) Protein inhibitors of proteinases. *Annual Review of Biochemistry*, **49**, 593–626.

Lee, J.L., Brown W., Graham J., Pearce G., Fox E., Dreher T. *et al.* (1986) Molecular characterization and phylogenetic studies of a wound-inducible proteinase inhibitor II gene in *Lycopersicon* species. *Proceedings of the National Academy Sciences, USA*, **83**, 7277–7281.

Lee M.K., Milne R.E., Ge A.Z. and Dean D.H. (1992) Location of a *Bombyx mori* receptor binding region on a *Bacillus thuringiensis* δ-endotoxin. *Journal of Biological Chemistry*, **267**, 3115–3121.

Li J., Carroll J. and Ellar D.J. (1991) Crystal structure of insecticidal δ-endotoxin from *Bacillus thuringiensis* at 2.5Å resolution. *Nature*, **353**, 815–821.

Lightner J., Pearce G., Ryan C.A. and Browse J. (1993) Isolation of signalling mutants of tomato (*Lycopersicon esculentum*). *Molecular and General Genetics*, **241**, 595–601.

Lorberth R., Dammann C., Ebneth M., Amati S. and Sanchez-Serrano J. (1992) Promoter elements involved in environmental and developmental control of potato proteinase inhibitor II expression. *The Plant Journal*, **2**, 477–486.

MacIntosh S.C., Kishore G., Perlack F., Marrone P., Stone T., Sims S. *et al.* (1990a) Potentiation of *Bacillus thuringiensis* insecticidal activity by serine protease inhibitors. *Journal of Agricultural and Food Chemistry*, **38**, 1145–1152.

MacIntosh S.C., Stone T., Sims S, Hunst P., Greenplate J., Marrore P. *et al.* (1990b) Specificity and efficacy of purified *Bacillus thuringiensis* proteins against agronomically important insects. *Journal of Invertebrate Pathology*, **56**, 258–266.

Mahoney W.C., Hermodson M.A., Jones B., Powers D.D., Corfman R.S. and Reeck G.R. (1984) Amino acid sequence and secondary structural analysis of the corn inhibitor of trypsin and activated Hageman factor. *Journal of Biological Chemistry*, **259**, 8412–8416.

Manen J.F., Simon P., Van Slooten J.C., Osteras M., Frutiger S. and Hughes G.J. (1991) A nodulin specifically expressed in senescent nodules of a winged bean is a protease inhibitor. *The Plant Cell*, **3**, 259–270.

Mares M., Meloun B., Pavlik M., Kosta V. and Bardys M. (1989) Primary structure of cathepsin D inhibitor from potatoes and its structure relationship to soybean trypsin inhibitor family. *FEBS Letters*, **251**, 94–98.

Margossian L.J., Federman A.D., Giovannoni J.J. and Fischer R.L. (1988) Ethylene regulated expression of a tomato fruit ripening gene encoding a proteinase inhibitor I with a glutamic residue at the reactive site. *Proceedings of the National Academy of Sciences, USA*, **85**, 8012–8016.

Martinez-Ramirez A.C., Escriche B., Real D.M., Silva F.J. and Ferré J. (1995) Inheritance of resistance to a *Bacillus thuringiensis* toxin in a field population of diamondback moth. *Pesticide Science*, **43**, 115–120.

Maruniak J.E., Fiesler S.E., and McGuire P.M. (1990) Susceptibility of insect cells to ricin. *Comparative Biochemistry and Physiology*, **96B**, 543–548.

Masson L., Mazza A., Brousseau R. and Tabashnik B. (1995) Kinetics of *Bacillus thuringiensis* toxin binding with brush border membrane vesicles from susceptible and resistant larvae of *Plutella xylostella*. *Journal of Biological Chemistry*, **270**, 11887–11896.

May R.M. (1993) Resisting resistance. *Nature*, **361**, 593–594.

McBride K.E., Svab Z., Schaaf D.J., Hogan P.S., Stalker D.M. and Maliga P. (1995) Amplification of a chimeric *Bacillus thuringiensis* gene in chloroplasts leads to an extraordinary level of an insecticidal protein in tobacco. *Bio/Technology*, **13**, 362–365.

McGaughey W.H. (1985) Insect resistance to the biological insecticide *Bacillus thuringiensis*. *Science*, **229**, 193–195.

McGaughey W.H. and Beeman R.W. (1988) Resistance to *Bacillus thuringiensis* in colonies of indianmeal moth and almond moth. *Journal of Economic Entomology*, **81**, 28–33.

McGaughey W.H. and Whalon M.E. (1992) Managing insect resistance to *Bacillus thuringiensis* toxins. *Science*, **258**, 1451–1455.

McGurl B., Orozco-Cárdenas M., Pearce G. and Ryan C.A. (1994) Overexpression of the prosystemin gene in transgenic tomato plants generates a systemic signal that constitutively induces proteinase inhibitor synthesis. *Proceedings of the National Academy of Sciences, USA*, **91**, 9799–9802.

McGurl B., Pearce G., Orozco-Cárdenas M. and Ryan C.A. (1992) Structure, expression, and antisense inhibition of the systemin precursor gene. *Science*, **255**, 1570–1573.

Menegatti E., Palmieri S., Walde P., and Luisi P.L. (1985) Isolation and characterization of a trypsin inhibitor from white mustard (*Sinapis alba*). *Journal of Agricultural and Food Chemistry*, **33**, 784–789.

Montfort W., Villafranca J., Mozingo A., Ernst S., Katzin B., Rurenber E. *et al.* (1987) The three dimensional structure of ricin at 2.8Å. *Journal of Biological Chemistry*, **11**, 5398–5403.

Moreno J. and Chrispeels M.J. (1989) A lectin gene encodes the α-amylase inhibitor of the common bean. *Proceedings of the National Academy of Sciences, USA*, **86**, 7885–7889.

Mundy J., Hejgaard J. and Suendsen I. (1984) Characterization of a bifunctional wheat inhibitor of endogenous α-amylase and subtilisin. *FEBS Letters*, **167**, 210–214.

Murdock L.L., Huesing J.E., Nielsen S.S., Pratt R.C. and Shade R.E. (1990) Biological effects of plant lectins on the cowpea weevil. *Phytochemistry*, **29**, 85–89.

Murray C. and Christeller J.T. (1994) Genomic sequence of a proteinase inhibitor II gene. *Plant Physiology*, **106**, 1681.

Nakanishi K. (1980) Insect antifeedants from plants. In *Insect Biology in the Future*, edited by M. Locke and D.S. Smith. pp 603–611. New York: Academic Press.

Narváez-Vásquez J., Orozco-Cárdenas M.L. and Ryan C.A. (1994) A sulfhydryl reagent modulates systemic signalling for wound-induced and systemin-induced proteinase inhibitor synthesis. *Plant Physiology*, **105**, 725–730.

Narváez-Vásquez J., Pearce G., Orozco-Cárdenas M.L., Franceschi V.R. and Ryan C.A. (1995) Autoradiographic and biochemical evidence for the systemic translocation of systemin in tomato plants. *Planta*, **195**, 593–600.

Nicol D., Copaja S.V., Wratten S.D. and Niemeyer H.M. (1992) A screen of worldwide wheat cultivars for hydroxamic acid levels and aphid antixenosis. *Annals of Applied Biology*, **21**, 11–18.

Odani S., Koide T., Ono T. and Ohnishi K. (1983) Structural relationship between barley (*Hordeum vulgare*) trypsin inhibitor and castor bean (*Ricinus communis*) storage protein. *The Biochemical Journal*, **213**, 543–545.

Odani S., Ono T., and Ikenaka T. (1979) Proteinase inhibitors from a Mimosoideae legume, *Albizzia julibrissin*. *Journal of Biochemistry*, **86**, 1795–1805.

Pang S.Z. Oberhaus S., Rasmussen J., Knipple D., Bloomquist J., Dean D. *et al.* (1992) Expression of a gene encoding a scorpion insectotoxin peptide in yeast, bacteria and plants. *Gene*, **116**, 165–172.

Pearce G., Johnson S. and Ryan C.A. (1993) Structure-activity of deleted and substituted systemin, an 18 amino acid polypeptide inducer of plant defensive genes. *Journal of Biological Chemistry*, **268**, 212–216.

Pearce G., Ryan C.A. and Liljegren D. (1988) Proteinase inhibitors I and II in fruit of wild tomato species: Transient components of a mechanism for defense and seed dispersal. *Planta*, **175**, 527–531.

Pearce G., Strydom D., Johnson S. and Ryan C.A. (1991) A polypeptide from tomato leaves induces wound-inducible proteinase inhibitor proteins. *Science*, **253**, 895–898.

Peferoen M., Jansens S., Reynaerts A. and Leemans J. (1990) Potato plants with engineered resistance against insect attack. In *Molecular and Cellular Biology of the Potato*. edited by M. Vayda and W. Park, pp 193–204. Wallingford, Oxon: CAB International.

Peña-Cortés H., Fisahn J. and Willmitzer L. (1995) Signals involved in wound-induced proteinase inhibitor II gene expression in tomato and potato plants. *Proceedings of the National Academy of Sciences, USA*, **92**, 4106–4113.

Peña-Cortés H., Willmitzer L. and Sanchez-Serrano J.J. (1991) Abscisic acid mediates wound induction but not developmental specific expression of the proteinase inhibitor II gene family. *The Plant Cell*, **3**, 963–972.

Perlak F.J., Deaton R.W., Armstrong T.A., Fuchs R.L., Sims S.R., Greenplate J.T. *et al.* (1990) Insect resistant cotton plants. *Bio/Technology*, **8**, 939–943.

Perlak F.J., Fuchs R.L., Dean D.A., McPherson S.L. and Fischhoff D.A. (1991) Modification of the coding sequences enhances plant expression of insect control protein genes. *Proceedings of the National Academy of Sciences, USA*, **88**, 3324–3328.

Purcell J.P., Greenplate J.T. and Sammons R.D. (1992) Examination of midgut luminal proteinase activities in six economically important insects. *Insect Biochemistry and Molecular Biology*, **22**, 41–47.

Reddy M.N., Keim P.S., Heinrikson R.L. and Kezdy F.J. (1975) Primary structure analysis of sulfhydryl protease inhibitors from pineapple stem. *Journal of Biological Chemistry*, **250**, 1741–1750.

Richardson M. (1977) The proteinase inhibitors of plants and microorganisms. *Phytochemistry*, **16**, 159–169.

Ritonja A. Krizaj I., Mesko P., Kopitar M., Lucovnik P., Strukelj B. *et al.* (1990). The amino acid sequence of a novel inhibitor of cathepsin D from potato. *FEBS Letters*, **267** 13–15.

Robinson J.F., Klun J.A. and Brindley T.A. (1978) European corn borer: A non-preference mechanism of leaf feeding resistance and its relationship to the 1,4 benzoxazin-3–one concentration in dent corn tissue. *Journal of Economic Entomology*, **71**, 461–465.

Rodis P. and Hoff J.E. (1984) Naturally occuring protein crystals in the potato. *Plant Physiology*, **74**, 907–911.

Rosenthal G.A. (1982) *Plant Non-protein Amino and Imino Acids*. New York: Academic Press.

Rosenthal G.A. (1991) The biochemical basis for the deleterious effects of L-canavinine. *Phytochemistry*, **30**, 1055–1058.

Rosenthal G.A., Berge M.A., Bleiler J.A. and Rudd T. (1987) Avoidance of aberrant protein production and an organism's ability to utilize or tolerate L-canavanine. *Experientia*, **43**, 558–561.

Rosenthal G.A. and Dahlman D.L. (1986) L-canavanine, insect protein synthesis, and higher plant chemical defense. *Proceedings of the National Academy of Sciences, USA*, **83**, 14–18.

Rudiger H. (1984) On the physiological role of plant lectins. *Bio Science*, **34**, 95–99.

Ryan C.A. (1989) Proteinase inhibitor gene families: Strategies for transformation to improve plant defenses against herbivores. *Bioessays*, **10**, 20–24.

Ryan C.A. (1992) The search for the proteinase inhibitor inducing factor, PIIF. *Plant Molecular Biology*, **19**, 123–133.

Sanders J.R., Winter J.A., Barnason A.R., Rogers S.G. and Fraley R.T. (1987) Comparison of cauliflower mosaic 35S and nopaline synthase promoters in transgenic plants. *Nucleic Acids Research*, **15**, 1543–1558.

Sangadala S., Walters F.S., English L.H. and Adang M.J. (1994) A mixture of *Manduca sexta* aminopeptidase and phosphatase enhance *Bacillus thuringiensis* insecticidal CryIA(c) toxin binding and ^{86}Rb-K^{+} efflux *in vitro*. *Journal of Biological Chemistry*, **269**, 10088–10092.

Schaller A., and Ryan C.A. (1994) Identification of a 50 kDa systemin-binding protein in tomato plasma membranes having kex2p-like properties. *Proceedings of the National Academy of Sciences, USA*, **91**, 11802–11806.

Schroeder H.E. Gollasch S., Moore A., Table L., Craig S., Hardie D. *et al.* (1995) Bean α-amylase inhibitor confers resistance to the pea weevil (*Bruchus pisorum*) in transgenic peas (*Pisum sativum* L.). *Plant Physiology*, **107**, 1233–1239.

Shade R.E. Schroeder H., Pueyo J., Tabe L., Murdock L., Higgins T. and Chrispeels M. 1994) Transgenic pea seeds expressing the α-amylase inhibitor of the common bean are resistant to bruchid beetles. *Bio/Technology*, **12**, 793–796.

Sick A., Gaertner F. and Wong A. (1990) Nucleotide sequence of a coleopteran active toxin gene from a new isolate of *Bacillus thuringiensis* subsp. *tolworthi*. *Nucleic Acids Research*, **18**, 1305.

Slaney A.C., Robbins H.L., and English L. (1992) Mode of action of *Bacillus thuringiensis* toxin CryIIIA: An analysis of toxicity in *Leptinotarsa decemlineata* and *Diabrotica undecimpunctata howardi*. *Insect Biochemistry and Molecular Biology*, **22**, 9–18.

Smith C.M. (1989) Plant Resistance to Insects. New York: Wiley-Interscience.

Stapleton A., Blankenship D., Ackermann B., Chen T.M., Gorder G., Manley G. *et al.* (1990) Curtatoxins: Neurotoxic insecticidal polypeptides isolated from the funnel-web spider *Hololena curta*. *Journal of Biological Chemistry*, **265**, 2054–2059.

Steele C.L., Lewinsohn E. and Croteau R. (1995) Induced oleoresin biosynthesis in grand fir as a defense against bark beetles. *Proceedings of the National Academy of Sciences, USA*, **92**, 4164–4168.

Stewart L.M.D., Hirst M., Ferber M.L., Merryweather A.T., Cayley P.J. and Possee R.D. (1991) Construction of an improved baculovirus insecticide containing an insect-specific toxin gene. *Nature*, **352**, 85–88.

Stoner K.A. (1992) Bibliography of plant resistance to arthropods in vegetables, 1977–1991. *Phytoparasitica*, **20**, 125–180.

Strukel B. Pungercar J., Ritonja A., Krizaj I., Gubensek F., Kregar I. and Turk V. (1990) Nucleotide and deduced amino acid sequence of an aspartic proteinase inhibitor homolog from potato tubers. *Nucleic Acids Research*, **18**, 4605.

Suh S.G., Stiekema W.J. and Hannapel D.J. (1991) Proteinase inhibitor activity and wound inducible gene expression of the 22 kDa potato tuber proteins. *Planta*, **184**, 423–430.

Tabashnik B.E. (1994) Evolution of resistance to *Bacillus thuringiensis*. *Annual Review of Entomology*, **39**, 47–79.

Tabashnik B.E., Cushing N.L., Finson N., and Johnson M.W. (1990) Field development of resistance to *Bacillus thuringiensis* in diamondback moth (Lepidoptera: Plutellidae). *Journal of Economic Entomology*, **83**, 1671–1676.

Tabashnik B.E., Finson N., Johnson M.W. and Heckel, D.G. (1994) Cross-resistance to *Bacillus thuringiensis* toxin CryIF in the diamondback moth (*Plutella xylostella*). *Applied and Environmental Microbiology*, **60**, 4627–4629.

Tallamy D.W. and McCloud E.S. (1991) Squash beetles cucumber beetles, and inducible cucurbit responses. In *Phytochemical Induction by Herbivores*, edited by D.W. Tallamy and M.J. Raupp, pp 155–181. New York: John Wiley and Sons

Tarczynski M.C., Jenson R.G. and Bohnert H.J. (1993) Stress protection of transgenic tobacco by production of the osmolyte mannitol. *Science*, **259**, 508–510.

Thomas J.C., Wasmann C.C., Echt C., Dunn R.L., Bohnert H.J. and McCoy T.J. (1994) Introduction and expression of an insect proteinase inhibitor in alfalfa (*Medicago sativa* L.). *Plant Cell Reports*, **14**, 31–36.

Thornburg R.W., An G., Cleveland T.E., Johnson R. and Ryan C.A. (1987) Wound-inducible expression of a potato inhibitor II-chloramphenicol acetyl transferase gene fusion in transgenic tobacco plants. *Proceedings of the National Academy of Sciences, USA*, **84**, 744–748.

Tomalski M.D. and Miller L.K. (1991) Insect paralysis by baculovirus-mediated expression of a mite neurotoxin gene. *Nature*, **352**: 82–85.

Turlings T.C.J., Loughrin J.H., McCall P.J., Rose U.S.R., Lewis W.J. and Tumlinson J.H. (1995) How caterpillar-damaged plants protect themselves by attracting parasitic wasps. *Proceedings of the National Academy of Sciences, USA*, **92**, 4169–4174.

Turlings T.C.J. and Tumlinson J.H. (1992) Systemic release of chemical signals by herbivore-injured corn. *Proceedings of the National Academy of Sciences, USA*, **89**, 8399–8402.

Vaeck M. Reynaerts A., Hofte H., Jansens S., De Beuckekeer M. Dean C. *et al.* (1987) Transgenic plants protected from insect attack. *Nature*, **328**, 33–37.

von Bodman S.B., Domier L.L. and Farrand S.K. (1995) Expression of multiple eukaryotic genes from a single promoter in *Nicotiana*. *Bio/Technology*, **13**, 587–591.

Walsh T.A. and Strickland J.A. (1993) Proteolysis of the 85 kilodalton crystalline cysteine proteinase inhibitor from potato releases functional cystatin domains. *Plant Physiology*, **103**, 1227–1234.

Warkentin T.D., Jordan M.C. and Hobbs S.L.A. (1992) Effect of promoter-leader sequences on transient reporter gene expression in particle bombarded pea tissues. *Plant Science*, **87**, 171–177.

Weselake R., MacGregor A. and Hill R. (1983) An endogenous α-amylase inhibitor in barley kernels. *Plant Physiology*, **72**, 809–812.

Widner W.R. and Whiteley H.R. (1989) Two highly related insecticidal crystal proteins of *Bacillus thuringiensis* subsp. *kurstaki* possess different host range specificities. *Journal of Bacteriology*, **171**, 965–974.

Wieczorek M., Otlewski J., Cook J., Parks K., Leluk J., Wilimowska-Pelc A. *et al.* (1985) The squash inhibitor family of serine proteinase inhibitors: Amino acid sequences and association equilibrium constants of inhibitors from squash, summer squash, zucchini, and cucumber seeds. *Biochemical and Biophysical Research Communications*, **126**, 646–652.

Wildon D.C. Thain J., Minchin P., Gubb I., Reilly A., Skipper Y. *et al.* (1992) Electrical signalling and systemic proteinase inhibitor induction in the wounded plant. *Nature*, **360**, 62–64.

Williamson M.S., Forde J., Buxton B. and Kreis, M. (1987) Nucleotide sequence of barley chymotrypsin inhibitor 2 (CL-2) and its expression in normal and high-lysine barley. *European Journal of Biochemistry*, **165**, 99–106.

Wingate V.P.M., Broadway R.M. and Ryan C.A. (1989) Isolation and characterization of novel, developmentally regulated proteinase inhibitor I protein and cDNA from the fruit of a wild species of tomato. *Journal of Biological Chemistry*, **264**, 17734–17738.

Wingate V.P.M. and Ryan C.A. (1991) A novel fruit expressed trypsin inhibitor I gene from a wild species of tomato. *Journal of Biological Chemistry*, **266**, 5814–5818.

Wolfson J.L. (1991) The effects of induced plant proteinase inhibitors on herbivorous insects. In *Phytochemical Induction by herbivores,* edited by D.W. Tallamy and M.J. Raupp, pp 223–243. New York: John Wiley and Sons.

Wolfson J.L. and Murdock L.L. (1987) Suppression of larval Colorado potato beetle growth and development by digestive proteinase inhibitors. Entomologia Experimentalis et Applicata, **44**, 235–240.

Wu C., and Whitaker J.R. (1991) Homology among trypsin/chymotrypsin inhibitors from red kidney bean, brazilian pink bean, lima bean, and soybean. *Journal of Agriculture and Food Chemistry*, **39**, 1583–1589.

Xu J. and Hukuhara T. (1992) Enhanced infection of a nuclear polyhedrosis virus in larvae of the armyworm, *Pseudaletia separata*, by a factor in the spheroids of an entomopoxvirus. *Journal of Invertebrate Pathology*, **60**, 259–264.

Yamamoto M., Hara S., and Ikenaka T. (1983) Amino acid sequences of two trypsin inhibitors from winged bean seeds (*Psophocarpus tetragonolobus*). *Journal of Biochemistry*, **94**, 849–863.

Young R.J., Scheuring C.F., Harris-Haller L. and Taylor B.H. (1994) An auxin-inducible proteinase inhibitor gene from tomato. *Plant Physiology*, **104**, 811–812.

Zemke K., Fahrnow A.M., Jany K.D., Pal G.P. and Saenger W. (1991) The three-dimensional structure of the bifunctional proteinase K/α-amylase inhibitor from wheat (PK13) at 2.5Å resolution. *FEBS Letters*, **279**, 240–242.

Zhang W., McElroy D. and Wu R. (1991) Analysis of rice 5' region activity in transgenic rice plants. *The Plant Cell*, **3**, 1155–1165.

INDEX